Parallel Algorithms for Optimal Control of Large Scale Linear Systems

Communications and Control Engineering Series

Editors: B.W. Dickinson · A. Fettweis · J.L. Massey · J.W. Modestino
E.D. Sontag · M. Thoma

Zoran Gajić and Xuemin Shen

Parallel Algorithms for Optimal Control of Large Scale Linear Systems

With 35 Figures

Springer-Verlag
London Berlin Heidelberg New York
Paris Tokyo Hong Kong
Barcelona Budapest

Zoran Gajić, PhD
Rutgers University,
Department of Electrical and Computer Engineering,
Piscataway, U.S.A.

Xuemin Shen, BSc, MSc, PhD
University of Alberta,
Department of Electrical Engineering,
Edmonton,
Canada, T6G 2G7

ISBN 978-1-4471-3221-9   ISBN 978-1-4471-3219-6 (eBook)
DOI 10.1007/978-1-4471-3219-6

British Library Cataloguing in Publication Data
A catalogue record for this book is available from the British Library

Library of Congress Cataloging-in-Publication Data
A catalog record for this book is available from the Library of Congress

Typesetting: Camera ready by authors

To Professors: P. Kokotović, H. Khalil
and J. Medanić

Zoran Gajić

To my father Zhong Shen

Xuemin Shen

Preface

This book is designed to be a comprehensive treatment of parallel algorithms for optimal control of large scale linear and bilinear systems. These algorithms were originally evolved in the context of the recursive reduced-order methods for singularly perturbed and weakly coupled linear systems. There are numerous examples of large scale singularly perturbed and weakly coupled dynamic systems that provide great challenges to engineers, mathematicians and computer scientists. Some of the examples of singularly perturbed and weakly coupled systems include electric power systems, aerospace systems, large electric and communication networks, robotics, and process control systems in the chemical and petroleum industries.

The parallel algorithms presented in this book are applicable to a wider classes of practical systems than the traditional methods for large scale singularly perturbed and weakly coupled systems based on the power-series expansion methods. The synchronous parallel reduced-order algorithms presented in this book offer several advantages: the higher order of accuracy can be easily achieved at low cost, the parallel processing of information can be used, results are obtained under much milder assumptions (no analyticity requirements imposed on the problem coefficients), the software and hardware implementation of the control algorithms is highly simplified due to complete parallelism in the design procedures.

This book is intended for a wide readership, including control engineers, applied mathematicians, computer scientists, and advanced graduate students who seek a comprehensive view of the current developments in the theory of large scale linear and bilinear singularly perturbed and weakly coupled control systems. The book emphasizes mathematical developments as well as their application to solving practical problems without requiring a strong mathematical background.

To demonstrate the usefulness of the presented methods for large scale singularly perturbed and weakly coupled linear and bilinear systems, and to point out its various advantages, we have included many real control system examples such as: F-8 aircraft, L-1011 fighter aircraft, fluid catalytic cracker, twelve plate absorption column, magnetic tape control system, power system composed of two interconnected areas, distillation column, steam power system, hydro power plant, chemical plants, gas absorber, supported beam control problem, induction motor drives (bilinear model), large space structure, optimal control of a paper making machine (bilinear model), satellite control problem, and synchronous machine connected to an infinite bus.

The authors hope that this book will reduce some of barriers that exist in recognizing the power and usefulness of the synchronous parallel algorithms for optimal control of large scale linear and bilinear systems, and that it will help to broaden their implementation in practice. Also, we hope that this book will motivate some researchers to develop the corresponding asynchronous parallel algorithms and extend the presented results to nonlinear control systems.

The authors are thankful for support and contributions from Professors V. Gourishankar, J. Momoh, D. Petkovski, B. Petrovic, P. Milojevic, and M. Rao, from colleagues Dr. D. Arnautovic, Dr. Q. Xia, and from our present and former graduate students Z. Aganovic, T. Grodt, N. Harkara, S. Hogan, M. Huey, A. Kolarov, M. Qureshi, V. Radisavljevic, N. Rayavarupu, I. Seskar, D. Skataric, W. Su, Y. Ying, and J. Zhuang.

Dr. Shen is particularly thankful to the Canadian NSERC for a two-year international postdoctoral fellowship at the University of Alberta, in the Departments of Chemical and Electrical Engineering, which helped him a lot of in the completion of this book.

Z. Gajic and X. Shen

Piscataway, NJ, USA
Edmonton, Canada
October 1992

Contents

CONTENTS

Chapter 1

Introduction

This book presents the parallel algorithms for optimal controllers of large scale linear dynamical systems inherently parallel in nature, namely for singularly perturbed and weakly coupled linear systems. The book is written in the spirit of parallel and distributed computations (Bertsekas and Tsitsiklis, 1989; 1991) and parallel processing of information in terms of the reduced-order controllers and filters (Gajic et al., 1990). It covers almost all important concepts of the optimal linear control theory and its applications, in the context of continuous and discrete, deterministic and stochastic linear systems. A generalization of the presented methods to the optimal control of singularly perturbed and weakly coupled bilinear systems is also considered.

The other classes of the general linear optimal control problems can be studied by using the parallel reduced-order algorithms presented in this book with the help of some standard control techniques like the pole placement, overlapping decomposition, and prescribed degree of stability requirement. With the pole placement by performance modification method (Medanic, 1988; Tharp, 1992), it is possible to separate the closed-loop eigenvalues into two disjoint sets, that is to introduce the slow and fast phenomena (singular perturbations). The overlapping methods of Siljak are very powerful tools in the system decomposition (Siljak, 1991). The overlapping decomposition technique documented also in (Ikeda and Siljak, 1980; Ohta and Siljak, 1985; Sezer and Siljak, 1986; Calvet and Titli, 1989) can influence weak coupling (Arabacioglu et al., 1986). The prescribed degree of stability requirement (Anderson and Moore, 1990), imposed on the system in order to assure given stability margin, can bring the system matrix into the block diagonally dominant form and make the system internally weakly coupled. Also, it is well known that the cheap control and high gain feedback optimal control problems can be studied by using the theory of singular perturbations (Kokotovic and Khalil, 1986).

The results of this book complement the theory and application of singularly perturbed and weakly coupled control systems, introduced to the control audience, and developed for practical implementations, by Professor Petar Kokotovic and his coworkers.

This book represents an improved and considerably extended version (from 202 to almost 450 pages) of our monograph titled "Singularly Perturbed and Weakly Coupled Linear Control Systems — A Recursive Approach" (Gajic et al., 1990), published in the Springer-Verlag Series Lecture Notes in Control and Information Sciences. The book is mostly based on the authors recent research papers and we have been following them very closely in many parts of this book.

The theory of singular perturbations has been a highly recognized and rapidly developing area of control in the last twenty five years (Kokotovic and Khalil, 1986; Kokotovic et al., 1986). It has been studied so far by using the Taylor series, asymptotic expansions, and power-series methods. Being nonrecursive in nature, these expansion methods become very cumbersome and computationally very expensive (the size of computations required can be considerable) when a high order of accuracy is required. In such a case, the advantage of using the expansion methods (the important theoretical tools) is questionable from the numerical point view, and sometimes these methods are almost not applicable for practical computations (Grodt and Gajic, 1988; Gajic et al., 1989). In the era of an increased application of the modern control theory results to the real world systems that might be a serious problem.

In addition, if a small perturbation parameter ϵ is not very small ("small enough"), then the $O(\epsilon)^1$ theory, used so far in the study of singularly perturbed problems, can not produce satisfactory results for the given class of problems. In order to broaden the class of applicable problems, the development of the $O(\epsilon^k)$ theory is a necessary requirement. Even more, it is pointed out in (Hemker, 1983) that the $O(\epsilon^k)$ theory is a trend in the modern numerical analysis of singularly perturbed problems: "numerical analysis of singular perturbation problems mainly concentrates on the following question: how to find a numerical approximation to the solution for small as well as intermediate values of ϵ, where no short asymptotic expansion is available. Or more general, how

---

[1] $O(\epsilon^k)$ stands for $C\epsilon^k$, where C is a bounded constant and k is any arbitrary constant.

to construct a single numerical method that can be applied both in the case of extremely small ϵ and for larger values of ϵ, when one wouldn't consider the problem as singularly perturbed any longer."

Furthermore, in the case of singularly perturbed structures induced by a high gain feedback (Kokotovic and Khalil, 1986; Kokotovic et al., 1986), the standard statement of the singular perturbation theory "it exists ϵ small enough" means it exists control input big enough. Thus, that assumption "it exists ϵ small enough" limits the practical implementation of the $O(\epsilon)$ singular perturbation theory quite a lot. In the recent paper (Gajic et al., 1989), a real world example demonstrates the failure of the $O(\epsilon)$ theory for the problem of the optimal static output feedback control of a linear singularly perturbed system. The same example is solved successfully in (Gajic et al., 1989) by using the $O(\epsilon^k)$ theory for $k \geq 2$.

The development of the recursive techniques (based on the fixed point reduced-order parallel algorithms) for singularly perturbed linear-quadratic steady state control problems has started recently (Gajic, 1986; Gajic et al., 1987; Grodt and Gajic, 1988; Shen, 1990; Gajic and Shen 1991a, 1991b; Qureshi, 1992; Qureshi et al., 1992; Su et al., 1992b; Skataric and Gajic, 1992). The recursive reduced-order numerical methods for finite time singularly perturbed control systems have been developed in (Grodt and Gajic, Su et al., 1992a; Shen, 1992).

The linear weakly coupled systems have been studied in different set-ups by many researchers (Kokotovic et al., 1969; Medanic and Avramovic, 1975; Ishimatsu et al., 1975; Ozguner and Perkins, 1977; Delacour et al., 1978; Mahmoud, 1978; Petkovski and Rakic, 1979; Washburn and Mendel, 1980; Sezer and Siljak, 1986; Khalil and Kokotovic, 1978; Srikant and Basar, 1992a). The solutions of the Riccati-type and/or Lyapunov-type equations are obtained in terms of the Taylor series and power-series expansions with respect to a small coupling parameter ϵ. Approximate feedback control laws were derived by truncating expansions of the feedback coefficients of the optimal control law (Kokotovic et al., 1969; Ozguner and Perkins, 1977; Delacour, 1978; Petkovski and Rakic, 1979). Such approximations have been shown to be near-optimal with performance that can made as close to the optimal performance as desired by including enough terms in the truncated expansions.

The recursive approach to weakly coupled control systems, based on the fixed point iterations, has been developed recently (Petrovic and Gajic, 1988; Harkara et al., 1989; Gajic and Shen, 1989; Shen and Gajic, 1990a, 1990b, 1990c; Shen, 1990; Su, 1990; Su and Gajic, 1991, 1992b; Qureshi, 1992). It has been shown that the recursive methods are particularly useful when the coupling parameter ϵ is not extremely small and/or when any desired order of accuracy is required, namely, $O\left(\epsilon^k\right)$, where $k = 2, 3, 4, \ldots$. In some applications a very good approximation is required, such as for a plant-filter augmented system (Shen and Gajic, 1990a), where the accuracy of $O\left(\epsilon^k\right)$, $k \geq 6$ is necessary to stabilize given closed-loop system.

This book consists of fourteen chapters. Chapter 1 comprises an introduction. In Chapter 2, we study the main algebraic equations of the linear steady state optimal control theory for both singularly perturbed and weakly coupled systems, namely, the algebraic Lyapunov and Riccati equations. We derive the corresponding recursive algorithms for the solution of these equations in the most general case when the problem matrices are functions of a small perturbation parameter. The numerical decomposition has been achieved so that only low-order systems are involved in algebraic computations. The introduced recursive methods are of the fixed point type and can be implemented as parallel synchronous algorithms (Bertsekas and Tsitsiklis, 1991). Both continuous-time and discrete-time versions of the algebraic Lyapunov and Riccati equations are studied. The partitioned expressions of the Riccati equations have very complicated forms in the discrete-time domain for both singularly perturbed and weakly coupled systems. We have overcome that problem by using corresponding bilinear transformations (Shen and Gajic, 1990b; Gajic and Shen 1991b), that are applicable under quite mild assumptions, so that the solution of the discrete algebraic Riccati equations of singularly perturbed and weakly coupled systems are obtained by using results from Sections 2.2.2 and 2.3.2, derived for the corresponding continuous-time algebraic Riccati equations. It is shown that the singular perturbation recursive methods converge with the rate of convergence of $O\left(\epsilon\right)$, whereas the recursive methods for weakly coupled linear systems converge faster, that is, with the rate of convergence of $O\left(\epsilon^2\right)$.

Having obtained approximate solutions of the algebraic Lyapunov and Riccati equations, the corresponding approximate linear-quadratic optimal control problems are solved in terms of these solutions. Some real world control examples are included in order to demonstrate the procedures: magnetic tape, F-8 aircraft, catalytic cracker, and chemical plant.

In Chapter 3, the Chang decoupling transformation (Chang, 1972) of singularly perturbed linear systems is introduced. We also present a version of the Chang transformation which is applicable to weakly coupled linear systems (Gajic and Shen, 1989). The transformation matrices are obtained from two coupled matrix equations in both cases.

Algorithms that efficiently generate solutions of these equations are derived. New versions of the Chang transformation for both singularly perturbed and weakly coupled linear systems, that lead to the decoupled matrix equations for the transformation matrices (Qureshi, 1992; Qureshi and Gajic, 1992) are derived as well. The presented transformations are utilized to completely and exactly decompose the Lyapunov differential equations of singularly perturbed and weakly coupled systems into the reduced-order ones. In this chapter, we also present results for the decomposition of the general boundary value problems of linear weakly coupled, continuous-time and discrete-time, systems.

In Chapter 4, the deterministic output feedback control of singularly perturbed and weakly coupled linear systems is studied. The well-defined recursive numerical technique for the solution of nonlinear algebraic matrix equations, associated with the output feedback control problem of singularly perturbed systems is developed. The numerical slow-fast decomposition is achieved so that only low-order systems are involved in algebraic computations. It is shown that each iteration step of the proposed algorithm improves the accuracy by an order of magnitude, that is, the accuracy of $O\left(\epsilon^k\right)$ can be obtained by performing only k iterations (Gajic et al., 1989). This represents the significant improvement since all results on the output feedback control problems for the singularly perturbed systems have been obtained so far with the accuracy of $O\left(\epsilon\right)$ only. A real world example, an industrial important reactor — fluid catalytic cracker — demonstrates the efficiency of the proposed algorithm and the failure of $O\left(\epsilon\right)$ theory.

Following similar lines a recursive algorithm is also developed for solving nonlinear algebraic equations comprising the solutions of the optimal static output feedback control problem of linear weakly coupled systems. The effectiveness of the proposed reduced-order algorithm and its advantages over the global full-order algorithm is demonstrated on a twelve-plate chemical absorption column (Harkara et al., 1989). Obtained results strongly support the necessity for the existence of reduced-order numerical techniques for solving corresponding nonlinear algebraic equations. In addition to the reduction in required computations, it can be easier to find a good initial guess and to handle the problem of nonuniqueness of the solution of these nonlinear equations — they represent the necessary conditions only.

For both singularly perturbed and weakly coupled linear output feedback control problems, the synchronous algorithms that solve in parallel six reduced-order algebraic Lyapunov equations are derived.

5

In Chapter 5, we present the approach to the decomposition and approximation of the linear-quadratic Gaussian estimation and control problems. The global Kalman filter of linear weakly coupled continuous systems is decomposed into separate reduced-order local filters via the use of a decoupling transformation, introduced in Chapter 3. A near-optimal control law is derived by approximating the coefficients of the optimal control law (Khalil and Gajic, 1984). The order of approximation of the optimal performance is $O\left(\epsilon^k\right)$, where k is the order of approximation of the coefficients. The electrical power system example demonstrates the failures of the $O\left(\epsilon^2\right)$ and $O\left(\epsilon^4\right)$ theories and the necessity for the existence of the $O\left(\epsilon^k\right)$ theory (Shen and Gajic, 1990a). The proposed method produces the reduction in both off-line and on-line computational requirements and converges under mild assumptions. Similarly, in this chapter we also study the linear-quadratic Gaussian control problem of singularly perturbed systems. In that context the reduced-order recursive algorithm is used to design a controller for an F-8 aircraft by using Kalman filters operating independently in slow and fast time scales (Khalil and Gajic, 1984; Gajic, 1986).

In the remaining part of Chapter 5, the obtained results for continuous systems are extended to the discrete-time domain. In that respect, the near-optimum steady state regulator is derived for the discrete stochastic weakly coupled system and applied to a fifth-order distillation column (Shen and Gajic, 1990c). The corresponding singularly perturbed discrete stochastic problem is studied for a steam power system (Shen, 1990; Gajic and Shen, 1991a). The proposed methods allow parallel processing of information and reduce considerably the size of required off-line and on-line computations, since they introduce full parallelism in the design procedures.

Chapter 6 is about the finite time open-loop control problems (linear two-point boundary value problem) for weakly coupled and singularly perturbed systems. The main idea of this chapter is to exploit the reduced-order subsystems to find efficiently the optimal open-loop control in the new coordinates. This change of coordinates is particularly important for singularly perturbed systems, where the original two-point boundary value problem is transformed in the pure-slow and pure-fast reduced-order completely decoupled initial value problems. By doing this, the stiffness of the singularly perturbed two-point boundary value problem is converted in the problem of an ill-defined system of linear algebraic equations (Su et al., 1992a). Two real world examples, a distillation column for weakly coupled continuous-time systems, and a magnetic tape control system for singularly perturbed continuous-time systems are included in order to demonstrate the efficiency of the considered methods. The study of the open-loop control problem presented for both singularly perturbed and weakly coupled continuous-time systems is extended in Sections 6.5–6.8 to the corresponding discrete-time problems.

In Chapter 7, the algebraic Riccati equation is considered in terms of the Hamiltonian matrix. For both weakly coupled and singularly perturbed systems, the Hamiltonian matrices retain the weakly coupled and singularly perturbed forms by interchanging some of the state and costate variables so that they can be block diagonalized via the decoupling transformations introduced in Chapter 3. The main idea of this chapter is to obtain the solutions of the global algebraic Riccati equations from two decoupled reduced-order subsystems — both leading to the nonsymmetric algebraic Riccati equations. It has been shown that the solutions exist under stablizability-detectability conditions imposed on subsystems.

The algebraic Riccati equation of singularly perturbed control systems is completely and exactly decomposed into two reduced-order algebraic Riccati equations corresponding to slow and fast time scales (Su et al., 1992b). The pure-slow and pure-fast algebraic Riccati equations are nonsymmetric ones, but their $O(\epsilon)$ perturbations are symmetric. It is shown that the Newton method is very efficient for solving the obtained nonsymmetric Riccati equations. The presented method might produce a new insight into the time scale optimal filtering and control problems.

Similar results are obtained for the weakly coupled algebraic Riccati equation (Su and Gajic, 1992). The use of the nonsymmetric reduced-order Riccati equations can produce a lot of savings; that is, $O(n)$, in the size of computations required. Furthermore, the proposed method is very suitable for parallel computations since it allows complete parallelism, on the contrary to Chapter 2, where intermediate results have to be interchanged after each iteration step. A satellite control problem is solved in order to demonstrate the procedure.

Chapter 8 deals with finite time optimal control problems. In that direction the recursive reduced-order numerical solution of the weakly coupled and singularly perturbed matrix differential and difference Riccati equations are obtained. The order-reductions are achieved in both cases via the use of decoupling transformations applied to the corresponding Hamiltonian matrices (Grodt and Gajic, 1988; Su and Gajic, 1991). It is shown that corresponding algorithms converge under stabilizability-observability conditions imposed on subsystems with the rate of convergence of $O(\epsilon^2)$ for weakly coupled and $O(\epsilon)$ for singularly perturbed systems. Corresponding results for the difference Riccati equations of singularly perturbed (Shen, 1992) and weakly coupled systems (Shen et al., 1991b) are also presented in this chapter. Several real world examples demonstrate the reduced-order recursive techniques for solving differential and difference Riccati equations: synchronous machine connected to an infinite bus, F-8 aircraft, and gas absorber.

Two large classes of small parameter systems have been studied independently so far in the context of control theory: singularly perturbed and weakly coupled systems. However, models of many real physical systems (for example, power systems, flexible space structures) are at the same time both singularly perturbed and weakly coupled. Even more, very often the structure of power systems cannot be put either in the standard singularly perturbed or standard weakly coupled form. Some of these structures, which we call quasi singularly perturbed and quasi weakly coupled control systems, are studied in Chapter 9.

In Chapter 9, we consider a special class of linear control systems represented by the standard singularly perturbed system matrix and with the control input matrix having three different nonstandard forms. The obtained results are quite simplified (comparing to the standard singularly perturbed control systems), and in one case the optimal solution of the algebraic Riccati equation is completely determined in terms of the reduced-order algebraic Lyapunov equations. The proposed method is successfully applied to the reduced-order design of optimal controllers for a hydro power plant of Serbian power system (Skataric and Gajic, 1992). It is important to point out that the solutions to the real 11th and 14th-order hydro power control systems are obtained by the presented reduced-order parallel algorithms, but the global method fails to produce the answers in both cases.

In this chapter, we also consider a special class of linear systems having block diagonally dominant system matrix and with the control input influencing only one of subsystems. The optimal reduced-order controllers are designed through the recursive reduced-order parallel algorithm, which converges quickly to the required optimal solution (Skataric et al., 1990). Many real world systems, such as power systems, chemical reactors, flexible space structures, and in general, systems with only few actuators, possess the control structures studied in this section of Chapter 9. The proposed method is demonstrated on the controller design of three real control systems: chemical reactor, F-4 fighter aircraft, and the design of the decentralized multivariable excitation controllers in a nearly weakly coupled multimachine system.

In addition, the reduced-order solution is obtained for a class of linear-quadratic optimal control problems having weakly interconnected system matrix, strongly coupled control input matrix, and with a special structure for the state penalty matrix. Thus, the choice of the penalty matrix sometimes plays an important role in the process of system decomposition and distributed and parallel computations (Skataric et al., 1991). The results for this kind of the control system decomposition are demonstrated on the models of L-1011 fighter aircraft and distillation column.

Chapter 10 presents a case of singularly perturbed and weakly coupled systems, which exhibit at the same time both the multiple time scale and weak coupling phenomena (Gajic and Skataric, 1991). On the contrary to the multimodeling concept (Khalil and Kokotovic, 1978; Khalil, 1980b; Saksena and Cruz, 1981a, 1981b; Saksena and Basar, 1982; Saksena et al., 1983; Gajic and Khalil, 1986; Gajic, 1988; Zhuang and Gajic, 1990), where the weak coupling is allowed between fast variables only, in this chapter we study the effect of weak coupling between slow and fast variables. The optimal results are obtained in terms of parallel synchronous reduced-order algorithms under milder assumptions than for the standard singularly perturbed systems. These kind of systems appear particularly in the case of linearized models of flexible space structures (Moerder and Calise, 1985b) and in the optimal control problems of systems described by partial differential equations and presented in modal coordinates (Meirovich, 1967; Meirovich and Baruh, 1983; Baruh and Choe, 1990). Several case studies are presented: supported beam, satellite, power system, and fluid catalytic cracker.

The static output feedback of discrete quasi weakly coupled stochastic systems is studied in Chapter 11. A parallel reduced-order algorithm is developed which is very efficient, since the algorithm decomposes a high order system into a low order system. The low order system is represented by six Lyapunov equations which may be solved in parallel to reduce computational time. The required solution is obtained up to an arbitrary order of accuracy, $O\left(\epsilon^{2k}\right)$, where ϵ is a weak coupling parameter and k represents the number of iterations. The efficiency of the proposed method is demonstrated on two real aircraft examples which possess the quasi weakly coupled structure under prescribed degree of stability assumption (Hogan and Gajic, 1992).

In this chapter, we also study the static output feedback control problem for discrete linear singularly perturbed stochastic systems. A recursive algorithm is presented to solve the corresponding nonlinear algebraic equations (Qureshi et al., 1992). The algorithm removes the ill-conditioning by decomposing the higher order equations into lower order equations corresponding to the fast and slow time scales. As a case study the discrete model of a steam power system is considered. The obtained numerical results support the theoretical findings.

The application of the recursive reduced-order approach to differential games is given in Chapter 12. The analysis is restricted to the weakly coupled linear-quadratic Nash games and to the solution of corresponding coupled algebraic Riccati equations (Petrovic and Gajic, 1988). These results can be extended to the other types of differential games either in the context of weakly coupled or singularly perturbed systems.

Chapter 13 deals with the problem of high gain feedback and cheap control. The singular perturbation methodology is used to describe the problems under consideration (Kokotovic et al., 1986; Kokotovic and Khalil, 1986). The reduced-order parallel algorithm producing any arbitrary order of accuracy is obtained under the control oriented assumptions (Huey, 1992). It is important to point out that in the presented methodology there is no need to study the high gain feedback and cheap control problems in the limit when a small parameter ϵ tends to zero. This avoids the impulsive behavior and the presence of singular controls. The presented results are demonstrated on an example of a flexible space structure. In addition, in this chapter the open-loop cheap control problem and the problem of complete decomposition of the algebraic "cheap" Riccati equation into the reduced-order pure-slow and pure-fast Riccati equations are studied.

The time varying singularly perturbed and weakly coupled systems have been studied in several sections of Chapter 3, where the corresponding structures are outlined and the decomposition techniques have been developed for their studies. These methods are utilized in Chapter 14 in the context of linear approach to the optimal control of bilinear singularly perturbed and weakly coupled systems.

In Chapter 14, the composite near-optimal control of singularly perturbed bilinear systems is obtained (Aganovic and Gajic, 1991a) by combining the ideas from (Chow and Kokotovic, 1976) and (Cebuhar and Constanza, 1984). Obtained results are demonstrated on a fourth-order induction motor drives. The extension of the near-optimal composite control to the optimal reduced-order control is also considered. The reduced-order open-loop optimal control of singularly perturbed bilinear systems is presented by following results of (Aganovic and Gajic, 1991b).

In the remaining part of Chapter 14, we study the weakly coupled continuous-time bilinear optimal control problem (Aganovic, 1992). Both the open-loop and closed-loop optimal control problems are considered. As a case study, we consider the control problem of a paper making machine.

The system decomposition in this book is presented for both singularly perturbed and weakly coupled control systems composed of two subsystems. Corresponding parallel algorithms are solved by using two, three, or six processors working in parallel. However, under certain assumptions, the presented methods can be extended to the singularly perturbed systems with n-time scales, and to the weakly coupled systems with n subsystems. In those cases the original systems would be decomposed in n subsystems. Apparently, the corresponding control algorithms would be solved by many parallel processors. The study in that direction remains an open research problem.

In this book we do not study the implementation of the presented algorithms on the parallel computers. This may be another topic for future research. We have demonstrated the advantages of the presented algorithms. In two cases, 11th and 14th-order real hydro power control systems, (Chapter 9), we have shown that the classical approach fails to produce the answers. The parallel reduced-order synchronous algorithms have solved these problems successfully. The development of the asynchronous versions of these algorithms is underway. The advantages of the parallel algorithms will come into full effect once we solve very high order control systems.

We hope that these results, based on the recursive reduced-order fixed-point approach, can be extended to the nonlinear singularly perturbed and nonlinear weakly coupled control problems. There are a lot of research papers published on the nonlinear singularly perturbed systems (Kokotovic et al., 1986; Kokotovic and Khalil, 1986; O'Malley, 1991). All of them are based on the expansion methods. Introducing the recursive, fixed-point approach in this study will be a challenging research task. The nonlinear weakly coupled systems were originally introduced to the control audience in (Kokotovic and Singh, 1991). Recently, they have been studied in the context of differential games by (Srikant and Basar, 1991; 1992b).

It is known that the linearized models of dynamical systems described by partial differential equations in the modal coordinates (Meirovich and Baruh, 1983; Baruh and Choe, 1990) consist of an infinite set of second order internally decoupled differential equations. The coupling comes externally through the control input components. In practical applications an infinite dimensional set of differential equations is approximated by a finite one of order $2n$. Using the techniques developed in this book we believe that these kind of control problems can be solved in terms of n parallel algorithms of order 2.

INTRODUCTION

The book is divided into two parts: theoretical concepts (Chapters 2-8) and applications (Chapters 9-14). Each chapter of the book is self-contained, so that the reader, after completion of Chapters 2 and 3, can go directly to the chapter of his/her interests.

The book contains several exercises, computer assignments, and formulations of the research problems to help the instructors who might be using this book as a graduate text on large scale systems and/or parallel design of controllers. The required background for this book is a graduate level course on optimal control (Kwakernaak and Sivan, 1972; Sage and White, 1977; Lewis, 1986; Anderson and Moore, 1990). For the related control theory concepts we refer the reader to the excellent books (Kailath, 1980; Chen, 1984; Sontag, 1990).

PART ONE — Theoretical Concepts

Linear-Quadratic Control Problems

Decoupling Transformations

Output Feedback Control

Linear Stochastic Systems

Open-Loop Optimal Control

Exact Decompositions of Algebraic Riccati Equations

Differential and Difference Riccati Equations

Chapter 2

Linear-Quadratic Control Problems

2.1 Introduction

In this chapter, we study the main algebraic equations of the linear steady state control theory: the Lyapunov and Riccati algebraic equations, for both singularly perturbed and weakly coupled systems. We derive the corresponding recursive, reduced-order parallel algorithms for the solution of these equations in the most general case when the problem matrices are functions of a small perturbation parameter. The numerical decomposition has been achieved, so that only low-order systems are involved in algebraic computations. The introduced recursive methods are of the fixed point type and can be implemented as synchronous parallel algorithms (Bertsekas and Tsitsiklis, 1989; 1991).

Both continuous-time and discrete-time versions of the algebraic Lyapunov and Riccati equations are studied. The partitioned expressions of the algebraic Riccati equations have very complicated forms in the discrete-time domain for both singularly perturbed and weakly coupled systems. We have overcome that problem by using corresponding bilinear transformations, which are applicable under quite mild assumptions, so that the the solutions of the discrete algebraic Riccati equations for both singularly perturbed systems and weakly coupled systems are obtained by using results for the corresponding continuous-time algebraic Riccati equations. It is shown that the singular perturbation recursive methods converge with the rate of convergence of $O(\epsilon)$, whereas the recursive methods for weakly coupled linear systems converge faster, that is, with the rate of convergence of $O(\epsilon^2)$.

Having obtained the approximate solutions of the algebraic Lyapunov and Riccati equations, the corresponding approximate linear-quadratic control problems are solved in terms of these solutions. Several real world examples are included in order to demonstrate the procedures: magnetic tape control problem, F-8 aircraft, catalytic cracker, and chemical plant.

2.2 Recursive Methods for Singularly Perturbed Linear Continuous Systems

Consider a linear dynamic system

$$\dot{x} = A(\epsilon) x + B(\epsilon) u, \qquad x(0) = x_0 \qquad (2.1)$$

with a performance index

$$J(\epsilon) = \frac{1}{2} \int_0^\infty \left[x^T Q(\epsilon) x + u^T R(\epsilon) u \right] dt, \quad Q(\epsilon) \geq 0, \quad R(\epsilon) > 0$$

$$(2.2)$$

which has to be minimized, where ϵ is a small positive parameter, $x \in \Re^n$, $u \in \Re^m$, are state and control variables, respectively, with appropriate dimensions of the corresponding matrices. The optimal control u that minimizes (2.2) along trajectories of (2.1) is given by the well-known expression

$$u(t) = -R^{-1}(\epsilon) B^T(\epsilon) P(\epsilon) x(t) \qquad (2.3)$$

where $P(\epsilon)$ is the positive semidefinite stabilizing solution of the algebraic Riccati equation

$$P(\epsilon) A(\epsilon) + A^T(\epsilon) P(\epsilon) + Q(\epsilon) - P(\epsilon) S(\epsilon) P(\epsilon) = 0$$
$$S(\epsilon) = B R^{-1} B^T \qquad (2.4)$$

For $S(\epsilon) = 0$, the equation (2.4) becomes the algebraic Lyapunov equation. In this section, we will also study a dual form of the algebraic Lyapunov equation that represents a variance equation of a linear system driven by white noise

$$\dot{x} = A(\epsilon) x + G(\epsilon) \omega \qquad (2.5)$$

where ω is a zero-mean Gaussian stationary white noise process with a unity intensity matrix. The algebraic Lyapunov equation corresponding to (2.5), and representing the variance equation of $x(t)$, is given by

$$K(\epsilon) A^T(\epsilon) + A(\epsilon) K(\epsilon) + G(\epsilon) G^T(\epsilon) = 0 \qquad (2.6)$$

16

According to the theory of singular perturbations (Kokotovic and Khalil, 1986; Kokotovic et al., 1986), the following partitions of the problem matrices are introduced

$$A(\epsilon) = \begin{bmatrix} A_1(\epsilon) & A_2(\epsilon) \\ A_3(\epsilon)/\epsilon & A_4(\epsilon)/\epsilon \end{bmatrix}, \quad B(\epsilon) = \begin{bmatrix} B_1(\epsilon) \\ B_2(\epsilon)/\epsilon \end{bmatrix}$$

$$Q(\epsilon) = \begin{bmatrix} q_1^T(\epsilon)\,q_1(\epsilon) & q_1^T(\epsilon)\,q_2(\epsilon) \\ q_2^T(\epsilon)\,q_1(\epsilon) & q_2^T(\epsilon)\,q_2(\epsilon) \end{bmatrix}, \quad G(\epsilon) = \begin{bmatrix} G_1(\epsilon) \\ G_2(\epsilon)/\epsilon \end{bmatrix} \quad (2.7)$$

$$P(\epsilon) = \begin{bmatrix} P_1(\epsilon) & \epsilon P_2(\epsilon) \\ \epsilon P_2^T(\epsilon) & \epsilon P_3(\epsilon) \end{bmatrix}, \quad K(\epsilon) = \begin{bmatrix} K_1(\epsilon) & K_2(\epsilon) \\ K_2^T(\epsilon) & K_3(\epsilon)/\epsilon \end{bmatrix}$$

Newly defined matrices are of dimensions $A_1, P_1, K_1, q_1^T q_1 \in \Re^{n_1 \times n_1}$; $A_4, P_3, K_3, q_2^T q_2 \in \Re^{n_2 \times n_2}$; $B_i, G_i \in \Re^{n_i \times m}$, $i = 1, 2$; with $n_1 + n_2 = n$. It is assumed that all matrices are continuous functions of ϵ.

2.2.1 Parallel Algorithm for Solving Algebraic Lyapunov Equation

The partitioned form of the Lyapunov equation given in (2.6) is

$$A_1(\epsilon) K_1(\epsilon) + K_1(\epsilon) A_1^T(\epsilon) + A_2(\epsilon) K_2^T(\epsilon)$$
$$+ K_2(\epsilon) A_2^T(\epsilon) + G_1(\epsilon) G_1^T(\epsilon) = 0$$

$$K_2(\epsilon) A_4^T(\epsilon) + \epsilon A_1(\epsilon) K_2(\epsilon) + K_1(\epsilon) A_3^T(\epsilon)$$
$$+ A_2(\epsilon) K_3(\epsilon) + G_1(\epsilon) G_2^T(\epsilon) = 0 \qquad (2.8)$$

$$K_3(\epsilon) A_4^T(\epsilon) + A_4(\epsilon) K_3(\epsilon) + \epsilon A_3(\epsilon) K_2(\epsilon)$$
$$+ \epsilon K_2^T(\epsilon) A_3^T(\epsilon) + G_2(\epsilon) G_2^T(\epsilon) = 0$$

Define the following $O(\epsilon)$ perturbation of (2.8)

$$\mathbf{A_1}(\epsilon) \mathbf{K_1}(\epsilon) + \mathbf{K_1}(\epsilon) A_1^T(\epsilon) + \mathbf{K_2}(\epsilon) A_2^T(\epsilon)$$
$$+ A_2(\epsilon) \mathbf{K_2^T}(\epsilon) + G_1(\epsilon) G_1^T(\epsilon) = 0$$
$$A_2(\epsilon) \mathbf{K_3}(\epsilon) + \mathbf{K_2}(\epsilon) A_4^T(\epsilon) + \mathbf{K_1}(\epsilon) A_3^T(\epsilon) + G_1(\epsilon) G_2^T(\epsilon) = 0$$
$$\mathbf{K_3}(\epsilon) A_4^T(\epsilon) + A_4(\epsilon) \mathbf{K_3}(\epsilon) + G_2(\epsilon) G_2^T(\epsilon) = 0$$
$$(2.9)$$

Note that we did not set $\epsilon = 0$ in $A_i's$ and $G_i's$. In the rest of the chapter we will assume that all matrices are functions of ϵ. However, the explicit dependence on ϵ of the problem matrices will be omitted in order to simplify notation. Solution of (2.9) is in fact given in terms of two lower order algebraic Lyapunov equations

$$A_0 K_1 + K_1 A_0^T + G_0 G_0^T = 0$$
$$A_4 K_3 + K_3 A_4^T + G_2 G_2^T = 0$$
(2.10)

and

$$K_2 = - \left[A_2 K_3 + K_1 A_3^T + G_1 G_2^T \right] A_4^{-T}$$
(2.11)

where

$$A_0 = A_1 - A_2 A_4^{-1} A_3, \quad G_0 = G_1 - A_2 A_4^{-1} G_2$$
(2.12)

Exercise 2.1: Derive expressions (2.10)-(2.12) from equation (2.9).

$\triangle$

Unique solutions of (2.10)-(2.11) exist under the following assumption.
Assumption 2.1 Matrices $A_0(\epsilon)$ and $A_4(\epsilon)$ are stable.

$\triangle$

This is a standard assumption in the theory of singular perturbations (Kokotovic and Khalil, 1986; Kokotovic et al., 1986). Defining approximation errors as

$$K_1 = \mathbf{K}_1 + \epsilon E_1$$
$$K_2 = \mathbf{K}_2 + \epsilon E_2$$
$$K_3 = \mathbf{K}_3 + \epsilon E_3$$
(2.13)

and subtracting (2.10)-(2.11) from (2.8), we get the error equations (after some algebra) in the form

$$A_0 E_1 + E_1 A_0^T = A_0 \left[\mathbf{K}_2 + \epsilon E_2 \right] A_4^{-T} A_2^T + A_2 A_4^{-1} \left[\mathbf{K}_2 + \epsilon E_2 \right]^T A_0^T$$
$$A_4 E_3 + E_3 A_4^T = -A_3 \left[\mathbf{K}_2 + \epsilon E_2 \right] - \left[\mathbf{K}_2 + \epsilon E_2 \right]^T A_3^T$$
$$A_2 E_3 + E_1 A_3^T + E_2 A_4^T + A_1 \left[\mathbf{K}_2 + \epsilon E_2 \right] = 0$$
(2.14)

These equations have very nice forms since the unknown quantity E_2 in equations for E_1 and E_3 is multiplied by a small parameter ϵ. This fact suggests the following reduced-order parallel algorithm for solving (2.14).

Algorithm 2.1:

$$A_0 E_1^{(i+1)} + E_1^{(i+1)} A_0^T = A_0 \left[K_2 + \epsilon E_2^{(i)} \right] A_4^{-T} A_2^T$$

$$+ A_2 A_4^{-1} \left[K_2 + \epsilon E_2^{(i)} \right]^T A_0^T$$

$$A_4 E_3^{(i+1)} + E_3^{(i+1)} A_4^T = -A_3 \left[K_2 + \epsilon E_2^{(i)} \right] - \left[K_2 + \epsilon E_2^{(i)} \right]^T A_3^T$$

$$E_2^{(i+1)} = - \left\{ A_2 E_3^{(i+1)} + E_1^{(i+1)} A_3^T + A_1 \left[K_2 + \epsilon E_2^{(i)} \right] \right\} A_4^{-T},$$

$$i = 0, 1, 2, \dots.$$

(2.15)

with the starting point $E_2^{(0)} = 0$.

$\triangle$

Using the stability property imposed in Assumption 2.1, it is easy to show ((2.15) is a contraction mapping, Luenberger, 1969) that

$$\left\| E_j^{(i)} - E_j \right\| = O\left(\epsilon^i\right), \qquad j = 1, 2, 3; \qquad i = 1, 2, \dots \qquad (2.16)$$

Note that (2.16) is valid in the case when the last equation of (2.15) is in the form

$$E_2^{(i+1)} = - \left\{ A_2 E_3^{(i)} + E_1^{(i)} A_3^T + A_1 \left[K_2 + \epsilon E_2^{(i)} \right] \right\} A_4^{-T},$$

$$i = 0, 1, 2, \dots; \quad E_1^{(0)} = 0, \quad E_3^{(0)} = 0$$

(2.17)

Thus, the algorithm (2.15) is convergent. Using $E_j^{(\infty)}$, j = 1, 2, 3, in (2.15) and comparing it to (2.14), imply that the algorithm (2.15) converges to the unique solution of (2.14). In summary, we have the following theorem.

Theorem 2.1 *Under stability assumptions imposed on $A_0(\epsilon)$ and $A_4(\epsilon)$, the algorithm (2.15) converges to the exact solution E with the rate of convergence of $O(\epsilon)$, and thus, the required solution K can be obtained with the accuracy of $O(\epsilon^i)$ from*

$$K_j^{(i)} = K_j + \epsilon E_j^{(i)} = K_j + O\left(\epsilon^i\right), \qquad j = 1, 2, 3; \qquad i = 1, 2, \dots. \quad (2.18)$$

$\diamond$

It is important to notice that in the proposed method we do not need to expand $A_i(\epsilon)$, i = 1,...,4, into the Taylor or power-series, and we do not require stability of $A_0(0)$ and $A_4(0)$, which make the important features of the presented method. However, both $\det A_0(\epsilon)$ and $\det A_4(\epsilon)$ must be $O(1)$. Assumption 2.1 is more natural and less binding than the stability assumption imposed on $A_0(0)$ and $A_4(0)$ (Kokotovic and Khalil, 1986; Kokotovic et al., 1986). Namely, the singularly perturbed structure of a system is the consequence of a strict inequality $\epsilon > 0$ (small positive parameter). The stability requirement imposed on $A_0(0)$ and $A_4(0)$ is based on the continuation argument, but it can not be indefinitely exploited.

It is known that the power-series expansion method leads to two reduced-order Lyapunov equations similar to those in (2.15) — they are of the same order, but the number of terms on the right-hand sides of these equations for the power-series expansion method is growing very quickly with an increase in the required accuracy. It can be seen from (2.15) that for the fixed point method the number of terms on the right-hand side is constant. The number of matrix multiplications required to form right-hand sides of the Lyapunov equations, corresponding to the fast variables, that is $E_3^{(i+1)}$, for the accuracy of $O(\epsilon^i)$, is given in Table 2.1.

i	1	2	3	4	5	6
Fixed points	1	1	1	1	1	1
Power series	3	6	9	12	15	18

Table 2.1: Required number of matrix multiplications per iteration

This table shows very strong support for the proposed fixed point method. In addition, an important advantage of the presented fixed point algorithm is in its parallel and distributed structure.

2.2.2 Parallel Algorithm for Solving Algebraic Riccati Equation

This approach has been developed first in (Gajic, 1986) for the non-parametrized case. In this section, we study the fixed point method to the solution of the algebraic Riccati equation for a more general case when the problem matrices are continuous functions of ϵ.

Consider the algebraic Riccati equation of singularly perturbed systems defined in (2.4) and (2.7). Partitioning (2.4) subject to (2.7) we get the following equations

$$P_1 A_1 + A_1^T P_1 + P_2 A_3 + A_3^T P_2^T - P_1 S_1 P_1 - P_1 S P_2^T - P_2 S P_1$$
$$- P_2 S_2 P_2^T + q_1^T q_1 = 0$$

$$(2.19)$$

$$P_1 A_2 + P_2 A_4 + \epsilon A_1^T P_2 + A_3^T P_3 - \epsilon P_1 S_1 P_2 - P_1 S P_3 - \epsilon P_2 S^T P_2$$
$$- P_2 S_2 P_3 + q_1^T q_2 = 0$$

$$(2.20)$$

$$P_3 A_4 + A_4^T P_3 + \epsilon P_2^T A_2 + \epsilon A_2^T P_2 - P_3 S_2 P_3 - \epsilon^2 P_2^T S_1 P_2 - \epsilon P_2^T S P_3$$
$$- \epsilon P_3 S^T P_2 + q_2^T q_2 = 0$$

$$(2.21)$$

where

$$S_i = B_i R^{-1} B_i^T, \quad i = 1, 2 ; \qquad S = B_1 R^{-1} B_2^T \qquad (2.22)$$

Let us define the following $O(\epsilon)$ perturbation of (2.19)-(2.21)

$$\mathbf{P}_1 A_1 + A_1^T \mathbf{P}_1 + \mathbf{P}_2 A_3 + A_3^T \mathbf{P}_2^T - \mathbf{P}_1 S_1 \mathbf{P}_1$$
$$- \mathbf{P}_1 S \mathbf{P}_2^T - \mathbf{P}_2 S^T \mathbf{P}_1 - \mathbf{P}_2 S_2 \mathbf{P}_2^T + q_1^T q_1 = 0$$

$$(2.23)$$

$$\mathbf{P}_1 A_2 + \mathbf{P}_2 A_4 + A_3^T \mathbf{P}_3 - \mathbf{P}_1 S \mathbf{P}_3 - \mathbf{P}_2 S_2 \mathbf{P}_3 + q_1^T q_2 = 0 \qquad (2.24)$$

$$\mathbf{P}_3 A_4 + A_4^T \mathbf{P}_3 - \mathbf{P}_3 S_2 \mathbf{P}_3 + q_2^T q_2 = 0 \qquad (2.25)$$

It is important to point out that ϵ in the coefficient matrices is not set to zero.

The Riccati equation (2.25) will produce the unique positive semidefinite stabilizing solution under the following assumption.

Assumption 2.2 The triple $(A_4(\epsilon), B_2(\epsilon), q_2(\epsilon))$ is stabilizable-detectable.

$$\triangle$$

From (2.24) we obtain

$$\mathbf{P}_2 = - \left(\mathbf{P}_1 A_2 + A_3^T \mathbf{P}_3 - \mathbf{P}_1 S \mathbf{P}_3 + q_1^T q_2 \right) (A_4 - S_2 \mathbf{P}_3)^{-1} \qquad (2.26)$$

which after a substitution in (2.23) and elimination of $\mathbf{P}_3$ produces the reduced-order slow algebraic Riccati equation, (Kokotovic and Khalil, 1986; Kokotovic et al., 1986), in the form

$$\mathbf{P}_1 A + A^T \mathbf{P}_1 - \mathbf{P}_1 S \mathbf{P}_1 + \mathbf{Q} = 0 \qquad (2.27)$$

where
$$A = A_0 - B_0 R_0^{-1} r^T Q_0, \quad S = B_0 R_0^{-1} B_0^T, \quad B_0 = B_1 - A_2 A_4^{-1} B_2$$
$$Q_0 = q_1 - q_2 A_4^{-1} A_3, \quad R_0 = R + r^T r, \quad r = -q_2 A_4^{-1} B_2$$
$$\mathbf{Q} = Q_0^T \left(I - r R_0^{-1} r^T \right) Q_0$$

$$(2.28)$$

Exercise 2.2: Derive the expression for the "slow" algebraic Riccati equation (2.27).

$$\triangle$$

The unique positive semidefinite stabilizing solution of (2.27) exists under the following assumption.

Assumption 2.3 The triple $\left(\mathbf{A}(\epsilon), B_0(\epsilon), \sqrt{\mathbf{Q}(\epsilon)} \right)$ is stabilizable-detectable.

$$\triangle$$

Therefore the zero-order solution has the form
$$\mathbf{P}(\epsilon) = \begin{bmatrix} \mathbf{P}_1(\epsilon) & \epsilon \mathbf{P}_2(\epsilon) \\ \epsilon \mathbf{P}_2^T(\epsilon) & \epsilon \mathbf{P}_3(\epsilon) \end{bmatrix}$$

$$(2.29)$$

The zero-order solution is $O(\epsilon)$ close to the exact one. We define errors as
$$P_j(\epsilon) = \mathbf{P}_j(\epsilon) + \epsilon E_j(\epsilon), \quad j = 1, 2, 3$$

$$(2.30)$$

The $O(\epsilon^k)$ approximation of $E_i's$ will produce the $O(\epsilon^{k+1})$ approximation of the required matrix P, which is why we are interested in finding equations for the error term and a convenient algorithm for their solutions. Subtracting (2.23)-(2.25) from (2.19)-(2.21) and using (2.30) we arrive at the following expression for the error equation
$$E_1 D_1 + D_1^T E_1 = D^T H_1 + H_1 D + D^T H_3 D + \epsilon H_2$$
$$E_2 D_3 + E_1 D_{21} + D_{22}^T E_3 = -H_1$$
$$E_3 D_3 + D_3^T E_3 = H_3$$

$$(2.31)$$

where
$$D_3 = A_4 - S_2 \mathbf{P}_3, \quad D_{22} = A_3 - S^T \mathbf{P}_1 - S_2 \mathbf{P}_2^T$$
$$D_{21} = A_2 - S \mathbf{P}_3, \quad D = D_3^{-1} D_{22}$$
$$D_{11} = A_1 - S_1 \mathbf{P}_1 - S \mathbf{P}_2^T, \quad D_1 = D_{11} - D_{21} D_3^{-1} D_{22}$$

$$(2.32)$$

and
$$H_1 = A_1^T \mathbf{P}_2 - \mathbf{P}_1 S_1 \mathbf{P}_2 - \mathbf{P}_2 S^T \mathbf{P}_2 - \epsilon (E_1 S E_3 + E_2 S_2 E_3)$$
$$H_2 = E_1 S_1 E_1 + E_1 S E_2^T + E_2 S^T E_1 + E_2 S_2 E_2^T$$
$$H_3 = -\mathbf{P}_2^T A_2 - A_2^T \mathbf{P}_2 + \epsilon \mathbf{P}_2^T S_1 \mathbf{P}_2 + \epsilon E_3 S_2 E_3 + \mathbf{P}_2^T S \mathbf{P}_3 + \mathbf{P}_3 S^T \mathbf{P}_2$$

$$(2.33)$$

Equations (2.31) have all cross-coupling terms and all nonlinear terms multiplied by a small parameter ϵ, which suggests that a fixed point algorithm can be efficient for their solution. We will propose the following algorithm, similar to one obtained in (Gajic, 1986), for the non-parametrized case.

Algorithm 2.2:

$$E_3^{(i+1)} D_3 + D_3^T E_3^{(i+1)} = H_3^{(i)}$$
$$E_2^{(i+1)} D_3 + E_1^{(i+1)} D_{21} + D_{22}^T E_3^{(i+1)} = -H_1^{(i)}$$
$$E_1^{(i+1)} D_1 + D_1^T E_1^{(i+1)} = D^T H_1^{(i)T} + H_1^{(i)} D + D^T H_3^{(i)} D + \epsilon H_2^{(i)}$$
$$E_1^{(0)} = 0, \quad E_2^{(0)} = 0, \quad E_3^{(0)} = 0, \quad i = 0, 1, 2, 3, \ldots$$

$$(2.34)$$

$$\triangle$$

The following theorem indicates the features of the algorithm (2.34).

Theorem 2.2 *Under stabilizability-detectability conditions, imposed in Assumptions 2.2 and 2.3, the algorithm (2.34) converges to the exact solution of E with the rate of convergence of $O(\epsilon)$, that is*

$$\left\| E - E^{(i+1)} \right\| = O(\epsilon) \left\| E - E^{(i)} \right\|, \quad i = 0, 1, 2, 3, \ldots. \quad (2.35)$$

or equivalently

$$\left\| E - E^{(i)} \right\| = O(\epsilon^i) \quad (2.36)$$

$$\diamond$$

Proof: As a staring point we need to show the existence of a bounded solution of $E_1, E_2,$ and E_3 in the neighborhood of ϵ^*, where $\epsilon^* \in [\epsilon_{min}; \epsilon_{max}]$. To prove that, by the implicit function theorem, it is enough to show, that the corresponding Jacobian is nonsingular at ϵ^*. The Jacobian is given by

$$J(\epsilon) = \begin{bmatrix} J_{11}(\epsilon) & 0 & 0 \\ J_{21}(\epsilon) & J_{22}(\epsilon) & J_{23}(\epsilon) \\ 0 & 0 & J_{33}(\epsilon) \end{bmatrix} + \begin{bmatrix} 0 & 0 & 0 \\ O(\epsilon) & O(\epsilon) & 0 \\ 0 & O(\epsilon) & O(\epsilon) \end{bmatrix} \quad (2.37)$$

Using the Kronecker product representation we have

$$J_{11} = D_{11}^T \oplus I_{n_1} + I_{n_1} \oplus D_{11}^T, \quad J_{22} = D_3^T \oplus I_{n_2}$$
$$J_{33} = D_3^T \oplus I_{n_2} + I_{n_2} \oplus D_3^T \quad (2.38)$$

For the Jacobian to be nonsingular J_{ii}, $i = 1, 2, 3$, have to be nonsingular. The matrix D_3 is the closed-loop matrix of the fast subsystem, and thus stable by the well-known properties of the solution of the algebraic Riccati equation. The matrix D_1 can be easily shown to be the closed-loop matrix of the reduced slow subsystem and it is stable also (Kokotovic and Khalil, 1986; Kokotovic et al., 1986). By known properties of the Kronecker product (Lancaster and Tismenetsky, 1985), matrices J_{ii} are then nonsingular. Thus, for ϵ^* small enough the Jacobian $J(\epsilon^*)$ is nonsingular.

The second part of the proof is to produce an estimate of the rate of convergence and to verify (2.35) and (2.36). That can be done similarly to the corresponding proof of Theorem 2.4 and thus, is omitted here.

●

Therefore we are able to find the exact solution of the full-order algebraic Riccati equation of singularly perturbed systems, by recursively solving two reduced-order Lyapunov equations and one linear equation in a parallel manner.

2.2.3 Case Study: Magnetic Tape Control Problem

In order to illustrate the efficiency of the proposed algorithm for solving the algebraic Riccati equation of singularly perturbed systems, we shall consider the magnetic tape control system example (Chow and Kokotovic, 1976) given by

$$
A(\epsilon) = \begin{bmatrix} 0 & 0.4 & 0 & 0 \\ 0 & 0 & 0.345 & 0 \\ 0 & \frac{-0.524}{\epsilon} & \frac{-0.465}{\epsilon} & \frac{0.262}{\epsilon} \\ 0 & 0 & 0 & -1/\epsilon \end{bmatrix}, \quad B = \begin{bmatrix} 0 \\ 0 \\ 0 \\ \frac{1}{\epsilon} \end{bmatrix}
$$

$$
R = diag\{1\ 0\ 1\ 0\}, \quad Q = 1, \quad \epsilon = 0.1
$$

With an accuracy of up to 6 decimal places, we obtained convergence to the exact solution in 4 iterations. Componentwise results are given in Table 2.2. It can be seen that the obtained numerical results are in agreement with the statement established in Theorem 2.2.

Using different values of ϵ in the same example, we can note very good convergence rates even with relatively large values of perturbation parameter. These results are displayed in Table 2.3

The advantages of the fixed point method presented are: (a) The size of the required computations is considerably less. Since this size does not grow per iteration, the method is extremely efficient for obtaining

	$P^{(0)}$	$P^{(1)}$	$P^{(2)}$	$P^{(3)}$	$P^{(4)}$ =exact
P_{11}	7.384024	7.540292	7.540064	7.540042	7.540043
P_{12}	5.904760	6.166314	6.170524	6.170445	6.170447
P_{13}	0.399308	0.405280	0.405345	0.405342	0.405342
P_{14}	0.100000	0.100000	0.100000	0.100000	0.100000
P_{22}	7.151604	7.452234	7.467309	7.467275	7.467278
P_{23}	0.379770	0.394804	0.395104	0.395100	0.395100
P_{24}	0.086123	0.089245	0.089202	0.089202	0.089202
P_{33}	0.104029	0.129797	0.130441	0.130441	0.130441
P_{34}	0.018036	0.024283	0.024398	0.024396	0.024396
P_{44}	0.004619	0.006183	0.006200	0.006200	0.006200

Table 2.2: Solution of the Riccati equation for magnetic tape control problem

ϵ	number of iterations for convergence
0.01	1
0.1	4
0.5	7
0.9	10

Table 2.3: Dependence of number of iterations on ϵ.

solutions of very high accuracy. (b) The fixed point method is recursive in nature (the power-series expansion method is not), and is thus much easier to implement. (c) In the more general case, when coefficients are functions of the small parameter ϵ, the power-series expansion method asks for analyticity of all coefficients with respect to ϵ at $\epsilon = 0$, whereas for application of the fixed-point method, we need only continuity of the same coefficients on the compact convex set, or continuously differentiable functions of ϵ on the compact set (Zangwill and Garcia, 1981).

Thus, the fixed point method demands much milder conditions than a power-series expansion method.

2.3 Recursive Methods for Weakly Coupled Linear Continuous Systems

The weakly coupled linear-quadratic control problem is defined by (2.1)-(2.4), subject to the following partition of the problem matrices (Kokotovic et al., 1969)

$$A(\epsilon) = \begin{bmatrix} A_1(\epsilon) & \epsilon A_2(\epsilon) \\ \epsilon A_3(\epsilon) & A_4(\epsilon) \end{bmatrix}, \qquad B(\epsilon) = \begin{bmatrix} B_1(\epsilon) & \epsilon B_2(\epsilon) \\ \epsilon B_3(\epsilon) & B_4(\epsilon) \end{bmatrix}$$

(2.39)

$$Q(\epsilon) = \begin{bmatrix} Q_1(\epsilon) & \epsilon Q_2(\epsilon) \\ \epsilon Q_2^T(\epsilon) & Q_3(\epsilon) \end{bmatrix}, \qquad R(\epsilon) = \begin{bmatrix} R_1(\epsilon) & 0 \\ 0 & R_2(\epsilon) \end{bmatrix}$$

where ϵ is a small parameter. Dimensions of partitioned matrices are compatible to those defined in (2.7). In addition, the existence of the weakly coupled systems of order n_1 and n_2 is conditioned by the following assumption (Chow and Kokotovic, 1983).

Assumption 2.4 The linear system (2.1) subject to (2.39) displays the weakly coupled structure under the assumption $det\ A_1(\epsilon) = O(1)$ and $det\ A_4(\epsilon) = O(1)$.

△

In this section, we will develop the recursive fixed point type parallel algorithms for solving the algebraic Lyapunov and Riccati equations of weakly coupled systems.

2.3.1 Parallel Algorithm for Solving Algebraic Lyapunov Equation

The algebraic Lyapunov equation of weakly coupled systems ("regulator type") is given by

$$A^T(\epsilon) P(\epsilon) + P(\epsilon) A(\epsilon) + Q(\epsilon) = 0 \qquad (2.40)$$

Due to block dominant structure of matrices A and Q, the required solution P is properly scaled as follows

$$P(\epsilon) = \begin{bmatrix} P_1(\epsilon) & \epsilon P_2(\epsilon) \\ \epsilon P_2^T(\epsilon) & P_3(\epsilon) \end{bmatrix} \qquad (2.41)$$

Partitioned form of (2.40) subject to (2.39) produces

$$\begin{aligned} P_1 A_1 + A_1^T P_1 + Q_1 + \epsilon^2 \left(P_2 A_3 + A_3^T P_2^T \right) &= 0 \\ P_1 A_2 + P_2 A_4 + A_1^T P_2 + A_3^T P_3 + Q_2 &= 0 \\ P_3 A_4 + A_4^T P_3 + Q_3 + \epsilon^2 \left(P_2^T A_2 + A_2^T P_2 \right) &= 0 \end{aligned} \qquad (2.42)$$

We define the $O\left(\epsilon^2\right)$ approximation of (2.42) as

$$\begin{aligned} \mathbf{P}_1 A_1 + A_1^T \mathbf{P}_1 + Q_1 &= 0 \\ \mathbf{P}_2 A_4 + A_1^T \mathbf{P}_2 &= -\mathbf{P}_1 A_2 - A_3^T \mathbf{P}_3 - Q_2 \\ \mathbf{P}_3 A_4 + A_4^T \mathbf{P}_3 + Q_3 &= 0 \end{aligned} \qquad (2.43)$$

Note that we did not set $\epsilon = 0$ in $A_i's$ and $Q_i's$, so that $\mathbf{P}_i$'s are functions of ϵ.

The unique solution of (2.43) exists under the following assumption.
Assumption 2.5 Matrices $A_1(\epsilon)$ and $A_4(\epsilon)$ are stable.

$\triangle$

Defining approximation errors as

$$P_j = \mathbf{P}_j + \epsilon^2 E_j, \qquad j = 1, 2, 3 \qquad (2.44)$$

and subtracting (2.43) from (2.42) we obtain the following expression for the errors

$$\begin{aligned} E_1 A_1 + A_1^T E_1 + \mathbf{P}_2 A_3 + A_3^T \mathbf{P}_2^T + \epsilon^2 \left(E_2 A_3 + A_3^T E_2^T \right) &= 0 \\ E_2 A_4 + A_1^T E_2 + E_1 A_2 + A_3^T E_3 &= 0 \\ E_3 A_4 + A_4^T E_3 + A_2^T \mathbf{P}_2 + \mathbf{P}_2^T A_2 + \epsilon^2 \left(A_2^T E_2 + E_2^T A_2 \right) &= 0 \end{aligned} \qquad (2.45)$$

We propose the following algorithm, having reduced-order and parallel structure, for solving (2.45).
Algorithm 2.3:

$$\begin{aligned} E_1^{(i+1)} A_1 + A_1^T E_1^{(i+1)} + P_2^{(i)} A_3 + A_3^T P_2^{(i)^T} &= 0 \\ E_3^{(i+1)} A_4 + A_4^T E_3^{(i+1)} + A_2^T P_2^{(i)} + P_2^{(i)^T} A_2 &= 0 \\ E_2^{(i+1)} A_4 + A_1^T E_2^{(i+1)} + E_1^{(i+1)} A_2 + A_3^T E_3^{(i+1)} &= 0, \qquad i = 0, 1, 2, ... \end{aligned} \qquad (2.46)$$

with the starting point $E_2^{(0)} = 0$ and with

$$P_j^{(i)} = P_j + \epsilon^2 E_j^{(i)}, \qquad j = 1, 2, 3; \quad i = 0, 1, 2, .. \tag{2.47}$$

$$\triangle$$

Using the same arguments like in Section 2.2, we can establish the following theorem.

Theorem 2.3 *Under stability assumptions imposed on matrices $A_1(\epsilon)$ and $A_4(\epsilon)$, the algorithm (2.46) converges to the exact solution E with the rate of convergence of $O(\epsilon^2)$, and thus, the required solution P can be obtained with the accuracy of $O(\epsilon^{2i})$ from (2.47), that is*

$$P_j = P_j^{(i)} + O(\epsilon^{2i}), \qquad j = 1, 2, 3; \quad i = 1, 2, 3.... \tag{2.48}$$

$$\diamond$$

2.3.2 Parallel Algorithm for Solving Algebraic Riccati Equation

The algebraic Riccati equation (2.4), subject to the weakly coupled structure given in (2.39), has the solution partitioned as in (2.41). Substitution of (2.39) and (2.41) in (2.4) will produce the following partitioned equations

$$P_1 A_1 + A_1^T P_1 + Q_1 - P_1 S_1 P_1 + \epsilon^2 \left(P_2 A_3 + A_3^T P_2^T \right)$$
$$-\epsilon^2 \left[(P_1 S_{12} + P_2 Z^T) P_1 + (P_1 Z + P_2 (S_2 + \epsilon^2 S_{21})) P_2^T \right] = 0 \tag{2.49}$$

$$P_3 A_4 + A_4^T P_3 + Q_3 - P_3 S_2 P_3 + \epsilon^2 \left(P_2^T A_2 + A_2^T P_2 \right)$$
$$-\epsilon^2 \left[(P_3 S_{21} + P_2^T Z) P_3 + (P_3 Z^T + P_2^T (S_1 + \epsilon^2 S_{12})) P_2 \right] = 0 \tag{2.50}$$

$$P_1 A_2 + P_2 A_4 + A_1^T P_2 + A_3^T P_3 + Q_2 - P_1 S_1 P_2 - P_1 Z P_3 - P_2 S_2 P_3$$
$$-\epsilon^2 \left[(P_1 S_{12} + P_2 Z^T) P_2 + P_2 S_{21} P_3 \right] = 0 \tag{2.51}$$

where

$$S_1 = B_1 R_1^{-1} B_1^T, \quad S_2 = B_4 R_2^{-1} B_4^T, \quad S_{12} = B_2 R_2^{-1} B_2^T$$
$$S_{21} = B_3 R_1^{-1} B_3^T, \qquad Z = B_1 R_1^{-1} B_3^T + B_2 R_2^{-1} B_4^T \tag{2.52}$$

The $O(\epsilon^2)$ approximation of (2.49)-(2.51) is defined as

$$P_1 A_1 + A_1^T P_1 - P_1 S_1 P_1 + Q_1 = 0$$
$$P_3 A_4 + A_4^T P_3 - P_3 S_2 P_3 + Q_3 = 0 \tag{2.53}$$

and

$$P_2 D_2 + D_1^T P_2 = -\left(P_1 A_2 + A_3^T P_3 + Q_2 - P_1 Z P_3\right) \tag{2.54}$$

where

$$D_1(\epsilon) = [A_1(\epsilon) - S_1(\epsilon) P_1(\epsilon)], \quad D_2(\epsilon) = [A_4(\epsilon) - S_2(\epsilon) P_3(\epsilon)] \tag{2.55}$$

The unique positive semidefinite stabilizing solutions of (2.53) exist under the following assumption.

Assumption 2.6 The triples $(A_1(\epsilon), B_1(\epsilon), \sqrt{Q_1(\epsilon)})$ and $(A_4(\epsilon), B_4(\epsilon), \sqrt{Q_3(\epsilon)})$ are stabilizable-detectable.

$$\triangle$$

Under Assumption 2.6 matrices $D_1(\epsilon)$ and $D_2(\epsilon)$ are stable so that the unique solution of (2.54) exists also.

If the errors are defined as

$$P_j = \mathbf{P}_j + \epsilon^2 E_j, \quad j = 1, 2, 3 \tag{2.56}$$

then the exact solution will be of the form

$$P = \begin{bmatrix} \mathbf{P}_1 + \epsilon^2 E_1 & \epsilon\left(\mathbf{P}_2 + \epsilon^2 E_2\right) \\ \epsilon\left(\mathbf{P}_2 + \epsilon^2 E_2\right)^T & \mathbf{P}_3 + \epsilon^2 E_3 \end{bmatrix} \tag{2.57}$$

Subtracting (2.53)-(2.54) from the corresponding equations (2.49)-(2.51) and using (2.56) produce the following equations for the errors

$$E_1 D_1 + D_1^T E_1 = P_1 S_{12} P_1 + P_2 Z^T P_1 + P_1 Z P_2^T + P_2 S_2 P_2^T$$
$$- P_2 A_3 - A_3^T P_2^T + \epsilon^2 \left(E_1 S_1 E_1 + P_2 S_{21} P_2^T\right) \tag{2.58}$$

$$E_3 D_2 + D_2^T E_3 = P_3 S_{21} P_3 + P_2^T S_1 P_2 + P_3 Z^T P_2 + P_2^T Z P_3$$
$$- P_2^T A_2 - A_2^T P_2 + \epsilon^2 \left(E_3 S_2 E_3 + P_2^T S_{12} P_2\right) \tag{2.59}$$

$$D_1^T E_2 + E_2 D_2 = P_1 S_{12} P_2 + P_2 Z^T P_2 + P_2 S_{21} P_3$$
$$- E_1 D_{12} - D_{21}^T E_3 + \epsilon^2 \left(E_1 S_1 E_2 + E_1 Z E_2 + E_2 S_2 E_3\right) \tag{2.60}$$

where

$$D_{12} = A_2 - S_1 \mathbf{P}_2 - Z \mathbf{P}_3, \quad D_{21} = A_3 - S_2 \mathbf{P}_2^T - Z^T \mathbf{P}_1 \tag{2.61}$$

It can be easily shown that the nonlinear equations (2.58)-(2.60) have the form

$$E_1 D_1 + D_1^T E_1 = const + \epsilon^2 f_1 \left(E_1, E_2, \epsilon^2 \right)$$
$$E_3 D_2 + D_2^T E_3 = const + \epsilon^2 f_3 \left(E_2, E_3, \epsilon^2 \right) \qquad (2.62)$$
$$E_2 D_2 + D_1^T E_2 = const + \epsilon^2 f_2 \left(E_1, E_2, E_3, \epsilon^2 \right)$$

We can see that all cross coupling terms and all nonlinear terms in (2.58)-(2.60) are multiplied by ϵ^2, so that we propose the following reduced-order parallel algorithm for solving (2.58)-(2.60).

Algorithm 2.4:

$$E_1^{(i+1)} D_1 + D_1^T E_1^{(i+1)} = P_1^{(i)} S_{12} P_1^{(i)} + P_2^{(i)} Z^T P_1^{(i)} + P_1^{(i)} Z P_2^{(i)^T}$$
$$-P_2^{(i)} A_3 - A_3^T P_2^{(i)^T} + \epsilon^2 \left(E_1^{(i)} S_1 E_1^{(i)} + P_2^{(i)} S_{21} P_2^{(i)^T} \right) + P_2^{(i)} S_2 P_2^{(i)^T}$$
$$(2.63)$$

$$E_3^{(i+1)} D_2 + D_2^T E_3^{(i+1)} = P_3^{(i)} S_{21} P_3^{(i)} + P_2^{(i)^T} Z P_3^{(i)} + P_3^{(i)} Z^T P_2^{(i)}$$
$$-P_2^{(i)^T} A_2 - A_2^T P_2^{(i)} + \epsilon^2 \left(E_3^{(i)} S_2 E_3^{(i)} + P_2^{(i)^T} S_{12} P_2^{(i)} \right) + P_2^{(i)^T} S_1 P_2^{(i)}$$
$$(2.64)$$

$$D_1^T E_2^{(i+1)} + E_2^{(i+1)} D_2 = P_1^{(i+1)} S_{12} P_2^{(i)} + P_2^{(i)} Z^T P_2^{(i)}$$
$$-E_1^{(i+1)} D_{12} - D_{21}^T E_3^{(i+1)} + P_2^{(i)} S_{21} P_3^{(i)}$$
$$+\epsilon^2 \left(E_1^{(i+1)} S_1 E_2^{(i)} + E_1^{(i+1)} Z E_2^{(i)} + E_2^{(i)} S_2 E_3^{(i+1)} \right)$$
$$(2.65)$$

with $E_1^{(0)} = 0, \; E_2^{(0)} = 0, \; E_3^{(0)} = 0$, where

$$P_j^{(i)} = P_j + \epsilon^2 E_j^{(i)}, \quad j = 1, 2, 3; \quad i = 1, 2, 3, ... \qquad (2.66)$$

$$\triangle$$

Exercise 2.3: Derive Algorithm 2.4 and write a MATLAB program (Hill, 1988) for solving equations (2.53)-(2.54) and (2.63)-(2.66).

$$\triangle$$

The following theorem indicates the features of the algorithm (2.63)-(2.66).

Theorem 2.4 *Under Assumption 2.6, the algorithm (2.63)-(2.66) converges to the exact solution of E with the rate of convergence of $O\left(\epsilon^2\right)$, that is*

$$\left\| E - E^{(i+1)} \right\| = O\left(\epsilon^2\right) \left\| E - E^{(i)} \right\|, \quad i = 0, 1, 2, .. \quad (2.67)$$

or equivalently

$$\left\| E - E^{(i)} \right\| = O\left(\epsilon^{2i}\right) \quad (2.68)$$

◇

Proof: The Jacobian of (2.49)-(2.51), at some $\epsilon = \epsilon^*$, is given by

$$J(\epsilon) = \begin{bmatrix} J_{11}(\epsilon) & 0 & 0 \\ J_{21}(\epsilon) & J_{22}(\epsilon) & J_{23}(\epsilon) \\ 0 & 0 & J_{33}(\epsilon) \end{bmatrix} + \begin{bmatrix} O\left(\epsilon^2\right) & O\left(\epsilon^2\right) & 0 \\ O\left(\epsilon^2\right) & O\left(\epsilon^2\right) & O\left(\epsilon^2\right) \\ 0 & O\left(\epsilon^2\right) & O\left(\epsilon^2\right) \end{bmatrix}$$
$$(2.69)$$

where

$$J_{11}(\epsilon) = I_{n_1} \oplus D_1^T(\epsilon) + D_1^T(\epsilon) \oplus I_{n_1}$$
$$J_{22}(\epsilon) = I_{n_2} \oplus D_2^T(\epsilon) + D_1^T(\epsilon) \oplus I_{n_1} \quad (2.70)$$
$$J_{33}(\epsilon) = I_{n_2} \oplus D_2^T(\epsilon) + D_2^T(\epsilon) \oplus I_{n_2}$$

Since $D_1(\epsilon)$ and $D_2(\epsilon)$ are stable matrices (by Assumption 2.6), $J_{ii}(\epsilon)$, $i = 1, 2, 3$, are nonsingular and hence the Jacobian will be nonsingular at $\epsilon = \epsilon^*$, assuming that ϵ is sufficiently small. Then, by the implicit function theorem, the existence of the unique bounded solution of (2.49)-(2.51) is guaranteed.

In the next step, we have to prove convergence of the algorithm (2.63)-(2.66) and to give an estimate of the rate of convergence. For $i = 0$, (2.58) and (2.63) imply

$$\left(E_1 - E_1^{(1)}\right) D_1 + D_1^T \left(E_1 - E_1^{(1)}\right) = \epsilon^2 f_1 \left(E_1, E_2, \epsilon^2\right) \quad (2.71)$$

Since $D_1(\epsilon)$ is stable and E_1 and E_2 are bounded it follows that

$$\left\| E_1 - E_1^{(1)} \right\| = O\left(\epsilon^2\right) \quad (2.72)$$

Similarly from (2.59) and (2.64) we have

$$\left(E_3 - E_3^{(1)}\right) D_2 + D_2^T \left(E_3 - E_3^{(1)}\right) = \epsilon^2 f_3 \left(E_2, E_3, \epsilon^2\right) \quad (2.73)$$

and

$$\left\| E_3 - E_3^{(1)} \right\| = O\left(\epsilon^2\right) \quad (2.74)$$

Using the same arguments in (2.60) and (2.65) will produce

$$\left\| E_2 - E_2^{(1)} \right\| = O\left(\epsilon^2\right) \qquad (2.75)$$

Continuing the same procedure and by induction we conclude that

$$\left\| E_1 - E_1^{(i)} \right\| = O\left(\epsilon^{2i}\right)$$
$$\left\| E_2 - E_2^{(i)} \right\| = O\left(\epsilon^{2i}\right) \qquad (2.76)$$
$$\left\| E_3 - E_3^{(i)} \right\| = O\left(\epsilon^{2i}\right)$$

with $i = 1, 2, 3, \dots$, which completes the proof of Theorem 2.4.

2.4 Approximate Linear Regulator for Continuous Systems

The positive semidefinite stabilizing solution of the algebraic Riccati equation (2.4), produces the answer to the optimal linear-quadratic steady state control problem. Namely, the quadratic criterion (2.2) is minimized along trajectories of the linear dynamic system (2.1) by using the control input in the form (2.3). It is proved in (Kokotovic and Cruz, 1969) that the near-optimal control given by

$$u^{(j)}(t) = -R^{-1}B^T P^{(j)} x^{(j)}(t) = -F^{(j)} x^{(j)}(t) \qquad (2.77)$$

where $P^{(j)}$ satisfies

$$P^{(j)} - P^{opt} = O\left(\epsilon^j\right) \qquad (2.78)$$

and

$$\dot{x}^{(j)}(t) = A\left(\epsilon\right) x^{(j)}(t) + B\left(\epsilon\right) u^{(j)}(t) \qquad (2.79)$$

is near-optimal in the sense

$$J^{(j)} - J^{opt} = O\left(\epsilon^{2j}\right) \qquad (2.80)$$

The approximate performance $J^{(j)}$ can be obtained from the algebraic Lyapunov equation

$$\left(A - BF^{(j)}\right)^T K^{(j)} + K^{(j)} \left(A - BF^{(j)}\right) + Q + F^{(j)^T} R F^{(j)} = 0 \quad (2.81)$$

so that

$$J^{(j)} = \frac{1}{2} x^T(0) K^{(j)} x(0) \qquad (2.82)$$

In the previous sections, we have developed very efficient techniques for generating $P^{(j)}$ for both singularly perturbed and weakly coupled systems. Thus, the proposed algorithms represent the methods for solving the linear-quadratic optimal control problems of both singularly perturbed and weakly coupled systems.

Exercise 2.4: Using results from Exercise 2.3 find the 5-th order approximations of the optimal criterion and the optimal trajectories for a chemical plant given in (Gomathi et al., 1980). Assume that the penalty matrices are identities.

$$\triangle$$

2.5 Recursive Methods for Singularly Perturbed Linear Discrete Systems

The linear singularly perturbed discrete systems have been studied in different set ups by many researchers. Two main structures of singularly perturbed linear discrete systems have been considered: the fast time scale version (Butuzov and Vasileva, 1971; Hoppensteadt and Miranker, 1977; Blankenship, 1981; Litkouhi and Khalil, 1984, 1985; Mahmoud, 1986; Oloomi and Sawan, 1987; Khorasani and Azimi-Sadjadi, 1987) and the slow time scale version (Phillips, 1980; Naidu and Rao, 1985). Discrete-time models of singularly perturbed linear systems, similar to (Phillips, 1980; Naidu and Rao, 1985), were studied also in (Othman et al., 1985; Mahmoud et al., 1986). Since the slow time scale version presupposes the asymptotic stability of the fast modes, it seems that in the design procedure of stabilizing feedback controllers, the fast time scale version is much more appropriate (Litkouhi and Khalil, 1985).

In this section, we will adopt the structure of singularly perturbed discrete linear systems defined by Litkouhi and Khalil, and study the corresponding linear-quadratic discrete control problems. We will take the approach based on a bilinear transformation (Kondo and Furuta, 1986). The main equations of the optimal linear control theory — the

Lyapunov and Riccati equations — are solved by recursive, reduced-order parallel algorithms in the most general case when the system matrices are functions of a small perturbation parameter. Since the Riccati equation has quite complicated form in the discrete-time domain, partitioning this equation, in the spirit of singular perturbation methodology, will produce a lot of terms and make corresponding problem numerically inefficient, even though the problem order-reduction is achieved. By applying a bilinear transformation, the solution of the discrete-time algebraic Riccati equation of singularly perturbed systems is obtained by using already known results for the corresponding continuous-time algebraic Riccati equation.

The proposed methods produce the reduced-order near-optimal solutions, up to an arbitrary order of accuracy, for both the Lyapunov and Riccati equations, that is $O\left(\epsilon^k\right)$, where ϵ is a small perturbation parameter. In addition, they reduce the size of required computations. The methods are very suitable for parallel and distributed computations. A real world example, an F-8 aircraft demonstrates the efficiency of the presented methods.

2.5.1 Parallel Algorithm for Solving Discrete Algebraic Lyapunov Equation

A discrete-time constant linear system with the zero-input

$$x\left(k+1\right) = Ax\left(k\right) \tag{2.83}$$

is asymptotically stable if and only if the solution of the algebraic discrete-time Lyapunov equation

$$A^T P A - P = -Q \tag{2.84}$$

is positive definite, where Q is any positive definite symmetric matrix. Equation (2.84) represents also a variance equation of a linear stochastic system driven by zero-mean stationary Gaussian white noise $\omega\left(k\right)$ with the intensity matrix Q

$$x\left(k+1\right) = Ax\left(k\right) + \omega\left(k\right) \tag{2.85}$$

Consider the algebraic discrete Lyapunov equation of the singularly perturbed linear discrete system represented by the matrix partitions (Litkouhi and Khalil, 1984; Shen et al., 1991a)

$$A = \begin{bmatrix} I + \epsilon A_1 & \epsilon A_2 \\ A_3 & A_4 \end{bmatrix}, \; Q = \begin{bmatrix} Q_1 & Q_2 \\ Q_2^T & Q_3 \end{bmatrix}, \; P = \begin{bmatrix} P_1/\epsilon & P_2 \\ P_2^T & P_3 \end{bmatrix} \quad (2.86)$$

A_i, i = 1, 2, 3, 4, and Q_j, j = 1, 2, 3, are assumed to be continuous functions of ϵ. Matrices P_1 and P_3 are of dimensions $n \times n$ and $m \times m$, respectively. Remaining matrices are of compatible dimensions.

The partitioned form of (2.84) subject to (2.86) is

$$P_1 A_1 + A_1^T P_1 + P_2 A_3 + A_3^T P_2^T + A_3^T P_3 A_3 + Q_1$$
$$+\epsilon \left(A_1^T P_1 A_1 + A_1^T P_2 A_3 + A_3^T P_2^T A_1 \right) = 0 \quad (2.87)$$

$$P_1 A_2 + P_2 A_4 + A_3^T P_3 A_4 - P_2 + Q_2$$
$$+\epsilon \left(A_1^T P_2 A_4 + A_3^T P_2^T A_2 \right) = 0 \quad (2.88)$$

$$A_4^T P_3 A_4 - P_3 + Q_3 + \epsilon \left(A_2^T P_1 A_2 + A_2^T P_2 A_4 + A_4^T P_2^T A_2 \right) = 0 \quad (2.89)$$

Define $O(\epsilon)$ perturbations of (2.87)-(2.89) by

$$P_1 A_1 + A_1^T P_1 + P_2 A_3 + A_3^T P_2^T + A_3^T P_3 A_3 + Q_1 = 0 \quad (2.90)$$

$$P_1 A_2 + P_2 A_4 + A_3^T P_3 A_4 - P_2 + Q_2 = 0 \quad (2.91)$$

$$A_4^T P_3 A_4 - P_3 + Q_3 = 0 \quad (2.92)$$

Note that we did not set $\epsilon = 0$ in $A_i's$ and $Q_j's$. From equation (2.91) the matrix P_2 can be expressed in terms of P_1 and P_3 as

$$P_2 = L_1 + P_1 L_2 \quad (2.93)$$

where

$$L_1 = \left(A_3^T P_3 A_4 - P_3 + Q_2 \right) (I - A_4)^{-1}$$
$$L_2 = A_2 (I - A_4)^{-1} \quad (2.94)$$

The invertibility of the matrix $(I - A_4)$ follows from the stability assumption that $| \lambda(A_4) | < 1$.

After doing some algebraic calculations we get

$$P_1 A + A^T P_1 + Q = 0 \quad (2.95)$$

where

$$A = A_1 + L_2 A_3 = A_1 + A_2 (I - A_4)^{-1} A_3$$
$$Q = L_1 A_3 + A_3^T L_1^T + A_3^T P_3 A_3 - Q_1 \qquad (2.96)$$

Thus, we can get solutions for P_1, P_2, and P_3 by solving one lower order continuous-time Lyapunov equation and lower order discrete-time Lyapunov equation. It is assumed that A and A_4 are stable matrices so that solutions of (2.90) and (2.92) exist. These are standard assumptions in the theory of singularly perturbed linear discrete-time systems (Litkouhi and Khalil, 1985).

Assumption 2.7 The matrix A is stable in the continuous-time domain and the matrix A_4 is stable in the discrete-time domain.

$$\triangle$$

Define errors as

$$P_1 = P_1 + \epsilon E_1$$
$$P_2 = P_2 + \epsilon E_2 \qquad (2.97)$$
$$P_3 = P_3 + \epsilon E_3$$

Subtracting (2.92)-(2.95) from (2.87)-(2.89) and doing some algebra, the following set of equations is obtained

$$E_1 A + A^T E_1 = -D A_3 - A_3^T D^T - A_1^T P_1 A_1 - A_1^T P_2 A_3$$
$$- A_3^T P_2^T A_1 - A_3^T E_3 A_4 - A_4^T E_3 A_3 - A_3^T E_3 A_3$$

$$A_4^T E_3 A_4 - E_3 = -A_2^T P_1 A_2 - A_2^T P_2 A_4 - A_4^T P_2^T A_2 \qquad (2.98)$$

$$E_2 = E_1 L_2 + D + A_3^T E_3 A_4$$

where

$$D = \left(A_3^T E_3 A_4 + A_1^T P_1 A_2 + A_1^T P_2 A_4 + A_3^T P_2^T A_2 \right) (I - A_4)^{-1} \qquad (2.99)$$

The solution of (2.98) of the given accuracy will produce the same accuracy for the solution of the Lyapunov equation (2.84). The proposed parallel synchronous algorithm for the numerical solution of (2.98) is as following.

Algorithm 2.5:

$$E_1^{(i+1)}A + A^T E_1^{(i+1)} = -D^{(i)}A_3 - A_3^T D^{T^{(i)}} - A_1^T P_1^{(i)} A_1$$
$$-A_1^T P_2^{(i)} A_3 - A_3^T P_2^{T^{(i)}} A_1 - A_3^T E_3^{(i)} A_4 - A_4^T E_3^{(i)} A_3 - A_3^T E_3^{(i)} A_3$$

$$A_4^T E_3^{(i+1)} A_4 - E_3^{(i+1)} = -A_2^T P_1^{(i)} A_2 - A_2^T P_2^{(i)} A_4 - A_4^T P_2^{T^{(i)}} A_2$$

$$E_2^{(i+1)} = E_1^{(i+1)} L_2 + D^{(i)} + A_3^T E_3^{(i+1)} A_4 \tag{2.100}$$

with starting points $E_1^{(0)} = E_2^{(0)} = E_3^{(0)} = 0$ and

$$P_j^{(i)} = P_j + \epsilon E_j^{(i)}, \quad j = 1, 2, 3; \quad i = 0, 1, 2... \tag{2.101}$$

$\triangle$

The main feature of Algorithm 2.5 is given in the following theorem.

Theorem 2.5 *Based on the stability assumptions imposed on* A *and* A_4 *the algorithm (2.100)-(2.101) converges to the exact solutions for* $E_j's$ *with the rate of convergence of* $O(\epsilon)$.

$\diamond$

The proof of this theorem is similar to the proof of the corresponding algorithm for the continuous-time algebraic singularly perturbed Lyapunov equation studied in Section 2.2.1. It uses the bilinear transformation from (Power, 1967) to transform the discrete-time Lyapunov equation into the continuous one and then follows the ideas of Section 2.2.1. A direct, discrete-time domain proof, will be presented in Chapter 11 in the context of the output feedback control problem for discrete stochastic systems.

2.5.2 Case Study: An F-8 Aircraft

A numerical example for a linearized model of an F-8 aircraft with the small positive parameter $\epsilon = 0.03333$ (Litkouhi, 1983) demonstrates the efficiency of the proposed method. The problem matrices A and Q are given by

$$A = 10^{-3} \begin{bmatrix} 998.51 & -8.044 & -0.10886 & -0.018697 \\ 0.15659 & 1000 & -0.76232 & 3.2272 \\ -213.94 & 0.88081 & 897.21 & 92.826 \\ 110.17 & -0.37821 & -445.56 & 929.68 \end{bmatrix}$$

$$Q = diag\,[0.1,\ 0.1,\ 0.1,\ 0.1]$$

Simulation results, obtained by using the software package L-A-S (West et al., 1985), are presented in Table 2.4. It can be seen that obtained numerical results are consistent with the established theoretical statements.

iteration	$P_1^{(i)}$	$P_2^{(i)}$	$P_3^{(i)}$
0	7.26070 0.17779 0.17779 8.02970	-1.4252 -0.81171 -7.3907 1.29590	2.10690 -0.25231 -0.25231 0.54983
1	8.15990 0.20340 0.20340 8.95420	-1.6239 -0.89268 -8.2950 1.47450	2.33520 -0.29196 -0.29196 0.55729
2	8.27660 0.20656 0.20656 9.07410	-1.6494 -0.90328 -8.4104 1.49760	2.36500 -0.29714 -0.29714 0.55884
3	8.28940 0.20696 0.20696 9.08960	-1.6526 -0.90465 -8.4253 1.50060	2.36880 -0.29781 -0.29781 0.55884
4	8.29130 0.20701 0.20701 9.09160	-1.6531 -0.90483 -8.4272 1.50100	2.36930 -0.29790 -0.29790 0.55887
5 = exact	8.29160 0.20702 0.20702 9.09180	-1.6531 -0.90485 -8.4275 1.50100	2.36940 -0.29791 -0.29791 0.55887

Table 2.4: Recursive solution of the singularly
perturbed discrete Lyapunov equation

2.5.3 Parallel Algorithm for Solving Discrete Algebraic Riccati Equation

The algebraic Riccati equation of singularly perturbed linear discrete systems is given by

$$P = A^T P A + Q - A^T P B \left(B^T P B + R \right)^{-1} B^T P A \quad R > 0, \quad Q \geq 0 \tag{2.102}$$

where (Litkouhi, 1983; Litkouhi and Khalil, 1984, 1985)

$$A = \begin{bmatrix} I + \epsilon A_1 & \epsilon A_2 \\ A_3 & A_4 \end{bmatrix}, \quad B = \begin{bmatrix} \epsilon B_1 \\ B_2 \end{bmatrix}, \quad Q = \begin{bmatrix} Q_1 & Q_2 \\ Q_2^T & Q_3 \end{bmatrix} \tag{2.103}$$

and ϵ is a small positive singular perturbation parameter. In addition, the following condition is satisfied (Litkouhi and Khalil, 1985)

$$det (I - A_4) \neq 0 \tag{2.104}$$

Due to the special structure of the problem matrices and its representation in the fast time scale, the required solution P has the form

(Litkouhi and Khalil, 1984)

$$P = \begin{bmatrix} P_1/\epsilon & P_2 \\ P_2^T & P_3 \end{bmatrix} \qquad (2.105)$$

The main goal in the theory of singular perturbations is to obtain the required solution in terms of the reduced-order problems, namely subsystems. In the case of the algebraic singularly perturbed discrete-time Riccati equation, the expansion of the partitioned form of (2.102) will produce a lot of terms and make corresponding approach computationally very involved even though one is faced with the reduced-order numerical problems. In order to overcome this problem, we have used a bilinear transformation introduced in (Kondo and Furuta, 1986) to transform the discrete-time Riccati equation (2.102) into the continuous-time algebraic Riccati equation of the form

$$A_c^T P_c + P_c A_c + Q_c - P_c S_c P_c = 0, \quad S_c = B_c R_c^{-1} B_c^T \qquad (2.106)$$

such that the solution of (2.102) is equal to the solution of (2.106).

It will be shown that the equation (2.106) preserves the structure of singularly perturbed systems. This equation (2.106) can be solved in terms of the reduced-order problems very efficiently by using the recursive method developed in Section 2.2.2, which converges with the rate of convergence of $O(\epsilon)$.

The bilinear transformation states that equations (2.102) and (2.106) have the same solution if the following hold (Kondo and Furuta, 1986)

$$A_c = I - 2D^{-T}$$
$$S_c = 2(I + A)^{-1} S_d D^{-1}, \quad S_d = BR^{-1}B^T$$
$$Q_c = 2D^T Q (I + A)^{-1}$$
$$D = (I + A)^T + Q (I + A)^{-1} S_d \qquad (2.107)$$

assuming that $(I + A)^{-1}$ exists. It can be easily seen that the matrix

$$I + A = \begin{bmatrix} 2I + \epsilon A_1 & \epsilon A_2 \\ A_3 & I + A_4 \end{bmatrix} \qquad (2.108)$$

is invertible for small values of ϵ if and only if the matrix $I + A_4$ is invertible. Using the standard result from (Stewart, 1973) the invertibility is assured if the matrix A_4 has no eigenvalues at -1. Thus, the method proposed in this section and used through out of this book will be applicable under the following assumption.

Assumption 2.8 The fast subsystem matrix has no eigenvalues located at −1.

$$\triangle$$

It is important to point out that, under given Assumption 2.8, the matrix D defined in (2.107) is nonsingular (Bar-Ness and Halbersberg, 1980).

Exercise 2.5: Prove that the matrix D defined in (2.107) is invertible.

$$\triangle$$

Let us show that applying the bilinear transformation, the system still preserves the singularly perturbed structure, namely, matrices defined in (2.107) should correspond to the linear-quadratic (LQ) singularly perturbed continuous-time control problem.

Using the formula for an inversion of block partitioned matrices, the following can be obtained from (2.103) and (2.107)

$$(I + A)^{-1} = \begin{bmatrix} I + O(\epsilon) & O(\epsilon) \\ O(1) & (1) \end{bmatrix}, \quad S_d^f = \begin{bmatrix} O(\epsilon^2) & O(\epsilon) \\ O(\epsilon) & O(1) \end{bmatrix}$$

$$D^f = \begin{bmatrix} I + O(\epsilon) & O(1) \\ O(\epsilon) & O(1) \end{bmatrix}, \quad D^{f-T} = \begin{bmatrix} I + O(\epsilon) & O(\epsilon) \\ O(1) & O(1) \end{bmatrix}$$
(2.109)

so that

$$A_c^f = \begin{bmatrix} O(\epsilon) & O(\epsilon) \\ O(1) & (1) \end{bmatrix}, \quad Q_c^f = \begin{bmatrix} O(1) & O(1) \\ O(1) & O(1) \end{bmatrix}$$

(2.110)

$$S_c^f = \begin{bmatrix} O(\epsilon^2) & O(\epsilon) \\ O(\epsilon) & O(1) \end{bmatrix} \Rightarrow B_c^f = \begin{bmatrix} O(\epsilon) \\ O(1) \end{bmatrix}$$

where f indicates the fast time scale version quantities.

It is the well-known fact that the structure of matrices obtained in (2.110) corresponds to the fast time scale representation of the continuous-time singularly perturbed LQ control problem (Litkouhi and Khalil, 1984; Kokotovic and Khalil, 1986; Kokotovic et al., 1986).

Since there is no difference in the use of either the slow or fast time scale representation for the continuous-time LQ control problem of singularly perturbed systems, we will adopt the slow time scale version for this problem. It is customary to represent continuous-time singularly perturbed systems by their slow time version (Kokotovic and Khalil, 1986; Kokotovic et al., 1986).

The slow time version of (2.110) can be obtained by multiplying the matrix A_c^f by $1/\epsilon$ and matrix S_c^f by $1/\epsilon^2$. Introducing a notation for the compatible partitions of these matrices we have

$$A_c = \begin{bmatrix} A_{11} & A_{12} \\ \frac{A_{21}}{\epsilon} & \frac{A_{22}}{\epsilon} \end{bmatrix}, \quad S_c = \begin{bmatrix} S_{11} & \frac{S_{12}}{\epsilon} \\ \frac{S_{12}^T}{\epsilon} & \frac{S_{22}}{\epsilon^2} \end{bmatrix} \qquad (2.111)$$

By doing this, the required solution P from (2.105), obtained now from (2.106), will be multiplied by ϵ, that is

$$\epsilon P = P_c = \begin{bmatrix} P_1 & \epsilon P_2 \\ \epsilon P_2^T & \epsilon P_3 \end{bmatrix} \qquad (2.112)$$

Going from the fast time version to the slow time version does not change the matrix Q_c. It is partitioned as

$$Q_c = \begin{bmatrix} Q_{11} & Q_{12} \\ Q_{12}^T & Q_{22} \end{bmatrix} = Q_c^f \qquad (2.113)$$

It is important to notice that partitions defined in (2.111)-(2.113) have to be performed by a computer only, in the process of calculations, and there is no need for the corresponding analytical expressions.

The solution of (2.106) can be found in terms of the reduced-order problems by imposing standard stabilizability-detectability assumptions on the slow and fast subsystems. The efficient recursive reduced-order algorithm for solving (2.106) is obtained in Section 2.2.2. It will be briefly summarized here taking into account the specific features of the problem under study.

First of all, we derive expressions for B_c and R_c so that the analogy between the discrete quantities (A, B, Q, R) and continuous ones (A_c, B_c, Q_c, R_c) is completed. By definition

$$S_c^f = B_c^f R_c^{-1} B_c^{fT} \qquad (2.114)$$

From (2.107) we have

$$S_c^f = 2 (I + A)^{-1} S_d D^{-1} (I + A^T) (I + A^T)^{-1} \qquad (2.115)$$

Since

$$S_d D^{-1} (I + A^T) = S_d \left[(I + A^T)^{-1} D \right]^{-1} =$$

$$= S_d \left[I + (I + A^T)^{-1} Q (I + A)^{-1} S_d \right]^{-1}$$
$$= B \left[R + B^T (I + A^T)^{-1} Q (I + A)^{-1} B \right]^{-1} B^T \qquad (2.116)$$

(the last step in this expression is justified in (Bar-Ness and Halbersberg, 1980)), we get

$$S_c^f = 2 (I + A)^{-1} B \left[R + B^T (I + A^T)^{-1} Q (I + A)^{-1} B \right]^{-1}$$
$$\times B^T (I + A^T)^{-1}$$

$$(2.117)$$

Comparing (2.114) and (2.116) we conclude

$$B_c^f = (I + A)^{-1} B = \begin{bmatrix} \epsilon B_1^f \\ B_2^f \end{bmatrix} \qquad (2.118)$$

and

$$R_c = 0.5 \left[R + B^T (I + A^T)^{-1} Q (I + A)^{-1} B \right] \qquad (2.119)$$

Note that R_c is positive definite.

Exercise 2.6: Verify formula (2.119).

$\triangle$

The slow time version of (2.118) is

$$B_c = \frac{1}{\epsilon} B_c^f = \begin{bmatrix} B_1^f \\ \frac{B_2^f}{\epsilon} \end{bmatrix} \qquad (2.120)$$

The $O(\epsilon)$ approximation of (2.106) subject to (2.111)-(2.113) and (2.119)-(2.120) can be obtained from the following reduced-order algebraic equations

$$0 = P_1 \underline{A} + \underline{A}^T P_1 + \underline{Q} - P_1 \underline{S} P_1, \quad \underline{S} = B_0 R_0^{-1} B_0^T$$
$$0 = P_3 A_{22} + A_{22}^T P_3 + Q_{22} - P_3 S_{22} P_3 \qquad (2.121)$$
$$P_2 = P_1 Z_1 - Z_2$$

where newly defined matrices can be obtained easily using results from Section 2.2.2.

The unique positive semidefinite stabilizing solution of (2.121) exists under the following assumption.

Assumption 2.9 The triples $(\underline{A}, B_0, \sqrt{\underline{Q}})$ and $(A_{22}, B_2, \sqrt{Q_{22}})$ are stabilizable-detectable.

$\triangle$

Defining the approximation errors as

$$P_i = \bar{P}_i + \epsilon E_i, \quad i = 1, 2, 3. \tag{2.122}$$

the recursive reduced-order algorithm, with the rate of convergence of $O(\epsilon)$, can be derived similarly to (2.34).

Algorithm 2.6:

$$E_1^{(j+1)} D_1 + D_1^T E_1^{(j+1)} = D^T H_1^{(j)^T} + H_1^{(j)} D + D^T H_3^{(j)} D + \epsilon H_2^{(j)}$$
$$E_2^{(j+1)} D_3 + E_1^{(j+1)} D_{21} + D_{22}^T E_3^{(j+1)} = H_1^{(j,j+1)}$$
$$E_3^{(j+1)} D_3 + D_3^T E_3^{(j+1)} = H_3^{(j)}$$
$$\tag{2.123}$$

with $j = 0, 1, 2, ...,$ and $E_1^{(0)} = 0, E_2^{(0)} = 0, E_3^{(0)} = 0$, where newly defined matrices are given by

$$D_1 = A_{11} - S_{11} P_1 - S_{12} P_2^T - D_{21} D_3^{-1} D_{22} = D_{11} - D_{21} D_3^{-1} D_{22}$$
$$D_3 = A_{22} - S_{22} P_3, \quad D = D_3^{-1} D_{22}$$
$$D_{21} = A_{12} - S_{12} P_3, \quad D_{22} = A_{21} - S_{12}^T P_1 - S_{22} P_2^T$$
$$\tag{2.124}$$

$$H_1^{(j,j+1)} = A_{21}^T P_2^{(j)} - P_1^{(j+1)} S_{11} P_2^{(j)} - P_2^{(j)} S_{12}^T P_2^{(j)}$$
$$-\epsilon \left(E_1^{(j+1)} S_{12} E_3^{(j+1)} + E_2^{(j)} S_{22} E_2^{(j)} \right)$$

$$H_2^{(j)} = E_1^{(j)} S_{11} E_1^{(j)} + E_1^{(j)} S_{12} E_2^{(j)^T} + E_2^{(j)} S_{12}^T E_2^{(j)} + E_2^{(j)} S_{22} E_2^{(j)^T}$$
$$H_3^{(j)} = -P_2^{(j)^T} A_{12} - A_{12}^T P_2^{(j)} + \epsilon \left(P_2^{(j)^T} S_{11} P_2^{(j)} + E_3^{(j)} S_{22} E_3^{(j)} \right)$$
$$+ P_2^{(j)^T} S_{12} P_3^{(j)} + P_3^{(j)} S_{12}^T P_2^{(j)}$$
$$\tag{2.125}$$
$$\triangle$$

It is important to point out that D_1 and D_3 are stable matrices (Gajic, 1986).

The rate of convergence of (2.123) is $O(\epsilon)$, that is

$$\left\| P_i - P_i^{(j)} \right\| = O(\epsilon^j), \quad i = 1, 2, 3; \quad j = 0, 1, 2, ... \tag{2.126}$$

where

$$P_i^{(j)} = \bar{P}_i + \epsilon E_i^{(j)} \quad i = 1, 2, 3; \quad j = 0, 1, 2, ... \tag{2.127}$$

In summary, the proposed algorithm for the reduced-order solution of the singularly perturbed discrete algebraic Riccati equation has the following form:

  1) Transform (2.102) into (2.106) by using the bilinear transformation defined in (2.107).

  2) Solve (2.106) by using the recursive reduced-order parallel algorithm defined by (2.121)-(2.127).

2.6 Approximate Linear Regulator Problem for Discrete Systems

The positive semidefinite stabilizing solution of the algebraic discrete Riccati equation (2.102), produces the answer to the optimal linear-quadratic steady state control problem. Namely, a quadratic criterion

$$J = \frac{1}{2} \sum_{k=0}^{\infty} \left(x^T(k) Q x(k) + u^T(k) R u(k) \right) \qquad (2.128)$$

is minimized along trajectories of a linear dynamic system

$$x(k+1) = A x(k) + B u(k) \qquad (2.129)$$

by using the control input of the form

$$u(k) = -\left(R + B^T P B \right)^{-1} B^T P A x(k) \qquad (2.130)$$

where P is obtained from (2.102), (Dorato and Levis, 1971). This problem has been studied in the context of singular perturbations in (Litkouhi and Khalil, 1984), where the fast time version has been adopted, so that (2.128) is multiplied by a small perturbation parameter, that is,

$$J_f = \epsilon J \qquad (2.131)$$

It is proved in (Litkouhi and Khalil, 1985) that the near-optimal control given by

$$u^{(j)}(k) = -\left(R + B^T P^{(j)} B \right)^{-1} B^T P^{(j)} A x(k) = -F^{(j)} x(k) \qquad (2.132)$$

where $P^{(j)}$ satisfies

$$P^{(j)} - P^{opt} = O\left(\epsilon^j \right) \qquad (2.133)$$

is near-optimal in the sense

$$J_f^{(j)} - J_f^{opt} = O\left(\epsilon^{2j}\right) \tag{2.134}$$

The approximate performance $J^{(j)}$ can be obtained from the discrete algebraic Lyapunov equation

$$K^{(j)} = \left(A - BF^{(j)}\right)^T K^{(j)} \left(A - BF^{(j)}\right) + Q + F^{(j)^T} RF^{(j)} \tag{2.135}$$

so that

$$J^{(j)} = \frac{1}{2} x^T(0) K^{(j)} x(0) \tag{2.136}$$

In the previous section we have developed a very efficient technique for generating $P^{(j)}$ by using the recursive reduced-order schemes (2.121)-(2.127), such that each iteration improves the accuracy by an order of magnitude. Thus, the proposed algorithm and the theoretical results obtained in (Litkouhi and Khalil, 1985) and given in (2.132)-(2.134) comprise an efficient method for solving the linear-quadratic control problem of singularly perturbed discrete systems.

The efficiency of this method is demonstrated on a real world example in the next section.

2.6.1 Case Study: Discrete Model of An F-8 Aircraft

A linearized model of an F-8 aircraft is considered in (Elliott, 1977). By a proper scaling this model was presented in the singularly perturbed continuous-time form (fast time version) in (Litkouhi, 1983), with the system matrix

$$\begin{bmatrix} -0.015 & -0.0805 & -0.0011666 & 0 \\ 0 & 0 & 0 & 0.03333 \\ -2.28 & 0 & -0.84 & 1 \\ 0.6 & 0 & -4.8 & -0.49 \end{bmatrix}$$

and the control matrix

$$\begin{bmatrix} -0.0000916 & 0.0007416 \\ 0 & 0 \\ -0.11 & 0 \\ -8.7 & 0 \end{bmatrix}$$

Small elements in the first two rows indicate two slow variables in contrary to relatively big elements in the third and forth rows corresponding to fast variables. The small perturbation parameter ϵ is chosen as $\epsilon = 1/30$. This model is discretized in (Litkouhi, 1983) by using the sampling period $T = 1$, leading to

$$A = \begin{bmatrix} 0.98475 & -0.079903 & 0.0009054 & -0.0010765 \\ 0.041588 & 0.99899 & -0.035855 & 0.012684 \\ -0.54662 & 0.044916 & -0.32991 & 0.19318 \\ 2.6624 & -0.10045 & -0.92455 & -0.26325 \end{bmatrix}$$

$$B = \begin{bmatrix} 0.0037112 & 0.00073610 \\ -0.087051 & 0.0000093411 \\ -1.19844 & -0.00041378 \\ -3.1927 & 0.00092535 \end{bmatrix}$$

The linear-quadratic control problem is solved for weighting matrices $R = I_2$, $Q = 10^{-2} I_4$ and the initial condition $x(0) = [1, 0, 0.008, 0]^T$.

The eigenvalues of the matrix A_4 are $-0.297 \pm j0.442$, so that Assumption 2.7 is satisfied.

Simulation results for the reduced-order solution for the approximate values of the criterion are presented in Table 2.5.

j	$J_{apr}^{(j)} - J_{opt}$
0	0.208 x 10^{-2}
1	0.885 x 10^{-5}
2	0.155 x 10^{-7}
3	0.534 x 10^{-10}

Table 2.5: Near optimality of the approximate criterion

2.7 Recursive Methods for Weakly Coupled Linear Discrete Systems

The main goal in the theory of weakly coupled control systems is to obtain the required solution in terms of reduced-order problems, namely subsystems. In the case of the weakly coupled algebraic discrete Riccati equation, the inversion of the partitioned matrix $B^T P B + R$ will produce a lot of terms and make the corresponding approach computationally very involved, even though one is faced with the reduced-order numerical problems. To solve this problem, we have used the bilinear transformation (2.107) to transform the discrete-time Riccati equation into the continuous-time algebraic Riccati equation of the form (2.106), such that the solution of (2.102) is equal to the solution of (2.106).

It will be shown that the equation (2.106) preserves the structure of weakly coupled systems. This equation can be solved in terms of the reduced-order problems very efficiently by using the recursive method developed in Section 2.3.2, which converges with the rate of convergence of $O\left(\epsilon^2\right)$. A model of a discrete chemical plant is considered as an illustrative example.

For the reason of completeness, we present first the results for the Lyapunov equation. Corresponding parallel reduced-order algorithm for solving discrete Lyapunov equation of weakly coupled systems is derived and demonstrated on a discrete catalytic cracker model.

As before, algorithms for both the Lyapunov and Riccati equations are implemented as synchronous ones. Their implementation as the asynchronous parallel algorithms is under investigation.

2.7.1 Parallel Algorithm for Solving Discrete Algebraic Lyapunov Equation

Consider the algebraic discrete Lyapunov equation (2.84). In the case of a weakly coupled linear discrete system corresponding matrices are partitioned as

$$A = \begin{bmatrix} A_1 & \epsilon A_2 \\ \epsilon A_3 & A_4 \end{bmatrix}, \quad Q = \begin{bmatrix} Q_1 & \epsilon Q_2 \\ \epsilon Q_2^T & Q_3 \end{bmatrix}, \quad P = \begin{bmatrix} P_1 & \epsilon P_2 \\ \epsilon P_2^T & P_3 \end{bmatrix}$$
$$(2.137)$$

where A_i, $i = 1, 2, 3, 4$, and Q_j, $j = 1, 2, 3$, are assumed to be continuous functions of ϵ. Matrices P_1 and P_3 are of dimensions $n \times n$ and $m \times m$, respectively. Remaining matrices are of compatible dimensions.

The partitioned form of (2.84) subject to (2.137) is

$$A_1^T P_1 A_1 - P_1 + Q_1 + \epsilon^2 \left(A_1^T P_2 A_3 + A_3^T P_2^T A_1 + A_3^T P_3 A_3 \right) = 0 \quad (2.138)$$

$$A_1^T P_1 A_2 - P_2 + Q_2 + A_1^T P_2 A_4 + A_3^T P_3 A_4 + \epsilon^2 A_3^T P_2^T A_2 = 0 \quad (2.139)$$

$$A_4^T P_3 A_4 - P_3 + Q_3 + \epsilon^2 \left(A_2^T P_1 A_2 + A_2^T P_2 A_4 + A_4^T P_2^T A_2 \right) = 0 \quad (2.140)$$

Define, like in Section 2.3.1, $O\left(\epsilon^2\right)$ perturbations of (2.138)-(2.140) by

$$A_1^T P_1 A_1 - P_1 + Q_1 = 0 \qquad (2.141)$$

$$A_1^T P_1 A_2 + A_1^T P_2 A_4 + A_3^T P_3 A_4 - P_2 + Q_2 = 0 \qquad (2.142)$$

$$A_4^T P_3 A_4 - P_3 + Q_3 = 0 \qquad (2.143)$$

Note that we did not set $\epsilon = 0$ in $A_i's$ and $Q_i's$. Assume that the matrices A_1 and A_4 are stable (Assumption 2.5). Then the unique solutions of (2.141)-(2.143) exist.

Define errors as

$$\begin{aligned} P_1 &= P_1 + \epsilon E_1 \\ P_2 &= P_2 + \epsilon E_2 \\ P_3 &= P_3 + \epsilon E_3 \end{aligned} \qquad (2.144)$$

Subtracting (2.141)-(2.143) from (2.138)-(2.140), the following error equations are obtained

$$\begin{aligned} A_1^T E_1 A_1 - E_1 &= -A_1^T P_2 A_3 - A_3^T P_2^T A_1 - A_3^T P_3 A_3 \\ A_4^T E_3 A_4 - E_3 &= -A_2^T P_1 A_2 - A_2^T P_2 A_4 - A_4^T P_2^T A_2 \\ A_1^T E_2 A_4 - E_2 &= -A_1^T E_1 A_2 - A_3^T P_2^T A_2 - A_3^T E_3 A_4 \end{aligned} \qquad (2.145)$$

The proposed parallel synchronous algorithm for the numerical solution of (2.145) is as follows (Shen, et al., 1991a).

Algorithm 2.7:

$$\begin{aligned} A_1^T E_1^{(i+1)} A_1 - E_1^{(i+1)} &= -A_1^T P_2^{(i)} A_3 - A_3^T P_2^{(i)^T} A_1 - A_3^T P_3^{(i)} A_3 \\ A_4^T E_3^{(i+1)} A_4 - E_3^{(i+1)} &= -A_2^T P_1^{(i)} A_2 - A_2^T P_2^{(i)} A_4 - A_4^T P_2^{(i)^T} A_2 \\ A_1^T E_2^{(i+1)} A_4 - E_2^{(i+1)} &= -A_1^T E_1^{(i+1)} A_2 - A_3^T P_2^{(i)^T} A_2 - A_3^T E_3^{(i+1)} A_4 \end{aligned}$$
$$(2.146)$$

with starting points $E_1^{(0)} = E_2^{(0)} = E_3^{(0)} = 0$ and

$$P_j^{(i)} = P_j + \epsilon^2 E_j^{(i)}, \quad j = 1, 2, 3; \quad i = 0, 1, 2... \qquad (2.147)$$

$\triangle$

Now we have the following theorem analogous to Theorem 2.3.

Theorem 2.6 *Under stability Assumption 2.5, the algorithm (2.146)-(2.147) converges to the exact solutions for E_j's with the rate of convergence of $O\left(\epsilon^2\right)$.*

$\diamond$

The proof of this theorem is similar to the proof of the corresponding algorithm for continuous-time algebraic singularly perturbed Lyapunov equation studied in Section 2.2.1. It uses the bilinear transformation from (Power, 1967) to transform the discrete-time Lyapunov equation into the continuous one and then follows the ideas of Section 2.2.1.

2.7.2 Case Study: Discrete Catalytic Cracker

A fifth-order model of a catalytic cracker (Kando et al., 1988), demonstrates the efficiency of the proposed method. The problem matrix A (after performing discretization with the sampling period $T = 1$) is given by

$$A_d = \begin{bmatrix} 0.011771 & 0.046903 & 0.096679 & 0.071586 & -0.019178 \\ 0.014096 & 0.056411 & 0.115070 & 0.085194 & -0.022806 \\ 0.066395 & 0.252260 & 0.580880 & 0.430570 & -0.11628 \\ 0.027557 & 0.104940 & 0.240400 & 0.178190 & -0.048104 \\ 0.000564 & 0.002644 & 0.003479 & 0.002561 & -0.000656 \end{bmatrix}$$

The small weak coupling parameter is $\epsilon = 0.21$ and the state penalty matrix is chosen as $Q = I$.

The simulation results are presented in Table 2.6.

2.7.3 Parallel Algorithm for Solving Algebraic Riccati Equation

The algebraic Riccati equation of weakly coupled linear discrete systems is given by (2.102) with

$$A = \begin{bmatrix} A_1 & \epsilon A_2 \\ \epsilon A_3 & A_4 \end{bmatrix}, \quad B = \begin{bmatrix} B_1 & \epsilon B_2 \\ \epsilon B_3 & B_4 \end{bmatrix}$$

i	$P_1^{(i)}$	$P_2^{(i)}$	$P_3^{(i)}$
0	1.00030 0.00135 0.00135 1.00540	0.54689 0.40537 -0.10944 2.08640 1.54650 -0.41752	1.93020 0.68954 -0.18620 0.68954 1.51110 -0.13802 -0.18620 -0.13802 1.03730
1	1.01390 0.05290 0.052897 1.20180	0.66593 0.49359 -0.13322 2.54040 1.88290 -0.50820	2.20320 0.89183 -0.24071 0.89183 1.66100 -0.17841 -0.24071 -0.17841 1.04820
2	1.01620 0.06184 0.06184 1.23600	0.69091 0.51209 -0.13821 2.63570 1.95350 -0.52722	2.26010 0.93400 -0.25208 0.93400 1.69230 -0.18683 -0.25208 -0.18683 1.05040
3	1.01670 0.06371 0.06371 1.24310	0.69604 0.51590 -0.13923 2.65520 1.96800 -0.53113	2.27170 0.94260 -0.25439 0.94260 1.69860 -0.18855 -0.25439 -0.18855 1.05090
4	1.01680 0.06409 0.06409 1.24450	0.69710 0.51668 -0.13944 2.65930 1.97100 -0.53193	2.27410 0.94437 -0.25487 0.94437 1.70000 -0.18891 -0.25487 -0.18891 1.05100
5	1.01680 0.06417 0.06417 1.24480	0.69731 0.51684 -0.13948 2.66010 1.97160 -0.53210	2.27460 0.94473 -0.25497 0.94473 1.70020 -0.18898 -0.25497 -0.18898 1.05100
6	1.01680 0.06418 0.06418 1.24490	0.69736 0.51687 -0.13949 2.66010 1.97170 -0.53213	2.27470 0.94481 -0.25499 0.94481 1.70030 -0.18899 -0.25499 -0.18899 1.05100
7	1.01680 0.06419 0.06419 1.24490	0.69737 0.51688 -0.13950 2.66030 1.97180 -0.53214	2.27470 0.94482 -0.25499 0.94482 1.70030 -0.18900 -0.25499 -0.18900 1.05100

Table 2.6: Reduced-order solution of discrete weakly coupled algebraic Lyapunov equation ($P^{(7)} = P_{exact}$)

$$Q = \begin{bmatrix} Q_1 & \epsilon Q_2 \\ \epsilon Q_2^T & Q_3 \end{bmatrix}, \quad R = \begin{bmatrix} R_1 & 0 \\ 0 & R_2 \end{bmatrix} \qquad (2.148)$$

and ϵ is a small weak coupling parameter. Due to block dominant structure of the problem matrices, the required solution P has the form

$$P = \begin{bmatrix} P_1 & \epsilon P_2 \\ \epsilon P_2^T & P_3 \end{bmatrix} \qquad (2.149)$$

The bilinear transformation states that equations (2.102) and (2.106) have the same solutions if the relation (2.107) holds, that is

$$A_c = I - 2D^{-T}$$

$$S_c = 2(I + A)^{-1} S_d D^{-1}, \quad S_d = BR^{-1}B^T$$
$$Q_c = 2D^{-1}Q(I + A)^{-1}$$
$$D = (I + A)^T + Q(I + A)^{-1} S_d$$

assuming that $(I + A)^{-1}$ exists. It can be seen that for weakly coupled systems the matrix

$$(I + A)^{-1} = \begin{bmatrix} O(1) & O(\epsilon) \\ O(\epsilon) & O(1) \end{bmatrix} \qquad (2.150)$$

is invertible for small values of ϵ. It can be verified that the weakly coupled structure of the matrices defined in (2.148) will produce the weakly coupled structure of the transformed continuous-time matrices defined in (2.107). It follows from the fact that S_d from (2.107) and Q from (2.148) have the same weakly coupled structure as (2.150), so does D in (2.107). The inverse of D is also in the weakly coupled form as defined in (2.150). From (2.107) the weakly coupled structure of matrices A_c and Q_c follows directly since they are given in terms of sums and/or products of weakly coupled matrices.

Using the standard result from (Stewart, 1973), it follows that the method proposed in this section is applicable under the following assumption.

Assumption 2.10 The system matrix A has no eigenvalues located at -1.
$$\triangle$$

It is important to point out that the eigenvalues located in the neighborhood of -1 will produce ill-conditioning with respect to matrix inversion and make the algorithm numerically unstable.

Let us introduce the following notation for the compatible partitions of the transformed weakly coupled matrices, that is

$$A_c = \begin{bmatrix} A_{11} & \epsilon A_{12} \\ \epsilon A_{21} & A_{22} \end{bmatrix}, \quad S_c = \begin{bmatrix} S_{11} & \epsilon S_{12} \\ \epsilon S_{12}^T & S_{22} \end{bmatrix} \qquad (2.151)$$

$$P_c = \begin{bmatrix} P_1 & \epsilon P_2 \\ \epsilon P_2^T & P_3 \end{bmatrix}, \quad Q_c = \begin{bmatrix} Q_{11} & \epsilon Q_{12} \\ \epsilon Q_{12}^T & Q_{22} \end{bmatrix} \qquad (2.152)$$

These partitions have to be performed by a computer only, in the process of calculations, and there is no need for the corresponding analytical expressions.

The solution of (2.106) can be found in terms of the reduced-order problems by imposing standard stabilizability-detectability assumptions on the subsystems. The efficient recursive reduced-order algorithm for solving (2.106) is obtained in Section 2.3.2. It will be briefly summarized here taking into account the specific features of the problem under study.

The $O\left(\epsilon^2\right)$ approximation of (2.106) subject to (2.151)-(2.152) can be obtained from the following decoupled set of reduced-order algebraic equations

$$P_1 A_{11} + A_{11}^T P_1 - P_1 S_{11} P_1 + Q_{11} = 0$$
$$P_3 A_{22} + A_{22}^T P_3 - P_3 S_{22} P_3 + Q_{22} = 0 \tag{2.153}$$

and

$$P_2 \Delta_2 + \Delta_1^T P_2 = -\left(P_1 A_{12} + A_{21}^T P_3 + Q_{12} - P_1 S_{12} P_3\right) \tag{2.154}$$

where

$$\Delta_1\left(\epsilon\right) = \left[A_{11}\left(\epsilon\right) - S_{11}\left(\epsilon\right) P_1\left(\epsilon\right)\right], \quad \Delta_2\left(\epsilon\right) = \left[A_{22}\left(\epsilon\right) - S_{22}\left(\epsilon\right) P_3\left(\epsilon\right)\right] \tag{2.155}$$

The unique positive semidefinite stabilizing solutions of (2.153) exist under the following assumption.

Assumption 2.11 The triples $\left(A_{ii}\left(\epsilon\right), \sqrt{S_{ii}\left(\epsilon\right)}, \sqrt{Q_{ii}\left(\epsilon\right)}\right)$, $i = 1, 2$, are stabilizable-detectable.

$$\triangle$$

Under Assumption 2.11 matrices $\Delta_1\left(\epsilon\right)$ and $\Delta_2\left(\epsilon\right)$ are stable so that the unique solution of (2.154) exists also.

If the errors are defined as

$$P_j = \mathbf{P}_j + \epsilon^2 E_j, \quad j = 1, 2, 3 \tag{2.156}$$

then the exact solution will be of the form

$$P = \begin{bmatrix} \mathbf{P}_1 + \epsilon^2 E_1 & \epsilon\left(\mathbf{P}_2 + \epsilon^2 E_2\right) \\ \epsilon\left(\mathbf{P}_2 + \epsilon^2 E_2\right)^T & \mathbf{P}_3 + \epsilon^2 E_3 \end{bmatrix} \tag{2.157}$$

The fixed point parallel reduced-order algorithm for the error terms, obtained by using results from Section 2.3.2, has the form.

Algorithm 2.8:

$$E_1^{(i+1)} \Delta_1 + \Delta_1^T E_1^{(i+1)} = P_1^{(i)} S_{12} P_2^{(i)^T} + P_2^{(i)} S_{12}^T P_1^{(i)}$$
$$+ P_2^{(i)} S_{22} P_2^{(i)^T} - P_2^{(i)} A_{21} - A_{21}^T P_2^{(i)^T} + \epsilon^2 E_1^{(i)} S_{11} E_1^{(i)} \tag{2.158}$$

$$E_3^{(i+1)}\Delta_2 + \Delta_2^T E_3^{(i+1)} = P_2^{(i)^T} S_{11} P_2^{(i)} + P_3^{(i)} S_{12}^T P_2^{(i)}$$
$$+ P_2^{(i)^T} S_{12} P_3^{(i)} - P_2^{(i)^T} A_{12} - A_{12}^T P_2^{(i)} + \epsilon^2 E_3^{(i)} S_{22} E_3^{(i)} \quad (2.159)$$

$$\Delta_1^T E_2^{(i+1)} + E_2^{(i+1)}\Delta_2 + E_1^{(i+1)}\Delta_{12} + \Delta_{21}^T E_3^{(i+1)}$$
$$= P_2^{(i)} S_{12}^T P_2^{(i)} + \epsilon^2 \left(E_1^{(i+1)} S_{11} E_2^{(i)} + E_1^{(i+1)} S_{12} E_3^{(i)} + E_2^{(i)} S_{22} E_3^{(i+1)} \right) \quad (2.160)$$

with $E_1^{(0)} = 0$, $E_2^{(0)} = 0$, $E_3^{(0)} = 0$, where

$$P_j^{(i)} = \mathbf{P_j} + \epsilon^2 E_j^{(i)}, \quad j = 1, 2, 3; \quad i = 1, 2, 3, \dots \quad (2.161)$$

and

$$\Delta_{12} = [A_{12} - S_{11}\mathbf{P_2} - S_{12}\mathbf{P_3}], \quad \Delta_{21} = [A_{21} - S_{22}\mathbf{P_2^T} - S_{12}^T\mathbf{P_1}]$$
$$(2.162)$$
$$\triangle$$

This algorithm satisfies all conditions given in Theorem 2.4, so that it converges to the exact solution of E with the rate of convergence of $O\left(\epsilon^2\right)$, that is

$$\left\| E - E^{(i+1)} \right\| = O\left(\epsilon^2\right) \left\| E - E^{(i)} \right\|, \quad i = 0, 1, 2, \dots \quad (2.163)$$

or equivalently

$$\left\| E - E^{(i)} \right\| = O\left(\epsilon^{2i}\right) \quad (2.164)$$

In summary, the proposed parallel algorithm for the reduced-order solution of the weakly coupled discrete algebraic Riccati equation has the following form:

1) Transform (2.102) into (2.106) by using the bilinear transformation defined in (2.107).

2) Solve (2.106) by using the recursive reduced-order parallel algorithm defined by (2.153)-(2.162).

2.7.4 Case Study: Discrete Model of a Chemical Plant

A real world physical example (a chemical plant model (Gomathi et al., 1980)) demonstrates the efficiency of the proposed method. The system matrices are obtained from (Gomathi et al., 1980) by performing a discretization with a sampling rate $T = 0.5$.

$$A = 10^{-2} \begin{bmatrix} 95.407 & 1.9643 & 0.3597 & 0.0673 & 0.0190 \\ 40.849 & 41.317 & 16.084 & 4.4679 & 1.1971 \\ 12.217 & 26.326 & 36.149 & 15.930 & 12.383 \\ 4.1118 & 12.858 & 27.209 & 21.442 & 40.976 \\ 0.1305 & 0.5808 & 1.8750 & 3.6162 & 94.280 \end{bmatrix}$$

$$B^T = 10^{-2} \begin{bmatrix} 0.0434 & 2.6606 & 3.7530 & 3.6076 & 0.4617 \\ -0.0122 & -1.0453 & -5.5100 & -6.6000 & -0.9148 \end{bmatrix}$$

$$Q = I_5, \quad R = I_2$$

The small weak coupling parameter ϵ is built into the problem and can be roughly estimated from the strongest coupled matrix (matrix B). The strongest coupling is in the third row, where

$$\epsilon = \frac{b_{31}}{b_{32}} = \frac{3.753}{5.510} = 0.68$$

Simulation results are obtained using the MATLAB package for computer aided control system design. The solution of the algebraic Riccati equation, obtained from Algorithm 2.8, is presented in Table 2.7.

For this specific real world example the proposed algorithm perfectly matches the presented theory since convergence, with the accuracy of 10^{-4}, is achieved after 9 iterations ($0.68^{18} = 10^{-4}$).

Note very dramatic changes in the element $P_1^{(j)}$ per iteration. Thus, in this example only higher order approximations produce satisfactory results. Corresponding differences between the optimal and approximate state trajectories for the corresponding components of the state vector are presented in Figures 2.1-2.5. The optimal and approximate control strategies are shown in Figures 2.6-2.7. In these figures, the solid lines represent the optimal quantities (state trajectories and controls); the dotted, dashed, and dash-dotted lines represent the approximate trajectories and controls. The obtained figures justify the necessity for the existence of the higher order approximations for both the approximate control strategies and the approximate trajectories.

2.8 Notes and Comments

The presented parallel algorithms are applicable to the large scale systems already in the singularly perturbed or weakly coupled forms. It will

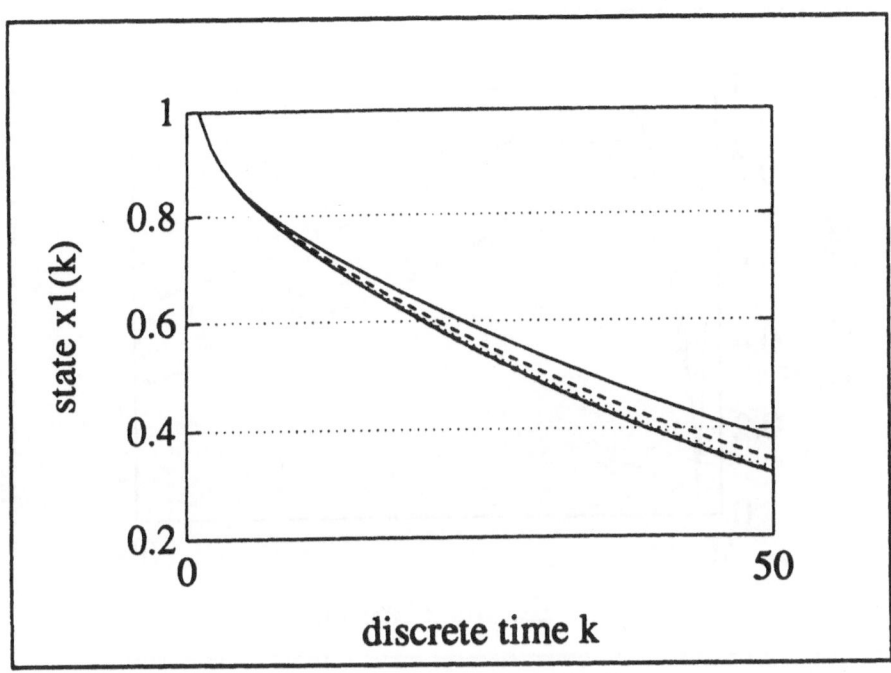

Figure 2.1: Approximate and optimal trajectories $x_1(k)$

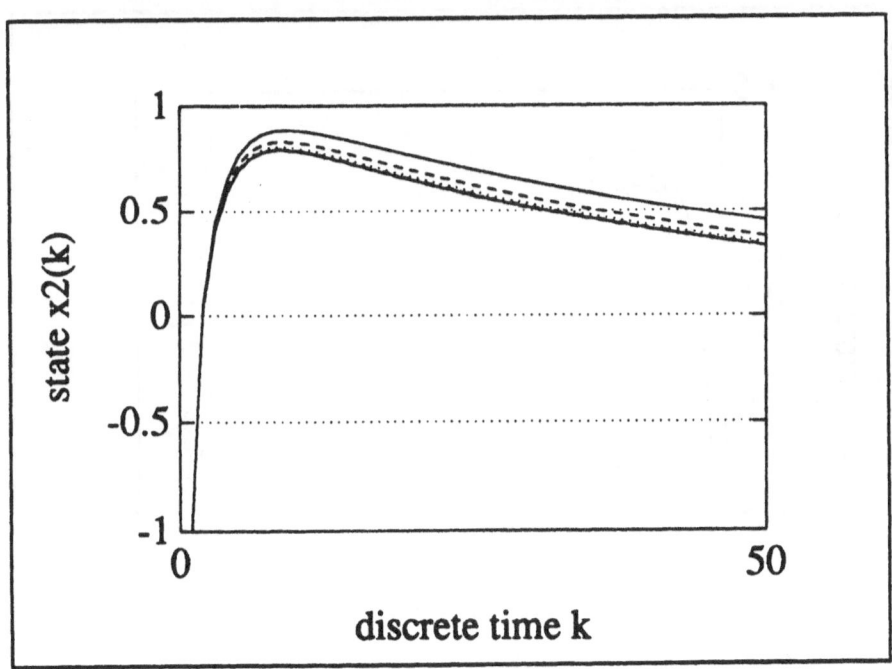

Figure 2.2: Approximate and optimal trajectories $x_2(k)$

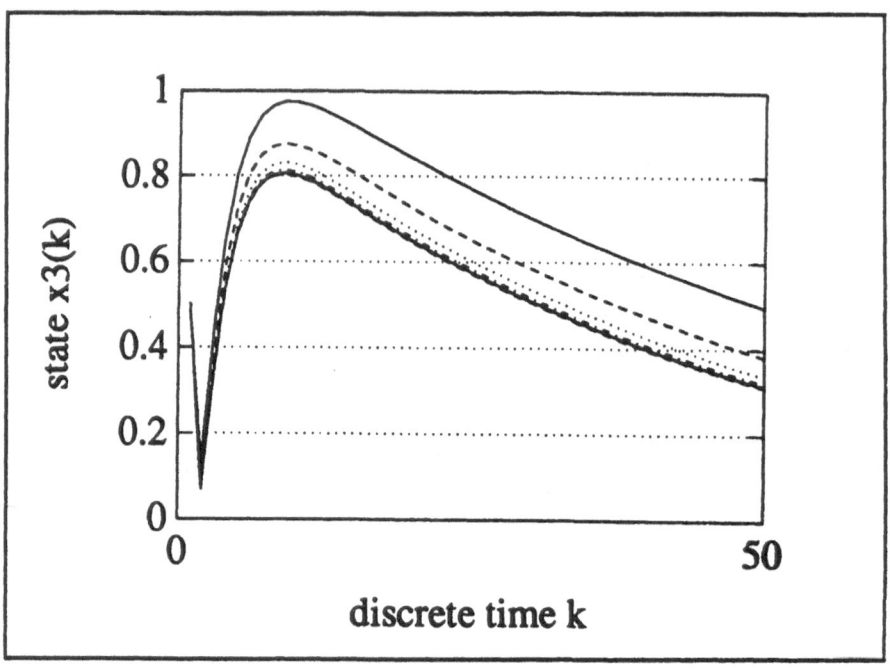

Figure 2.3: Approximate and optimal trajectories $x_3(k)$

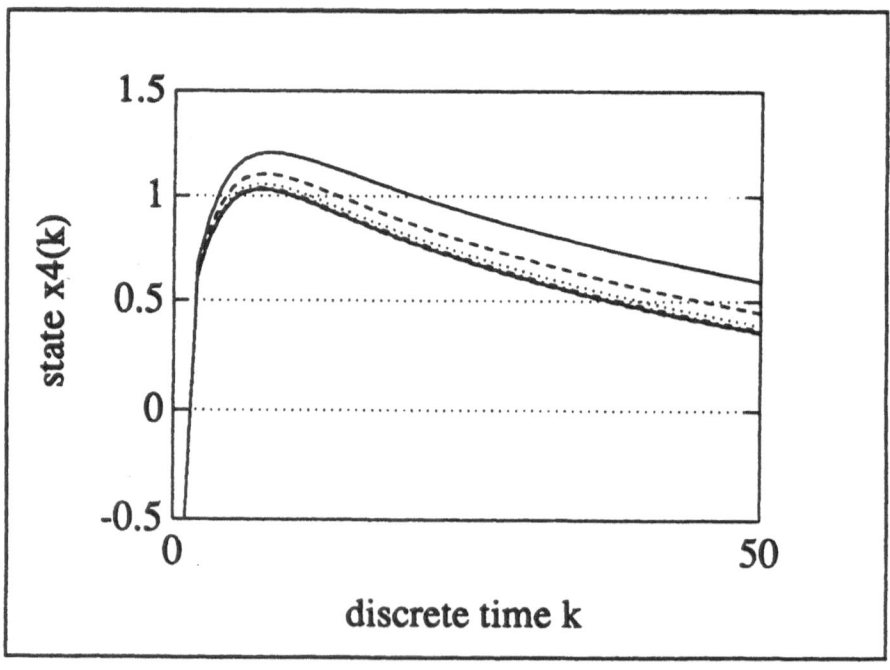

Figure 2.4: Approximate and optimal trajectories $x_4(k)$

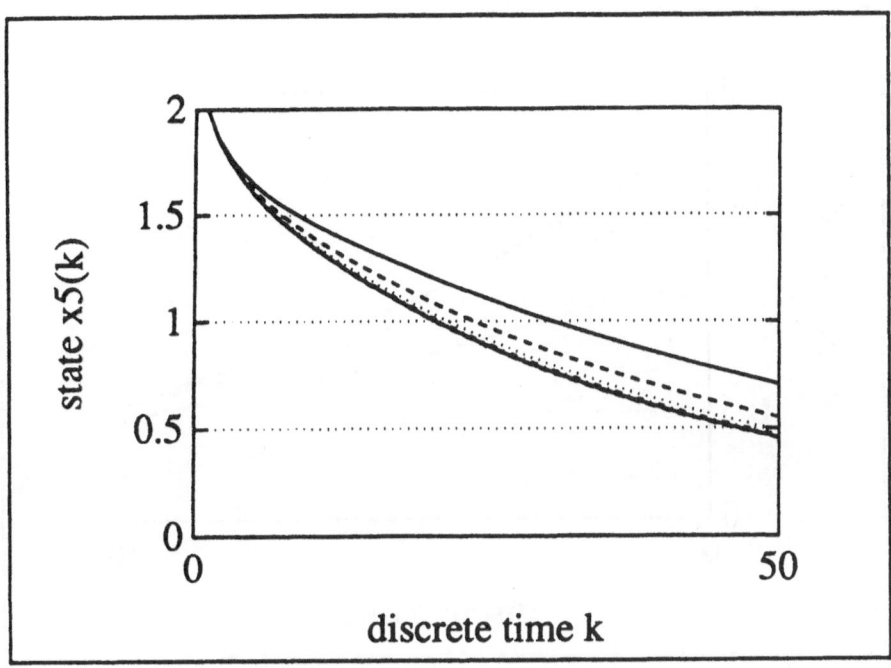

Figure 2.5: Approximate and optimal trajectories $x_5(k)$

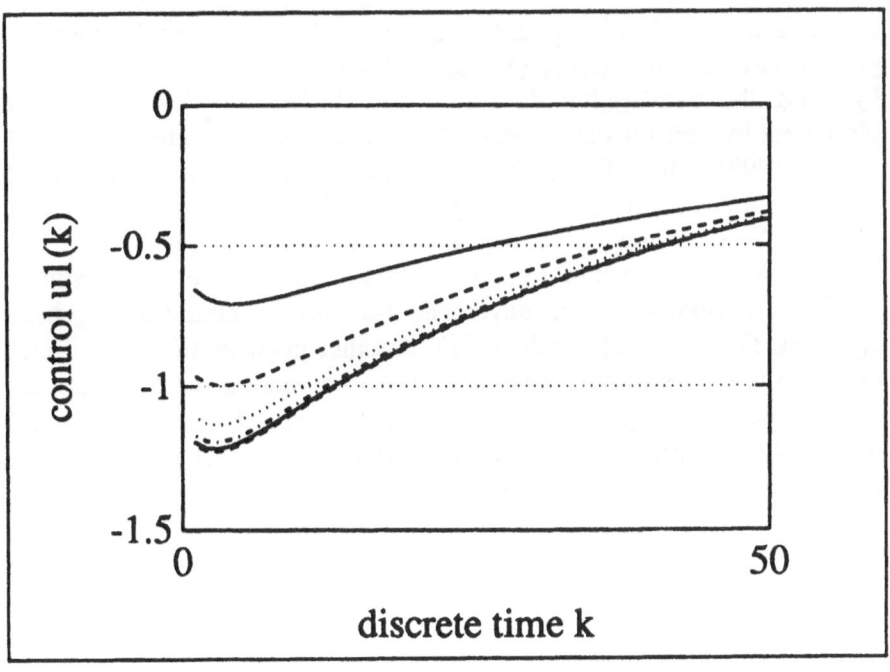

Figure 2.6: Approximate and optimal control strategies $u_1(k)$

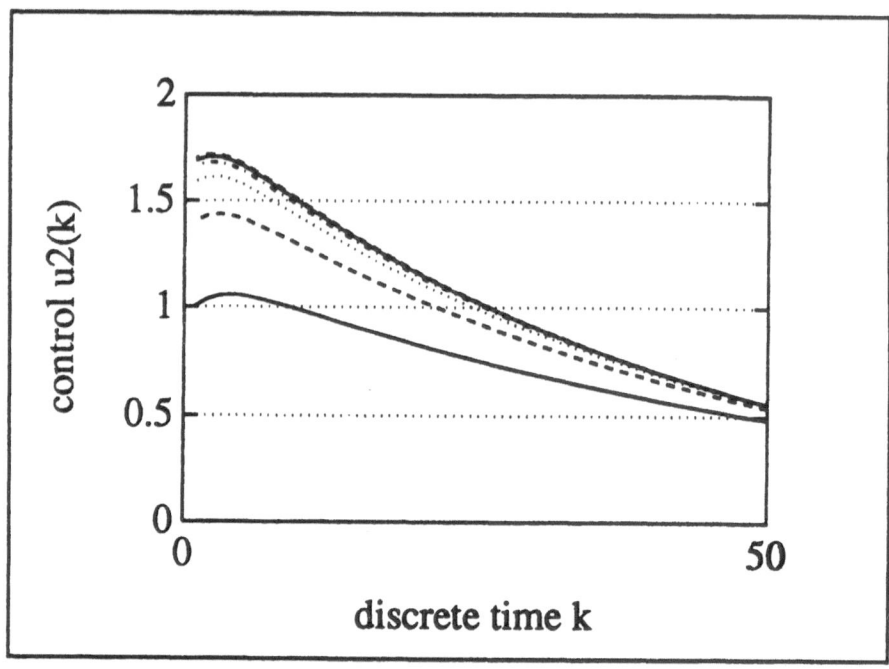

Figure 2.7: Approximate and optimal control strategies $u_2(k)$

be interesting to develop parallel algorithms for the optimal control of general large scale systems (Siljak, 1978) brought in the above forms by using the overlapping decomposition (Siljak, 1991) and the pole placement by performance modification technique (Medanic et al., 1988; Tharp, 1992). It will be also interesting to extend these results to the linear algebra problems of singularly perturbed and weakly coupled systems. Some results in that direction are already obtained by using the overlapping decomposition (Sezer and Siljak, 1991). Even more, similar type of algorithms can be developed for solving nonlinear algebraic equations (Zecevic and Siljak, 1992). Another important future research topic is the development of the asynchronous versions of the presented algorithms. The importance of the asynchronous algorithms for block diagonally dominant systems (weakly coupled systems) is documented in (Kaszkurewicz et al., 1990).

j	$P_1^{(j)}$	$P_2^{(j)}$	$P_3^{(j)}$
0	20.9061 0.9202 0.9202 1.2382	1.8865 1.4365 18.5536 0.5259 0.3219 2.1852	1.2937 0.1971 1.2516 0.1971 1.1514 1.2887 1.2516 1.2887 21.0090
1	39.2244 2.5453 2.5453 1.5406	3.4212 2.3932 28.8267 0.7575 0.4428 3.3277	1.4754 0.2982 2.0621 0.2982 1.2067 1.7456 2.0621 1.7456 25.1919
2	50.6375 3.6481 3.6481 1.6827	4.2746 2.8594 32.9119 0.8637 0.5006 3.8272	1.5558 0.3423 2.4450 0.3423 1.2304 1.9451 2.4450 1.9451 26.7777
3	56.1732 4.2167 4.2167 1.7492	4.6785 3.0634 34.4250 0.9111 0.5250 4.0179	1.5911 0.3609 2.5959 0.3423 1.2399 2.0161 2.5959 2.0161 27.2107
4	58.6366 4.4773 4.4773 1.7788	4.8566 3.1498 34.9986 0.9314 0.5351 4.0888	1.6063 0.3686 2.6519 0.3423 1.2436 2.0416 2.6519 2.0416 27.3486
5	59.6956 4.5906 4.5906 1.7915	4.9327 3.1858 35.2222 0.9400 0.5392 4.155	1.6127 0.3717 2.6727 0.3423 1.2451 2.0510 2.6727 2.0510 27.3982
6	60.1433 4.6387 4.6387 1.7969	4.9646 3.2008 35.3112 0.9436 0.5409 4.1258	1.6154 0.3729 2.6800 0.3729 1.2451 2.0546 2.6800 2.0546 27.4171
9	60.4410 4.6707 4.6707 1.8004	4.9857 3.2106 35.3676 0.9459 0.5420 4.1321	1.6171 0.3737 2.6853 0.3729 1.2461 2.0567 2.6853 2.0567 27.4288
12	60.4621 4.6730 4.6730 1.8006	4.9872 3.2113 35.3715 0.9461 0.5420 4.1326	1.6172 0.3738 2.6857 0.3729 1.2461 2.0569 · 2.6857 2.0569 27.4295
16	60.4636 4.6732 4.6732 1.8006	4.9873 3.2113 35.3717 0.9461 0.5420 4.1326	1.6172 0.3738 2.6857 0.3729 1.2461 2.0569 2.6857 2.0569 27.4296
	$P_1 = P_1^{(15)}$	$P_2 = P_2^{(15)}$	$P_3 = P_3^{(15)}$

Table 2.7: Reduced-order solution of the discrete
weakly coupled algebraic Riccati equation

Chapter 3

Decoupling Transformations

3.1 Introduction

Decoupling transformations play very important roles in the control theory of systems containing small parameters. Under certain, usually very mild conditions, these transformations allow the linear system decomposition into independent reduced-order subsystems. The decoupling transformation for linear singularly perturbed continuous-time varying systems is introduced in (Chang, 1972). Corresponding transformation for weakly coupled linear systems is presented in (Gajic and Shen, 1989). In the recent paper (Qureshi and Gajic, 1992) a new version of the Chang transformation is obtained for singularly perturbed continuous-time varying linear systems. In (Qureshi, 1992) a new version of the transformation obtained by (Gajic and Shen, 1989) for weakly coupled linear systems is derived.

In this chapter, we present the main results of (Chang, 1972) specialized for continuous-time invariant systems, Section 3.2. In Section 3.3, Chang's methodology is extended to continuous-time invariant linear weakly coupled systems. The numerical techniques for solving algebraic equations comprising the transformation for weakly coupled systems (Gajic and Shen, 1989) are also discussed in this section. Numerical techniques for solving corresponding algebraic equations for singularly perturbed systems will be presented in Chapter 8. The new versions of the decoupling transformations (Qureshi and Gajic, 1992; Qureshi, 1992) are presented in Section 3.4. These transformations, applied to the Lyapunov differential equations of both singularly perturbed and weakly coupled linear systems, produce the complete decompositions of these equations into three Lyapunov equations of the reduced-order — Section 3.5.

In Section 3.6, we have solved the general boundary value problem of continuous-time varying weakly coupled linear systems producing its decomposition into two reduced-order initial value problems (Qureshi and Gajic, 1991). The problem formulation and used methodology are dual to the original work of (Chang, 1972), done for the general boundary value problem of linear continuous-time varying singularly perturbed systems. In the last section of this chapter, the discrete-time version of the results obtained in Section 3.6 is presented.

3.2 Decoupling Transformation for Singularly Perturbed Linear Systems

Consider the singularly perturbed continuous-time invariant linear system defined by

$$\dot{x}_1 = A_1 x_1 + A_2 x_2 + B_1 u \tag{3.1}$$

$$\epsilon \dot{x}_2 = A_3 x_1 + A_4 x_2 + B_2 u \tag{3.2}$$

where $x_1 \in \Re^{n_1}, x_2 \in \Re^{n_2}$, and $u \in \Re^m$, are slow state variables, fast state variables, and a control input, respectively, and ϵ is a small positive parameter. The Chang transformation was derived in (Chang, 1972) for the general boundary value problem of singularly perturbed continuous-time varying systems. In this section, we give a simplified derivation of this transformation applicable to the continuous-time, time invariant initial value problem. This transformation is applicable to the standard singularly perturbed systems (Kokotovic et al., 1986) satisfying the following assumption.

Assumption 3.1. The fast subsystem matrix A_4 is nonsingular.

$\triangle$

If for any linear singularly perturbed system this assumption is not satisfied, the system is in a nonstandard singularly perturbed form. A methodology, presented in (Khalil, 1989), treats the nonstandard linear singularly perturbed systems. Another general approach, applicable for nonlinear singularly perturbed systems can be found in (Kokotovic et al., 1986).

The Chang transformation is defined by

$$\begin{bmatrix} \eta_1 \\ \eta_2 \end{bmatrix} = \begin{bmatrix} I_{n_1} - \epsilon H_1 L_1 & -\epsilon H_1 \\ L_1 & I_{n_2} \end{bmatrix} \begin{bmatrix} x_1 \\ x_2 \end{bmatrix} = \mathbf{T}_1 \begin{bmatrix} x_1 \\ x_2 \end{bmatrix} \tag{3.3}$$

where L_1 and H_1 satisfy the following matrix equations

$$A_4 L_1 - \epsilon L_1 A_1 + \epsilon L_1 A_2 L_1 - A_3 = 0 \tag{3.4}$$

$$H_1 (A_4 + \epsilon L_1 A_2) + \epsilon (A_1 - A_2 L_1) H_1 - A_2 = 0 \tag{3.5}$$

Note that (3.3) represents a nonsingular transformation. Numerical methods for solving efficiently the algebraic equations (3.4)-(3.5) in terms of the reduced-order algebraic equations will be presented in Chapter 8.

The transformation (3.3) completely decouples the original system (3.1)-(3.2) into pure-slow and pure-fast subsystems so that in the new coordinates we have

$$\dot{\eta}_1 = A_s \eta_1 + B_s u \tag{3.6}$$

$$\epsilon \dot{\eta}_2 = A_f \eta_2 + B_f u \tag{3.7}$$

where

$$A_s = A_1 - A_2 L_1, \quad A_f = A_4 + \epsilon L_1 A_2$$
$$B_s = B_1 - H_1 B_2 - \epsilon H_1 L_1 B_1, \quad B_f = B_2 + \epsilon L_1 B_1 \tag{3.8}$$

The inverse Chang transformation is given by

$$\begin{bmatrix} x_1 \\ x_2 \end{bmatrix} = \begin{bmatrix} I_{n_1} & \epsilon H_1 \\ -L_1 & I_{n_2} - \epsilon L_1 H_1 \end{bmatrix} \begin{bmatrix} \eta_1 \\ \eta_2 \end{bmatrix} = \mathbf{T}_1^{-1} \begin{bmatrix} \eta_1 \\ \eta_2 \end{bmatrix} \tag{3.9}$$

Exercise 3.1: By using the matrix multiplication and algebraic equations (3.4)-(3.5) verify that

$$\mathbf{T}_1 A \mathbf{T}_1^{-1} = \mathbf{T}_1 \begin{bmatrix} A_1 & A_2 \\ \frac{A_3}{\epsilon} & \frac{A_4}{\epsilon} \end{bmatrix} \mathbf{T}_1^{-1} = \begin{bmatrix} A_s & 0 \\ 0 & \frac{A_f}{\epsilon} \end{bmatrix}$$

and

$$\mathbf{T}_1 B = \begin{bmatrix} B_s \\ \frac{B_f}{\epsilon} \end{bmatrix}$$

$\triangle$

More details about the derivation of this transform will be given in the next section where we extend Chang's methodology to linear weakly coupled continuous-time invariant systems. Time varying version of the Chang transformation will come into account in Section 3.4.

3.3 Decoupling Transformation for Weakly Coupled Linear Systems

The linear weakly coupled system is represented by

$$\dot{x} = A_1 x + \epsilon A_2 z + B_1 u_1 + \epsilon B_2 u_2 \qquad (3.10)$$

$$\dot{z} = \epsilon A_3 x + A_4 z + \epsilon B_3 u_1 + B_4 u_2 \qquad (3.11)$$

where $x \in \Re^{n_1}$, $z \in \Re^{n_2}$, are subsystem states, $u_i \in \Re^{m_i}$ are subsystem controls, $i = 1, 2$, and ϵ is a small coupling parameter. In this section, we derive a nonsingular transformation that completely decouples linear weakly coupled systems (filters or estimators first of all).

Introducing the change of variables

$$x = \eta + \epsilon L_2 z \qquad (3.12)$$

the original system (3.10) is transformed into

$$\dot{\eta} = A_{10}\eta + \epsilon F_1(L_2) z + B_{10}u_1 + \epsilon B_{20}u_2 \qquad (3.13)$$

where

$$A_{10} = A_1 - \epsilon^2 L_2 A_3,$$
$$B_{10} = B_1 - \epsilon^2 L_2 B_3, \quad B_{20} = B_2 - L_2 B_4 \qquad (3.14)$$

and

$$F_1(L_2) = A_1 L_2 - L_2 A_4 + A_2 - \epsilon^2 L_2 A_3 L_2 \qquad (3.15)$$

Assuming that a matrix L_2 can be chosen such that $F_1(L_2) = 0$, then equation (3.13) represents a completely independent (decoupled) subsystem

$$\dot{\eta} = A_{10}\eta + B_{10}u_1 + \epsilon B_{20}u_2 \qquad (3.16)$$

As a matter of fact, equations (3.11) and (3.16) form a triangular system (after elimination of x from (3.11) by using (3.12)).

Introducing the second change of variables as

$$\zeta = z + \epsilon H_2 \eta \qquad (3.17)$$

the equation (3.11) becomes

$$\dot{\zeta} = A_{40}\zeta + \epsilon F_2(H_2)\eta + \epsilon B_{30}u_1 + B_{40}u_2 \qquad (3.18)$$

where

$$A_{40} = A_4 + \epsilon^2 A_3 L_2,$$
$$B_{30} = B_3 + H_2 B_{10}, \qquad B_{40} = B_4 + \epsilon^2 H_2 B_{20} \qquad (3.19)$$

and

$$F_2(H_2) = H_2 A_{10} - A_{40} H_2 + A_3 \qquad (3.20)$$

In addition, if a matrix H_2 can be chosen such that $F_2(H_2) = 0$, then we have

$$\dot{\zeta} = A_{40}\zeta + \epsilon B_{30}u_1 + B_{40}u_2 \qquad (3.21)$$

so that (3.16) and (3.21) represent two completely decoupled linear subsystems. Notice that the weakly coupled structure of the control inputs in (3.10) and (3.11) is preserved in the new coordinates, that is in (3.16) and (3.21). This means that the proposed transformation is applicable to the feedback structure of (3.10) and (3.11) also. Thus, applying the nonsingular transformation

$$\begin{bmatrix} \eta \\ \zeta \end{bmatrix} = \begin{bmatrix} I_{n_1} & -\epsilon L_2 \\ \epsilon H_2 & I_{n_2} - \epsilon^2 H_2 L_2 \end{bmatrix} \begin{bmatrix} x \\ z \end{bmatrix} = T_2 \begin{bmatrix} x \\ z \end{bmatrix} \qquad (3.22)$$

where

$$T_2^{-1} = \begin{bmatrix} I_{n_1} - \epsilon^2 L_2 H_2 & \epsilon L_2 \\ -\epsilon H_2 & I_{n_2} \end{bmatrix} \qquad (3.23)$$

the linear weakly coupled system (3.10)-(3.11) is completely decoupled and uniquely determined by its subsystems (3.16) and (3.21).

Obviously, the transformation T_2 is uniquely obtained if unique solutions of the following two algebraic equations exist

$$A_1 L_2 - L_2 A_4 + A_2 - \epsilon^2 L_2 A_3 L_2 = 0 \qquad (3.24)$$

$$H_2(A_1 - \epsilon^2 L_2 A_3) - (A_4 + \epsilon^2 A_3 L_2)H_2 + A_3 = 0 \qquad (3.25)$$

It is important to notice that at $\epsilon = 0$ we have

$$A_1 L_2^{(0)} - L_2^{(0)} A_4 + A_2 = 0 \qquad (3.26)$$

$$H_2^{(0)} A_1 - A_4 H_2^{(0)} + A_3 = 0 \qquad (3.27)$$

65

so that

$$L_2 = L_2^{(0)} + O\left(\epsilon^2\right)$$
$$H_2 = H_2^{(0)} + O\left(\epsilon^2\right)$$

(3.28)

Equations (3.26)-(3.27) are Sylvester equations and their unique solutions exist if matrices A_1 and A_4 have no eigenvalues in common (Lancaster and Tismenetsky, 1985). Thus, the presented results will be valid under the following assumption.

Assumption 3.2 Matrices A_1 and A_4 have no eigenvalues in common.

$\triangle$

By the implicit function theorem (Ortega and Rheinboldt, 1970), for a sufficiently small $\epsilon \in (0, \epsilon_1]$ there exists a unique solution of a weakly nonlinear algebraic equation (3.24). Under the assumption that A_1 and A_4 have no eigenvalues in common and by the fact that the eigenvalues are continuous functions of the matrix elements (Kato, 1980), there exists ϵ_2 small enough such that for any $\epsilon \in (0, \epsilon_2]$ matrices A_{10} and A_{40} will not have eigenvalues in common and thus, the unique solution of (3.25) will exist.

In summary, we have established the following theorem.

Theorem 3.1 *Under Assumption 3.2 there exists a small parameter* $\epsilon \in (0, min\,(\epsilon_1, \epsilon_2)]$ *such that the unique solutions of (3.24) and (3.25) exist.*

$\diamond$

Trajectories of the transformed (decoupled) system are $O\,(\epsilon)$ close to the trajectories of the original system. If the coupling parameter ϵ is extremely small, or if in the design procedure the accuracy of $O\,(\epsilon)$ is sufficient, there is no need for the decomposition. However, if $O\,(\epsilon)$ is not very small, or if the high accuracy is required, then one needs methods that will produce any desired accuracy, that is the accuracy of $O\,(\epsilon^k)$ where $k = 2, 3, 4, \ldots$. Thus, the method proposed in this section is very useful for the intermediate values of ϵ and for the systems with the high accuracy requirements. In addition, the importance of the proposed transformation is in the design of linear filters and observers — dynamical systems built by the designer. Apparently, it is much easier and less expensive to build two dynamical systems of order n_1 and n_2, than one dynamical system of order $n_1 + n_2$.

Note that transformations (3.12) and (3.17) can be used independently to put the system in either lower or upper triangular form. For some applications, this might be sufficient.

Numerical solutions for L_2 and H_2 can be obtained by using the fixed point type recursive algorithms similar to those developed by (Gajic, 1986; Petrovic and Gajic, 1988; Harkara et al., 1989). In the case of equations (3.24) and (3.25) the corresponding algorithm is given by

$$A_1 L_2^{(i+1)} - L_2^{(i+1)} A_4 + A_2 - \epsilon^2 L_2^{(i)} A_3 L_2^{(i)} = 0 \quad (3.29)$$

with $i = 0, 1, 2, ..., N\text{-}1$, and $L_2^{(0)}$ obtained from (3.26)

$$H_2^{(N)} A_{10}^{(N)} - A_{40}^{(N)} H_2^{(N)} + A_3 = 0 \quad (3.30)$$

where

$$A_{10}^{(N)} = A_1 - \epsilon^2 L_2^{(N)} A_3, \quad A_{40}^{(N)} = A_4 + \epsilon^2 A_3 L_2^{(N)} \quad (3.31)$$

Using the results of the references given above, it can be shown that

$$L_2 = L_2^{(N)} + O\left(\epsilon^{2N}\right) \quad (3.32)$$

and

$$H_2 = H_2^{(N)} + O\left(\epsilon^{2N}\right) \quad (3.33)$$

hence, the algorithm (3.17) converges with the rate of convergence of $O\left(\epsilon^2\right)$.

Example 3.1
In order to demonstrate the efficiency of the proposed algorithm (3.29), we have run a sixth-order example. Matrices A_i, $i = 1, 2, 3, 4$, are chosen randomly (standard deviation = 1 and mean value = 0 for A_1, A_2, and A_3; standard deviation = 2 and mean value = 0 for A_4).

$$A_1 = \begin{bmatrix} -1.720 & -0.999 & -0.592 \\ -1.434 & 0.799 & 0.856 \\ -0.729 & 0.105 & 0.867 \end{bmatrix}, \quad A_2 = \begin{bmatrix} -1.614 & -1.429 & 0.516 \\ 0.225 & 1.928 & 0.310 \\ -0.332 & 0.067 & 0.329 \end{bmatrix}$$

$$A_3 = \begin{bmatrix} -1.398 & 1.039 & 0.557 \\ 1.298 & 1.349 & -0.891 \\ -0.472 & -0.610 & -0.873 \end{bmatrix}, \quad A_4 = \begin{bmatrix} -2.956 & 1.219 & 2.269 \\ -0.038 & -2.240 & 2.296 \\ -0.873 & -2.020 & 2.344 \end{bmatrix}$$

The simulation results for different values of the coupling parameters ϵ are given in Table 3.1.

The results of Table 3.1 strongly support the necessity for the existence of the recursive scheme for the solution of (3.24), since unless ϵ is very small, the zeroth and first-order approximations are far from the optimal solution.

ϵ	Number of required iterations such that $\|L_2 - L_2^{(i)}\|_\infty < 10^{-10}$
0.8	*
0.7	28
0.6	17
0.5	12
0.3	9
0.1	5
0.05	3
0.01	2

Table 3.1: Number of iterations for the fixed point method

i $\epsilon = 0.1$	$\|L_2 - L_2^{(i)}\|_\infty$
0	4.129 x 10^{-2}
1	7.4645 x 10^{-4}
2	1.6401 x 10^{-5}
3	2.1149 x 10^{-7}
4	2.0989 x 10^{-10}

Table 3.2: Error propagation for the fixed point method

In Table 3.2, we show the propagations of the error per iteration when $\epsilon = 0.1$. We notice that the rate of convergence of the proposed algorithm (3.29) is $O\left(\epsilon^2\right) = O\left(10^{-2}\right)$.

$\triangle$

The algorithm (3.29) is based on the fixed point iterations, and it will converge as long as the small parameter ϵ is small enough so that the radius of convergence is $\rho(\epsilon) < 1$ at each iteration.

An alternative way of solving (3.24) is by using the Newton method where solution of (3.26) plays the role of the initial condition. The Newton method for the similar type of algebraic equations has been presented by (Grodt and Gajic, 1988). The Newton algorithm for (3.24) can be constructed by setting $L_2^{(i+1)} = L_2^{(i)} + \Delta L_2^{(i)}$ and neglecting $O\left((\Delta L_2)^2\right)$ terms. This will produce a Sylvester-type equation of the form

$$D_1^{(i)} L_2^{(i+1)} + L_2^{(i+1)} D_2^{(i)} = Q^{(i)}, \quad i = 0, 1, 2, \ldots \quad (3.34)$$

where

$$\begin{aligned}
D_1^{(i)} &= A_1 - \epsilon^2 L_2^{(i)} A_3 \\
D_2^{(i)} &= -\left(A_4 + \epsilon^2 A_3 L_2^{(i)}\right) \\
Q^{(i)} &= -\left(A_2 + \epsilon^2 L_2^{(i)} A_3 L_2^{(i)}\right)
\end{aligned} \quad (3.35)$$

with the initial condition $L_2^{(0)}$ obtained from (3.26).

Example 3.2

The Newton method is demonstrated by solving the same example. For the different values of ϵ the results are presented in Table 3.3.

ϵ	Number of required iterations such that $\|\|L_2 - L_2^{(i)}\|\|_\infty < 10^{-10}$
0.8	5
0.7	5
0.6	4
0.5	4
0.3	3
0.1	2
0.05	2
0.01	1

Table 3.3: Number of iterations for the Newton method

It can be seen, that for this particular example, the Newton method converges much faster than the fixed point iteration algorithm. It is the well-known fact that the Newton method converges quadratically in the neighborhood of the sought solution and that its main problem lies in the choice of the initial guess. For the algebraic equation (3.24) the initial guess is easily obtained with the accuracy of $O\left(\epsilon^2\right)$, and the Newton method, if it converges, will produce a sequence $O\left(\epsilon^4\right), O\left(\epsilon^8\right), O\left(\epsilon^{16}\right)$, close to the exact solution. However, in some cases the Newton method does not converge at all (bad initial guess) and one needs to have some other efficient techniques available.

The fixed point method presented earlier in this section is one of them, since its rate of convergence of $O\left(\epsilon^2\right)$ is remarkable.

The simulation results are obtained by using the software package L-A-S (West et al., 1985) for computer-aided control system design.

$$\triangle$$

The importance of the decoupling transformation introduced for weakly coupled linear systems is in the decomposition of the linear Kalman filters and/or observers. Namely, they are dynamical systems built by the control engineers. It is much easier and cheaper to build two filters of order n_1 and n_2, than one filter of order $n_1 + n_2$. The reduced order filters are much faster. Due to parallelism, the on-line computations are considerably reduced at every time instant.

3.4 New Versions of Decoupling Transformations

In this section, we present new versions of the decoupling transformations for both singularly perturbed and weakly coupled linear systems.

The singularly perturbed system under consideration is represented in the form

$$\dot{x}_1 = A_1 x_1 + A_2 x_2 \qquad (3.36)$$

$$\epsilon \dot{x}_2 = A_3 x_1 + A_4 x_2 \qquad (3.37)$$

where $x_1 \in \Re^{n_1}$, $x_2 \in \Re^{n_2}$ are slow and fast state variables, respectively, and matrices A_1, A_2, A_3, and A_4 are of appropriate dimensions, which are constant in the case of time invariant systems, and functions of time in the case of time varying systems. A small parameter ϵ is positive.

The linear weakly coupled system under consideration is represented by

$$\dot{x}_1 = A_1 x_1 + \epsilon A_2 x_2 \qquad (3.38)$$

$$\dot{x}_2 = \epsilon A_3 x_1 + A_4 x_2 \qquad (3.39)$$

where $x_1 \in \Re^{n_1}$, $x_2 \in \Re^{n_2}$ are subsystems state variables, and matrices A_1, A_2, A_3, and A_4 are of appropriate dimensions. In the case of time invariant systems these matrices are constant, while for time varying systems they are functions of time t. A small constant parameter ϵ of arbitrary sign couples the states x_1 and x_2.

The common approach to study these systems is to first transform them into new coordinates such that the states are independent (decoupled) from each other. This leads to a block diagonal form which is easier to solve. Chang (1972) developed a decoupling transformation for singularly perturbed linear systems (3.3), while Gajic and Shen (1989) proposed a similar decoupling transformation (3.22) for linear weakly coupled systems.

The continuous-time varying version for weakly coupled systems of the transformation defined in (3.22) has the following form

$$\mathbf{T}_2(t) = \begin{bmatrix} I_{n_1} & -\epsilon L_2(t) \\ \epsilon H_2(t) & I_{n_2} - \epsilon^2 H_2(t) L_2(t) \end{bmatrix} \qquad (3.40)$$

where $L_2(t)$ and $H_2(t)$ can be obtained from the following two coupled differential equations

$$\dot{L}_2 = A_1 L_2 - L_2 A_4 + A_2 - \epsilon^2 L_2 A_3 L_2 \qquad (3.41)$$

$$\dot{H}_2 = H_2(A_1 - \epsilon^2 L_2 A_3) - (A_4 + \epsilon^2 A_3 L_2) H_2 + A_3 \qquad (3.42)$$

For the singularly perturbed system, the continuous-time varying version of (3.3) is (Chang, 1972)

$$\mathbf{T}_1(t) = \begin{bmatrix} I_{n_1} - \epsilon H_1(t) L_1(t) & -\epsilon H_1(t) \\ L_1(t) & I_{n_2} \end{bmatrix} \qquad (3.43)$$

where $L_1(t)$ and $H_1(t)$ are the solutions of the following two equations

$$\epsilon \dot{L}_1 = A_4 L_1 - \epsilon L_1 A_1 + \epsilon L_1 A_2 L_1 - A_3, \qquad (3.44)$$

$$\epsilon \dot{H}_1 = \epsilon(A_1 - A_2 L_1) H_1 - H_1(A_4 + \epsilon L_1 A_2) - A_2 \qquad (3.45)$$

Note that for the time invariant cases, the derivatives $\dot{L}_1, \dot{H}_1, \dot{L}_2$, and $\dot{H}_2$ are zero. It is important to point out that the initial conditions for (3.41)-(3.42) and (3.44)-(3.45) are not specified so that they may be chosen arbitrarily.

The difficulty in solving equations (3.41)-(3.42) or (3.43)-(3.44) is that (3.42) or (3.44) can only be solved after the results of (3.41) or (3.43) are available. Therefore, computation must be done sequentially. Furthermore, two different algorithms are needed: one for (3.41) or (3.43) and the other for (3.42) or (3.44). In this section, this difficulty is overcome by introducing other transformations that decouple the original systems as well as the transformation equations for both weakly coupled and singularly perturbed linear systems. This will enable us to compute $L_1(t)$ and $H_1(t)$ or $L_2(t)$ and $H_2(t)$ in parallel, and by using only one algorithm. The proposed transformations are extremely efficient, from the numerical point of view, in the case of time varying systems since corresponding differential equations are completely decoupled. This is extremely important for singularly perturbed systems where both transformation equations (3.44) and (3.45) are stiff, and thus numerically ill-defined (Kokotovic and Khalil, 1986).

3.4.1 New Decoupling Transformation for Linear Weakly Coupled Systems

Introducing the change of variables like in (Gajic and Shen, 1989)

$$\eta(t) = x_1(t) - \epsilon L_3(t) x_2(t) \tag{3.46}$$

and differentiating both sides, we obtain

$$\dot{\eta} = \dot{x}_1 - \epsilon \dot{L}_3 x_2 - \epsilon L_3 \dot{x}_2 \tag{3.47}$$

Substituting for $\dot{x}_1$ and $\dot{x}_2$ and simplifying, we get

$$\dot{\eta} = A_{10}\eta - \epsilon F_3(L_3) x_2 \tag{3.48}$$

where

$$A_{10} = A_1 - \epsilon^2 L_3 A_3 \tag{3.49}$$

and

$$F_3(L_3) = \dot{L}_3 - A_1 L_3 + L_3 A_4 - A_2 + \epsilon^2 L_3 A_3 L_3 \tag{3.50}$$

Assuming that a matrix L_3 can be chosen such that $F_3(L_3) = 0$, the equation (3.48) will represent a completely independent (decoupled) system (Gajic and Shen, 1989).

Introducing the second change of variables as

$$\zeta(t) = -\epsilon H_3(t) x_1(t) + x_2(t) \qquad (3.51)$$

and following similar calculations, we get

$$\dot{\zeta} = A_{40}\zeta - \epsilon F_4(H_3) x_1 \qquad (3.52)$$

where

$$A_{40} = A_4 - \epsilon^2 H_3 A_2 \qquad (3.53)$$

and

$$F_4(H_3) = \dot{H}_3 - A_4 H_3 + H_3 A_1 - A_3 + \epsilon^2 H_3 A_2 H_3 \qquad (3.54)$$

Again assuming that a matrix $H_3(t)$ can be chosen such that $F_4(H_3) = 0$, then equations (3.48) and (3.52) will represent a completely decoupled linear system. Note that the initial conditions for differential equations (3.50) and (3.54) are arbitrary. Thus, applying the transformation

$$\begin{bmatrix} \eta \\ \zeta \end{bmatrix} = \begin{bmatrix} I_{n_1} & -\epsilon L_3(t) \\ -\epsilon H_3(t) & I_{n_2} \end{bmatrix} \begin{bmatrix} x_1 \\ x_2 \end{bmatrix} = T_3(t) \begin{bmatrix} x_1 \\ x_2 \end{bmatrix} \qquad (3.55)$$

where

$$T_3^{-1}(t) = \begin{bmatrix} I_{n_1} + \epsilon^2 L_3(t) M(t) H_3(t) & \epsilon L_3(t) M(t) \\ \epsilon M(t) H_3(t) & M(t) \end{bmatrix} \qquad (3.56)$$

with $M(t) = \left(I_{n_1} - \epsilon^2 H_3(t) L_3(t)\right)^{-1}$, the linear weakly coupled system is completely decoupled and uniquely determined by its subsystems (3.48) and (3.52). The nonsingularity of the transformation $T_3(t)$ can be noticed by the fact that the off-diagonal elements are of the order of ϵ, while the blocks on the main diagonal are indentity matrices. Therefore, for a sufficiently small ϵ, $T_3(t)$ is strictly diagonally dominant and hence nonsingular.

Note that exactly the same transformation can be applied to the continuous-time invariant systems. In that case differential equations (3.50) and (3.54) reduce to the algebraic ones, so that the matrix T_3 is constant.

We can apply the transformation (3.55) to the discrete-time invariant weakly coupled linear system defined in Chapter 2, and represented by

$$x_1(n+1) = A_{11}x_1(n) + \epsilon A_{12}x_2(n)$$
$$x_2(n+1) = \epsilon A_{21}x_1(n) + A_{22}x_2(n)$$

(3.57)

Following similar calculations, the transformed block diagonal system is obtained as follows

$$\eta(n+1) = \left[A_{11} - \epsilon^2 L_3 A_{21}\right] \eta(n)$$
$$\zeta(n+1) = \left[A_{22} - \epsilon^2 H_3 A_{12}\right] \zeta(n)$$

(3.58)

where L_3 and H_3 satisfy the following decoupled algebraic equations

$$A_{11}L_3 - L_3 A_{22} - \epsilon^2 L_3 A_{21} L_3 + A_{12} = 0$$

(3.59)

$$A_{22}H_3 - H_3 A_{11} - \epsilon^2 H_3 A_{12} H_3 + A_{21} = 0$$

(3.60)

Note that unique solutions of (3.59) and (3.60) exist for sufficiently small values of ϵ under the assumption that matrices A_{11} and A_{22} have no eigenvalues in common (Assumption 3.2).

Exercise 3.2: Derive equations (3.58)-(3.60). Then show that a similar transformation is applicable to the decomposition of the discrete-time varying linear control system represented by

$$x_1(n+1) = A_{11}(n)x_1(n) + \epsilon A_{12}(n)x_2(n) + B_{11}(n)u_1(n)$$
$$+\epsilon B_{12}(n)u_2(n)$$
$$x_2(n+1) = \epsilon A_{12}(n)x_1(n) + A_{22}(n)x_2(n) + \epsilon B_{21}(n)u_1(n)$$
$$+B_{22}(n)u_2(n)$$

Find expressions for the system matrices in new coordinates and the difference equations whose solutions comprise the required transformation.

$\triangle$

The advantage of this transformation over the previous one (Gajic and Shen, 1989) is that the transformation equations for L_3 and H_3, namely (3.50) and (3.54), respectively, have exactly the same form, and they are independent of each other. Therefore, we can use the same algorithm to solve both L_3 and H_3. Moreover, due to the fact that they are independent from each other, the computations can be done in parallel. However, a price is paid for this convenience when we go back to the original variables x_1 and x_2. This step requires the computation of an inverse of a matrix $\left(I_{n_1} - \epsilon^2 L_3 H_3\right)^{-1}$.

3.4.2 New Decoupling Transformation for Linear Singularly Perturbed Systems

Introducing the change of variables for the linear singularly perturbed system (3.36)-(3.37) as

$$\alpha = x_1 - \epsilon L_4 x_2$$
$$\beta = -H_4 x_1 + x_2 \tag{3.61}$$

and differentiating, we get

$$\dot{\alpha} = \dot{x}_1 - \epsilon L_4 \dot{x}_2 - \epsilon \dot{L}_4 x_2$$
$$\dot{\beta} = -H_4 \dot{x}_1 - \dot{H}_4 x_1 + \dot{x}_2 \tag{3.62}$$

Substituting for $\dot{x}_1$ and $\dot{x}_2$ from the original system, and simplifying, we get

$$\dot{\alpha} = A_{10}\alpha - G_1(L_4) x_2 \tag{3.63}$$

where

$$A_{10} = A_1 - L_4 A_3 \tag{3.64}$$

and

$$G_1(L_4) = \epsilon \dot{L}_4 + L_4 B_4 - \epsilon B_1 L_4 - B_2 + \epsilon L_4 B_3 L_4 \tag{3.65}$$

Also

$$\epsilon \dot{\beta} = A_{40}\beta - G_2(H_4) x_1 \tag{3.66}$$

where

$$A_{40} = A_4 - \epsilon H_4 A_2 \tag{3.67}$$

and

$$G_2(H_4) = \epsilon \dot{H}_4 + \epsilon H_4 A_1 - A_4 H_4 - A_3 + \epsilon H_4 A_2 H_4 \tag{3.68}$$

By setting $G_1(L_4) = 0$, and $G_2(H_4) = 0$, we get the decoupled system

$$\dot{\alpha} = A_{10}\alpha = (A_1 - L_4 A_3)\alpha \tag{3.69}$$

$$\epsilon \dot{\beta} = A_{40}\beta = (A_4 - \epsilon H_4 A_2)\beta \tag{3.70}$$

where L_4 and M_4 can be calculated from the following two stiff differential equations

$$\epsilon \dot{L}_4 = -L_4 A_4 + A_2 + \epsilon(A_1 L_4 - L_4 A_3 L_4) \tag{3.71}$$

75

$$\epsilon \dot{H}_4 = A_4 H_4 + A_3 - \epsilon (H_4 A_1 + H_4 A_2 H_4) \qquad (3.72)$$

The initial conditions for differential equations (3.71) and (3.72) are arbitrary (Chang, 1972; Smith, 1987). For time invariant systems equations (3.71)-(3.72) become algebraic ones. Efficient numerical methods for solving corresponding algebraic equations are discussed in (Grodt and Gajic, 1988). Thus, the introduced decoupling transformation is

$$\begin{bmatrix} \alpha \\ \beta \end{bmatrix} = \begin{bmatrix} I_{n_1} & -\epsilon L_4 \\ -H_4 & I_{n_2} \end{bmatrix} \begin{bmatrix} x_1 \\ x_2 \end{bmatrix} = \mathbf{T}_4(t) \begin{bmatrix} x_1 \\ x_2 \end{bmatrix} \qquad (3.73)$$

with

$$\mathbf{T}_4^{-1} = \begin{bmatrix} I_{n_1} + \epsilon L_4 N H_4 & \epsilon L_4 N \\ N H_4 & N \end{bmatrix} \qquad (3.74)$$

where $N = (I_{n_2} - \epsilon H_4 L_4)^{-1}$.

It is important to notice that in (3.44) and (3.45) one has to solve one Riccati and one Lyapunov equation sequentially. The total processing time in that case is greater than t_R, where t_R is the time for solving the Riccati equation. However, in (3.71) and (3.72) solutions of two Riccati equations are required, but due to parallelism the total processing time is t_R.

We can apply the same transformation as one given in (3.73) to the linear discrete-time singularly perturbed system introduced in Chapter 2, given by

$$\begin{aligned} x_1(n+1) &= (I + \epsilon A_{11}) x_1(n) + \epsilon A_{12} x_2(n) \\ x_2(n+1) &= A_{21} x_1(n) + A_{22} x_2(n) \end{aligned} \qquad (3.75)$$

Following similar calculation, the transformed block diagonal system is obtained as follows

$$\begin{aligned} \alpha(n+1) &= [I + \epsilon A_{11} - \epsilon L_4 A_{21}] \alpha(n) \\ \beta(n+1) &= [A_{22} - \epsilon H_4 A_{12}] \beta(n) \end{aligned} \qquad (3.76)$$

where L_4 and H_4 satisfy the following decoupled algebraic matrix equations

$$L_4 (I - A_{22}) + \epsilon A_{11} L_4 - \epsilon L_4 A_{21} L_4 + A_{12} = 0 \qquad (3.77)$$

$$(A_{22} - I) H_4 - \epsilon H_4 A_{11} - \epsilon H_4 A_{12} H_4 + A_{21} = 0 \qquad (3.78)$$

Note that unique solutions of (3.77)-(3.78) exist for sufficiently small values of ϵ under the assumption that the matrix A_{22} has no eigenvalues at -1, which is the standard condition imposed on singularly perturbed discrete systems (see formula 2.104).

It is left as an exercise to readers to derive the corresponding transformation for time-varying discrete singularly perturbed systems.

Exercise 3.3: Given the singularly perturbed time-varying discrete control system

$$x_1 (n + 1) = (I + \epsilon A_{11} (n)) x_1 (n) + \epsilon A_{12} (n) x_2 (n) + \epsilon B_1 (n) u (n)$$
$$x_2 (n + 1) = A_{21} (n) x_1 (n) + A_{22} (n) x_2 (n) + B_2 (n) u (n)$$

Find the transformation which completely decouples slow and fast variables.

$\triangle$

3.5 Decompositions of the Differential Lyapunov Equations

In the following, we first show that the introduced decoupling transformation for weakly coupled linear systems also completely decouples the Lyapunov matrix differential equations corresponding to weakly coupled systems. Then, we derive the similar result for the singularly perturbed differential Lyapunov equation.

Consider the Lyapunov matrix differential equation of weakly coupled systems

$$\dot{P} = A^T P + PA + Q, \quad Q = Q^T, \quad P(t_0) = P_0 \qquad (3.79)$$

where the given matrices A and Q are partitioned as

$$A = \begin{bmatrix} A_1 & \epsilon A_2 \\ \epsilon A_3 & A_4 \end{bmatrix}, \quad Q = \begin{bmatrix} Q_1 & \epsilon Q_2 \\ \epsilon Q_2^T & Q_3 \end{bmatrix} \qquad (3.80)$$

Due to assumed structure for A and Q, the matrix P is properly scaled as (Kokotovic et al., 1969)

$$P = \begin{bmatrix} P_1 & \epsilon P_2 \\ \epsilon P_2^T & P_3 \end{bmatrix} \qquad (3.81)$$

Multiplying (3.79) from the left by T_2^{-T} and from the right-hand side by T_2^{-1}, we get

$$T_2^{-T} \dot{P} T_2^{-1} = T_2^{-T} A^T P T_2^{-1} + T_2^{-T} P A T_2^{-1} + T_2^{-T} Q T_2^{-1} \qquad (3.82)$$

which can be written as

$$\dot{K} = a^T K + Ka + q, \qquad K(t_0) = K_0 \qquad (3.83)$$

where

$$a = T_2 A T_2^{-1} = \begin{bmatrix} A_{10} & 0 \\ 0 & A_{40} \end{bmatrix}$$

$$q = T_2^{-T} Q T_2 = \begin{bmatrix} q_1 & \epsilon q_2 \\ \epsilon q_2^T & q_3 \end{bmatrix} \qquad (3.84)$$

$$K = T_2^{-T} P T_2^{-1} = \begin{bmatrix} K_1 & \epsilon K_2 \\ \epsilon K_2^T & K_3 \end{bmatrix}, \quad K(t_0) = T_2^{-T} P_0 T_2^{-1}$$

Partitioning (3.83) we can note a completely decoupled form among elements of K, that is

$$\dot{K}_1 = K_1 A_{10} + A_{10}^T K_1 + q_1 \qquad (3.85)$$

$$\dot{K}_2 = K_2 A_{40} + A_{10}^T K_2 + q_2 \qquad (3.86)$$

$$\dot{K}_3 = K_3 A_{40} + A_{40}^T K_3 + q_3 \qquad (3.87)$$

Having obtained $K_i's$ from (3.85)-(3.87), we can get the solution of the Lyapunov differential equation in the original coordinates as

$$P = T_2^T K T_2 \qquad (3.88)$$

For the singularly perturbed Lyapunov matrix differential equation given by (3.79) the problem matrices are partitioned as

$$A = \begin{bmatrix} A_1 & A_2 \\ \frac{A_3}{\epsilon} & \frac{A_4}{\epsilon} \end{bmatrix}, \qquad Q = \begin{bmatrix} Q_1 & Q_2 \\ Q_2^T & Q_3 \end{bmatrix} \qquad (3.89)$$

Due to assumed structure for A and Q, the matrix P is properly scaled as (Kokotovic et al., 1986)

$$P = \begin{bmatrix} P_1 & \epsilon P_2 \\ \epsilon P_2^T & \epsilon P_3 \end{bmatrix} \qquad (3.90)$$

Multiplying (3.79) from the left by T_1^{-T} and from the right-hand side by T_1^{-1}, we get

$$T_1^{-T} \dot{P} T_1^{-1} = T_1^{-T} A^T P T_1^{-1} + T_1^{-T} P A T_1^{-1} + T_1^{-T} Q T_1^{-1} \quad (3.91)$$

which can be written as

$$\dot{K} = a^T K + K a + q, \qquad K(t_0) = K_0 \qquad (3.92)$$

where

$$a = \mathbf{T}_1 A \mathbf{T}_1^{-1} = \begin{bmatrix} A_s & 0 \\ 0 & \frac{A_f}{\epsilon} \end{bmatrix}$$

$$q = \mathbf{T}_1^{-T} Q \mathbf{T}_1^{-1} = \begin{bmatrix} q_1 & q_2 \\ q_2^T & q_3 \end{bmatrix} \qquad (3.93)$$

$$K = \mathbf{T}_1^{-T} P \mathbf{T}_1^{-1} = \begin{bmatrix} K_1 & \epsilon K_2 \\ \epsilon K_2^T & \epsilon K_3 \end{bmatrix}, \quad K(t_0) = \mathbf{T}_1^{-T} P_0 \mathbf{T}_1^{-1}$$

Partitioning (3.92) we can note a completely decoupled form among elements of K

$$\dot{K}_1 = K_1 A_s + A_s^T K_1 + q_1 \qquad (3.94)$$

$$\epsilon \dot{K}_2 = K_2 A_f + A_s^T K_2 + q_2 \qquad (3.95)$$

$$\epsilon \dot{K}_3 = K_3 A_f + A_f^T K_3 + q_3 \qquad (3.96)$$

Having obtained $K_i's$ from (3.94)-(3.96), we can get the solution of the singularly perturbed Lyapunov differential equation in the original coordinates as

$$\begin{bmatrix} P_1 & \epsilon P_2 \\ \epsilon P_2^T & \epsilon P_3 \end{bmatrix} = \mathbf{T}_1^T \begin{bmatrix} K_1 & \epsilon K_2 \\ \epsilon K_2^T & \epsilon K_3 \end{bmatrix} \mathbf{T}_1 \qquad (3.97)$$

It is left to the reader to find the corresponding decompositions of the difference Lyapunov equations for both weakly coupled and singularly perturbed systems.

Exercise 3.4: Following the methodology presented in Section 3.5 find the decomposition for both singularly perturbed and weakly coupled matrix difference Lyapunov equations.

$\triangle$

3.6 Boundary Value Problem of Linear Continuous Weakly Coupled Systems

In this section, we study the general boundary value problem of linear time varying weakly coupled systems. The existence of the solution in terms of the reduced-order completely decoupled dynamical systems is established.

Consider the general boundary value problem for the linear weakly coupled system

$$\dot{x} = A_1(t) x + \epsilon A_2(t) z \qquad (3.98)$$

$$\dot{z} = \epsilon A_3(t) x + A_4(t) z \qquad (3.99)$$

with boundary conditions

$$M(\epsilon) \begin{bmatrix} x(0, \epsilon) \\ z(0, \epsilon) \end{bmatrix} + N(\epsilon) \begin{bmatrix} x(1, \epsilon) \\ z(1, \epsilon) \end{bmatrix} = \begin{bmatrix} c_1(\epsilon) \\ c_2(\epsilon) \end{bmatrix} \qquad (3.100)$$

on the interval $0 \leq t \leq 1$, where $x \in \Re^{n_1}$, $z \in \Re^{n_2}$, $c_1 \in \Re^{n_1}$, $c_2 \in \Re^{n_2}$, and $A_1(t)$, $A_2(t)$, $A_3(t)$, $A_4(t)$, $M(\epsilon)$, $N(\epsilon)$ are matrices of appropriate dimensions. A small parameter ϵ couples the states x and z. Note that for weakly coupled systems the following standard assumption is imposed (Chow and Kokotovic, 1983).

Assumption 3.3

$$det A_1(t) = O(1) \quad and \quad det A_4(t) = O(1), \quad \forall t \in [0, 1]$$

$\triangle$

In addition, the following assumption is made in this section.

Assumption 3.4 The matrices $A_1(t)$, $A_2(t)$, $A_3(t)$, and $A_4(t)$ are continuous function for $0 \leq t \leq 1$, and $M(\epsilon) = M(0) + O(\epsilon)$, $N(\epsilon) = N(0) + O(\epsilon)$, $c_i(\epsilon) = c_i(0) + O(\epsilon)$, $i = 1, 2$.

$\triangle$

The corresponding boundary value problem for singularly perturbed systems is studied in (Chang, 1972). In the following, we will use the transformation derived in Section 3.4.1 to block diagonalize (3.98)-(3.99). The obtained results simplify the analytical and computational treatment of (3.98)-(3.99) in terms of the reduced-order completely decoupled dynamical systems.

It is shown in Section 3.4.1 that the following transformation

$$\begin{bmatrix} \eta \\ \zeta \end{bmatrix} = \begin{bmatrix} I_{n_1} & -\epsilon L_3\left(t\right) \\ -\epsilon H_3\left(t\right) & I_{n_2} \end{bmatrix} \begin{bmatrix} x_1 \\ x_2 \end{bmatrix} = \mathbf{T_3}\left(t,\ \epsilon\right) \begin{bmatrix} x_1 \\ x_2 \end{bmatrix} \qquad (3.101)$$

produces the block diagonal form of (3.98)-(3.99), that is,

$$\dot{\eta} = A_{10}\left(t\right)\eta \qquad (3.102)$$

$$\dot{\zeta} = A_{40}\left(t\right)\zeta \qquad (3.103)$$

where

$$A_{10}\left(t\right) = A_1\left(t\right) - \epsilon^2 L_3\left(t\right) A_3\left(t\right) \qquad (3.104)$$

$$A_{40}\left(t\right) = A_4\left(t\right) - \epsilon^2 H_3\left(t\right) A_2\left(t\right) \qquad (3.105)$$

The transformation matrices $L_3\left(t\right)$ and $H_3\left(t\right)$ satisfy the following Riccati-type differential equations

$$\dot{L}_3\left(t\right) = A_1\left(t\right) L_3\left(t\right) - L_3\left(t\right) A_4\left(t\right) + A_2\left(t\right) - \epsilon^2 L_3\left(t\right) A_3\left(t\right) L_3\left(t\right) \qquad (3.106)$$

$$\dot{H}_3\left(t\right) = A_4\left(t\right) H_3\left(t\right) - H_3\left(t\right) A_1\left(t\right) + A_3\left(t\right) - \epsilon^2 H_3\left(t\right) A_2\left(t\right) H_3\left(t\right) \qquad (3.107)$$

Note that for sufficiently small ϵ, the matrix $\mathbf{T_3}\left(t,\ \epsilon\right)$ in (3.101) is diagonally dominant, and hence, nonsingular. In addition, no initial or terminal conditions are imposed on equations (3.106)-(3.107).

Since the differential Riccati equation, in general, has a finite escape time (Sasagawa, 1982), we have to establish the existence of bounded solutions for (3.106)-(3.107). The only assumption we impose is that $A_1\left(t\right)$, $A_2\left(t\right)$, $A_3\left(t\right)$, and $A_4\left(t\right)$ are continuous function for $0 \leq t \leq 1$. Due to the same structure of (3.106)-(3.107), it suffices to show that either one of them possesses a bounded solution. The following lemma is proved for the existence of such a solution.

Lemma 3.1 *There exists $\epsilon_0 > 0$ such that equations (3.106)-(3.107) have solutions $L_3\left(t\right)$ and $H_3\left(t\right)$, respectively, which are uniformly bounded for $0 \leq t \leq 1$ in the region $0 < \epsilon \leq \epsilon_0$.*

$\diamond$

Proof: Let $X\left(t\right)$ be the fundamental matrix of $\dot{x} = A_1\left(t\right)x$, and $Z\left(t\right)$ be the fundamental matrix of $\dot{z} = A_4\left(t\right)z$. Since $A_1\left(t\right)$ and $A_4\left(t\right)$ are continuous on $[0, 1]$, and therefore bounded on $[0, 1]$, there exist $\sigma_1 > 0$ and $\sigma_2 > 0$ such that

$$\| A_1(t) \| < \sigma_1$$
$$\| A_4(t) \| < \sigma_2$$

which implies (Chang, 1972)

$$\| X(t) X^{-1}(s) \| \leq K_1 exp(\sigma_1 |t - s|) \quad 0 \leq t, s \leq 1$$
$$\| Z(t) Z^{-1}(s) \| \leq K_2 exp(\sigma_2 |t - s|) \quad 0 \leq t, s \leq 1 \tag{3.108}$$

where K_1 and K_2 are positive constants.

It can be verified by differentiation that

$$L_3(t) = \int\limits_0^t X(t) X^{-1}(s) \left[A_2(s) - \epsilon^2 L_3(s) A_3(s) L_3(s) \right] Z(s) Z^{-1}(t) \, ds \tag{3.109}$$

is a solution of (3.106). We will show that this is a bounded solution, that is, $\| L_3(t) \| \leq \rho$, for some ρ. By (3.109) we get

$$\| L_3(t) \| \leq \int\limits_0^t K_1 K_2 e^{\sigma_1 |t-s|} e^{\sigma_2 |t-s|} ds \, (\| A_2(t) \| + \epsilon^2 \rho^2 \| A_3(t) \|)$$

Since $s \leq t$, then $|t - s| = t - s$, so that we obtain

$$\| L_3(t) \| \leq \frac{K_1 K_2}{\sigma_1 + \sigma_2} \left[e^{(\sigma_1 + \sigma_2)t} - 1 \right] (\| A_2(t) \| + \epsilon^2 \rho^2 \| A_3(t) \|)$$

This inequality is valid for $t = 1$, giving

$$\| L_3(t) \| \leq \alpha \, (\| A_2(t) \| + \epsilon^2 \rho^2 \| A_3(t) \|)$$

where

$$\alpha = \frac{K_1 K_2}{\sigma_1 + \sigma_2} \left[e^{(\sigma_1 + \sigma_2)t} - 1 \right]$$

Now we have to find ρ and ϵ_0 such that $\| L_3(t) \| \leq \rho$. Pick

$$\rho = 2\alpha \| A_2(t) \|$$

so that

$$\| L_3(t) \| \le \frac{\rho}{2} + \alpha \epsilon^2 \rho^2 \| A_3(t) \|$$

Pick ϵ_0 such that

$$\alpha \epsilon^2 \rho^2 \| A_3(t) \| \le \frac{\rho}{2}$$

or

$$\epsilon_0^2 \le \frac{1}{4\alpha^2 \| A_2(t) \| \| A_3(t) \|}$$

Thus

$$\| L_3(t) \| \le \frac{\rho}{2} + \frac{\rho}{2} < \rho$$

Therefore, $L_3(t)$ has a bounded solution, that is $\| L_3(t) \| \le \rho$. Since $[0,1]$ is a compact interval, the solution is uniformly bounded. On the same lines, it can be proved that $H_3(t)$ also has a uniformly bounded solution.

Consequently, the change of variables (3.101) transforms the system (3.98)-(3.99) to (3.102)-(3.103). Applying (3.101) to the boundary conditions (3.100), we obtain

$$\hat{M}(\epsilon) \begin{bmatrix} \eta(0; \epsilon) \\ \zeta(0; \epsilon) \end{bmatrix} + \hat{N}(\epsilon) \begin{bmatrix} \eta(1; \epsilon) \\ \zeta(1; \epsilon) \end{bmatrix} = \begin{bmatrix} c_1(\epsilon) \\ c_2(\epsilon) \end{bmatrix} \tag{3.110}$$

where $\hat{M}(\epsilon) = M(\epsilon) T_3^{-1}(0, \epsilon)$, $\hat{N}(\epsilon) = N(\epsilon) T_3^{-1}(1, \epsilon)$, and $T_3^{-1}(t, \epsilon)$ is given by

$$T_3^{-1}(t, \epsilon) = \begin{bmatrix} I_{n_1} + \epsilon^2 L_3(t) M(t) H_3(t) & \epsilon L_3(t) M(t) \\ \epsilon M(t) H(t) & M(t) \end{bmatrix}$$

where $M(t) = (I_{n_2} - \epsilon^2 H(t) L(t))^{-1}$. Therefore, the transformed boundary value problem is described by (3.102)-(3.103) and (3.110).

Let $\xi(t, \epsilon)$ be the transition matrix of (3.102), and $\chi(t, \epsilon)$ be the transition matrix of (3.103), then we have the following solution for (3.102)-(3.103)

$$\begin{bmatrix} \eta(t, \epsilon) \\ \zeta(t, \epsilon) \end{bmatrix} = \begin{bmatrix} \xi(t, \epsilon) & 0 \\ 0 & \chi(t, \epsilon) \end{bmatrix} \begin{bmatrix} \alpha_1(\epsilon) \\ \alpha_2(\epsilon) \end{bmatrix} \tag{3.111}$$

where $\alpha_1(\epsilon)$ and $\alpha_2(\epsilon)$ are arbitrary constants. It remains to choose $\alpha_1(\epsilon)$ and $\alpha_2(\epsilon)$ to satisfy the boundary condition (3.110). Substituting (3.111) into (3.110) yields

$$\Delta(\epsilon) \begin{bmatrix} \alpha_1(\epsilon) \\ \alpha_2(\epsilon) \end{bmatrix} = \begin{bmatrix} c_1(\epsilon) \\ c_2(\epsilon) \end{bmatrix}$$

where

$$\Delta(\epsilon) = \hat{M}(\epsilon) + \hat{N}(\epsilon) \begin{bmatrix} \xi(1; \epsilon) & 0 \\ 0 & \chi(1, \epsilon) \end{bmatrix} \qquad (3.112)$$

If $\Delta^{-1}(\epsilon)$ exist, then

$$\begin{bmatrix} \alpha_1(\epsilon) \\ \alpha_2(\epsilon) \end{bmatrix} = \Delta^{-1}(\epsilon) \begin{bmatrix} c_1(\epsilon) \\ c_2(\epsilon) \end{bmatrix}$$

thus (3.111) represents a solution to the boundary value problem.

Note that as $\epsilon \to 0$, $T_3^{-1}(t, \epsilon) \to I$, and therefore $\lim_{\epsilon \to 0} \hat{M}(\epsilon) = M(0)$ and $\lim_{\epsilon \to 0} \hat{N}(\epsilon) = N(0)$

$$\lim_{\epsilon \to 0} \Delta(\epsilon) = M(0) + N(0) \begin{bmatrix} \xi(1; 0) & 0 \\ 0 & \chi(1, 0) \end{bmatrix}$$

where $\xi(t, 0) = \lim_{\epsilon \to 0} \xi(t, \epsilon)$ and $\chi(t, 0) = \lim_{\epsilon \to 0} \chi(t, \epsilon)$. Hence for sufficiently small ϵ, the inverse $\Delta^{-1}(\epsilon)$ exists under the following assumption.

Assumption 3.5 The matrix $\Delta(0) = \lim_{\epsilon \to 0} \Delta(\epsilon)$ in nonsingular.

$\triangle$

Note that Assumption 3.5 is always satisfied for the linear filtering and control problems (Qureshi, 1992).

Consequently, the following theorem summarizes the results.

Theorem 3.2 *Let Assumptions 3.4 and 3.5 hold, then for sufficiently small ϵ the original weakly coupled boundary value problem (3.98)-(3.100) has the solution given by*

$$\begin{bmatrix} x(t, \epsilon) \\ z(t, \epsilon) \end{bmatrix} = T_3^{-1}(t, \epsilon) \begin{bmatrix} \xi(t, \epsilon) & 0 \\ 0 & \chi(t, \epsilon) \end{bmatrix} \Delta^{-1}(\epsilon) \begin{bmatrix} c_1(\epsilon) \\ c_2(\epsilon) \end{bmatrix}$$

for $0 \le t \le 1$.

$\diamond$

3.7 Boundary Value Problem of Discrete Linear Weakly Coupled Systems

In the procedure of solving the general continuous-time boundary value problem of linear weakly coupled systems the transition matrices of two time-varying subsystems are required. Since, in general, these matrices can not be found analytically, it would be interesting, from the practical point of view, to develop the corresponding results for the weakly coupled time-varying discrete boundary value problem, where the corresponding discrete-time transition matrices can be found easily.

Consider the general boundary value problem of the discrete-time linear weakly coupled system

$$x(k+1) = A_1(k) x(k) + \epsilon A_2(k) z(k) \qquad (3.113)$$

$$z(k+1) = \epsilon A_3(k) x(k) + A_4(k) z(k) \qquad (3.114)$$

with boundary conditions

$$M(\epsilon) \begin{bmatrix} x(0,\epsilon) \\ z(0,\epsilon) \end{bmatrix} + N(\epsilon) \begin{bmatrix} x(n,\epsilon) \\ z(n,\epsilon) \end{bmatrix} = \begin{bmatrix} c_1(\epsilon) \\ c_2(\epsilon) \end{bmatrix} \qquad (3.115)$$

on the interval $0 \le k \le n$, where $x \in \Re^{n_1}, z \in \Re^{n_2}, c_1 \in \Re^{n_1}; c_2 \in \Re^{n_2}$, and $A_1(k), A_2(k), A_3(k), A_4(k), M(\epsilon), N(\epsilon)$ are of appropriate dimensions. A small parameter ϵ couples the states x and z. It is assumed that $M(\epsilon) = M(0) + O(\epsilon), \ N(\epsilon) = N(0) + O(\epsilon), \ c_i(\epsilon) = c_i(0) + O(\epsilon), \ i = 1, 2$. Note that for weakly coupled systems $\det A_1(k) = O(1)$ and $\det A_4(k) = O(1)$ for all $0 \le k \le n$ (see Assumption 3.3).

The corresponding boundary value problem for continuous weakly coupled systems is studied in the previous section. In this section, we will use the same transformation to block diagonalize (3.113)-(3.114), which simplifies the analytical and computational treatment of (3.113)-(3.114) in terms of the reduced-order completely decoupled dynamical systems.

The following transformation

$$\begin{bmatrix} \eta(k) \\ \zeta(k) \end{bmatrix} = \begin{bmatrix} I & -\epsilon L(k) \\ -\epsilon H(k) & I \end{bmatrix} \begin{bmatrix} x(k) \\ z(k) \end{bmatrix}$$
$$= T_3(k, \epsilon) \begin{bmatrix} x(k) \\ z(k) \end{bmatrix} \qquad (3.116)$$

produces a block diagonal form of (3.113)-(3.114), that is

$$\eta(k+1) = A_{10}(k) \eta(k) \qquad (3.117)$$

$$\zeta(k+1) = A_{40}(k)\zeta(k) \tag{3.118}$$

where

$$A_{10}(k) = A_1(k) - \epsilon^2 L(k) A_3(k) \tag{3.119}$$

$$A_{40}(k) = A_4(k) - \epsilon^2 H(k) A_2(k) \tag{3.120}$$

The transformation matrices $L(k)$ and $H(k)$ satisfy the following difference equations

$$\begin{aligned} &A_1(k)L(k) - L(k+1)A_4(k) + A_2(k) \\ &\quad -\epsilon^2 L(k+1) A_3(k) L(k) = 0 \end{aligned} \tag{3.121}$$

$$\begin{aligned} &A_4(k)H(k) - H(k+1)A_1(k) + A_3(k) \\ &\quad -\epsilon^2 H(k+1) A_2(k) H(k) = 0 \end{aligned} \tag{3.122}$$

Note that for sufficiently small ϵ, the matrix $T_3(k,\epsilon)$ in (3.116) is diagonally dominant, and hence nonsingular. In addition, no initial or terminal conditions are imposed on equations (3.121)-(3.122).

In order to have a bounded solution for the block diagonalized system (3.117)-(3.118), $\|A_{10}(k)\|$ and $\|A_{40}(k)\|$ must be bounded. First of all, we make the following assumption on the matrices $A_1(k), A_2(k), A_3(k)$, and $A_4(k)$.

Assumption 3.6 The matrices $A_i(k)$, $i = 1, 2, 3, 4$, are bounded, that is, $\|A_i(k)\| \le K_i$, $i = 1, 2, 3, 4$, where K_i are bounded constants. Furthermore, $A_1(k)$ and $A_4(k)$ are invertible.

$$\triangle$$

Under Assumption 3.6, it follows that (3.121)-(3.122) have bounded solutions, and hence, the systems (3.117)-(3.118) have bounded solutions. Thus, we have a dual lemma to Lemma 3.1 stated as following.

Lemma 3.2 *Under Assumptions 3.3 and 3.6, there exists $\epsilon_0 > 0$ such that equations (3.121) and (3.122) have solutions $L(k)$ and $H(k)$, respectively, which are bounded for $0 \le k \le n$ in the region $0 < \epsilon \le \epsilon_0$.*

$$\diamond$$

Consequently, the change of variable (3.116) transforms the system (3.113)-(3.114) to (3.117)-(3.118). Applying (3.116) to the boundary conditions (3.115), we obtain

$$\hat{M}(\epsilon) \begin{bmatrix} \eta(0,\epsilon) \\ \zeta(0,\epsilon) \end{bmatrix} + \hat{N}(\epsilon) \begin{bmatrix} \eta(n,\epsilon) \\ \zeta(n,\epsilon) \end{bmatrix} = \begin{bmatrix} c_1(\epsilon) \\ c_2(\epsilon) \end{bmatrix} \qquad (3.123)$$

where $\hat{M}(\epsilon) = M(\epsilon)\mathbf{T}_3^{-1}(0,\epsilon)$, $\hat{N}(\epsilon) = N(\epsilon)\mathbf{T}_3^{-1}(n,\epsilon)$ with $\mathbf{T}_3(k,\epsilon)$ defined in (3.116). Therefore the transformed boundary value problem is described by (3.117), (3.118), and (3.123).

Let $\mathcal{E}(k,\epsilon)$ be the transition matrix of (3.117), and $\mathcal{Z}(k,\epsilon)$ be the transition matrix of (3.118), then we have the following solutions for (3.117)-(3.118)

$$\begin{bmatrix} \eta(k,\epsilon) \\ \zeta(k,\epsilon) \end{bmatrix} = \begin{bmatrix} \mathcal{E}(k,\epsilon) & 0 \\ 0 & \mathcal{Z}(k,\epsilon) \end{bmatrix} \begin{bmatrix} \alpha_1(\epsilon) \\ \alpha_2(\epsilon) \end{bmatrix} \qquad (3.124)$$

where $\alpha_1(\epsilon)$ and $\alpha_2(\epsilon)$ are arbitrary constants. It remains to choose $\alpha_1(\epsilon)$ and $\alpha_2(\epsilon)$ to satisfy the boundary conditions (3.123). Substituting (3.124) into (3.123) yields

$$\Delta(\epsilon) \begin{bmatrix} \alpha_1(\epsilon) \\ \alpha_2(\epsilon) \end{bmatrix} = \begin{bmatrix} c_1(\epsilon) \\ c_2(\epsilon) \end{bmatrix}$$

where

$$\Delta(\epsilon) = \hat{M}(\epsilon) + \hat{N}(\epsilon) \begin{bmatrix} \mathcal{E}(n,\epsilon) & 0 \\ 0 & \mathcal{Z}(k,\epsilon) \end{bmatrix} \qquad (3.125)$$

If $\Delta^{-1}(\epsilon)$ exists then

$$\begin{bmatrix} \alpha_1(\epsilon) \\ \alpha_2(\epsilon) \end{bmatrix} = \Delta^{-1}(\epsilon) \begin{bmatrix} c_1(\epsilon) \\ c_2(\epsilon) \end{bmatrix}$$

and thus, (3.124) represents a solution of the considered boundary value problem.

It is easy to see that as $\epsilon \to 0$, $\mathbf{T}_3^{-1}(k,\epsilon) \to I$, $\lim_{\epsilon \to 0} \hat{M}(\epsilon) = M(0)$, and $\lim_{\epsilon \to 0} \hat{N}(\epsilon) = N(0)$. Therefore,

$$\lim_{\epsilon \to 0} \Delta(\epsilon) = M(0) + N(0) \begin{bmatrix} \mathcal{E}(n,0) & 0 \\ 0 & \mathcal{Z}(n,0) \end{bmatrix} \qquad (3.126)$$

where $\mathcal{E}(k,0) = \lim_{\epsilon \to 0} \mathcal{E}(k,\epsilon)$, and $\mathcal{Z}(k,0) = \lim_{\epsilon \to 0} \mathcal{Z}(k,\epsilon)$. Hence for sufficiently small ϵ, the unique solutions for $\alpha_1(\epsilon)$ and $\alpha_2(\epsilon)$ exist under the following assumption.

Assumption 3.7 The matrix (3.126) is nonsingular.

$$\triangle$$

Note that as opposed to the continuous-time systems, the transition

DECOUPLING TRANSFORMATIONS

matrices and hence, $\Delta(\epsilon)$ for discrete systems can be found analytically. This gives us the explicit solution for the boundary value problem of discrete weakly coupled systems.

Consequently, the following theorem summarizes the results.

Theorem 3.3 *Let Assumption 3.7 holds, then under conditions stated in Lemma 3.2, for sufficiently small ϵ, the original boundary value problem (3.113)-(3.114) has the solution given by*

$$\begin{bmatrix} x(k,\epsilon) \\ z(k,\epsilon) \end{bmatrix} = \mathbf{T}_3^{-1}(k,\epsilon) \begin{bmatrix} \mathcal{E}(k,\epsilon) & 0 \\ 0 & \mathcal{Z}(k,\epsilon) \end{bmatrix} \begin{bmatrix} \alpha_1(\epsilon) \\ \alpha_2(\epsilon) \end{bmatrix} \qquad (3.127)$$

for $0 \leq k \leq n$.

◊

Research Problem 3.1: In the continuous-time varying boundary value problem for singularly perturbed systems (Chang, 1972), the transition matrices of continuous-time varying slow and fast subsystems are required. These transitions matrices can not be found analytically. Derive the discrete-time version of the results reported in (Chang, 1972), such that the analytical expressions for the required transition matrices are obtained.

△

Chapter 4

Output Feedback Control

4.1 Introduction

The design of the optimal linear full state regulator requires the measurement of all system states. In many practical applications, this is not feasible due to either the high cost of the state measurements or the inaccessibility for measurement of some of the system states. The standard way to overcome these difficulties is to reconstruct the full state vector from the available measurements by the Luenberger observer or, if the measurements are noisy, by the Kalman filter. However, these state reconstruction methods will introduce an additional dynamical system. That is why, in the early 1970's, increasing attention was given to the problem of designing output constrained regulators where a very limited number of state measurements are available for control implementation (Levine and Athans, 1970; Levine et al., 1971; Mendel, 1974; Petkovski and Rakic, 1979). The optimal solution to this control problem is obtained in terms of high-order nonlinear matrix algebraic equations. The convergence complexities of the algorithms suggested for the solution of these equations have hindered for quite a long time a wider application of this technique. The convergence problem was solved in (Moerder and Calise, 1985a; Toivonen, 1985). Since then, the static output feedback control problem has become a very fruitful research area (Makila and Toivonen, 1987).

In this chapter, the output feedback control of singularly perturbed and weakly coupled linear systems is studied. The output feedback control problem attracted the attention of the researchers from the field of singular perturbations in the 1980's (Chemouil and Wahdam, 1980;

Fossard and Magni, 1980; Khalil, 1981; Arkun and Ramakrishnan, 1983; Moerder and Calise 1985b; Calise and Moerder, 1985; Khalil, 1987). It is well known that the singularly perturbed systems belong to the class of systems with ill-conditioned dynamics which make corresponding numerical problems stiff. Thus, in addition to the high-order nonlinear matrix algebraic equations, one is faced with the ill-defined numerical problems also.

Motivated by the results of (Gajic, 1986; Gajic et al., 1990) and (Moerder and Calise, 1985a), we have developed the well-defined recursive numerical technique for the solution of nonlinear algebraic matrix equations associated with the output feedback control problem of linear-quadratic singularly perturbed systems. Moreover, the numerical slow-fast decomposition has been achieved so that only low-order systems are involved in algebraic computations. It is shown that each iteration step of the proposed algorithm improves the accuracy by an order of magnitude, that is, the accuracy of $O\left(\epsilon^k\right)$, where ϵ is a small perturbation parameter, can be obtained by performing only k iterations. This represents a significant improvement, since all results on the output feedback control problems for the singularly perturbed systems have been obtained so far with an accuracy of $O\left(\epsilon\right)$ only. The real world example, an industrially important reactor, which demonstrates the efficiency of the proposed algorithm and the failure of $O\left(\epsilon\right)$ theory is included in the section.

The output feedback control problem for weakly coupled linear systems has been studied in (Petkovski and Rakic, 1979) by using a series expansion approach. This approach is not recursive in application and it is numerically inefficient when a high order of accuracy is required or when the coupling parameter ϵ is not very small.

Following the results presented in Section 4.2, a recursive algorithm is developed for solving nonlinear algebraic equations comprising the solution of the optimal static output feedback control problem of linear weakly coupled systems. The numerical decomposition has been achieved so that only low-order systems are involved in algebraic computations. The effectiveness of the proposed reduced-order algorithm and its advantages over the global full-order algorithm are demonstrated on a twelve-plate chemical absorption column. Obtained results strongly support the necessity of the existence of reduced-order numerical techniques for solving corresponding nonlinear algebraic equations. In addition to reduction in required computations, it would be easier to find a good initial guess and to handle the problem of nonuniqueness of the solution of these nonlinear equations — they represent the necessary conditions only.

In this chapter, we have limited our attention to the deterministic continuous-time output feedback control problem. Stochastic output feedback makes no sense in the continuous-time domain since it is not rational to feedback the continuous-time white noise. In Chapter 11, we will study the stochastic output feedback in the discrete-time domain.

4.2 Output Feedback for Singularly Perturbed Linear Systems

Consider the singularly perturbed linear system (Kokotovic and Khalil, 1986)

$$\dot{x}_1 = A_1 x_1 + A_2 x_2 + B_1 u, \qquad x_1(t_0) = x_{10} \tag{4.1}$$

$$\epsilon \dot{x}_2 = A_3 x_1 + A_4 x_2 + B_2 u, \qquad x_2(t_0) = x_{20} \tag{4.2}$$

$$y = C_1 x_1 + C_2 x_2 \tag{4.3}$$

where $x_1 \in \Re^{n_1}$ and $x_2 \in \Re^{n_2}$ are state vectors, $u \in \Re^m$ is a control input and $y \in \Re^r$ is a measured output. In the following, A_i, B_j, and C_j, $i = 1, \ldots 4$, $j = 1, 2$, are constant matrices of compatible dimensions; in general they are continuous functions of a small positive parameter ϵ (Gajic, 1986). With (4.1)-(4.3), consider the performance criterion

$$J = \frac{1}{2} \int\limits_0^{\infty} \left\{ \begin{bmatrix} x_1 \\ x_2 \end{bmatrix}^T Q \begin{bmatrix} x_1 \\ x_2 \end{bmatrix} + u^T R u \right\} dt \tag{4.4}$$

with positive definite R and positive semidefinite Q, which has to be minimized. In addition, the control input $u(t)$ is constrained to

$$u = -Fy \tag{4.5}$$

The optimal constant output feedback gain F is given by (Levine and Athans, 1970)

$$F = R^{-1} B^T K L C^T \left(C L C^T \right)^{-1} \tag{4.6}$$

where matrices K and L satisfy high-order nonlinear coupled algebraic equations

$$(A - BFC)L + L(A - BFC)^T + x_0 x_0^T = 0 \tag{4.7}$$

$$(A - BFC)^T K + K (A - BFC) + Q + C^T F^T RFC = 0 \quad (4.8)$$

and newly defined matrices are

$$A = \begin{bmatrix} A_1 & A_2 \\ \frac{A_3}{\epsilon} & \frac{A_4}{\epsilon} \end{bmatrix}, \quad B = \begin{bmatrix} B_1 \\ \frac{B_2}{\epsilon} \end{bmatrix}, \quad C = [C_1 \quad C_2], \quad x_0 = \begin{bmatrix} x_{10} \\ x_{20} \end{bmatrix} \quad (4.9)$$

Compatible to the nature of their solutions, matrices K and L are partitioned as follows

$$K = \begin{bmatrix} K_1 & \epsilon K_2 \\ \epsilon K_2^T & \epsilon K_3 \end{bmatrix}, \quad L = \begin{bmatrix} L_1 & L_2 \\ L_2^T & L_3 \end{bmatrix} \quad (4.10)$$

It is shown in (Moerder and Calise, 1985a) that the algorithm proposed for the numerical solution of (4.6)-(4.8), defined by Algorithm 4.1:

Choose $F^{(0)}$ such that $A - BF^{(0)}C$ is a stable matrix \quad (4.11)

$$\left(A - BF^{(i)}C\right) L^{(i+1)} + L^{(i+1)} \left(A - BF^{(i)}C\right)^T + x_0 x_0^T = 0 \quad (4.12)$$

$$\left(A - BF^{(i)}C\right)^T K^{(i+1)} + K^{(i+1)} \left(A - BF^{(i)}C\right) + Q$$
$$+ C^T F^{(i)^T} RF^{(i)}C = 0 \quad (4.13)$$

$$F^{(i+1)} = R^{-1} B^T K^{(i+1)} L^{(i+1)} C^T \left(CL^{(i+1)}C^T\right)^{-1} \quad (4.14)$$

with $i = 1, 2, ...$, converges to a local minimum under the nonrestrictive assumption. As a matter of fact, the updated value for F is defined in (Moerder and Calise, 1985a) as

$$F_N^{(i+1)} = F^{(i)} + \alpha \left(F^{(i+1)} - F^{(i)}\right) \quad (4.15)$$

where $\alpha \in (0, 1]$ is chosen at each iteration to ensure that the minimum is not overshoot, that is,

$$J_{i+1} = tr \left\{ K^{(i+1)} x_0 x_0^T \right\} < J_i = \left\{ K^{(i)} x_0 x_0^T \right\} \quad (4.16)$$

$\triangle$

Note that in (Toivonen and Makila, 1987) the Newton method was derived for solving the algebraic equations (4.6)-(4.8).

Exercise 4.1: Multiply equation (4.8) from the left by CL and from the right by $L^T C^T$. Show that this procedure reduces equation (4.8) to the standard algebraic Riccati equation. Comment on the procedure and obtained result.

$$\triangle$$

It has been customary in the control literature on the output feedback to assume that the initial conditions are uniformly distributed on the unit sphere, that is,

$$E\left\{x_0 x_0^T\right\} = I_{(n_1+n_2)} \tag{4.17}$$

Notice that applying the slow-fast decomposition transform of Chang (Chang, 1972) to problem (4.1)-(4.5) and finding the optimal gains for the slow and fast subsystems is producing the accuracy of $O\left(\epsilon\right)$ only, (Calise and Moerder, 1985; Moerder and Calise, 1985b, 1988). This leads to a well-posed problem, but there is no way to improve the approximation to any desired order of accuracy, that is, $O\left(\epsilon^k\right)$. In this section, we will achieve that goal through the numerical slow-fast decomposition of the algebraic equations (4.11)-(4.15).

In order to simplify derivations, we introduce the notation

$$A - BFC = \begin{bmatrix} D_1 & D_2 \\ \frac{D_3}{\epsilon} & \frac{D_4}{\epsilon} \end{bmatrix} = \begin{bmatrix} A_1 - B_1 F C_1 & A_2 - B_1 F C_2 \\ \frac{A_3 - B_2 F C_1}{\epsilon} & \frac{A_4 - B_2 F C_2}{\epsilon} \end{bmatrix} \tag{4.18}$$

$$Q + C^T F^T R F C = \begin{bmatrix} q_1 & q_2 \\ q_3 & q_4 \end{bmatrix}$$
$$= \begin{bmatrix} Q_1 + C_1^T F^T R F C_1 & Q_2 + C_1^T F^T R F C_2 \\ Q_2^T + C_2^T F^T R F C_1 & Q_3 + C_2^T F^T R F C_2 \end{bmatrix} \tag{4.19}$$

with obvious definitions for $D_i's$ and $q_i's$, $i = 1, 2, 3, 4$.

Partitioning (4.12)-(4.13) compatible to (4.9)-(4.10) and using (4.17)-(4.19) will produce the following set of equations

$$D_1^{(i)} L_1^{(i+1)} + L_1^{(i+1)} D_1^{(i)^T} + D_2^{(i)} L_2^{(i+1)^T} + L_2^{(i+1)} D_2^{(i)^T} + I = 0 \tag{4.20}$$

$$L_2^{(i+1)} D_4^{(i)^T} + \epsilon D_1^{(i)} L_2^{(i+1)} + L_1^{(i+1)} D_3^{(i)^T} + \epsilon D_2^{(i)} L_3^{(i+1)} = 0 \tag{4.21}$$

$$L_3^{(i+1)} D_4^{(i)^T} + D_4^{(i)} L_3^{(i+1)} + D_3^{(i)} L_2^{(i+1)} + L_2^{(i+1)^T} D_3^{(i)^T} + \epsilon I = 0 \tag{4.22}$$

and

$$D_1^{(i)^T} K_1^{(i+1)} + K_1^{(i+1)} D_1^{(i)} + D_3^{(i)^T} K_2^{(i+1)^T} + K_2^{(i+1)} D_3^{(i)} + q_1^{(i)} = 0$$
(4.23)

$$K_2^{(i+1)} D_4^{(i)} + \epsilon D_1^{(i)^T} K_2^{(i+1)} + D_3^{(i)^T} K_3^{(i+1)} + K_1^{(i+1)} D_2^{(i)} + q_2^{(i)} = 0 \quad (4.24)$$

$$K_3^{(i+1)} D_4^{(i)} + D_4^{(i)^T} K_3^{(i+1)} + \epsilon D_2^{(i)^T} K_2^{(i+1)} + \epsilon K_2^{(i+1)^T} D_2^{(i)} + q_3^{(i)} = 0$$
(4.25)

where

$$D_1^{(i)} = A_1 - B_1 F^{(i)} C_1, \quad D_2^{(i)} = A_2 - B_1 F^{(i)} C_2$$

$$D_3^{(i)} = A_3 - B_2 F^{(i)} C_1, \quad D_4^{(i)} = A_4 - B_2 F^{(i)} C_2$$

and

$$q_1^{(i)} = Q_1 + C_1^T F^{(i)^T} R F^{(i)} C_1$$

$$q_2^{(i)} = Q_2 + C_1^T F^{(i)^T} R F^{(i)} C_2$$

$$q_3^{(i)} = Q_3 + C_2^T F^{(i)^T} R F^{(i)} C_2, \quad i = 0, 1, 2, 3, \dots.$$

Since the matrix $A - BF^{(i)}C$ has n_1 slow eigenvalues of $O(1)$ and n_2 fast eigenvalues of $O\left(\frac{1}{\epsilon}\right)$, then $det\left(A - BF^{(i)}C\right)$ is of $O\left(\frac{1}{\epsilon^{n_2}}\right)$, which makes (4.12) and (4.13) numerically ill-defined. However, the partitioned forms of (4.12) and (4.13), given by (4.20)-(4.22) and (4.23)-(4.25), obtained after multiplying equations for $L_2(K_2)$ and $L_3(K_3)$ by ϵ, comprise the well-defined numerical problems, but there are no available methods for their solution. In what follows, we will derive the efficient numerical scheme for solving (4.20)-(4.22) and (4.23)-(4.25). Even more, the slow-fast decomposition will be achieved, and the required solutions will be obtained in terms of low-order problems of dimensions n_1 and n_2 — the original problems (4.20)-(4.22) and (4.23)-(4.25) are of dimensions $n_1 + n_2$.

Equations (4.23)-(4.25) form the standard Lyapunov equation of singularly perturbed linear systems. It is a special case of the more general Lyapunov equation studied in (Gajic et al., 1987). Its zeroth-order solution is obtained by setting $\epsilon = 0$ in (4.23)-(4.25), which after some algebra produces

$$\mathbf{K}_1^{(i+1)} D_0^{(i)} + D_0^{(i)^T} \mathbf{K}_1^{(i+1)} + G_0^{(i)^T} G_0^{(i)} = 0 \qquad (4.26)$$

$$K_3^{(i+1)} D_4^{(i)} + D_4^{(i)^T} K_3^{(i+1)} + q_3^{(i)} = 0 \qquad (4.27)$$

$$K_2^{(i+1)} = - \left(K_1^{(i+1)} D_2^{(i)} + D_3^{(i)^T} K_3^{(i+1)} + q_2^{(i)} \right) D_4^{(i)^{-1}} \qquad (4.28)$$

where

$$D_0^{(i)} = D_1^{(i)} - D_2^{(i)} D_4^{(i)^{-1}} D_3^{(i)}$$

$$G_0^{(i)} = G_1^{(i)} - G_3^{(i)} D_4^{(i)^{-1}} D_3^{(i)}, \qquad G_p^{(i)} = \sqrt{q_p^{(i)}}, \quad p = 1, 3$$

Note that there is no need to calculate the square roots of $q_p^{(i)}$'s. The expression for $G_0^{(i)}$ is used in (4.26) only to simplify notation, but not for real calculations since

$$q_1^{(i)} = G_1^{(i)^T} G_1^{(i)}, \qquad q_2^{(i)} = G_1^{(i)^T} G_3^{(i)}, \qquad q_3^{(i)} = G_3^{(i)^T} G_3^{(i)}$$

The zeroth-order solution

$$\mathbf{K}_1^{(i+1)} = \begin{bmatrix} \mathbf{K}_1^{(i+1)} & \epsilon \mathbf{K}_2^{(i+1)} \\ \epsilon \mathbf{K}_2^{(i+1)^T} & \epsilon \mathbf{K}_3^{(i+1)} \end{bmatrix} \qquad (4.29)$$

is $O(\epsilon)$ close to the required one $K^{(i+1)}$. We can relate them through the error term E

$$\epsilon E = K^{(i+1)} - \mathbf{K}^{(i+1)} \qquad (4.30)$$

or by using a compatible partition:

$$\begin{bmatrix} \epsilon E_1 & \epsilon^2 E_2 \\ \epsilon^2 E_2^T & \epsilon^2 E_3 \end{bmatrix} = \begin{bmatrix} K_1^{(i+1)} - \mathbf{K}_1^{(i+1)} & \epsilon \left(K_2^{(i+1)} - \mathbf{K}_2^{(i+1)} \right) \\ \epsilon \left(K_2^{(i+1)} - \mathbf{K}_2^{(i+1)} \right)^T & \epsilon \left(K_3^{(i+1)} - \mathbf{K}_3^{(i+1)} \right) \end{bmatrix} \qquad (4.31)$$

Clearly, the $O(\epsilon^k)$ approximation of E will produce the $O(\epsilon^{k+1})$ approximation of the sought solution, which is why we are interested in finding a convenient form for the error equation and an appropriate algorithm for its solution. It is shown in (Gajic et al., 1987) that the error equation is given by

$$D_1^{(i)^T} E_1 + E_1 D_1^{(i)} + D_3^{(i)^T} E_2^T + E_2 D_3^{(i)} = 0 \qquad (4.32)$$

$$E_2 D_4^{(i)} + \epsilon D_1^{(i)^T} E_2 + D_1^{(i)^T} K_2^{(i+1)} + D_3^{(i)^T} E_3 = 0 \qquad (4.33)$$

$$E_3 D_4^{(i)} + D_4^{(i)^T} E_3 + D_2^{(i)^T} K_2^{(i+1)} + K_2^{(i+1)^T} D_2^{(i)}$$
$$+\epsilon \left(D_2^{(i)^T} E_2 + E_2^T D_2^{(i)} \right) = 0 \qquad (4.34)$$

and that the following algorithm
Algorithm 4.2:

$$D_0^{(i)^T} E_1^{(j+1)} + E_1^{(j+1)} D_0^{(i)} = D_0^{(i)^T} \left(K_2^{(i+1)} + \epsilon E_2^{(j)} \right) D_4^{(i)^{-1}} D_3^{(i)}$$
$$+ D_3^{(i)^T} D_4^{(i)^{-T}} \left(K_2^{(i+1)} + \epsilon E_2^{(j)} \right)^T D_0^{(i)} = 0$$
$$(4.35)$$

$$E_3^{(j+1)} D_4^{(i)} + D_4^{(i)^T} E_3^{(j+1)} = -D_2^{(i)^T} \left(K_2^{(i+1)} + \epsilon E_2^{(j)} \right)$$
$$- \left(K_2^{(i+1)} + \epsilon E_2^{(j)} \right) D_2^{(i)} \qquad (4.36)$$

$$E_2^{(j+1)} = - \left[D_1^{(i)^T} \left(K_2^{(i+1)} + \epsilon E_2^{(j)} \right) + D_3^{(i)^T} E_3^{(i+1)} \right] D_4^{(i)^{-1}} \qquad (4.37)$$
$$j = 1, 2, 3, ...$$

with initial conditions chosen as $E_1^{(0)} = 0$, $E_2^{(0)} = 0$, and $E_3^{(0)} = 0$, converges to required solution E with the rate of convergence of $O(\epsilon)$ that is,

$$\left\| E - E^{(j)} \right\| = O\left(\epsilon^j\right), \quad j = 1, 2, 3, ... \qquad (4.38)$$

That implies

$$\left\| K^{(i+1)} - \left(K^{(i+1)} + \epsilon E^{(j)} \right) \right\| = O\left(\epsilon^j\right), \quad j = 1, 2, 3, ... \qquad (4.39)$$

$$\triangle$$

Note that the complete slow-fast decomposition is achieved, that is, the solution of the Lyapunov equations (4.23)-(4.25) of order $n_1 + n_2$ is obtained in terms of low-order Lyapunov equations, the slow one (4.35) of order n_1, and the fast one (4.36) of order n_2.

Equations (4.20)-(4.22) do not represent the standard Lyapunov equations of singularly perturbed systems due to the fact that the initial conditions satisfy (4.17). In the following, we apply the methodology of (Gajic et al., 1987) to (4.20)-(4.22) subject to (4.17), and derive the recursive algorithm for its solution in terms of the reduced-order problems.

Setting $\epsilon = 0$ in (4.20)-(4.22) will produce, after some algebra, the zeroth-order approximation of (4.20)-(4.22) as

$$\mathbf{L}_1^{(i+1)} D_0^{(i)^T} + D_0^{(i)} \mathbf{L}_1^{(i+1)} + I = 0 \tag{4.40}$$

$$\mathbf{L}_2^{(i+1)} = -\mathbf{L}_1^{(i+1)} D_3^{(i)^T} D_4^{(i)^{-T}} \tag{4.41}$$

$$\mathbf{L}_3^{(i+1)} D_4^{(i)^T} + D_4^{(i)} \mathbf{L}_3^{(i+1)} + \mathbf{L}_2^{(i+1)^T} D_3^{(i)^T} + D_3^{(i)} \mathbf{L}_2^{(i+1)} = 0 \tag{4.42}$$

Even though the complete slow-fast decomposition is not achieved (in contrary to (4.26)-(4.28)), these equations can be solved in terms of reduced-order problems in a sequential manner, namely, first solve (4.40), then (4.41), and finally solve (4.42).

Defining the error as

$$
\begin{aligned}
L^{(i+1)} - \mathbf{L}^{(i+1)} &= \epsilon M = \epsilon \begin{bmatrix} M_1 & M_2 \\ M_2^T & M_3 \end{bmatrix} \\
&= \begin{bmatrix} L_1^{(i+1)} - \mathbf{L}_1^{(i+1)} & L_2^{(i+1)} - \mathbf{L}_2^{(i+1)} \\ \left(L_2^{(i+1)} - \mathbf{L}_2^{(i+1)} \right)^T & L_3^{(i+1)} - \mathbf{L}_3^{(i+1)} \end{bmatrix}
\end{aligned}
\tag{4.43}
$$

and subtracting (4.40)-(4.42) from (4.20)-(4.22), we get the error equation as

$$D_1^{(i)} M_1 + M_1 D_1^{(i)^T} + D_2^{(i)} M_2^T + M_2 D_2^{(i)^T} = 0 \tag{4.44}$$

$$
\begin{aligned}
M_2 D_4^{(i)^T} &+ \epsilon D_1^{(i)} M_2 + M_1 D_3^{(i)^T} + \epsilon D_2^{(i)} M_3 \\
&+ D_1^{(i)} \mathbf{L}_2^{(i+1)} + D_2^{(i)} \mathbf{L}_3^{(i+1)} = 0
\end{aligned}
\tag{4.45}
$$

$$M_3 D_4^{(i)^T} + D_4^{(i)} M_3 + D_3^{(i)} M_2 + M_2^T D_3^{(i)^T} + I = 0 \tag{4.46}$$

Note that (4.45) is a weakly linear Lyapunov equation. At this point, we will ignore that fact and solve it with respect to M_2 as follows

$$
\begin{aligned}
M_2 = - &\left[D_1^{(i)} \left(\mathbf{L}_2^{(i+1)} + \epsilon M_2 \right) + M_1 D_3^{(i)^T} + D_2^{(i)} \left(\mathbf{L}_3^{(i+1)} + \epsilon M_3 \right) \right] \\
&\times D_4^{(i)^{-T}}
\end{aligned}
\tag{4.47}
$$

Using (4.47) in (4.44) yields

$$D_0^{(i)} M_1 + M_1 D_0^{(i)^T} - D_2^{(i)} D_4^{(i)^{-1}} H_2^T - H_2 D_4^{(i)^{-T}} D_2^{(i)^T} = 0 \tag{4.48}$$

where

$$H_2 = D_1^{(i)} L_2^{(i+1)} + D_2^{(i)} L_3^{(i+1)} \tag{4.49}$$

Thus, the weakly coupled and hierarchical structure of (4.44)-(4.46) can be exploited by proposing the following recursive scheme, which leads to the two low-order completely decoupled Lyapunov equations:

Algorithm 4.3:

$$D_0^{(i)} M_1^{(j+1)} + M_1^{(j+1)} D_0^{(i)^T} - D_2^{(i)} D_4^{(i)^{-1}} H_2^{(j)^T} - H_2^{(j)} D_4^{(i)^{-T}} D_2^{(i)^T} = 0 \tag{4.50}$$

$$M_2^{(j+1)} = - \left[H^{(j)} + M_1^{(j+1)} D_3^{(i)^T} \right] D_4^{(i)^{-T}} \tag{4.51}$$

$$M_3^{(j+1)} D_4^{(i)^T} + D_4^{(i)} M_3^{(j+1)} + D_3^{(i)} M_2^{(j+1)} + M_2^{(j+1)^T} D_3^{(i)^T} + I = 0 \tag{4.52}$$

where

$$H^{(j)} = D_1^{(i)} \left(\mathbf{L}_2^{(i+1)} + \epsilon M_2^{(j)} \right) + D_2^{(i)} \left(\mathbf{L}_3^{(i+1)} + \epsilon M_3^{(j)} \right) \tag{4.53}$$

with $j = 0, 1, 2, 3, ...,$ with initial conditions chosen as $M_1^{(0)} = 0$, $M_2^{(0)} = 0$, and $M_3^{(0)} = 0$.

△

The following theorem summarizes the features of the proposed scheme (Gajic et al., 1989).

Theorem 4.1 *The algorithm (4.50)-(4.53) converges, for sufficiently small values of ϵ, to the exact solution of the error terms, and thus to the solution $L^{(i+1)}$, with the rate of convergence of $O(\epsilon)$, that is,*

$$\left\| M_k - M_k^{(j)} \right\| = O\left(\epsilon^j \right), \quad k = 1, 2, 3; \quad j = 1, 2, 3, ...$$

◇

Research Problem 4.1: Define asynchronous versions of the synchronous parallel Algorithms 4.2 and 4.3 and study the convergence properties of the proposed algorithms.

△

4.3 Case Study: Fluid Catalytic Cracker

In order to demonstrate the efficiency of the proposed algorithm and the failure of the $O(\epsilon)$ theory, we have run a fifth-order real world example, an industrially important reactor (Arkun and Ramakrishnan, 1983). Matrices A, B, C, Q, and R are given by

$$A = \begin{bmatrix} -16.11 & -0.39 & 27.2 & 0 & 0 \\ 0.01 & -16.99 & 0 & 0 & 12.47 \\ 15.11 & 0 & -53.6 & -16.57 & 71.78 \\ -53.36 & 0 & 0 & -107.2 & 232.11 \\ 2.27 & 69.1 & 0 & 0 & -102.99 \end{bmatrix}$$

$$B^T = \begin{bmatrix} 11.12 & -3.61 & -21.91 & -53.6 & 69.1 \\ -12.6 & 3.36 & 0 & 0 & 0 \end{bmatrix}$$

$$C = \begin{bmatrix} 0 & 0 & 0 & 0 & 1 \\ 0 & 1 & 0 & 0 & 0 \end{bmatrix}, \quad Q = I_5, \quad R = I_2$$

The eigenvalues of the matrix A are -2.8, -7.7, -74, -82, -129. Thus, we have two slow and three fast variables. The small parameter is chosen as $\epsilon = 0.1$, which is roughly the ratio of 7.7 and 74.

The theory of singularly perturbed optimal output feedback problems is derived so far for the $O(\epsilon)$ approximation. Using the $O(\epsilon)$ approximation of the equations comprising the solution of the optimal output feedback, namely of (4.26)-(4.28) and (4.40)-(4.42), will fail to produce the desired approximation for this example. Even more, the algorithm does not converge to the near-optimum solution for the extremely small values of the parameter α such as 0.001. The cause of the trouble is the inversion of the quantity CLC^T. Its determinant for the optimal value of L is very small, that is, 0.9736 x 10^{-4}, and thus, this problem is very sensitive to $O(\epsilon)$ perturbations, which can be seen from Table 4.1.

The results from Table 4.1 strongly support the necessity for the existence of the recursive schemes which can produce any desired accuracy, that is, the development of the $O(\epsilon^k)$ theory.

In Table 4.2, we have presented results for the criterion and the gain error for the global algorithm (Moerder and Calise, 1985a), and the corresponding quantities for the proposed reduced-order recursive algorithm. The initial value for the gain is obtained from (Petkovski and Rakic, 1978). It can be seen that the initial guess is quite good, but the global algorithm converges very slowly to the optimal solution. As far as the criterion is concerned, it takes 28 iterations to achieve an

α	$det[CL^{(1)}C^T]$ $j = 6$	$det[CL^{(1)}C^T]$ $j = 1$	$det[CL^{(0)}C^T]$	$det[CL^{(1)}C^T]$
0.5	0.86846 x 10^{-4}	0.14432 x 10^{-3}	0.38943 x 10^{-6}	0.24392 x 10^{-10}
0.1	0.12749 x 10^{-3}	"	"	0.57491 x 10^{-9}
0.01	0.14244 x 10^{-3}	"	"	0.31742 x 10^{-7}
0.001	0.14413 x 10^{-3}	"	"	0.24904 x 10^{-6}

Table 4.1: Determinant of CLC^T

accuracy of up to five decimal digits, where $J_{opt} = 0.28573$. On the other hand, the trajectories of the approximate system after 30 iterations are still far apart from the optimal trajectories since the approximate gain is only $O\left(10^{-2}\right)$ close to the optimal one. Thus, this algorithm demands a lot of iterations in order to achieve high accuracy. This fact justifies even more the necessity for the existence of algorithms which will reduce computational requirements. In the proposed algorithm, only low-order Lyapunov equations are involved in algebraic computations. Even more, at the very beginning, they can be solved with reduced accuracy ($j = 1$ or 2), and once we approach the optimum, the accuracy can be increased to the desired one. The third column of Table 4.2 is obtained with $j = 2$, for $i \leq 16$, and $j = 6$ for $i > 16$. The second and fifth columns of Table 4.2 are obtained for $j = 6$ for all $i's$. The parameter α is chosen as $\alpha = 0.5$ since the global algorithm does not converge for $\alpha \geq 0.6$.

4.4 Output Feedback for Linear Weakly Coupled Systems

Consider the weakly coupled linear system (Petkovski and Rakic, 1979)

$$\dot{x}_1 = A_1 x_1 + \epsilon A_2 x_2 + B_1 u_1 + \epsilon B_2 u_2 \qquad (4.54)$$

$$\dot{x}_2 = \epsilon A_3 x_1 + A_4 x_2 + \epsilon B_3 u_1 + B_4 u_2 \qquad (4.55)$$

100

i	$J_{opt}^{(i)}$	$J_{app}^{(i)}$ $j=6$	$J_{app}^{(i)}$ $j=2,\ i<17$ $i>16,\ j=6$	$\|F_{opt}^{(i)} - F_{opt}\|_\infty$	$\|F_{opt}^{(i)} - F_{opt}\|_\infty$ $j=6$
1	0.30487	0.30488	0.30427	2.1520	2.1480
2	0.28733	0.28738	0.28879	0.1635	0.1684
4	0.28615	0.28619	0.28745	0.1296	0.1328
6	0.28595	0.28599	0.28710	0.1093	0.1120
8	0.28588	0.28591	0.28691	0.0913	0.0936
10	0.28583	0.28586	0.28676	0.0764	0.0783
12	0.28580	0.28583	0.28664	0.0638	0.0654
14	0.28578	0.28580	0.28654	0.0533	0.0550
16	0.28577	0.28578	0.28646	0.0446	0.0456
18	0.28575	0.28577	0.28584	0.0373	0.0380
20	0.28575	0.28576	0.28581	0.0311	0.0317
22	0.28574	0.28576	0.28579	0.0260	0.0256
24	0.28574	0.28575	0.28577	0.0217	0.0219
26	0.28574	0.28575	0.28577	0.0181	0.0181
28	0.28573	0.28575	0.28576	0.0150	0.0149
30	0.28573	0.28575	0.28575	0.0125	0.0122

Table 4.2: Optimal and approximate criteria

$$y = \begin{bmatrix} C_1 & \epsilon C_2 \\ \epsilon C_3 & C_4 \end{bmatrix} \begin{bmatrix} x_1 \\ x_2 \end{bmatrix} = \begin{bmatrix} y_1 \\ y_2 \end{bmatrix} \tag{4.56}$$

where $x_1 \in \Re^{n_1}$ and $x_2 \in \Re^{n_2}$ are state vectors, $u_i \in \Re^{m_i}$, $i = 1, 2$, are control inputs and $y_i \in \Re^{r_i}$, $i = 1, 2$, are measured outputs. In the following, A_i, B_i, and C_i, $i = 1, ..., 4$, are constant matrices of compatible dimensions; in general they are continuous functions of a small parameter (Petrovic and Gajic, 1988).

With (4.54)-(4.56), consider the performance criterion given in (4.4), with

$$Q = \begin{bmatrix} Q_1 & \epsilon Q_2 \\ \epsilon Q_2^T & Q_4 \end{bmatrix}, \quad R = \begin{bmatrix} R_1 & 0 \\ 0 & R_4 \end{bmatrix} \tag{4.57}$$

with positive definite R and positive semidefinite Q, which has to be minimized.

In addition, the control input $u(t)$ is constrained to be a direct feedback from the output $y(t)$ as given by equation (4.5).

The optimal constant output feedback gain F is given by (Levine and Athans, 1970)

$$F = R^{-1} B^T K L C^T \left(C L C^T \right)^{-1} \tag{4.58}$$

where matrices K and L satisfy high-order nonlinear coupled algebraic equations (4.7) and (4.8) and newly defined matrices A, B, and C are

$$A = \begin{bmatrix} A_1 & \epsilon A_2 \\ \epsilon A_3 & A_4 \end{bmatrix}, \quad B = \begin{bmatrix} B_1 & \epsilon B_4 \\ \epsilon B_3 & B_4 \end{bmatrix}, \quad C = \begin{bmatrix} C_1 & \epsilon C_2 \\ \epsilon C_3 & C_4 \end{bmatrix} \tag{4.59}$$

Compatible with the nature of their solutions, matrices K and L are partitioned as

$$K = \begin{bmatrix} K_1 & \epsilon K_2 \\ \epsilon K_2^T & K_3 \end{bmatrix}, \quad L = \begin{bmatrix} L_1 & \epsilon L_2 \\ \epsilon L_2^T & L_3 \end{bmatrix} \tag{4.60}$$

The algorithm given by equations (4.11)-(4.14), described in Section 4.2, can be applied for the numerical solution of the corresponding nonlinear matrix equations (4.6)-(4.8) with A, B, and C now defined by (4.59).

In order to simplify derivations, the following notation is introduced:

$$\begin{aligned} [A - BFC] &= \begin{bmatrix} D_1 & \epsilon D_2 \\ \epsilon D_3 & D_4 \end{bmatrix} \\ &= \begin{bmatrix} A_1 - B_1 F_1 C_1 & \epsilon(A_2 - B_1 F_2 C_4) \\ \epsilon(A_3 - B_4 F_3 C_1) & A_4 - B_4 F_4 C_4 \end{bmatrix} \end{aligned} \tag{4.61}$$

$$[Q + C^T F^T R F C] = \begin{bmatrix} q_1 & \epsilon q_2 \\ \epsilon q_3 & q_4 \end{bmatrix} \tag{4.62}$$

where

$$\begin{aligned} q_1 &= Q_1 + C_1^T F_1^T R_1 F_1 C_1 + C_1^T F_3^T R_4 F_3 C_1 \\ q_2 &= Q_2 + C_1^T F_1^T R_1 F_2 C_4 + C_1^T F_3^T R_4 F_4 C_4 \\ q_3 &= Q_4 + C_4^T F_2^T R_1 F_2 C_4 + C_4^T F_4^T R_4 F_4 C_4 \end{aligned}$$

In addition, without loss of generality, it has been assumed that matrices B_2, B_3, C_2, and C_3 are zeros.

Partitioning (4.12) and (4.13) compatible to (4.59) and (4.60) and using (4.61) and (4.62), we get the following set of equations

$$D_1^{(i)} L_1^{(i+1)} + L_1^{(i+1)} D_1^{(i)^T} + \epsilon^2 \left(D_2^{(i)} L_2^{(i+1)^T} + L_2^{(i+1)} D_2^{(i)^T} \right) + I = 0 \tag{4.63}$$

$$L_2^{(i+1)} D_4^{(i)^T} + D_1^{(i)} L_2^{(i+1)} + L_1^{(i+1)} D_3^{(i)^T} + D_2^{(i)} L_3^{(i+1)} = 0 \tag{4.64}$$

$$L_3^{(i+1)} D_4^{(i)^T} + D_4^{(i)} L_3^{(i+1)} + \epsilon^2 \left(D_3^{(i)} L_2^{(i+1)} + L_2^{(i+1)^T} D_3^{(i)^T} \right) + I = 0 \tag{4.65}$$

and

$$\begin{aligned} D_1^{(i)^T} K_1^{(i+1)} + K_1^{(i+1)} D_1^{(i)} + q_1^{(i)} \\ + \epsilon^2 \left(D_3^{(i)^T} K_2^{(i+1)^T} + K_2^{(i+1)} D_3^{(i)} \right) = 0 \end{aligned} \tag{4.66}$$

$$K_2^{(i+1)} D_4^{(i)} + D_1^{(i)^T} K_2^{(i+1)} + D_3^{(i)^T} K_3^{(i+1)} + K_1^{(i+1)} D_2^{(i)} + q_2^{(i)} = 0 \tag{4.67}$$

$$K_3^{(i+1)} D_4^{(i)} + D_4^{(i)^T} K_3^{(i+1)} + q_3^{(i)} + \epsilon^2 \left(D_2^{(i)^T} K_2^{(i+1)} + K_2^{(i+1)^T} D_2^{(i)} \right) = 0 \tag{4.68}$$

where

$$D_1^{(i)} = A_1 - B_1 F_1^{(i)} C_1, \qquad D_2^{(i)} = A_2 - B_1 F_2^{(i)} C_4$$
$$D_3^{(i)} = A_3 - B_3 F_3^{(i)} C_1, \qquad D_4^{(i)} = A_4 - B_4 F_4^{(i)} C_2$$

and

$$q_1^{(i)} = Q_1 + C_1^T F_1^{(i)^T} R_1 F_1^{(i)} C_1 + C_1^T F_3^{(i)^T} R_4 F_3^{(i)} C_1$$
$$q_2^{(i)} = Q_2 + C_1^T F_1^{(i)^T} R_1 F_2^{(i)} C_4 + C_1^T F_3^{(i)^T} R_4 F_4^{(i)} C_4$$
$$q_3^{(i)} = Q_4 + C_4^T F_2^{(i)^T} R_1 F_2^{(i)} C_4 + C_4^T F_4^{(i)^T} R_4 F_4^{(i)} C_4$$

with $i = 0, 1, 2, 3, \ldots$.

Equations (4.63)-(4.65) and (4.66)-(4.68) have been studied by (Petkovski and Rakic, 1979) by using a series expansion method. Their approach is not recursive in application and they numerically justify it for $O(\epsilon)$ accuracy only. In this section, we develop a recursive scheme which will efficiently extend the main results of their work to any arbitrary order of accuracy, namely $O(\epsilon^{2k})$, where k represents the number of required iterations of the proposed recursive scheme.

Note that all of matrices D_i, K_i, L_i, and q_i are functions of a small parameter ϵ. However, dependence on ϵ is suppressed in order to simplify notation.

103

Stability of the matrix given by

$$D^{(i)}(\epsilon) = \begin{bmatrix} D_1^{(i)}(\epsilon) & \epsilon D_2^{(i)}(\epsilon) \\ \epsilon D_3^{(i)}(\epsilon) & D_4^{(i)}(\epsilon) \end{bmatrix}$$

is guaranteed by (Moerder and Calise, 1985a). Due to block diagonal dominance of $D^{(i)}(\epsilon)$, matrices $D_1^{(i)}(\epsilon)$ and $D_4^{(i)}(\epsilon)$ are stable for a sufficiently small values of ϵ for $\forall i$.

In what follows equations (4.63)-(4.68) will be numerically solved in terms of the reduced-order Lyapunov and Sylvester equations by using results from (Gajic et al., 1987).

Notice that equations (4.63)-(4.68) represent standard Lyapunov equations of weakly coupled systems. An efficient recursive algorithm for their numerical solution, with $O(\epsilon^2)$ rate of convergence, has been derived in (Gajic et al., 1987). The zeroth-order solutions of these equations are obtained by setting $\epsilon = 0$ in (4.63)-(4.68)

$$D_1^{(i)}L_1^{(i+1)} + L_1^{(i+1)}D_1^{(i)^T} + I = 0 \tag{4.69}$$

$$L_2^{(i+1)}D_4^{(i)^T} + D_1^{(i)}L_2^{(i+1)} + L_1^{(i+1)}D_3^{(i)^T} + D_2^{(i)}L_3^{(i+1)} = 0 \tag{4.70}$$

$$L_3^{(i+1)}D_4^{(i)^T} + D_4^{(i)}L_3^{(i+1)} + I = 0 \tag{4.71}$$

and

$$D_1^{(i)^T}K_1^{(i+1)} + K_1^{(i+1)}D_1^{(i)} + q_1^{(i)} = 0 \tag{4.72}$$

$$K_2^{(i+1)}D_4^{(i)} + D_1^{(i)^T}K_2^{(i+1)} + D_3^{(i)^T}K_3^{(i+1)} + K_1^{(i+1)}D_2^{(i)} + q_2^{(i)} = 0 \tag{4.73}$$

$$K_3^{(i+1)}D_4^{(i)} + D_4^{(i)^T}K_3^{(i+1)} + q_3^{(i)} = 0 \tag{4.74}$$

It can be seen that the complete reduced-order decomposition is achieved in (4.69)-(4.74), that is, one needs to solve four reduced-order Lyapunov and two reduced-order Sylvester equations.

The existence of the unique and bounded solutions of (4.63)-(4.68) is guaranteed by the stability of $D^{(i)}(\epsilon)$, (Moerder and Calise, 1985a). Due to stability of $D_1^{(i)}(\epsilon)$ and $D_4^{(i)}(\epsilon)$ the unique solutions of (4.69)-(4.74) exist as well.

The zeroth-order solutions

$$L^{(i+1)} = \begin{bmatrix} L_1^{(i+1)} & \epsilon L_2^{(i+1)} \\ \epsilon L_2^{(i+1)^T} & L_3^{(i+1)} \end{bmatrix}, \quad K^{(i+1)} = \begin{bmatrix} K_1^{(i+1)} & \epsilon K_2^{(i+1)} \\ \epsilon K_2^{(i+1)^T} & K_3^{(i+1)} \end{bmatrix}$$
$$\tag{4.75}$$

are $O\left(\epsilon^2\right)$ close to the required ones $L^{(i+1)}$ and $K^{(i+1)}$ (see equations (4.63)-(4.68)). We can relate them through the error terms

$$L^{(i+1)} - \mathbf{L}^{(i+1)} = \epsilon^2 M = \epsilon^2 \begin{bmatrix} M_1 & \epsilon M_2 \\ \epsilon M_2^T & M_3 \end{bmatrix} \tag{4.76}$$

and

$$K^{(i+1)} - \mathbf{K}^{(i+1)} = \epsilon^2 E = \epsilon^2 \begin{bmatrix} E_1 & \epsilon E_2 \\ \epsilon E_2^T & E_3 \end{bmatrix} \tag{4.77}$$

Clearly, the $O\left(\epsilon^k\right)$ approximations for M and E will produce the $O\left(\epsilon^{k+2}\right)$ approximations of the required solutions. This is why we are interested in finding a convenient form for these error terms and the appropriate algorithm for their solution.

Using the results of (Gajic et al., 1987), it can be shown that these error equations are given by

$$M_1 D_1^{(i)^T} + D_1^{(i)} M_1 + D_2^{(i)} \left(\mathbf{L}_2^{(i+1)} + \epsilon^2 M_2 \right)^T$$
$$+ \left(\mathbf{L}_2^{(i+1)} + \epsilon^2 M_2 \right) D_2^{(i)^T} = 0 \tag{4.78}$$

$$M_2 D_4^{(i)^T} + D_1^{(i)} M_2 + D_2^{(i)} M_3 + M_1 D_3^{(i)^T} = 0 \tag{4.79}$$

$$M_3 D_4^{(i)^T} + D_4^{(i)} M_3 + D_3^{(i)} \left(\mathbf{L}_2^{(i+1)} + \epsilon^2 M_2 \right)$$
$$+ \left(\mathbf{L}_2^{(i+1)} + \epsilon^2 M_2 \right)^T D_3^{(i)^T} = 0 \tag{4.80}$$

and

$$E_1 D_1^{(i)} + D_1^{(i)^T} E_1 + D_3^{(i)^T} \left(\mathbf{K}_2^{(i+1)} + \epsilon^2 E_2 \right)^T$$
$$+ \left(\mathbf{K}_2^{(i+1)} + \epsilon^2 E_2 \right) D_3^{(i)} = 0 \tag{4.81}$$

$$E_2 D_4^{(i)} + D_1^{(i)^T} E_2 + E_1 D_2^{(i)} + D_3^{(i)^T} E_3 = 0 \tag{4.82}$$

$$E_3 D_4^{(i)} + D_4^{(i)^T} E_3 + \left(\mathbf{K}_2^{(i+1)} + \epsilon^2 E_2 \right)^T D_2^{(i)}$$
$$+ D_2^{(i)^T} \left(\mathbf{K}_2^{(i+1)} + \epsilon^2 E_2 \right) = 0 \tag{4.83}$$

The weakly coupled and hierarchical structure of (4.78)-(4.83) can be exploited by proposing the recursive scheme, which leads, after some

OUTPUT FEEDBACK CONTROL

algebra, to the six low-order completely decoupled recursive equations
Algorithm 4.4:

$$M_1^{(j+1)}D_1^{(i)^T} + D_1^{(i)}M_1^{(j+1)} + D_2^{(i)}\left(\mathbf{L}_2^{(i+1)} + \epsilon^2 M_2^{(j)}\right)^T$$
$$+ \left(\mathbf{L}_2^{(i+1)} + \epsilon^2 M_2^{(j)}\right)D_2^{(i)^T} = 0 \quad (4.84)$$

$$M_2^{(j+1)}D_4^{(i)^T} + D_1^{(i)}M_2^{(j+1)} + D_2^{(i)}M_3^{(j+1)} + M_1^{(j+1)}D_3^{(i)^T} = 0 \quad (4.85)$$

$$M_3^{(j+1)}D_4^{(i)^T} + D_4^{(i)}M_3^{(j+1)} + D_3^{(i)}\left(\mathbf{L}_2^{(i+1)} + \epsilon^2 M_2^{(j)}\right)$$
$$+ \left(\mathbf{L}_2^{(i+1)} + \epsilon^2 M_2^{(j)}\right)^T D_3^{(i)^T} = 0 \quad (4.86)$$

and

$$E_1^{(j+1)}D_1^{(i)} + D_1^{(i)^T}E_1^{(j+1)} + D_3^{(i)^T}\left(\mathbf{K}_2^{(i+1)} + \epsilon^2 E_2^{(j)}\right)^T$$
$$+ \left(\mathbf{K}_2^{(i+1)} + \epsilon^2 E_2^{(j)}\right)D_3^{(i)} = 0 \quad (4.87)$$

$$E_2^{(j+1)}D_4^{(i)} + D_1^{(i)^T}E_2^{(j+1)} + E_1^{(j+1)}D_2^{(i)} + D_3^{(i)^T}E_3^{(j+1)} = 0 \quad (4.88)$$

$$E_3^{(j+1)}D_4^{(i)} + D_4^{(i)^T}E_3^{(j+1)} + \left(\mathbf{K}_2^{(i+1)} + \epsilon^2 E_2^{(j)}\right)^T D_2^{(i)}$$
$$+ D_2^{(i)^T}\left(\mathbf{K}_2^{(i+1)} + \epsilon^2 E_2^{(j)}\right) = 0 \quad (4.89)$$

with $j = 1, 2, 3, ...$, and with initial conditions being chosen as $M_1^{(0)} = E_1^{(0)} = 0$, $M_2^{(0)} = E_2^{(0)} = 0$, and $M_3^{(0)} = E_3^{(0)} = 0$.

$\triangle$

Observe the decoupled structure of (4.84)-(4.89): the reduced-order Lyapunov equations (4.84), (4.86), (4.87), and (4.89) are solved first and then the Sylvester equations (4.85) and (4.88) are solved.

The following theorem is proved in (Harkara et al., 1989).

Theorem 4.2 *The algorithm (4.84)-(4.89) converges, for sufficiently small value of ϵ, to the solution of the error terms, and thus to the required solutions $L^{(i+1)}$ and $K^{(i+1)}$, with the rate of convergence of $O\left(\epsilon^2\right)$.*

$\diamond$

Research Problem 4.2: Derive an asynchronous version of the parallel synchronous Algorithms 4.4. Establish a set of necessary and sufficient conditions for convergence.

$\triangle$

4.5 Case Study: Twelve Plate Absorption Column

In order to illustrate the efficiency of the proposed algorithm for weakly coupled systems, the method is applied to the mathematical model of a twelve-plate absorption column originally derived in (Lapidus and Amundson, 1950; Lapidus et al., 1961) — see also (Petkovski and Rakic, 1979; Petkovski, 1981). The system matrix is given by

$$A = \begin{bmatrix} A_1 & A_2 \\ A_3 & A_4 \end{bmatrix}$$

where

$$A_1 = A_4 = \begin{bmatrix} a_1 & a_2 & 0 & 0 & 0 & 0 \\ a_3 & a_1 & a_2 & 0 & 0 & 0 \\ 0 & a_3 & a_1 & a_2 & 0 & 0 \\ 0 & 0 & a_3 & a_1 & a_2 & 0 \\ 0 & 0 & 0 & a_3 & a_1 & a_2 \\ 0 & 0 & 0 & 0 & a_3 & a_1 \end{bmatrix}$$

with

$$a_1 = -1.73058, \quad a_2 = 0.634231, \quad a_3 = 0.538827$$

Here A_2 has all entries equal to zero except for $(A_2)_{6,1} = a_2$, and A_3 has all entries equal to zero except for $(A_3)_{1,6} = a_3$. The control matrix is

$$[B_1 \quad \epsilon B_2] = \begin{bmatrix} b_1 & 0 \\ 0 & 0 \\ 0 & 0 \\ 0 & 0 \\ 0 & 0 \\ 0 & 0 \end{bmatrix}, \quad [\epsilon B_3 \quad B_4] = \begin{bmatrix} 0 & 0 \\ 0 & 0 \\ 0 & 0 \\ 0 & 0 \\ 0 & 0 \\ 0 & b_2 \end{bmatrix}$$

with

$$b_1 = 0.538827, \quad b_2 = 0.8809$$

and the input matrix is

$$C = \begin{bmatrix} 1 & 0 & 0 & 0 & 0 & 0 & 0 & 0 & 0 & 0 & 0 & 0 \\ 0 & 1 & 0 & 0 & 0 & 0 & 0 & 0 & 0 & 0 & 0 & 0 \\ 0 & 0 & 0 & 0 & 0 & 0 & 0 & 0 & 0 & 0 & 1 & 0 \\ 0 & 0 & 0 & 0 & 0 & 0 & 0 & 0 & 0 & 0 & 0 & 1 \end{bmatrix}$$

The initial conditions are

$$x_1^T(0) = [-0.036 \quad -0.066 \quad -0.092 \quad -0.113 \quad -0.132 \quad -0.148]$$
$$x_2^T(0) = [-0.161 \quad -0.173 \quad -0.182 \quad -0.190 \quad -0.197 \quad -0.203]$$

i	$J_{opt}^{(i)}$	$J_{red}^{(i)}$	$\|F_{opt}^{(i)} - F_{opt}\|_\infty$	$\|F_{red}^{(i)} - F_{opt}^{(i)}\|_\infty$ $j = 6$
1	0.97305	0.97289	13.038	13.086
2	0.27731	0.27778	4.050	3.975
3	0.24112	0.24109	2.527	2.616
4	0.22316	0.22308	1.677	2.207
5	0.21596	0.21604	1.834	1.574
6	0.21355	0.21372	7.861	0.908
7	0.21286	0.21301	11.759	0.477
8	0.21277	0.21281	3.003	0.242
9	0.21274	0.21275	3.157	0.123
10	0.21274	0.21274	4.625	0.064
12	0.21273	0.21273	6.626	0.019
16	0.21273	0.21273	62.600	0.002
18	0.21273	0.21273	36.207	0.000
20	0.21273	0.21273	26.833	0.000
22	*	0.21273	*	0.000

* = the global algorithm fails to produce solution for $i > 21$

Table 4.3: Optimal and approximate criteria and gains

The penalty matrices in the performance index are $Q = I_{12}$, $R = I_2$. The small coupling parameter ϵ is equal to 0.5.

In Table 4.3, we present results for the criterion and the gain error for the global algorithm (Moerder and Calise, 1985a), and the corresponding quantities for the proposed reduced-order recursive algorithm. The parameter α is chosen as $\alpha = 0.5$.

In order to facilitate finding the solution to the problem under study by using the global algorithm and to avoid problems of system instability, smaller values of α were used. The entries in Table 4.4 show the results obtained by using $\alpha = 0.05$ and $\alpha = 0.01$. The global algorithm fails to produce a unique value for the solution even though convergence to the optimal value of the criterion is achieved at $i = 116$ and $i = 55$ for $\alpha = 0.05$ and $\alpha = 0.1$, respectively.

The initial value for the gain $F^{(0)}$ is obtained by using the method proposed in (Petkovski and Rakic, 1978). The global algorithm takes 11 iterations to achieve the accuracy of up to 5 decimal digits, where $J_{opt} = 0.21273$.

It is important to note that the non-uniqueness of the solution of equations (4.7), (4.8), and (4.58) is shown by the entries in the fourth column of Table 4.3 which are obtained by using the global algorithm. It is seen that there are several possible solutions to the optimal control problem even though convergences to the optimal value of the criterion is achieved at $i = 11$. Furthermore, for $i \geq 22$, with $\alpha = 0.5$ the global algorithm fails to produce the solution so that it can not converge to the unique value of the gain. From the entries in the fifth column of Table 4.3, it is clear that by using the reduced-order algorithm proposed, the difficulty of non-uniqueness of the solution to the optimal output control problem is resolved since the reduced-order algorithm produces a unique value of the feedback gain F. In addition, there are no problems with system instability when the reduced-order algorithm is used.

The example clearly shows the superiority of the reduced-order algorithm over the global algorithm.

All simulation results in this chapter are obtained by using the software package L-A-S for computer aided control system design (West et al., 1985).

Research Problem 4.3: Study the output feedback problem of the singularly perturbed and weakly coupled linear systems in the context of the projective control theory (Medanic, 1979; Hopkins et al., 1981; Medanic and Uskokovic, 1983, 1988; Medanic et al., 1985; Mea et al., 1986; Arnautovic and Medanic, 1987, 1990; Ramaker et al., 1990; Arnautovic and Skataric, 1991). The projective control technique represents a method for designing low-order controllers for high-order systems by retaining a given subset of eigenvalues and eigenvectors.

$\triangle$

$\epsilon = 0.5$	$\|F_{opt}^{(i)} - F_{opt}\|_\infty$	$\|F_{opt}^{(i)} - F_{opt}\|_\infty$
i	$\alpha = 0.05$	$\alpha = 0.1$
1	13.028	13.028
10	3.175	1.587
20	1.626	1.107
30	1.058	2.071
40	1.025	2.666
50	1.249	0.608
60	1.811	1.110
70	2.527	1.753
80	2.631	2.019
90	1.632	2.069
100	0.826	2.009
110	0.714	1.903
120	0.949	1.783
130	1.253	1.666
140	1.510	1.555
150	1.692	1.456
160	1.804	1.366
170	1.864	1.283
180	1.889	1.210
190	1.888	1.144
200	1.867	1.082

Table 4.4: Nonuniqueness of the global algorithm

Chapter 5

Linear Stochastic Systems

In this chapter, we study the stochastic optimal control of linear time invariant singularly perturbed and weakly coupled systems in both continuous and discrete-time domains. The main issue in the linear optimal stochastic control is the design of the optimal Kalman filter which has the same order as a dynamical system under consideration. In the case of large scale systems composed of slow, fast, and weakly coupled state variables, it is possible to replace the design of the global Kalman filters in terms of the reduced-order Kalman filters. In addition to these on-line simplifications, we also present the simplified calculations of the regulator and filter gains by using the corresponding parallel algorithms derived in Chapters 2 and 3.

5.1 Recursive Approach to Singularly Perturbed Linear Stochastic Systems

Singularly perturbed linear stochastic continuous-time estimation and control problems have been studied in the past by a few researchers (Haddad, 1976; Haddad and Kokotovic, 1977; Khalil and Gajic, 1984; Teneketzis and Sandell, 1977). The paper (Khalil and Gajic, 1984) seems to be the most complete one; it alleviates the difficulties of the previous approaches and is conceptually simple. We shall briefly summarize the main results of (Khalil and Gajic, 1984). Consider the singularly perturbed system

$$\dot{x}_1 = A_1 x_1 + A_2 x_2 + B_1 u + G_1 w \tag{5.1}$$

$$\epsilon \dot{x}_2 = A_3 x_1 + A_4 x_2 + B_2 u + G_2 w \qquad (5.2)$$

with corresponding measurements $y(t) \in \Re^{r_2}$

$$y = C_1 x_1 + C_2 x_2 + v \qquad (5.3)$$

where $x_1 \in \Re^{n_1}$ and $x_2 \in \Re^{n_2}$ are state vectors, $u \in \Re^m$ is a control input, $w \in \Re^{r_1}$ and $v \in \Re^{r_2}$ are zero-mean, stationary, white Gaussian noise with intensities $W > 0$ and $V > 0$, respectively, and ϵ is a small positive parameter. In the following, A_i, B_j, G_j, C_j, $i = 1, ..., 4$, $j = 1, 2$, are constant matrices; in general, they are analytic functions of ϵ (Khalil and Gajic, 1984). With (5.1)-(5.3), consider the performance criterion

$$J = \lim_{\substack{t_0 \to -\infty \\ t_1 \to \infty}} \frac{1}{t_1 - t_0} E \left\{ \int_{t_0}^{t_1} \left[\begin{pmatrix} x_1 \\ x_2 \end{pmatrix}^T R_1 \begin{pmatrix} x_1 \\ x_2 \end{pmatrix} + u^T R_2 u \right] dt \right\} \qquad (5.4)$$

with positive definite R_2 and positive semidefinite R_1, which has to be minimized.

The optimal control is given by

$$u = -F_1(\epsilon) \hat{x}_1 - F_2(\epsilon) \hat{x}_2 \qquad (5.5)$$

where $\hat{x}_1$ and $\hat{x}_2$ are optimal estimates of the state vectors x_1 and x_2

$$\dot{\hat{x}}_1 = A_1 \hat{x}_1 + A_2 \hat{x}_2 + B_1 u + K_1(\epsilon)(y - C_1 \hat{x}_1 - C_2 \hat{x}_2) \qquad (5.6a)$$

$$\epsilon \dot{\hat{x}}_2 = A_3 \hat{x}_1 + A_4 \hat{x}_2 + B_2 u + K_2(\epsilon)(y - C_1 \hat{x}_1 - C_2 \hat{x}_2) \qquad (5.6b)$$

The matrices F_1, F_2 and K_1, K_2 are regulator and filter gains, respectively

$$F_1 = R_2^{-1} \left(B_1^T P_1 + B_2^T P_2^T \right), \qquad F_2 = R_2^{-1} \left(\epsilon B_1^T P_2 + B_2^T P_3 \right) \qquad (5.7a)$$

$$K_1 = \left(Q_1 C_1^T + Q_2 C_2^T \right) V^{-1}, \qquad K_2 = \left(\epsilon Q_2^T C_1^T + Q_3 C_2^T \right) V^{-1} \qquad (5.7b)$$

where P_i, Q_i, $i = 1, 2, 3$, are solutions of the corresponding regulator and filter Riccati equations

$$A^T(\epsilon) P(\epsilon) + P(\epsilon) A(\epsilon) - P(\epsilon) S_R(\epsilon) P(\epsilon) + R_1 = 0 \qquad (5.8a)$$

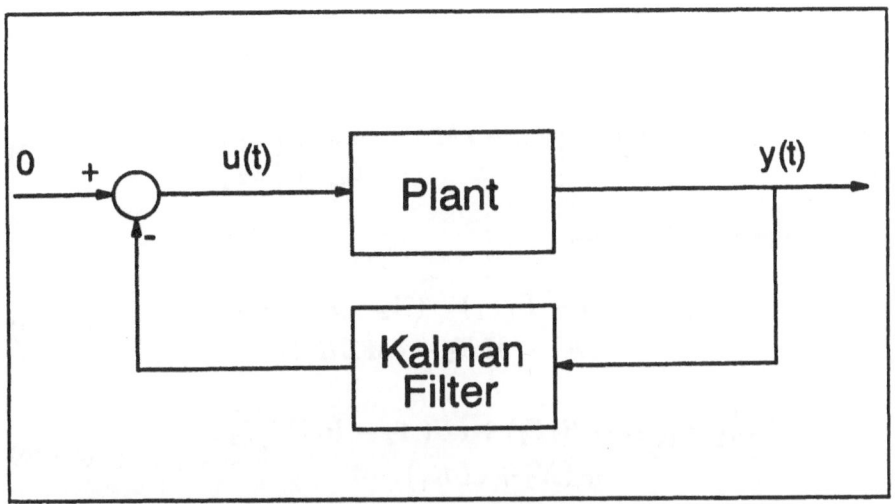

Figure 5.1: Optimal feedback control law with global Kalman filter

$$A(\epsilon)Q(\epsilon) + Q(\epsilon)A^T(\epsilon) - Q(\epsilon)S_F(\epsilon)Q(\epsilon) + G(\epsilon)WG^T(\epsilon) = 0$$
$$(5.8b)$$

with scaling compatible to the nature of their solutions

$$P(\epsilon) = \begin{bmatrix} P_1 & \epsilon P_2 \\ \epsilon P_2^T & \epsilon P_3 \end{bmatrix}, \qquad Q(\epsilon) = \begin{bmatrix} Q_1 & Q_2 \\ Q_2^T & \frac{1}{\epsilon}Q_3 \end{bmatrix} \qquad (5.9)$$

and newly defined matrices as

$$A(\epsilon) = \begin{bmatrix} A_1 & A_2 \\ \frac{A_3}{\epsilon} & \frac{A_4}{\epsilon} \end{bmatrix}, \quad B(\epsilon) = \begin{bmatrix} B_1 \\ \frac{B_2}{\epsilon} \end{bmatrix}, \quad G(\epsilon) = \begin{bmatrix} G_1 \\ \frac{G_2}{\epsilon} \end{bmatrix} \qquad (5.10)$$
$$C = [C_1, C_2], \quad S_R(\epsilon) = B(\epsilon)R_2^{-1}B^T(\epsilon), \quad S_F = C^T V^{-1} C$$

Eliminating u from (5.6), by using (5.5), the optimal filter can be represented as a system driven by the innovation process $\nu = y - C_1\hat{x}_1 - C_2\hat{x}_2$

$$\dot{\hat{x}}_1 = (A_1 - B_1 F_1)\hat{x}_1 + (A_2 - B_1 F_2)\hat{x}_2 + K_1\nu \qquad (5.11a)$$
$$\epsilon\dot{\hat{x}}_2 = (A_3 - B_2 F_1)\hat{x}_1 + (A_4 - B_2 F_2)\hat{x}_2 + K_2\nu \qquad (5.11b)$$

The block diagram for the optimal control law given in terms of the global Kalman filter is represented in Figure 5.1.

As was shown in (Khalil and Gajic, 1984), for the purpose of achieving decomposition on the slow and fast variables, this filter is transformed via the use of a nonsingular transformation (Chang, 1972) into new coordinates

$$\begin{bmatrix} \hat{\eta}_1 \\ \hat{\eta}_2 \end{bmatrix} = \begin{bmatrix} I_{n_1} - \epsilon ML & -\epsilon M \\ L & I_{n_2} \end{bmatrix} \begin{bmatrix} \hat{x}_1 \\ \hat{x}_2 \end{bmatrix} \tag{5.12}$$

so that the filter becomes

$$\dot{\hat{\eta}}_1 = [(A_1 - B_1 F_1) - (A_2 - B_1 F_2) L] \hat{\eta}_1 \\ + (K_1 - MK_2 - \epsilon MLK_1) \nu \tag{5.13a}$$

$$\epsilon \dot{\hat{\eta}}_2 = [(A_4 - B_2 F_2) + \epsilon L (A_2 - B_1 F_2)] \hat{\eta}_2 \\ + (K_2 + \epsilon LK_1) \nu \tag{5.13b}$$

with the innovation process

$$\nu = y - (C_1 - C_2 L) \hat{\eta}_1 - [C_2 + \epsilon (C_1 - C_2 L) M] \hat{\eta}_2$$

The optimal control is now given by

$$u = -(F_1 - F_2 L) \hat{\eta}_1 - [F_2 + \epsilon (F_1 - F_2 L) M] \hat{\eta}_2 \tag{5.14}$$

Matrices L and M satisfy

$$(A_4 - B_2 F_2) L - (A_3 - B_2 F_1) \\ -\epsilon L [(A_1 - B_1 F_1) - (A_2 - B_1 F_2) L] = 0 \tag{5.15a}$$

$$-M (A_4 - B_2 F_2) + (A_2 - B_1 F_2) - \epsilon ML (A_2 - B_1 F_2) \\ +\epsilon [(A_1 - B_1 F_1) - (A_2 - B_1 F_2) L] M = 0 \tag{5.15b}$$

Thus, in order to find the optimal solution in the decomposed form above, we have to solve two Riccati equations (5.8a)-(5.8b), a weakly nonlinear equation (5.15a), and a linear equation (5.15b).

The following lemma is summarized from (Chow and Kokotovic, 1976; Khalil and Gajic, 1984).

Lemma 5.1 *If A_4 is nonsingular and the triples (A_0, B_0, ρ_0), (A_0, G_0, C_0), (A_4, B_2, ρ_2), (A_4, G_2, C_2) are stabilizable and detectable, then for a sufficiently small ϵ, equations (5.8a)-(5.8b) will have unique stabilizing solutions which possess power series expansions at $\epsilon = 0$.*

◇

Matrices appearing in Lemma 5.1 are given by

$$A_0 = A_1 - A_2 A_4^{-1} A_3, \quad B_0 = B_1 - A_2 A_4^{-1} B_2$$
$$G_0 = G_1 - A_2 A_4^{-1} G_2, \quad C_0 = C_1 - C_2 A_4^{-1} A_3$$
$$R_1 = (\rho_1 \ \rho_2)^T (\rho_1 \ \rho_2), \quad \rho_0 = \rho_1 - \rho_2 A_4^{-1} A_3$$

Using the results of Lemma 5.1 the approximate stabilizing control is defined as

$$
\begin{aligned}
u_{apr}^{(k)} = &- \left(F_1^{(k)} - F_2^{(k)} L^{(k)} \right) \hat{\eta}_1^{(k)} \\
&- \left[F_2^{(k)} + \epsilon \left(F_1^{(k)} - F_2^{(k)} L^{(k)} \right) M^{(k)} \right] \hat{\eta}_2^{(k)}
\end{aligned}
\tag{5.16}
$$

with approximative filters

$$
\begin{aligned}
\dot{\hat{\eta}}_1^{(k)} = &\left[\left(A_1 - B_1 F_1^{(k)} \right) - \left(A_2 - B_1 F_2^{(k)} \right) L^{(k)} \right] \hat{\eta}_1^{(k)} \\
&+ \left(K_1^{(k)} - M^{(k)} K_2^{(k)} - \epsilon M^{(k)} L^{(k)} K_1^{(k)} \right) \nu^{(k)}
\end{aligned}
\tag{5.17a}
$$

$$
\begin{aligned}
\epsilon \dot{\hat{\eta}}_2^{(k)} = &\left[\left(A_4 - B_2 F_2^{(k)} \right) + \epsilon L^{(k)} \left(A_2 - B_1 F_2^{(k)} \right) \right] \hat{\eta}_2^{(k)} \\
&+ \left(K_2^{(k)} + \epsilon L^{(k)} K_1^{(k)} \right) \nu^{(k)}
\end{aligned}
\tag{5.17b}
$$

where

$$
\begin{aligned}
\nu^{(k)} = &y - \left(C_1 - C_2 L^{(k)} \right) \hat{\eta}_1^{(k)} \\
&- \left[C_2 + \epsilon \left(C_1 - C_2 L^{(k)} \right) M^{(k)} \right] \hat{\eta}_2^{(k)}
\end{aligned}
$$

and

$$
\begin{aligned}
F_1^{(k)} &= R_2^{-1} \left(B_1^T P_1^{(k)} + B_2^T P_2^{T(k)} \right) = F_1 + O\left(\epsilon^k \right) \\
F_2^{(k)} &= R_2^{-1} \left(\epsilon B_1^T P_2^{(k)} + B_2^T P_3^{(k)} \right) = F_2 + O\left(\epsilon^k \right) \\
K_1^{(k)} &= \left(Q_1^{(k)} C_1^T + Q_2^{(k)} C_2^T \right) V^{-1} = K_1 + O\left(\epsilon^k \right) \\
K_2^{(k)} &= \left(\epsilon Q_2^{T(k)} C_1^T + Q_3^{(k)} C_2^T \right) V^{-1} = K_2 + O\left(\epsilon^k \right)
\end{aligned}
$$

Corresponding block diagram for the approximate (near-optimal) control law is shown in Figure 5.2.

The main result from (Khalil and Gajic, 1984) can be summerized in the following theorem.

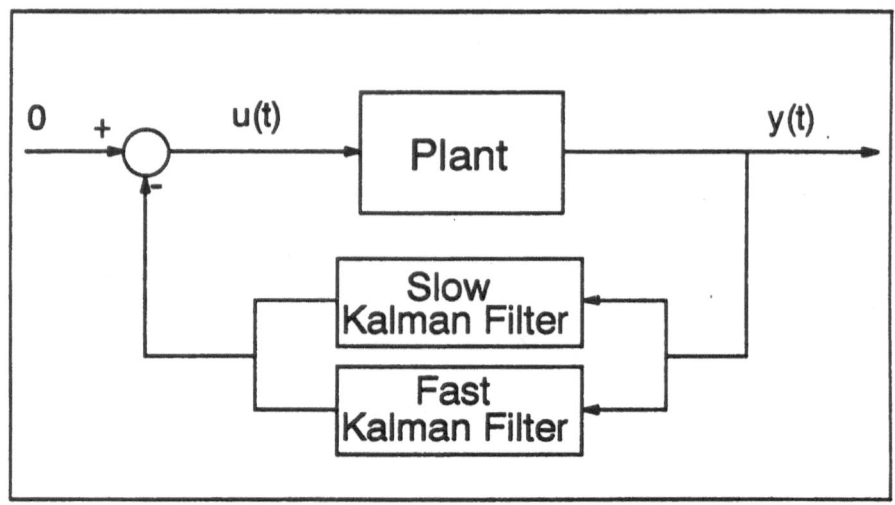

Figure 5.2: Approximate feedback control
law with slow and fast Kalman filters

Theorem 5.1 *Suppose that conditions of Lemma 5.1 hold. Let x_1 and x_2 be the optimal trajectories and J be the optimal value of the performance criterion. Let $x_1^{(k)}$, $x_2^{(k)}$, and $J^{(k)}$ be the corresponding quantities under the approximative control law $u_{apr}^{(k)}$ then*

$$\frac{J^{(k)} - J}{J} = O\left(\epsilon^k\right) \qquad\qquad (5.18)$$

$$var\left(x_1 - x_1^{(k)}\right) = O\left(\epsilon^{2k}\right) \qquad (as\ t \to \infty) \qquad (5.19a)$$
$$var\left(x_2 - x_2^{(k)}\right) = O\left(\epsilon^{2k-1}\right) \qquad (as\ t \to \infty) \qquad (5.19b)$$

◇

Note that by choosing appropriate initial conditions for (5.17) as

$$\hat{\eta}_1^{(k)}(0) = \left(I_{n_1} - \epsilon M^{(k)} L^{(k)}\right) \hat{x}_1(0) - \epsilon M^{(k)} \hat{x}_2(0)$$
$$\hat{\eta}_2^{(k)}(0) = L^{(k)} \hat{x}_1(0) + \hat{x}_2(0)$$

then (5.19) holds for all $t \geq 0$.

Exercise 5.1: Prove Theorem 5.1. Hint: (Khalil and Gajic, 1984).

△

Using standard techniques (Jamshidi, 1980; Bittanti et al., 1991) for the direct solution of the Riccati equation (5.8) can be inappropriate since one would be faced with the stiff numerical problem of the full-order. The well-known singular perturbation techniques (Kokotovic and Yackel, 1972), based on the power-series expansion with respect to ϵ, will convert given full-order stiff problem (5.8) to the family of well-defined reduced-order problems for which direct methods (Jamshidi, 1980; Bittanti et al., 1991) are very well suited. However, the power-series expansion method is not recursive in its nature. When we are interested in a high degree of accuracy or when ϵ is not very small, which can often be the case, the size of the computations required can be considerable, even though we are solving low-order problems. In such cases, the advantage of doing series expansion method is questionable. The presence of a small parameter ϵ can be exploited from a different point of view, which has been done in Chapter 2.

In summary, we can notice that the complete solution of the linear-quadratic Gaussian (LQG) control problem for singularly perturbed system is based on the solution of the algebraic Riccati equations (5.8) and algebraic equations for the transformation matrices L and M, namely, (5.15). Thus, the nature of their solutions determines the nature of the overall LQG problem. The algebraic Riccati equations can be solved using results from Section 2.2.2.

The filter equation (5.8b) can be solved by the same algorithm taking into account the following analogies

$$A_1 \to A_1^T, \quad A_2 \to A_3^T, \quad A_3 \to A_2^T, \quad A_4 \to A_4^T;$$
$$B_1 \to C_1^T, \quad B_2 \to C_2^T, \quad R_2 \to V$$
$$\rho_1^T \rho_1 \to G_1 W G_1^T, \quad \rho_1^T \rho_2 \to G_1 W G_2^T, \quad \rho_2^T \rho_2 \to G_2 W G_2^T$$

Thus, the remaining problem is to find a corresponding procedure for solving (5.15).

Solutions of the L and M equations (5.15) that are needed for our LQG optimal control can be sought through the iterative form proposed in (Kokotovic et al., 1980) — see also Sections 2.2 and 8.1.

$$L^{(i+1)} = \left(A_4 - B_2 F_2^{(N)}\right)^{-1} \left(A_3 - B_2 F_1^{(N)}\right) + \epsilon \left(A_4 - B_2 F_2^{(N)}\right)^{-1}$$
$$\times L^{(i)} \left[\left(A_1 - B_1 F_1^{(N)}\right) - \left(A_2 - B_1 F_2^{(N)}\right)^{-1} L^{(i)}\right]$$

$$(5.20)$$

$$M^{(i+1)} = \left(A_2 - B_1 F_2^{(N)}\right)\left(A_4 - B_2 F_2^{(N)}\right)^{-1}$$
$$-\epsilon M^{(i)} L^{(N)}\left(A_2 - B_1 F_2^{(N)}\right)\left(A_4 - B_2 F_2^{(N)}\right)^{-1}$$
$$+\epsilon\left[\left(A_1 - B_1 F_1^{(N)}\right) - \left(A_2 - B_1 F_2^{(N)}\right) L^{(N)}\right]$$
$$\times M^{(i)}\left(A_4 - B_2 F_2^{(N)}\right)^{-1}, \quad i = 1, 2, ..., N - 1 \tag{5.21}$$

with initial values

$$L^{(0)} = \left(A_4 - B_2 F_2^{(0)}\right)^{-1}\left(A_3 - B_2 F_1^{(0)}\right) \tag{5.22}$$

$$M^{(0)} = \left(A_2 - B_1 F_2^{(0)}\right)\left(A_4 - B_2 F_2^{(0)}\right)^{-1} \tag{5.23}$$

Note that in (5.20) $F_1^{(N)}$ and $F_2^{(N)}$ have already been determined using (5.7a) and the solution of the Riccati equation (5.8) which can be obtained using the algorithm from Section 2.2. The same holds for $L^{(N)}$ in equation (5.21).

Equations (5.20) and (5.21) take the forms

$$D_3 L = \gamma_1 + \epsilon \Im_1 (L, \epsilon) \tag{5.24}$$

$$M D_3 = \gamma_2 + \epsilon \Im_2 (M, L, \epsilon) \tag{5.25}$$

where D_3 is a stable matrix, γ_1 and γ_2 are known constants, $\Im_1$ and $\Im_2$ are nonlinear functions. Then the following theorem holds for L and M.

Theorem 5.2 *Under the conditions of Theorem 5.1, algorithm (5.20)-(5.21) converges to the exact solution L and M with the rate of convergence of $O(\epsilon)$, that is*

$$\| L - L^{(i+1)} \| = O(\epsilon) \| L - L^{(i)} \|$$
$$\| M - M^{(i+1)} \| = O(\epsilon) \| M - M^{(i)} \|$$

or equivalently

$$\| L - L^{(i)} \| = O(\epsilon^i)$$
$$\| M - M^{(i)} \| = O(\epsilon^i)$$

◇

Proof of this theorem uses the same arguments as the proof of Theorem 2.2 and thus is omitted here.

So far we have developed the iterative procedures that generate in a very efficient way all coefficients for the approximate solution of LQG (5.16)-(5.17). Those coefficients for an $O\left(\epsilon^k\right)$ accuracy of the approximation of LQG are obtained by doing $k - 1$ iterations on the same set of equations. Contrary to the power-series expansions, where we have to solve k-different sets of equations, our iterative scheme is very useful in the case where a high order of accuracy is required. Instead of defining and solving a new set of equations, we perform just one additional iteration on the already existing set of equations. On the other hand in both cases we are faced with low-order numerical problems (in fact they are of the same order).

Equations (5.16)-(5.17) can be written in the following composite forms

$$u_{opt}^{(k)} = -f_1^{(k)}\hat{\eta}_1^{(k)} - f_2^{(k)}\hat{\eta}_2^{(k)} \qquad (5.26)$$

$$\dot{\hat{\eta}}_1^{(k)} = a_1^{(k)}\hat{\eta}_1^{(k)} + g_1^{(k)}\nu^{(k)} \qquad (5.27)$$

$$\epsilon\dot{\hat{\eta}}_2^{(k)} = a_2^{(k)}\hat{\eta}_2^{(k)} + g_2^{(k)}\nu^{(k)} \qquad (5.28)$$

where

$$\nu^{(k)} = y - \xi_1^{(k)}\hat{\eta}_1^{(k)} - \xi_2^{(k)}\hat{\eta}_2^{(k)}$$

with obvious expressions for $f_i^{(k)}$, $a_i^{(k)}$, $g_i^{(k)}$, $\xi_i^{(k)}$, $i = 1, 2$. The suboptimal criterion

$$J^{(k)} = \lim_{\substack{t_0 \to -\infty \\ t_1 \to \infty}} \frac{1}{t_1 - t_0} E\left\{\int_{t_0}^{t_1}\left[\begin{pmatrix}x_1^{(k)} \\ x_2^{(k)}\end{pmatrix}^T R_1 \begin{pmatrix}x_1^{(k)} \\ x_2^{(k)}\end{pmatrix} + u^{(k)^T} R_2 u^{(k)}\right] dt\right\} \qquad (5.29)$$

is then given by

$$J^{(k)} = tr\left\{R_1 q_{11}^{(k)} + f^{(k)^T} R_2 f^{(k)} q_{22}^{(k)}\right\}, \quad f^{(k)} = \left(f_1^{(k)} \; f_2^{(k)}\right) \qquad (5.30)$$

where

LINEAR STOCHASTIC SYSTEMS

$$q_{11}^{(k)} = Var\left(\begin{matrix} x_1^{(k)} \\ x_2^{(k)} \end{matrix}\right) \quad and \quad q_{22}^{(k)} = Var\left(\begin{matrix} \hat{\eta}_1^{(k)} \\ \hat{\eta}_2^{(k)} \end{matrix}\right)$$

Quantities $q_{11}^{(k)}$ and $q_{22}^{(k)}$ can be obtained by studying the variance equation of the following system

$$\begin{bmatrix} \dot{x}_1^{(k)} \\ \epsilon\dot{x}_2^{(k)} \\ \dot{\hat{\eta}}_1^{(k)} \\ \epsilon\dot{\hat{\eta}}_2^{(k)} \end{bmatrix} = \begin{bmatrix} A_1 & A_2 & -B_1 f_1^{(k)} & -B_1 f_2^{(k)} \\ A_3 & A_4 & -B_2 f_1^{(k)} & -B_2 f_2^{(k)} \\ g_1^{(k)} C_1 & g_1^{(k)} C_2 & a_1^{(k)} - g_1^{(k)}\xi_1^{(k)} & -g_1^{(k)}\xi_2^{(k)} \\ g_2^{(k)} C_1 & g_2^{(k)} C_2 & -g_2^{(k)} C_2 & a_2^{(k)} - g_2^{(k)}\xi_2^{(k)} \end{bmatrix}$$

$$\times \begin{bmatrix} x_1^{(k)} \\ x_2^{(k)} \\ \hat{\eta}_1^{(k)} \\ \hat{\eta}_2^{(k)} \end{bmatrix} + \begin{bmatrix} G_1 & 0 \\ G_2 & 0 \\ 0 & g_1^{(k)} \\ 0 & g_2^{(k)} \end{bmatrix}\begin{bmatrix} w \\ v \end{bmatrix}$$

(5.31)

or, in a composite form

$$\dot{Z}^{(k)} = A^{(k)}Z^{(k)} + G^{(k)}\tilde{w} \quad \tilde{w} = (w^T \ v^T)^T \tag{5.32}$$

with obvious definitions for $A^{(k)}$ and $G^{(k)}$.

The variance of $Z^{(k)}$ denoted by $q^{(k)}$ is given by the well-known Lyapunov equation

$$A^{(k)}q^{(k)} + q^{(k)}A^{(k)^T} + G^{(k)}\tilde{W}G^{(k)^T} = 0, \quad \tilde{W} = diag\,(W, V) \tag{5.33}$$

where $q^{(k)}$ is partitioned as

$$q^{(k)} = \begin{bmatrix} q_{11}^{(k)} & q_{12}^{(k)} \\ q_{12}^{(k)^T} & \frac{1}{\epsilon}q_{22}^{(k)} \end{bmatrix} \tag{5.34}$$

This procedure is demonstrated in the next section by a numerical example, showing the required convergence properties

$$f_1^{(k)} \to f_1^{opt}$$
$$a_1^{(k)} \to a_1^{opt}$$
$$g_1^{(k)} \to g_1^{opt}$$
$$\xi_i^{(k)} \to \xi_i^{opt}$$
$$J^{(k)} \to J^{opt}$$
$$k = 0, 1, ...; \quad i = 1, 2.$$

Exercise 5.2: Find the value for the optimal performance criterion in the new coordinates, namely, in terms of $Var\,(\eta_1, \eta_2, \hat{\eta}_1, \hat{\eta}_2)$. Are white noise processes $w\,(t)$ and $\nu\,(t)$ correlated in the new coordinates?

$\triangle$

5.2 Case Study: F-8 Aircraft LQG Controller

In order to demonstrate the numerical behavior of the near-optimum design of singularly perturbed LQG regulators, we present results for an LQG controller of an F-8 aircraft which was considered in (Teneketzis and Sandell, 1977). The controller is designed to produce elevator commands to keep the aircraft in steady level flight in the face of wind disturbances. For simplicity, the wind disturbance is modeled as white. The aircraft's longitudinal variables are

$$x = \begin{bmatrix} V \\ \gamma \\ \alpha \\ q \end{bmatrix} \qquad u = \delta_e$$

where

  V   horizontal-velocity deviation (feet/second)

  γ flight-path angle (radians)

  α angle of attack (radians)

  q pitch rate (radians/second)

  δ_e elevator deflection (radians)

The equations of motion of the airplane are a set of coupled nonlinear equations in the longitudinal and lateral state variables. If the equations are linearized about the nominal state and control variables, the resulting linear equations are found to approximately decouple into separate sets of the longitudinal and lateral dynamics. In our case the system model is given by

$$\begin{bmatrix} \dot{\tilde{x}}_1 \\ \dot{\tilde{x}}_2 \\ \dot{\tilde{x}}_3 \\ \dot{\tilde{x}}_4 \end{bmatrix} = \begin{bmatrix} -1.357 \times 10^{-2} & -32.2 & -46.3 & 0 \\ 1.2 \times 10^{-4} & 0 & 1.214 & 0 \\ -1.212 \times 10^{-4} & 0 & -1.214 & 1 \\ 5.7 \times 10^{-4} & 0 & -9.01 & -0.6696 \end{bmatrix} \begin{bmatrix} \tilde{x}_1 \\ \tilde{x}_2 \\ \tilde{x}_3 \\ \tilde{x}_4 \end{bmatrix} +$$

$$+ \begin{bmatrix} -0.433 \\ 0.1394 \\ -0.1394 \\ -0.1577 \end{bmatrix} u + \begin{bmatrix} -46.3 \\ 1.214 \\ -1.214 \\ -9.01 \end{bmatrix} w$$

$$y = \begin{bmatrix} 0 & 0 & 0 & 1 \\ 1 & 0 & 0 & 0 \end{bmatrix} \begin{bmatrix} \tilde{x}_1 \\ \tilde{x}_2 \\ \tilde{x}_3 \\ \tilde{x}_4 \end{bmatrix} + v$$

where the white noise processes w and v are independent and have intensities $W = 3.15 \times 10^{-4}$ and $V = diag\, [6.859 \times 10^{-4}, \, 40]$. The performance criterion is

$$J = \lim_{\substack{t_0 \to -\infty \\ t_f \to \infty}} \frac{1}{t_f - t_0} E\left\{ \int_{t_0}^{t_f} [0.01\tilde{x}_1^2 + 3260\,(\tilde{x}_3^2 + \tilde{x}_4^2 + u^2)]\, dt \right\}$$

The reader is referred to (Teneketzis and Sandell, 1977) for a discussion of the modelling aspects and the choice of J.

The open-loop eigenvalues are $-0.94 \pm j2.98$ and $-0.0075 \pm j0.0076$ which shows clearly the two-time-scale property of the system. The choice of the state variables adopted in (Teneketzis and Sandell, 1977) led to a formulation in which the first two variables are slow variables. A logical choice of the parameter ϵ is $\epsilon = 0.025$, which is roughly the ratio of the magnitude of the slow eigenvalues to the magnitude of the fast eigenvalues. The singularly perturbed nature of this system becomes more evident (Chow, 1982) by using a state transformation $x = T\tilde{x}$ where

$$T = \begin{bmatrix} 1 & 1618 & 133.92 & 200 \\ 0 & 500 & 40.8 & 61 \\ 0 & 0 & 600 & 0 \\ 0 & 0 & 0 & 200 \end{bmatrix}$$

Introducing ϵ artificially by multiplying the left-hand sides by 0.025, the system takes the singularly perturbed form of (5.1)-(5.4) with

$$A_1 = \begin{bmatrix} 0.278386 & -0.965256 \\ 0.089833 & -0.290700 \end{bmatrix}, \quad A_2 = \begin{bmatrix} -0.074210 & 0.016017 \\ 0.012815 & -0.001398 \end{bmatrix}$$

$$A_3 = \begin{bmatrix} -0.001815 & 0.005873 \\ 0.002850 & -0.009223 \end{bmatrix}, \quad A_4 = \begin{bmatrix} -0.030344 & 0.075024 \\ -0.075092 & -0.016777 \end{bmatrix}$$

$$B_1 = \begin{bmatrix} 174.907714 \\ 54.392760 \end{bmatrix}, \quad B_2 = \begin{bmatrix} -2.091000 \\ -0.780500 \end{bmatrix}$$

$$\rho_1^T \rho_1 = \begin{bmatrix} 0.010000 & -0.032360 \\ -0.032360 & 0.104717 \end{bmatrix}$$

$$\rho_1^T \rho_2 = \begin{bmatrix} -0.000032 & -0.000130 \\ 0.000102 & 0.000421 \end{bmatrix}$$

$$\rho_2^T \rho_2 = \begin{bmatrix} 0.009056 & 0.000000 \\ 0.000000 & 0.018502 \end{bmatrix}$$

$$R_2 = 3260, \quad W = 0.000315, \quad V = diag\{0.000686,\ 40\}$$

$$C_1 = \begin{bmatrix} 0 & 0 \\ 1 & -3.236000 \end{bmatrix}, \quad C_2 = \begin{bmatrix} 0 & 0.005000 \\ -0.003152 & 0.013020 \end{bmatrix}$$

$$G_1 = \begin{bmatrix} -46.626960 \\ 7.858776 \end{bmatrix}, \quad G_2 = \begin{bmatrix} -18.210002 \\ -45.049998 \end{bmatrix}$$

Corresponding results are shown in Table 5.1

5.3 Recursive Approach to Weakly Coupled Linear Stochastic Systems

The linear-quadratic Gaussian control problem of the weakly coupled continuos-time systems is studied in this section by using the results reported in (Shen and Gajic, 1990a). Corresponding result for another class of small parameter systems — singularly perturbed systems (Kokotovic and Khalil, 1986), has been presented in Section 5.1 — by using the fixed point theory. Although the duality of the regulator and filter Riccati equations can be used, together with the results presented in Chapter 2, to obtain corresponding approximations to the regulator and filter gains, such approximations will not be sufficient, because they only reduce the off-line computations and do not help the on-line computations of implementing the Kalman filter which will be of the same order as the overall weakly coupled system. The weakly coupled structure of the global Kalman filter is exploited in this section such that it may be replaced by two lower order local filters. This has been achieved via the use of a decoupling transformation introduced in (Gajic and Shen, 1989).

In this section, we present the approach to the decomposition and approximation of the linear-quadratic Gaussian control problem of weakly

	First	*Second*	*Third*	*Forth = Optimal*
f_1	-0.000448 0.002551	-0.000449 0.002554	-0.000449 0.002554	-0.000449 0.002554
f_2	-0.000493 -0.000061	-0.000496 -0.000065	-0.000496 -0.000065	-0.000496 -0.000065
ζ_1	0.000230 -0.000863 0.999329 -3.233571	0.000234 -0.000881 0.999314 -3.233517	0.000235 -0.000882 0.999314 -3.233516	0.000235 -0.000882 0.999314 -3.233516
ζ_2	0.000004 0.005009 0.010988 0.019571	0.000004 0.005009 0.011113 0.019609	0.000004 0.005009 0.011143 0.019625	0.000004 0.005009 0.011143 0.019626
a_1	0.355782 -1.409892 0.114433 0.429934	0.355852 -1.410409 0.114482 -0.430155	0.355843 -1.410384 0.114480 -0.430148	0.355843 -1.410384 0.114480 -0.430149
a_2	-0.031306 0.074993 -0.075317 -0.016826	-0.031313 0.074936 -0.075317 -0.016829	-0.031314 0.074936 -0.075317 -0.016829	-0.031314 0.074036 -0.075317 -0.016829
g_1	25.459425 0.000494 7.792644 0.001299	25.898515 0.003132 7.945627 0.000704	25.897536 0.004113 7.930356 0.000723	25.920266 0.004134 7.936231 0.000723
g_2	9.084229 0.001998 22.483368 0.0018733	9.104124 0.001986 22.486045 0.001870	9.103953 0.001972 22.486804 0.001875	9.103760 0.001971 22.486923 0.001875
J	25.066942	25.066604	25.066597	25.066597

Table 5.1: A successive approximation solution of LQG for an aircraft F-8

coupled systems by treating the decomposition and approximation tasks separately from each other. The decoupling transformation of (Gajic and Shen, 1989) is used for the exact block diagonalization of the global

Kalman filter. The approximate feedback control law is then obtained by approximating the coefficients of the optimal local filters with the accuracy of $O\left(\epsilon^N\right)$. The resulting feedback control law is shown to be a near-optimal solution of the LQG by studying the corresponding closed-loop system as a system driven by white noise. It is shown that the order of approximation of the optimal performance is $O\left(\epsilon^N\right)$, and the order of approximation of the optimal system trajectories is $O\left(\epsilon^{2N}\right)$. All required coefficients of the desired accuracy are easily obtained by using the recursive fixed point type numerical techniques developed in Chapter 2. Given numerical algorithms converge to the required coefficients with the rate of convergence of $O\left(\epsilon^2\right)$. In addition, only low-order subsystems are involved in the algebraic computations and no analyticity requirements are imposed on the system coefficients — which is the standard assumption in the power-series expansion method. As a consequence of these properties, under very mild conditions (coefficients are bounded functions of a small coupling parameter over $\epsilon \in [0, \epsilon_1]$), in addition to the standard stabilizability-detectability subsystem assumptions, we have achieved the reduction in both off-line and on-line computational requirements.

This section is organized as follows. At beginning, we study the approximation of weakly coupled systems driven by white noise. It is shown that an *Nth*-order approximation in which the system coefficients are $O\left(\epsilon^N\right)$ close to the exact ones is a valid approximation in the sense that the differences between the exact and approximate solutions are $O\left(\epsilon^{2N}\right)$. Then, we use these results in the study of the LQG problem. A decoupling nonsingular transformation is used to represent the Kalman filter in new coordinates in which local filters are completely decoupled. An *Nth*-order approximate feedback control law is defined by approximating coefficients by $O\left(\epsilon^N\right)$. A study of the corresponding closed-loop system, shows that the absolute increase in the performance criterion over its optimal value is $O\left(\epsilon^N\right)$.

Consider the linear time-invariant weakly coupled system driven by white noise

$$\begin{bmatrix} \dot{x}_1 \\ \dot{x}_2 \end{bmatrix} = \begin{bmatrix} A_{11}(\epsilon) & \epsilon A_{12}(\epsilon) \\ \epsilon A_{21}(\epsilon) & A_{22}(\epsilon) \end{bmatrix} \begin{bmatrix} x_1 \\ x_2 \end{bmatrix} + \begin{bmatrix} G_{11}(\epsilon) & \epsilon G_{12}(\epsilon) \\ \epsilon G_{21}(\epsilon) & G_{22}(\epsilon) \end{bmatrix} \begin{bmatrix} w_1 \\ w_2 \end{bmatrix}$$
$$(5.35)$$

where $x_i \in \Re^{n_i}$, $w_i \in \Re^{r_i}$, $i = 1, 2$, and ϵ is a small parameter. The system matrices are bounded functions of ϵ, (Gajic et al., 1990; Harkara et. al, 1989; Petrovic and Gajic, 1988) of appropriate dimensions. The inputs $w_i(t)$ are zero mean, stationary, Gaussian uncorrelated white noise processes with intensities $W_i > 0$, $i = 1, 2$. It is well known that the variance of the linear systems driven by white noise is given by the

Lyapunov equation (Kwakernaak and Sivan, 1972). In order to assure the existence of its solution we assume that $A_{ii}(\epsilon)$, $i = 1, 2$, are stable matrices. The purpose of this section is to study approximations of $x_i(t)$, $i = 1, 2$, when ϵ is small. We are interested in approximations $x_i^N(t)$ which are defined by the following equations

$$\begin{bmatrix} \dot{x}_1^N \\ \dot{x}_2^N \end{bmatrix} = \begin{bmatrix} A_{11}^N(\epsilon) & \epsilon A_{12}^N(\epsilon) \\ \epsilon A_{21}^N(\epsilon) & A_{22}^N(\epsilon) \end{bmatrix} \begin{bmatrix} x_1^N \\ x_2^N \end{bmatrix} + \begin{bmatrix} G_{11}^N(\epsilon) & \epsilon G_{12}^N(\epsilon) \\ \epsilon G_{21}^N(\epsilon) & G_{22}^N(\epsilon) \end{bmatrix} \begin{bmatrix} w_1 \\ w_2 \end{bmatrix}$$
(5.36)

where

$$A_{ij}(\epsilon) - A_{ij}^N(\epsilon) = O(\epsilon^N),$$
$$G_{ij}(\epsilon) - G_{ij}^N(\epsilon) = O(\epsilon^N), \quad i, j = 1, 2$$
(5.37)

The quantities of interest are the variances of the errors

$$e_i(t) = x_i(t) - x_i^N(t), \quad i = 1, 2$$
(5.38)

at steady state. We study the impact of the steady state errors on a quadratic form given by

$$\sigma = tr\{ \begin{bmatrix} H^T(\epsilon) H(\epsilon) & H^T(\epsilon) J(\epsilon) \\ J^T(\epsilon) H(\epsilon) & J^T(\epsilon) J(\epsilon) \end{bmatrix}$$
$$\times E \begin{bmatrix} x_1(t) x_1^T(t) & x_1(t) x_2^T(t) \\ x_2(t) x_1^T(t) & x_2(t) x_2^T(t) \end{bmatrix} \}$$
(5.39)

where $H(\epsilon)$ and $J(\epsilon)$ are bounded functions of ϵ also. Such a quadratic form appears in the steady state LQG control problem. We examine the approximation of σ by σ^N defined by

$$\sigma^N = tr\{ \begin{bmatrix} H^{NT}(\epsilon) H^N(\epsilon) & H^{NT}(\epsilon) J^N(\epsilon) \\ J^{NT}(\epsilon) H^N(\epsilon) & J^{NT}(\epsilon) J^N(\epsilon) \end{bmatrix}$$
$$\times E \begin{bmatrix} x_1^N(t) x_1^{NT}(t) & x_1^N(t) x_2^{NT}(t) \\ x_2^N(t) x_1^{NT}(t) & x_2^N(t) x_2^{NT}(t) \end{bmatrix} \}$$
(5.40)

where

$$H^N(\epsilon) - H(\epsilon) = O(\epsilon^N), \quad J^N(\epsilon) - J(\epsilon) = O(\epsilon^N)$$
(5.41)

In the following we will suppress the ϵ-dependence of the problem matrices in order to simplify notation.

The main results of this section are given in the following two theorems.

126

Theorem 5.3 *Under stability assumptions imposed on A_{ii}, $i = 1, 2$; the approximation errors at steady state satisfy*

$$Var\{e_i\} = Var\{x_i - x_i^N\} = O(\epsilon^{2N}), \quad i = 1, 2$$
$$Cov\{e_1, e_2\} = O(\epsilon^{2N}) \tag{5.42}$$

◇

Theorem 5.4 *Under conditions stated in Theorem 5.3, the quadratic forms (5.39) and (5.40) at steady state satisfy*

$$\Delta\sigma = \sigma - \sigma^N = O(\epsilon^N) \tag{5.43}$$

◇

Proof: The proof of these two theorems can be obtained by studying the following augmented system driven by white noise

$$\begin{bmatrix} \dot{x}_1 \\ \dot{e}_1 \\ \dot{x}_2 \\ \dot{e}_2 \end{bmatrix} = \begin{bmatrix} A_{11} & 0 & \epsilon A_{11} & 0 \\ O(\epsilon^N) & A_{11}^N & O(\epsilon^{N+1}) & \epsilon A_{12}^N \\ \epsilon A_{21} & 0 & A_{22} & 0 \\ O(\epsilon^{N+1}) & \epsilon A_{21}^N & O(\epsilon^N) & A_{22}^N \end{bmatrix} \begin{bmatrix} x_1 \\ e_1 \\ x_2 \\ e_2 \end{bmatrix}$$

$$+ \begin{bmatrix} G_{11} & \epsilon G_{12} \\ O(\epsilon^N) & O(\epsilon^{N+1}) \\ \epsilon G_{21} & G_{22} \\ O(\epsilon^{N+1}) & O(\epsilon^N) \end{bmatrix} \begin{bmatrix} w_1 \\ w_2 \end{bmatrix} \tag{5.44}$$

For shorthand notation (5.44) is written as

$$\dot{z} = \Lambda z + \Gamma w \tag{5.45}$$

with obvious definitions of z, ω, Λ, and Γ. The variance of z at steady state is given by the algebraic Lyapunov equation (Kwakernaak and Sivan, 1972)

$$0 = \Lambda Q + Q\Lambda^T + \Gamma W \Gamma^T \tag{5.46}$$

where

$$W = \begin{bmatrix} W_1 & 0 \\ 0 & W_2 \end{bmatrix}$$

The variance of z is partitioned as

$$Q = \begin{bmatrix} Q_{11} & Q_{12} & \epsilon Q_{13} & \epsilon Q_{14} \\ Q_{12}^T & Q_{22} & \epsilon Q_{23} & \epsilon Q_{24} \\ \epsilon Q_{13}^T & \epsilon Q_{23}^T & Q_{33} & Q_{34} \\ \epsilon Q_{14}^T & \epsilon Q_{24}^T & Q_{34}^T & Q_{44} \end{bmatrix}$$

Studying the partitioned form of (5.46) will produce (after lengthy calculations)

$$Q_{ij} = O(1), \qquad ij = 11, 13, 33 \tag{5.47a}$$
$$Q_{ij} = O(\epsilon^{2N}), \qquad ij = 22, 24, 44 \tag{5.47b}$$
$$Q_{ij} = O(\epsilon^N), \qquad ij = 12, 14, 23, 34 \tag{5.47c}$$

which proves Theorem 5.3.

Quadratic forms defined in (5.39) and (5.40) can be now expressed in terms of the elements of the matrix Q as

$$\sigma = tr\left\{ H^T H Q_{11} + J^T J Q_{33} + 2\epsilon J^T H Q_{13} \right\} \tag{5.48}$$

and

$$\sigma^N = tr\left\{ 2\epsilon J^{N^T} H^N (Q_{13} - Q_{23} - Q_{14} + Q_{24}) \right\}$$
$$+ tr\left\{ H^{N^T} H^N (Q_{11} - 2Q_{12} + Q_{22}) + J^{N^T} J^N (Q_{33} - 2Q_{34} + Q_{44}) \right\} \tag{5.49}$$

From (5.48)-(5.49) and estimates for Q_{ij}, $i, j = 1, 2, 3, 4$, one has

$$\Delta\sigma = \sigma - \sigma^N$$
$$= O(\epsilon^N) tr\left\{ (Q_{11} + Q_{33}) + \left(H^{N^T} H^N + J^{N^T} J^N \right) \right\} + O(\epsilon^{N+1}) \tag{5.50}$$

Since Q_{11}, Q_{33} and H^N, J^N are $O(1)$ quantities, one can conclude that $\Delta\sigma = O(\epsilon^N)$, which completes the proof of Theorem 5.4.

At this point we can introduce the linear-quadratic Gaussian control problem of weakly coupled systems and study its approximation and decomposition by utilizing results from Theorems 5.3 and 5.4.

Consider the weakly coupled linear system

$$\begin{bmatrix} \dot{x}_1 \\ \dot{x}_2 \end{bmatrix} = \begin{bmatrix} A_{11}(\epsilon) & \epsilon A_{12}(\epsilon) \\ \epsilon A_{21}(\epsilon) & A_{22}(\epsilon) \end{bmatrix} \begin{bmatrix} x_1 \\ x_2 \end{bmatrix} + \begin{bmatrix} B_{11}(\epsilon) & \epsilon B_{12}(\epsilon) \\ \epsilon B_{21}(\epsilon) & B_{22}(\epsilon) \end{bmatrix} \begin{bmatrix} u_1 \\ u_2 \end{bmatrix}$$
$$+ \begin{bmatrix} G_{11}(\epsilon) & \epsilon G_{12}(\epsilon) \\ \epsilon G_{21}(\epsilon) & G_{22}(\epsilon) \end{bmatrix} \begin{bmatrix} w_1 \\ w_2 \end{bmatrix} \tag{5.51}$$

$$\begin{bmatrix} y_1 \\ y_2 \end{bmatrix} = \begin{bmatrix} C_{11}(\epsilon) & \epsilon C_{12}(\epsilon) \\ \epsilon C_{21}(\epsilon) & C_{22}(\epsilon) \end{bmatrix} \begin{bmatrix} x_1 \\ x_2 \end{bmatrix} + \begin{bmatrix} v_1 \\ v_2 \end{bmatrix} \qquad (5.52)$$

where $x_i \in \Re^{n_i}$, $u_i \in \Re^{m_i}$, $y_i \in \Re^{r_i}$, $i = 1, 2$, are state, control and measurement vectors, respectively, and $w_i \in \Re^{s_i}$, $v_i \in \Re^{r_i}$, $i = 1, 2$, are independent zero-mean stationary white Gaussian noise processes with intensities W_i, V_i, $i = 1, 2$. The degree of interaction between subsystems is measured by a small parameter ϵ. With (5.51)-(5.52), consider the performance criterion

$$J = tr \left\{ D^T D E \begin{bmatrix} x_1 x_1^T & x_1 x_2^T \\ x_2 x_1^T & x_2 x_2^T \end{bmatrix} + RE \begin{bmatrix} u_1 u_1^T & u_1 u_2^T \\ u_2 u_1^T & u_2 u_2^T \end{bmatrix} \right\} \qquad (5.53)$$

with positive definite R. In the following all matrices are bounded functions of ϵ, (Gajic et al., 1990; Harkara et al., 1989; Petrovic and Gajic, 1988), of the appropriate dimensions. In addition, matrices $D^T D$ and R have the weakly coupled structure. We assume that they are given by

$$D^T D = \begin{bmatrix} D_1^T D_1 & \epsilon D_1^T D_2 \\ \epsilon D_2^T D_1 & D_2^T D_2 \end{bmatrix} \qquad R = \begin{bmatrix} R_1 & 0 \\ 0 & R_2 \end{bmatrix}$$

where $R_i \in \Re^{m_i \times m_i}$ and $D_i^T D_i \in \Re^{n_i \times n_i}$, $i = 1, 2$.

The optimal control law has the very well-known form (Kwakernaak and Sivan, 1972)

$$\begin{bmatrix} u_1(t) \\ u_2(t) \end{bmatrix} = - \begin{bmatrix} F_{11} & \epsilon F_{12} \\ \epsilon F_{21} & F_{22} \end{bmatrix} \begin{bmatrix} \hat{x}_1(t) \\ \hat{x}_2(t) \end{bmatrix} \qquad (5.54)$$

$$\begin{bmatrix} \dot{\hat{x}}_1(t) \\ \dot{\hat{x}}_2(t) \end{bmatrix} = \begin{bmatrix} A_{11} & \epsilon A_{12} \\ \epsilon A_{21} & A_{22} \end{bmatrix} \begin{bmatrix} \hat{x}_1(t) \\ \hat{x}_2(t) \end{bmatrix} + \begin{bmatrix} B_{11} & \epsilon B_{12} \\ \epsilon B_{21} & B_{22} \end{bmatrix} \begin{bmatrix} u_1(t) \\ u_2(t) \end{bmatrix}$$

$$+ \begin{bmatrix} K_{11} & \epsilon K_{12} \\ \epsilon K_{21} & K_{22} \end{bmatrix} \begin{bmatrix} y_1(t) - C_{11} \hat{x}_1(t) - \epsilon C_{12} \hat{x}_2(t) \\ y_2(t) - \epsilon C_{21} \hat{x}_1 - C_{22} \hat{x}_2(t) \end{bmatrix}$$

$$(5.55)$$

Introducing the notation

$$A = \begin{bmatrix} A_{11} & \epsilon A_{12} \\ \epsilon A_{21} & A_{22} \end{bmatrix}, \quad B = \begin{bmatrix} B_{11} & \epsilon B_{12} \\ \epsilon B_{21} & B_{22} \end{bmatrix}, \quad G = \begin{bmatrix} G_{11} & \epsilon G_{12} \\ \epsilon G_{21} & G_{22} \end{bmatrix}$$

$$C = \begin{bmatrix} C_{11} & \epsilon C_{12} \\ \epsilon C_{21} & C_{22} \end{bmatrix}, \quad F = \begin{bmatrix} F_{11} & \epsilon F_{12} \\ \epsilon F_{21} & F_{22} \end{bmatrix}, \quad K = \begin{bmatrix} K_{11} & \epsilon K_{12} \\ \epsilon K_{21} & K_{22} \end{bmatrix}$$

$$W = \begin{bmatrix} W_1 & 0 \\ 0 & W_2 \end{bmatrix}, \quad V = \begin{bmatrix} V_1 & 0 \\ 0 & V_2 \end{bmatrix}.$$

the regulator and filter gains are obtained from

$$F = R^{-1}B^T P, \quad K = QC^T V^{-1} \tag{5.56}$$

where P and Q are positive semidefinite stabilizing solutions of the algebraic Riccati equations

$$A^T P + PA - PS_R P + D^T D = 0, \quad S_R = BR^{-1}B^T \tag{5.57}$$

$$AQ + QA^T - QS_F Q + GWG^T = 0, \quad S_F = C^T V^{-1} C \tag{5.58}$$

Due to the weakly coupled structure of all coefficients in (5.57)-(5.58), solutions of these equations have the form

$$P = \begin{bmatrix} P_1 & \epsilon P_2 \\ \epsilon P_2^T & P_3 \end{bmatrix}, \quad Q = \begin{bmatrix} Q_1 & \epsilon Q_2 \\ \epsilon Q_2^T & Q_3 \end{bmatrix} \tag{5.59}$$

Solutions of (5.57)-(5.58) can be found in terms of the reduced-order problems by imposing standard stabilizability-detectability assumptions on subsystems. The efficient fixed point algorithms for solving (5.57) and (5.58) are obtained in Section 2.3.2. The algorithms for solving regulator and filter algebraic Riccati equations of weakly coupled systems are convergent under the following assumptions.

Assumption 5.1 Triples (A_{ii}, B_{ii}, D_{ii}), $i = 1, 2$, are stabilizable and detectable.

$\triangle$

Assumption 5.2 Triples (A_{ii}, C_{ii}, G_{ii}), $i = 1, 2$, are stabilizable and detectable.

$\triangle$

Getting approximate solutions for P and Q in terms of the reduced-order problems will produce savings in off-line computations. However, in the case of stochastic systems, where an additional dynamic system — filter — has to be built, one is particularly interested in the reduction of on-line computations. We will achieve that by using the decoupling transformation presented in Section 3.3.

The Kalman filter (5.55) is viewed as a system driven by the innovation process. However, one might study the filter form when it is driven by both measurements and controls. The filter form under consideration is obtained from (5.55) as

$$
\begin{bmatrix} \dot{\hat{x}}_1 \\ \dot{\hat{x}}_2 \end{bmatrix} = \begin{bmatrix} K_{11} & \epsilon K_{12} \\ \epsilon K_{21} & K_{22} \end{bmatrix} \begin{bmatrix} \nu_1 \\ \nu_2 \end{bmatrix} +
$$
$$
\begin{bmatrix} (A_{11} - B_{11}F_{11} - \epsilon^2 B_{12}F_{21}) & \epsilon(A_{12} - B_{11}F_{12} - B_{12}F_{22}) \\ \epsilon(A_{21} - B_{21}F_{11} - B_{22}F_{21}) & (A_{22} - B_{22}F_{22} - \epsilon^2 B_{21}F_{12}) \end{bmatrix} \begin{bmatrix} \hat{x}_1 \\ \hat{x}_2 \end{bmatrix} \tag{5.60}
$$

with innovation processes

$$
\begin{aligned}
\nu_1 &= y_1 - C_{11}\hat{x}_1 - \epsilon C_{12}\hat{x}_2 \\
\nu_2 &= y_2 - \epsilon C_{21}\hat{x}_1 - C_{22}\hat{x}_2
\end{aligned} \tag{5.61}
$$

The nonsingular state transformation from Section 3.3 will block diagonalize (5.60) under condition that matrices $(A_{11} - B_{11}F_{11} - \epsilon^2 B_{12}F_{12})$ and $(A_{22} - B_{22}F_{22} - \epsilon^2 B_{21}F_{12})$ have no eigenvalues in common. This transformation is given by

$$
\begin{bmatrix} \hat{\eta}_1 \\ \hat{\eta}_2 \end{bmatrix} = \begin{bmatrix} I_{n_1} - \epsilon^2 LH & -\epsilon L \\ \epsilon H & I_{n_2} \end{bmatrix} \begin{bmatrix} \hat{x}_1 \\ \hat{x}_2 \end{bmatrix} = T_2^{-1} \begin{bmatrix} \hat{x}_1 \\ \hat{x}_2 \end{bmatrix} \tag{5.62}
$$

with

$$
T_2 = \begin{bmatrix} I_{n_1} & \epsilon L \\ -\epsilon H & I_{n_2} - \epsilon^2 HL \end{bmatrix} \tag{5.63}
$$

where matrices L and H satisfy equations given in Section 3.3. The optimal feedback control expressed in the new coordinates has the form

$$
u_1 = -f_{11}\hat{\eta}_1 - \epsilon f_{12}\hat{\eta}_2 \tag{5.64a}
$$
$$
u_2 = -\epsilon f_{21}\hat{\eta}_1 - f_{22}\hat{\eta}_2 \tag{5.64b}
$$

with

$$
\dot{\hat{\eta}}_1 = \alpha_1 \hat{\eta}_1 + \beta_{11}\nu_1 + \epsilon\beta_{12}\nu_2 \tag{5.65a}
$$
$$
\dot{\hat{\eta}}_2 = \alpha_2 \hat{\eta}_2 + \epsilon\beta_{21}\nu_1 + \beta_{22}\nu_2 \tag{5.65b}
$$

where

$$
\begin{aligned}
f_{11} &= F_{11} - \epsilon^2 F_{12}H, & f_{12} &= F_{12} + (F_{11} - \epsilon^2 F_{12}H)L \\
f_{21} &= F_{21} - F_{22}H, & f_{22} &= F_{22} + \epsilon^2 (F_{21} - F_{22}H)L \\
\alpha_1 &= a_{11} - \epsilon^2 a_{12}H, & \alpha_2 &= a_{22} + \epsilon^2 H a_{12} \\
\beta_{11} &= K_{11} - \epsilon^2 (LH + LK_{21}), & \beta_{12} &= K_{12} - LK_{22} - \epsilon^2 LHK_{12} \\
\beta_{21} &= HK_{11} + K_{21}, & \beta_{22} &= K_{22} + \epsilon^2 HK_{12}
\end{aligned}
$$

$$\tag{5.66}$$

and

$$a_{11} = A_{11} - B_{11}F_{11} - \epsilon^2 B_{12}F_{21}, \quad a_{12} = A_{12} - B_{11}F_{12} - B_{12}F_{22}$$
$$a_{21} = A_{21} - B_{21}F_{11} - B_{22}F_{21}, \quad a_{22} = A_{22} - B_{22}F_{22} - \epsilon^2 B_{21}F_{12}$$

The innovation processes ν_1 and ν_2 are now given by

$$\nu_1 = y_1 - d_{11}\hat{\eta}_1 - \epsilon d_{12}\hat{\eta}_2 \tag{5.67a}$$
$$\nu_2 = y_2 - \epsilon d_{21}\hat{\eta}_1 - d_{22}\hat{\eta}_2 \tag{5.67b}$$

where

$$d_{11} = C_{11} - \epsilon^2 C_{12}H, \quad d_{12} = C_{11}L + C_{12} - \epsilon^2 C_{12}HL$$
$$d_{21} = C_{21} - C_{22}H, \quad d_{22} = C_{22} + \epsilon^2 (C_{21} - C_{22}H)L$$

Approximate control laws are defined by perturbing coefficients F_{ij}, K_{ij}, $i,j = 1, 2$; L and H by $O\left(\epsilon^k\right)$, $k = 1, 2, ...$, in other words by using k-th approximations for these coefficients, where k stands for the required order of accuracy, that is,

$$u_1^{(k)} = -f_{11}^{(k)}\hat{\eta}_1^{(k)} - \epsilon f_{12}^{(k)}\hat{\eta}_2^{(k)} \tag{5.68a}$$
$$u_2^{(k)} = -\epsilon f_{21}^{(k)}\hat{\eta}_1^{(k)} - f_{22}^{(k)}\hat{\eta}_2^{(k)} \tag{5.68b}$$

with

$$\dot{\hat{\eta}}_1^{(k)} = \alpha_1^{(k)}\hat{\eta}_1^{(k)} + \beta_{11}^{(k)}\nu_1^{(k)} + \epsilon\beta_{12}^{(k)}\nu_2^{(k)} \tag{5.69a}$$
$$\dot{\hat{\eta}}_2^{(k)} = \alpha_2^{(k)}\hat{\eta}_2^{(k)} + \epsilon\beta_{21}^{(k)}\nu_1^{(k)} + \beta_{22}^{(k)}\nu_2^{(k)} \tag{5.69b}$$

where

$$\nu_1^{(k)} = y_1 - d_{11}^{(k)}\hat{\eta}_1^{(k)} - \epsilon d_{12}^{(k)}\hat{\eta}_2^{(k)} \tag{5.70a}$$
$$\nu_2^{(k)} = y_2 - \epsilon d_{21}^{(k)}\hat{\eta}_1^{(k)} - d_{22}^{(k)}\hat{\eta}_2^{(k)} \tag{5.70b}$$

and

$$f_{ij}^{(k)} = f_{ij} + O\left(\epsilon^k\right), \quad d_{ij}^{(k)} = d_{ij} + O\left(\epsilon^k\right)$$
$$\alpha_{ij}^{(k)} = \alpha_{ij} + O\left(\epsilon^k\right), \quad \beta_{ij}^{(k)} = \beta_{ij} + O\left(\epsilon^k\right) \tag{5.71}$$
$$i, j = 1, 2$$

The near-optimality of the proposed control law (5.68) is established in the following theorem.

Theorem 5.5 *Let x_1 and x_2 be the optimal trajectories and J be the optimal value of the performance criterion. Let $x_1^{(k)}$, $x_2^{(k)}$, and $J^{(k)}$ be the corresponding quantities under the approximate control law $u^{(k)}$, then*

$$J - J^{(k)} = O\left(\epsilon^k\right) \tag{5.72}$$

$$Var\left\{\left(x_i - x_i^{(k)}\right)\right\} = O\left(\epsilon^{2k}\right), \quad k = 0, 1, 2, \ldots \tag{5.73}$$

◇

Proof: The results of Theorems 5.3 and 5.4 are employed by studying a system of equations driven by white noise. For the truly optimal control consider the equations

$$\begin{bmatrix} \dot{\hat{\eta}}_1 \\ \dot{e}_1 \\ \dot{\hat{\eta}}_2 \\ \dot{e}_2 \end{bmatrix} = \begin{bmatrix} A_{11} & \epsilon A_{12} \\ \epsilon A_{21} & A_{22} \end{bmatrix} \begin{bmatrix} \hat{\eta}_1 \\ e_1 \\ \hat{\eta}_2 \\ e_2 \end{bmatrix} + \begin{bmatrix} \Theta_{11} & \epsilon\Theta_{12} \\ \epsilon\Theta_{21} & \Theta_{22} \end{bmatrix} \begin{bmatrix} v_1 \\ w_1 \\ v_2 \\ w_2 \end{bmatrix} \tag{5.74}$$

where $e_i = \eta_i - \hat{\eta}_i$, $i = 1, 2$, are estimation errors. The corresponding equation for the approximate control is

$$\begin{bmatrix} \dot{\hat{\eta}}_1^N \\ \dot{e}_1^N \\ \dot{\hat{\eta}}_2^N \\ \dot{e}_2^N \end{bmatrix} = \begin{bmatrix} A_{11}^N & \epsilon A_{12}^N \\ \epsilon A_{21}^N & A_{22}^N \end{bmatrix} \begin{bmatrix} \hat{\eta}_1^N \\ e_1^N \\ \hat{\eta}_2^N \\ e_2^N \end{bmatrix} + \begin{bmatrix} \Theta_{11}^N & \epsilon\Theta_{12}^N \\ \epsilon\Theta_{21}^N & \Theta_{22}^N \end{bmatrix} \begin{bmatrix} v_1 \\ w_1 \\ v_2 \\ w_2 \end{bmatrix} \tag{5.75}$$

where $e_i^N = \eta_i - \hat{\eta}_i^N$ are corresponding estimation errors. The matrices A_{ij}, Θ_{ij} and A_{ij}^N, Θ_{ij}^N in (5.74) and (5.75) are obtained in an obvious way. It can be verified that

$$A_{ij} - A_{ij}^N = O\left(\epsilon^N\right), \quad \Theta_{ij} - \Theta_{ij}^N = O\left(\epsilon^N\right), \quad i, j = 1, 2$$

and $A_{ii}(0)$, $i = 1, 2$, are given by

$$A_{ii}(0) = \begin{bmatrix} A_{ii} - B_{ii}F_{ii} & K_{ii}C_{ii} \\ 0 & A_{ii} - K_{ii}C_{ii} \end{bmatrix} \tag{5.76}$$

which by stabilizability-detectability assumptions imposed on the triples (A_{ii}, B_{ii}, D_i) and (A_{ii}, C_{ii}, G_{ii}), $i = 1, 2$, guarantees the stability of matrices $A_{ii}(0)$. The results of Theorems 5.3 and 5.4 can be now directly used to establish (5.72) and (5.73).

Results obtained in in this section are along the lines of those obtained in (Kokotovic and Cruz, 1969). It is shown in (Kokotovic and Cruz, 1969) that an $O\left(\epsilon^{N}\right)$ approximation of coefficients for a deterministic linear-quadratic regulator implies the $O\left(\epsilon^{2N}\right)$ approximation of the corresponding performance criterion. The same problem for the singularly perturbed linear-quadratic stochastic regulator produces the relative error of $O\left(\epsilon^{N}\right)$ — (Theorem 5.1). In this section, we show that for the weakly coupled LQG problem an $O\left(\epsilon^{N}\right)$ approximation of coefficients implies the absolute error of the performance criterion of $O\left(\epsilon^{N}\right)$ — (Theorem 5.4).

5.4 Case Study: Electric Power System

In order to demonstrate the numerical behavior of the near-optimum design of weakly coupled LQG regulator, we present results for an LQG controller of a power system composed of two interconnected areas (Geromel and Peres, 1985). The system model is given by

$$
A = \begin{bmatrix}
0 & 0.55 & 0.0 & 0.0 & 0.0 & -0.55 & 0.0 & 0.0 & 0.0 \\
0 & 0.0 & 1.0 & 0.0 & 0.0 & 0.0 & 0.0 & 0.0 & 0.0 \\
0 & -3.3 & -0.05 & 6.0 & 0.0 & 3.3 & 0.0 & 0.0 & 0.0 \\
0 & 0.0 & 0.0 & -3.3 & 3.3 & 0.0 & 0.0 & 0.0 & 0.0 \\
0 & 0.0 & -5.2 & 0.0 & -13 & 0.0 & 0.0 & 0.0 & 0.0 \\
0 & 0.0 & 0.0 & 0.0 & 0.0 & 0.0 & 1.0 & 0.0 & 0.0 \\
0 & 3.3 & 0.0 & 0.0 & 0.0 & -3.3 & -0.05 & 6.0 & 0.0 \\
0 & 0.0 & 0.0 & 0.0 & 0.0 & 0.0 & 0.0 & -3.3 & 3.3 \\
0 & 0.0 & 0.0 & 0.0 & 0.0 & 0.0 & -5.2 & 0.0 & -13
\end{bmatrix}
$$

$$
B = \begin{bmatrix}
0 & 0 & 0 & 0 & 13 & 0 & 0 & 0 & 0 \\
0 & 0 & 0 & 0 & 0 & 0 & 0 & 0 & 13
\end{bmatrix}
$$

$$
C = \begin{bmatrix}
1 & 0.43 & 0 & 0 & 0 & 0.0 & 0 & 0 & 0 \\
0 & 0.0 & 0 & 1 & 0 & 0.0 & 0 & 0 & 0 \\
-1 & 0.0 & 0 & 0 & 0 & 0.43 & 0 & 0 & 0 \\
0 & 0.0 & 0 & 0 & 0 & 0.0 & 0 & 1 & 0
\end{bmatrix}
$$

$$D^T D = \begin{bmatrix} 1.0 & 0.0 & 0.0 & 0.0 & 0.0 & 0.0 & 0.0 & 0.0 & 0.0 \\ 0.0 & 1.3 & 0.0 & 0.0 & 0.0 & -0.3 & 0.0 & 0.0 & 0.0 \\ 0.0 & 0.0 & 1.0 & 0.0 & 0.0 & 0.0 & 0.0 & 0.0 & 0.0 \\ 0.0 & 0.0 & 0.0 & 0.0 & 0.0 & 0.0 & 0.0 & 0.0 & 0.0 \\ 0.0 & 0.0 & 0.0 & 0.0 & 0.0 & 0.0 & 0.0 & 0.0 & 0.0 \\ 0.0 & -0.3 & 0.0 & 0.0 & 0.0 & 1.3 & 0.0 & 0.0 & 0.0 \\ 0.0 & 0.0 & 0.0 & 0.0 & 0.0 & 0.0 & 1.0 & 0.0 & 0.0 \\ 0.0 & 0.0 & 0.0 & 0.0 & 0.0 & 0.0 & 0.0 & 0.0 & 0.0 \\ 0.0 & 0.0 & 0.0 & 0.0 & 0.0 & 0.0 & 0.0 & 0.0 & 0.0 \end{bmatrix}$$

$$R = \begin{bmatrix} 1.0 & 0.0 \\ 0.0 & 1.0 \end{bmatrix}$$

It is assumed that $G = B$, and that the noise intensity matrices are given by

$$W_1 = 0.1, \quad W_2 = 0.1, \quad V_1 = I_2, \quad V_2 = I_2$$

We can note relatively big elements in the cross coupling matrices A_{12}, A_{21}, and C_{21}. The small parameter ϵ is built in the problem. The value for ϵ should be estimated from the problem strongest coupled matrix — in this case matrix C. It seems from our experience that the formula

$$\epsilon = \frac{max \left(\| C_{12} \|, \| C_{21} \| \right)}{max \left(\| C_{11} \|, \| C_{22} \| \right)} = \frac{1}{1.43} = 0.699 \qquad (5.77)$$

produces quite good estimate for ϵ, where $\| \; \|$ is any suitable norm. In this example we have used the infinity norm.

It is important to notice that there is no known method in the literature which produces an upper bound for the small parameter ϵ. This is true for the entire theory of small parameters (weak coupling and singular perturbations). It happens that in this particular example, despite the relatively large value for the small parameter ϵ, the proposed method converges, since the radius of convergence of all algorithms used is less than one at each iteration. The simulation results are presented in the following table.

The small parameter ϵ is relatively big in this example, that is $\epsilon =$ 0.7. Since $O\left(0.7^{26}\right) \approx 10^{-4}$, it will require 24 terms in order to get the accuracy of 10^{-4} if the power-series expansion method is used — which is not feasible. On the other hand, the fixed pointed method scheme used in this section will demand 12 iterations (rate of convergence is $O\left(\epsilon^2\right)$)

k	$J^{(k)}$	$J^{(k)} - J$	$(0.7)^k$
2	∞	∞	*
4	∞	∞	*
6	5.9415	0.9645	0.11765
10	5.1111	0.1341	0.02825
18	4.9788	0.0018	0.00163
26	4.9770	$< 10^{-4}$	9.4×10^{-5}
Optimal	4.9770	*	*

Table 5.2: Approximate values for criterion

of the presented algorithms — which can be easily achieved. Even more, it happens in this problem that the $O\left(\epsilon^2\right)$ and $O\left(\epsilon^4\right)$ approximate filters do not stabilize the plant-filter augmented system, and the approximate filter has to be found with the accuracy of at least $O\left(\epsilon^6\right)$.

Table 5.2 verifies the result of Theorem 5.5, namely $J - J^{(k)} = O\left(\epsilon^k\right)$, and support the formula (5.77) for the estimate of the weak coupling parameter.

The modeling issue for the megawatt-frequency control problem of multiarea electric energy systems was considered in (Elgerd and Fosha, 1970; Fosha and Elgerd, 1970). A summary of their results can be found in (Gajic et al., 1990). The same model was used in (Geromel and Peres, 1985) for decentralized load-frequency control.

5.5 Parallel Reduced-Order Controller for Stochastic Linear Discrete Singularly Perturbed Systems

The continuous-time LQG problem of singularly perturbed systems is solved in (Khalil and Gajic, 1984) by using the power-series expansion approach, and later on in (Gajic, 1986) by using the fixed point theory. In this section, we will solve the discrete-time LQG problem of singularly perturbed system by using the results obtained in Sections 2.5–2.6 and Section 3.2.

This section presents the approach to the decomposition and approximation of the linear-quadratic Gaussian control problem of singularly

perturbed discrete systems by treating the decomposition and approximation tasks separately from each other. The decoupling transformation of
(Chang, 1972) is used for the exact block diagonalization of the global
Kalman filter. The approximate feedback control law is then obtained
by approximating the coefficients of the optimal regulator and the optimal local filters with the accuracy of $O\left(\epsilon^N\right)$. The resulting feedback
control law is shown to be a near-optimal solution of the LQG by studying the corresponding closed-loop system as a system driven by white
noise. It is shown that the order of approximation of the optimal system
trajectories is $O\left(\epsilon^{N+1/2}\right)$ in the case of slow variables and $O\left(\epsilon^N\right)$ in
the case of fast variables. All required coefficients of desired accuracy
are easily obtained by using the recursive reduced-order fixed point type
numerical techniques developed in Chapter 2 and Section 5.1. Obtained
numerical algorithms converge to the required optimal coefficients with
the rate of convergence of $O\left(\epsilon\right)$. In addition, only low-order subsystems
are involved in the algebraic computations and no analyticity requirements are imposed on the system coefficients — which is the standard
assumption in the power-series expansion method. As a consequence of
these, under very mild conditions (coefficients are bounded functions of a
small perturbation parameter), in addition to the standard stabilizability-
detectability subsystem assumptions, we have achieved the reduction in
both off-line and on-line computational requirements.

The results presented in this section are mostly based on the doctoral
dissertation (Shen, 1990) and the recent research papers of (Gajic and
Shen, 1991a, 1991b).

Consider the discrete linear singularly perturbed stochastic system
represented in the fast time scale by (this structure is justified in Appendix
5.1)

$$\begin{aligned}
x_1\left(n+1\right) &= \left(I_{n_1} + \epsilon A_{11}\right)x_1\left(n\right) + \epsilon A_{12}x_2\left(n\right) \\
&\quad + \epsilon B_1 u\left(n\right) + \epsilon G_1 w\left(n\right) \\
x_2\left(n+1\right) &= A_{21}x_1\left(n\right) + A_{22}x_2\left(n\right) + B_2 u\left(n\right) + G_2 w\left(n\right)
\end{aligned} \qquad (5.78)$$

$$y\left(n\right) = C_1 x_1\left(n\right) + C_2 x_2\left(n\right) + v\left(n\right)$$

with the performance criterion

$$J = \frac{1}{2}E\left\{\sum_{n=0}^{\infty}\left[z^T\left(n\right)z\left(n\right) + u^T\left(n\right)Ru\left(n\right)\right]\right\}, \quad R > 0 \qquad (5.79)$$

where $x_i \in \Re^{n_i}$, $i = 1$, 2, comprise slow and fast state vectors, respectively, $u \in \Re^m$ is the control input, $y \in \Re^l$ is the observed output,

$w \in \Re^r$ and $v \in \Re^l$ are independent zero-mean stationary Gaussian mutually uncorrelated white noise processes with intensities $W > 0$ and $V > 0$, respectively, and $z \in \Re^s$ is the controlled output given by

$$z(n) = D_1 x_1(n) + D_2 x_2(n) \tag{5.80}$$

All matrices are bounded functions of a small positive parameter ϵ (Gajic et al., 1990) having appropriate dimensions.

The optimal control law is given by (Kwakernaak and Sivan, 1972)

$$u(n) = -F\hat{x}(n) \tag{5.81}$$

with

$$\hat{x}(n+1) = A\hat{x}(n) + Bu(n) + K[y(n) - C\hat{x}(n)] \tag{5.82}$$

where

$$A = \begin{bmatrix} I_{n_1} + \epsilon A_{11} & \epsilon A_{12} \\ A_{21} & A_{22} \end{bmatrix}, \quad B = \begin{bmatrix} \epsilon B_1 \\ B_2 \end{bmatrix}, \quad C = [C_1 \quad C_2]$$
$$K = \begin{bmatrix} \epsilon K_1 \\ K_2 \end{bmatrix}, \quad F = [F_1 \quad F_2] \tag{5.83}$$

The regulator gain F and filter gain K are obtained from

$$F = (R + B^T P B)^{-1} B^T P A \tag{5.84}$$

$$K = A Q C^T (V + C Q C^T)^{-1} \tag{5.85}$$

where P and Q are positive semidefinite stabilizing solutions of the discrete-time algebraic regulator and filter Riccati equations, respectively given by

$$P = D^T D + A^T P A - A^T P B (R + B^T P B)^{-1} B^T P A \tag{5.86}$$

$$Q = A Q A^T - A Q C^T (V + C Q C^T)^{-1} C Q A^T + G W G^T \tag{5.87}$$

where

$$D = [D_1 \quad D_2], \quad G = \begin{bmatrix} \epsilon G_1 \\ G_2 \end{bmatrix} \tag{5.88}$$

Due to the singularly perturbed structure of the problem matrices the required solutions P and Q in the fast time scale version have the forms

$$P = \begin{bmatrix} P_{11}/\epsilon & P_{12} \\ P_{12}^T & P_{22} \end{bmatrix}, \quad Q = \begin{bmatrix} \epsilon Q_{11} & \epsilon Q_{12} \\ \epsilon Q_{12}^T & Q_{22} \end{bmatrix} \qquad (5.89)$$

In order to obtain required solutions of (5.86)-(5.87) in terms of the reduced-order problems and overcome the complicated partitioned form of the discrete-time algebraic Riccati equation, we have used the method developed in Section 2.5 (based on a bilinear transformation), to transform the discrete-time algebraic Riccati equations (5.86)-(5.87) into continuous-time algebraic Riccati equations of the forms

$$A_R^T \mathbf{P} + \mathbf{P} A_R - \mathbf{P} S_R \mathbf{P} + D_R^T D_R = 0, \quad S_R = B_R R_R^{-1} B_R^T \qquad (5.90)$$

$$A_F \mathbf{Q} + \mathbf{Q} A_F^T - \mathbf{Q} S_F \mathbf{Q} + G_F W_F G_F^T = 0, \quad S_F = C_F^T V_F^{-1} C_F \qquad (5.91)$$

such that the solutions of (5.86)-(5.87) are equal to the solutions of (5.90)-(5.91), that is

$$P = \mathbf{P}, \quad Q = \mathbf{Q} \qquad (5.92)$$

where

$$\begin{aligned} A_R &= I - 2\left(\Delta_R^{-1}\right)^T \\ B_R R_R^{-1} B_R^T &= 2\left(I + A\right)^{-1} B R^{-1} B^T \Delta_R^{-1} \\ D_R^T D_R &= 2\Delta_R^{-1} D^T D \left(I + A\right)^{-1} \\ \Delta_R &= \left(I + A^T\right) + D^T D \left(I + A\right)^{-1} B R^{-1} B^T \end{aligned} \qquad (5.93)$$

and

$$\begin{aligned} A_F &= I - 2\left(\Delta_F^{-1}\right) \\ C_F^T V_F^{-1} C_F &= 2\left(I + A^T\right)^{-1} C^T V^{-1} C \Delta_F^{-1} \\ G_F W_F G_F^T &= 2\Delta_F^{-1} G W G^T \left(I + A^T\right)^{-1} \\ \Delta_F &= \left(I + A\right) + G W G^T \left(I + A^T\right)^{-1} C^T V^{-1} C \end{aligned} \qquad (5.94)$$

It is shown in Section 2.5 that the equations (5.90)-(5.91) preserve the structure of singularly perturbed systems. These equations can be solved in terms of the reduced-order problems very efficiently by using the recursive method developed in Chapter 2, which converges with the rate of convergence of $O(\epsilon)$ under the following assumption.

LINEAR STOCHASTIC SYSTEMS

Assumption 5.3 The matrix A_{22} has no eigenvalues located at -1.

$\triangle$

Under this assumption matrices $\triangle_R$ and $\triangle_F$ are invertible.

Solutions of (5.90) and (5.91) are found in terms of the reduced-order problems by imposing standard stabilizability-detectability assumptions on subsystems.

Getting approximate solutions for P and Q in terms of reduced-order problems will produce savings in off-line computations. However, in the case of stochastic systems, where an additional dynamical system — filter — has to be built, one is particularly interested in the reduction of on-line computations. In this section, the savings of on-line computation will be achieved by using a decoupling transformation introduced in (Chang, 1972). The Kalman filter (5.82) is viewed as a system driven by the innovation process (Khalil and Gajic, 1984). However, one might study the filter form when it is driven by both measurements and control. The filter form under consideration is obtained from (5.82) as

$$\hat{x}_1(n+1) = (I_{n_1} + \epsilon A_{11} - \epsilon B_1 F_1)\hat{x}_1(n)$$
$$+\epsilon(A_{12} - B_1 F_2)\hat{x}_2(n) + \epsilon K_1 v(n)$$
$$\hat{x}_2(n+1) = (A_{21} - B_2 F_1)\hat{x}_1(n) + (A_{22} - B_2 F_2)\hat{x}_2(n) + K_2 v(n)$$
$$(5.95)$$

with the innovation process

$$v(n) = y(n) - C_1\hat{x}_1(n) - C_2\hat{x}_2(n) \tag{5.96}$$

The nonsingular state transformation of (Chang, 1972) will block diagonalize (5.95). The transformation is given by (see Chapter 3)

$$\begin{bmatrix} \hat{\eta}_1(n) \\ \hat{\eta}_2(n) \end{bmatrix} = \begin{bmatrix} I_{n_1} - \epsilon HL & -\epsilon H \\ L & I_{n_2} \end{bmatrix} \begin{bmatrix} \hat{x}_1(n) \\ \hat{x}_2(n) \end{bmatrix} = T_1 \begin{bmatrix} \hat{x}_1(n) \\ \hat{x}_2(n) \end{bmatrix} \tag{5.97}$$

with

$$T_1^{-1} = \begin{bmatrix} I_{n_1} & \epsilon H \\ -L & I_{n_2} - \epsilon LH \end{bmatrix} \tag{5.98}$$

where matrices L and H satisfy equations

$$\epsilon L a_{11} + (I - a_{22})L + a_{21} - \epsilon L a_{12}L = 0 \tag{5.99}$$

$$H(I - a_{22} - \epsilon L a_{12}) + \epsilon(a_{11} - a_{12}L)H + a_{12} = 0 \tag{5.100}$$

140

with

$$a_{11} = A_{11} - B_1 F_1, \quad a_{12} = A_{12} - B_1 F_2$$
$$a_{21} = A_{21} - B_2 F_1, \quad a_{22} = A_{22} - B_2 F_2$$

(5.101)

The optimal feedback control, expressed in the new coordinates, has the form

$$u(n) = -f_1 \hat{\eta}_1(n) - f_2 \hat{\eta}_2(n)$$

(5.102)

with

$$\hat{\eta}_1(n+1) = \alpha_1 \hat{\eta}_1(n) + \epsilon \beta_1 v(n)$$
$$\hat{\eta}_2(n+1) = \alpha_2 \hat{\eta}_2(n) + \beta_2 v(n)$$

(5.103)

where

$$f_1 = F_1 - F_2 L, \quad f_2 = F_2 + \epsilon (F_1 - F_2 L) H$$
$$\alpha_1 = I_{n_1} + \epsilon (a_{11} - a_{12} L), \quad \alpha_2 = a_{22} + \epsilon L a_{12}$$
$$\beta_1 = K_1 - H (K_2 + \epsilon L K_1), \quad \beta_2 = K_2 + \epsilon L K_1$$

(5.104)

The innovation process $v(n)$ is now given by

$$v(n) = y(n) - d_1 \hat{\eta}_1(n) - d_2 \hat{\eta}_2(n)$$

(5.105)

where

$$d_1 = C_1 - \epsilon C_2 L, \quad d_2 = C_2 + \epsilon (C_1 - C_2 L) H$$

(5.106)

Near-optimum control law is defined by perturbing coefficients F_i, K_i, $i, j = 1, 2$, L and H by $O(\epsilon^k)$, $k = 1, 2, ...,$ in other words by using k-th approximations for these coefficients, where k stands for the required order of accuracy, that is

$$u^{(k)}(n) = -f_1^{(k)} \hat{\eta}_1^{(k)}(n) - f_2^{(k)} \hat{\eta}_2^{(k)}(n)$$

(5.107)

with

$$\hat{\eta}_1^{(k)}(n+1) = \alpha_1^{(k)} \hat{\eta}_1^{(k)}(n) + \epsilon \beta_1^{(k)} v^{(k)}(n)$$
$$\hat{\eta}_2^{(k)}(n+1) = \alpha_2^{(k)} \hat{\eta}_2^{(k)}(n) + \beta_2^{(k)} v^{(k)}(n)$$

(5.108)

where

$$v^{(k)}(n) = y(n) - d_1^{(k)} \hat{\eta}_1^{(k)}(n) - d_2^{(k)} \hat{\eta}_2^{(k)}(n)$$

(5.109)

LINEAR STOCHASTIC SYSTEMS

and

$$f_i^{(k)} = f_i + O\left(\epsilon^k\right), \quad d_i^{(k)} = d_i + O\left(\epsilon^k\right)$$
$$\alpha_i^{(k)} = \alpha_i + O\left(\epsilon^k\right), \quad \beta_i^{(k)} = \beta_i + O\left(\epsilon^k\right) \tag{5.110}$$
$$i = 1, 2$$

The approximate values of $J^{(k)}$ are obtained from the following expression

$$J^{(k)} = \frac{1}{2}E\left\{\sum_{n=0}^{\infty}\left[x^{(k)^T}(n)D^TDx^{(k)}(n) + u^{(k)^T}(n)Ru^{(k)}(n)\right]\right\}$$
$$= \frac{1}{2}tr\left\{D^TDq_{11}^{(k)} + f^{(k)^T}Rf^{(k)}q_{22}^{(k)}\right\} \tag{5.111}$$

where

$$q_{11}^{(k)} = Var\left\{\left(x_1^{(k)} \ x_2^{(k)}\right)^T\right\}, \quad q_{22}^{(k)} = Var\left\{\left(\hat{\eta}_1^{(k)} \ \hat{\eta}_2^{(k)}\right)^T\right\}$$
$$f^{(k)} = \left[f_1^{(k)} \ f_2^{(k)}\right] \tag{5.112}$$

Quantities $q_{11}^{(k)}$ and $q_{22}^{(k)}$ can be obtained by studying the variance equation of the following system driven by white noise

$$\begin{bmatrix} x^{(k)}(n+1) \\ \hat{\eta}^{(k)}(n+1) \end{bmatrix} = \begin{bmatrix} A & -Bf^{(k)} \\ \beta^{(k)}C & \alpha^{(k)} - \beta^{(k)}d^{(k)} \end{bmatrix}\begin{bmatrix} x^{(k)}(n) \\ \hat{\eta}^{(k)}(n) \end{bmatrix}$$
$$+ \begin{bmatrix} G & 0 \\ 0 & \beta^{(k)} \end{bmatrix}\begin{bmatrix} w(n) \\ v(n) \end{bmatrix} \tag{5.113}$$

where

$$\alpha^{(k)} = \begin{bmatrix} \alpha_1^{(k)} & 0 \\ 0 & \alpha_2^{(k)} \end{bmatrix}, \quad \beta^{(k)} = \begin{bmatrix} \epsilon\beta_1^{(k)} \\ \beta_2^{(k)} \end{bmatrix}, \quad d^{(k)} = \begin{bmatrix} d_1^{(k)} & d_2^{(k)} \end{bmatrix} \tag{5.114}$$

Equation (5.113) can be represented in a composite form

$$\Gamma^{(k)}(n+1) = \Lambda^{(k)}\Gamma^{(k)}(n) + \Pi^{(k)}w(n) \tag{5.115}$$

with obvious definitions for $\Lambda^{(k)}$, $\Pi^{(k)}$, $\Gamma^{(k)}(n)$, and $w(n)$. The variance of $\Gamma^{(k)}(n)$ at steady state denoted by $q^{(k)}$, is given by the discrete algebraic Lyapunov equation (Kwakernaak and Sivan, 1972)

$$q^{(k)}(n+1) = \Lambda^{(k)}q^{(k)}\Lambda^{(k)^T} + \Pi^{(k)}\overline{W}\Pi^{(k)^T}, \quad \overline{W} = diag(W, V) \tag{5.116}$$

142

with $q^{(k)}$ partitioned as

$$q^{(k)} = \begin{bmatrix} q_{11}^{(k)} & q_{12}^{(k)} \\ q_{12}^{(k)^T} & q_{22}^{(k)} \end{bmatrix} \tag{5.117}$$

On the other hand, the optimal value of J has the very well-known form, (Kwakernaak and Sivan, 1972)

$$J^{opt} = \frac{1}{2} tr \left[D^T D \mathbf{Q} + \mathbf{P} K \left(C \mathbf{Q} C^T + V \right) K^T \right] \tag{5.118}$$

where $\mathbf{P}$, $\mathbf{Q}$, F, and K are obtained from (5.84)-(5.87).

The near-optimality of the proposed approximate control law (5.107) is established in the following theorem.

Theorem 5.6 *Let x_1 and x_2 be optimal trajectories and J be the optimal value of the performance criterion. Let $x_1^{(k)}$, $x_2^{(k)}$, and $J^{(k)}$ be corresponding quantities under the approximate control law $u^{(k)}$ given by (5.107). Under the condition stated in Assumption 5.3 and the stabilizability-detectability subsystem assumptions, the following hold*

$$J^{opt} - J^{(k)} = O\left(\epsilon^k\right)$$
$$Var\left\{x_1 - x_1^{(k)}\right\} = O\left(\epsilon^{2k+1}\right) \tag{5.119}$$
$$Var\left\{x_2 - x_2^{(k)}\right\} = O\left(\epsilon^{2k}\right) \qquad k = 0, 1, 2, \dots.$$

◇

The proof of this theorem is rather lengthly and is omitted. It follows the ideas of Theorems 1 and 2 from (Khalil and Gajic, 1984). In addition, due to the discrete nature of the problem, the proof of Theorem 5.6 utilizes the bilinear transformation from (Power, 1967) which transforms the discrete Lyapunov equation (5.116) into the continuous one and compares it with the corresponding equation under the optimal control law. More about the proof can be found in (Shen, 1990).

Exercise 5.3: Find the values of the optimal and approximate criterion (of the LQG problem studied in Section 5.5) in the new coordinates, namely, in terms of $Var\left(\eta_1, \eta_2, \hat{\eta}_1, \hat{\eta}_2\right)$ and $Var\left(\eta_1^{(k)}, \eta_2^{(k)}, \hat{\eta}_1^{(k)}, \hat{\eta}_2^{(k)}\right)$, respectively.

△

5.6 Case Study: Discrete Steam Power System

A real world physical example, a fifth-order discrete model of a steam power system (Mahmoud, 1982) demonstrates the efficiency of the proposed method. The problem matrices A and B are given by

$$A = \begin{bmatrix} 0.9150 & 0.0510 & 0.0380 & 0.015 & 0.038 \\ -0.030 & 0.889 & -0.0005 & 0.046 & 0.111 \\ -0.006 & 0.468 & 0.247 & 0.014 & 0.048 \\ -0.715 & -0.022 & -0.0211 & 0.240 & -0.024 \\ -0.148 & -0.003 & -0.004 & 0.090 & 0.026 \end{bmatrix}$$

$$B^T = [0.0098 \quad 0.122 \quad 0.036 \quad 0.562 \quad 0.115]$$

Remaining matrices are chosen as

$$C = \begin{bmatrix} 1 & 1 & 0 & 0 & 0 \\ 0 & 0 & 1 & 1 & 1 \end{bmatrix}, \quad D^T D = diag\{5 \quad 5 \quad 5 \quad 5 \quad 5\}, \quad R = 1$$

It is assumed that $G = B$ and that white noise intensity matrices are given by

$$W = 5, \quad V_1 = 5, \quad V_2 = 5$$

It is shown (Mahmoud, 1982) that this model possesses the singularly perturbed property with $n_1 = 2$, $n_2 = 3$, and $\epsilon = 0.264$.

The simulation results are presented in the following table.

k	$J^{(k)}$	$J^{(k)} - J$
0	13.4918	0.229×10^{-1}
1	13.4825	0.136×10^{-1}
2	13.4700	0.110×10^{-2}
3	13.4695	0.600×10^{-3}
4	13.4690	1.000×10^{-4}
5	13.4689	$< 10^{-4}$
optimal	13.4689	

Table 5.3: Approximate values for the criterion

It can be seen from this table that the approximate solution has quite rapid convergence to the optimal solution. This table justifies the result of Theorem 5.6, that $J^{(k)} - J^{opt} = O(\epsilon)$. Notice that $(0.246)^6 = 3 \times 10^{-4}$.

5.7 Linear-Quadratic Gaussian Control of Discrete Weakly Coupled Systems at Steady State

In this section, we study the linear-quadratic Gaussian control problem of weakly coupled discrete-time systems. The partitioned form of the main equation of the optimal linear control theory — the Riccati equation, has a very complicated form in the discrete-time domain. In Chapter 2, that problem is overcome by using a bilinear transformation which is applicable under quite mild assumption, so that the reduced-order solution of the discrete algebraic Riccati equation of weakly coupled systems can be obtained up to any order of accuracy, by using known reduced-order results for the corresponding continuous-time algebraic Riccati equation.

Although the duality of the filter and regulator Riccati equations can be used together with results reported in (Shen and Gajic, 1990b) to obtain corresponding approximations to the filter and regulator gains, such approximations will not be sufficient because they only reduce the off-line computations of implementing the Kalman filter which will be of the same order as the overall weakly coupled system. The weakly coupled structure of the global Kalman filter is exploited in this section such that it may be replaced by two lower order local filters. This has been achieved via the use of a decoupling transformation introduced in (Gajic and Shen, 1989).

The decoupling transformation of (Gajic and Shen, 1989) is used for the exact block diagonalization of the global Kalman filter. The approximate feedback control law is then obtained by approximating the coefficients of the optimal local filters and regulators with the accuracy of $O(\epsilon^N)$. The resulting feedback control law is shown to be a near-optimal solution of the LQG by studying the corresponding closed-loop system as a system driven by white noise. It is shown that the order of approximation of the optimal performance is $O(\epsilon^N)$, and the order of approximation of the optimal system trajectories is $O(\epsilon^{2N})$. All required coefficients of desired accuracy are easily obtained by using the recursive reduced-order fixed point type numerical techniques developed in Chapter 2. The obtained numerical algorithms converge to the required optimal coefficients with the rate of convergence of

$O\left(\epsilon^2\right)$. In addition, only low-order subsystems are involved in the algebraic computations and no analyticity requirements are imposed on the system coefficients — which is the standard assumption in the power-series expansion method. As a consequence of these properties, under very mild conditions (coefficients are bounded functions of a small coupling parameter), in addition to the standard stabilizability-detectability subsystem assumptions, we have achieved the reduction in both off-line and on-line computational requirements.

The results presented in this section are mostly based on the doctoral dissertation (Shen, 1990) and on the recent research papers (Shen and Gajic, 1990b, 1990c).

Consider the linear discrete weakly coupled stochastic system

$$x_1\left(n+1\right) = A_{11}x_1\left(n\right) + \epsilon A_{12}x_2\left(n\right) + B_{11}u_1\left(n\right) + \epsilon B_{12}u_2\left(n\right)$$
$$+G_{11}w_1\left(n\right) + \epsilon G_{12}w_2\left(n\right)$$
$$x_2\left(n+1\right) = \epsilon A_{21}x_1\left(n\right) + A_{22}x_2\left(n\right) + \epsilon B_{21}u_1\left(n\right) + B_{22}u_2\left(n\right)$$
$$+\epsilon G_{21}w_1\left(n\right) + G_{22}w_2\left(n\right)$$

$$(5.120)$$

$$y_1\left(n\right) = C_{11}x_1\left(n\right) + \epsilon C_{12}x_2\left(n\right) + v_1\left(n\right)$$
$$y_2\left(n\right) = \epsilon C_{21}x_1\left(n\right) + C_{22}x_2\left(n\right) + v_2\left(n\right)$$

with the performance criterion

$$J = \frac{1}{2}E\left\{\sum_{n=0}^{\infty}\left[z^T\left(n\right)z\left(n\right) + u_1^T\left(n\right)R_1u_1\left(n\right) + u_2^T\left(n\right)R_2u_2\left(n\right)\right]\right\}$$

$$(5.121)$$

where $x_i \in \Re^{n_i}$, $i = 1, 2$, comprise state vectors, $u_i \in \Re^{m_i}$, $i = 1, 2$, are control inputs, $y_i \in \Re^{l_i}$, $i = 1, 2$, are observed outputs, $w_i \in \Re^{r_i}$ and $v_i \in \Re^{l_i}$ are independent zero-mean stationary Gaussian mutually uncorrelated white noise processes with intensities $W_i > 0$ and $V_i > 0$, respectively, and $z_i \in \Re^{s_i}$, $i = 1, 2$, are the controlled outputs given by

$$z_1\left(n\right) = D_{11}x_1\left(n\right) + \epsilon D_{12}x_2\left(n\right)$$
$$z_2\left(n\right) = \epsilon D_{21}x_1\left(n\right) + D_{22}x_2\left(n\right)$$

$$(5.122)$$

All matrices are bounded functions of a small coupling parameter ϵ and have appropriate dimensions. In addition, it is assumed that R_i, $i = 1, 2$, are positive definite matrices.

The optimal control law is given by (Kwakernaak and Sivan, 1972)

$$u\left(n\right) = -F\hat{x}\left(n\right) \qquad (5.123)$$

with

$$\hat{x}(n+1) = A\hat{x}(n) + Bu(n) + K\left[y(n) - C\hat{x}(n)\right] \qquad (5.124)$$

where

$$A = \begin{bmatrix} A_{11} & \epsilon A_{12} \\ \epsilon A_{21} & A_{22} \end{bmatrix}, \quad B = \begin{bmatrix} B_{11} & \epsilon B_{12} \\ \epsilon B_{21} & B_{22} \end{bmatrix}, \quad C = \begin{bmatrix} C_{11} & \epsilon C_{12} \\ \epsilon C_{21} & C_{22} \end{bmatrix}$$

$$K = \begin{bmatrix} K_{11} & \epsilon K_{12} \\ \epsilon K_{21} & K_{22} \end{bmatrix}, \quad F = \begin{bmatrix} F_{11} & \epsilon F_{12} \\ \epsilon F_{21} & F_{22} \end{bmatrix}$$

$$(5.125)$$

The regulator gain F and filter gain K are obtained from

$$F = \left(R + B^T P B\right)^{-1} B^T P A \qquad (5.126)$$

$$K = AQC^T \left(V + CQC^T\right)^{-1} \qquad (5.127)$$

where P and Q are positive semidefinite stabilizing solutions of the discrete-time algebraic regulator and filter Riccati equations, respectively, given by

$$P = D^T D + A^T P A - A^T P B \left(R + B^T P B\right)^{-1} B^T P A \qquad (5.128)$$

$$Q = AQA^T - AQC^T \left(V + CQC^T\right)^{-1} CQA^T + GWG^T \qquad (5.129)$$

with

$$R = diag\left(R_1 \ R_2\right), \quad W = diag\left(W_1 \ W_2\right), \quad V = diag\left(V_1 \ V_2\right)$$

$$(5.130)$$

and

$$D = \begin{bmatrix} D_{11} & \epsilon D_{12} \\ \epsilon D_{21} & D_{22} \end{bmatrix}, \quad G = \begin{bmatrix} G_{11} & \epsilon G_{12} \\ \epsilon G_{21} & G_{22} \end{bmatrix} \qquad (5.131)$$

Due to the block dominant structure of the problem matrices the required solutions P and Q have the form

$$P = \begin{bmatrix} P_{11} & \epsilon P_{12} \\ \epsilon P_{12}^T & P_{22} \end{bmatrix}, \quad Q = \begin{bmatrix} Q_{11} & \epsilon Q_{12} \\ \epsilon Q_{12}^T & Q_{22} \end{bmatrix} \qquad (5.132)$$

In order to obtain the required solutions of (5.128) and (5.129) in terms of the reduced-order problems and to overcome the complicated partitioned form of the discrete-time algebraic Riccati equation, we have used the method developed in the previous section, to transform the discrete-time algebraic Riccati equations (5.128) and (5.129) into the continuous-time algebraic Riccati equations of the form

$$A_R^T \mathbf{P} + \mathbf{P} A_R - \mathbf{P} S_R \mathbf{P} + D_R^T D_R = 0, \quad S_R = B_R R_R^{-1} B_R^T \quad (5.133)$$

$$A_F \mathbf{Q} + \mathbf{Q} A_F^T - \mathbf{Q} S_F \mathbf{Q} + G_F W_F G_F^T = 0, \quad S_F = C_F^T V_F^{-1} C_F \quad (5.134)$$

such that the solutions of (5.128) and (5.129) are equal to the solutions of (5.133) and (5.134), that is

$$P = \mathbf{P}, \quad Q = \mathbf{Q} \tag{5.135}$$

where

$$\begin{aligned} A_R &= I - 2\left(\Delta_R^{-1}\right)^T \\ B_R R_R^{-1} B_R^T &= 2\left(I + A\right)^{-1} B R^{-1} B^T \Delta_R^{-1} \end{aligned} \tag{5.136}$$

$$\begin{aligned} D_R^T D_R &= 2\Delta_R^{-1} D^T D \left(I + A\right)^{-1} \\ \Delta_R &= \left(I + A^T\right) + D^T D \left(I + A\right)^{-1} B R^{-1} B^T \end{aligned}$$

and

$$\begin{aligned} A_F &= I - 2\left(\Delta_F^{-1}\right) \\ C_F^T V_F^{-1} C_F &= 2\left(I + A^T\right)^{-1} C^T V^{-1} C \Delta_F^{-1} \\ G_F W_F G_F^T &= 2\Delta_F^{-1} G W G^T \left(I + A^T\right)^{-1} \\ \Delta_F &= \left(I + A\right) + G W G^T \left(I + A^T\right)^{-1} C^T V^{-1} C \end{aligned} \tag{5.137}$$

It is shown in Section 2.7 that the equations (5.133) and (5.134) preserve the structure of weakly coupled systems. These equations can be solved in terms of the reduced-order problems very efficiently by using the recursive method developed in Chapter 2, which converges with the rate of convergence of $O\left(\epsilon^2\right)$. Solutions of (5.133) and (5.134) are found in terms of the reduced-order problems by imposing standard stabilizability-detectability assumptions on subsystems (see Section 2.3.2).

Getting approximate solution for P and Q in terms of the reduced-order problems will produce savings in off-line computations. However, in the case of stochastic systems, where the additional dynamical system — filter — has to be built, one is particularly interested in the reduction of on-line computations. In this section, the savings of on-line computation will be achieved by using a decoupling transformation introduced in (Gajic and Shen, 1989). The basic properties of that transformation in the discrete-time domain are given in Chapter 3.

The Kalman filter (5.124) is viewed as a system driven by the innovation process. However, one might study the filter form when it is driven by both measurements and control. The filter form under consideration is obtained from (5.124) as

$$\hat{x}_1(n+1) = \left(A_{11} - B_{11}F_{11} - \epsilon^2 B_{12}F_{12}\right)\hat{x}_1(n)$$
$$+ \epsilon\left(A_{12} - B_{11}F_{12} - B_{12}F_{22}\right)\hat{x}_2(n) + K_{11}v_1(n) + \epsilon K_{12}v_2(n)$$

$$\hat{x}_2(n+1) = \epsilon\left(A_{21} - B_{21}F_{11} - B_{22}F_{21}\right)\hat{x}_1(n)$$
$$+ \left(A_{22} - \epsilon^2 B_{21}F_{12} - B_{22}F_{22}\right)\hat{x}_2(n) + \epsilon K_{21}v_1(n) + K_{22}v_2(n) \tag{5.138}$$

with the innovation process

$$v_1(n) = y_1(n) - C_{11}\hat{x}_1(n) - \epsilon C_{12}\hat{x}_2(n)$$
$$v_2(n) = y_2(n) - \epsilon C_{21}\hat{x}_1(n) - C_{22}\hat{x}_2(n) \tag{5.139}$$

The nonsingular state transformation of (Gajic and Shen, 1989) will block diagonalize (5.138) under condition that the subsystem feedback matrices $\left(A_{11} - B_{11}F_{11} - \epsilon^2 B_{12}F_{21}\right)$ and $\left(A_{22} - B_{22}F_{22} - \epsilon^2 B_{21}F_{12}\right)$ have no eigenvalues in common (see Chapter 3). The transformation is given by

$$\begin{bmatrix} \hat{\eta}_1(n) \\ \hat{\eta}_2(n) \end{bmatrix} = \begin{bmatrix} I_{n_1} - \epsilon^2 LH & -\epsilon L \\ \epsilon H & I_{n_2} \end{bmatrix} \begin{bmatrix} \hat{x}_1(n) \\ \hat{x}_2(n) \end{bmatrix} = \mathbf{T}_2^{-1} \begin{bmatrix} \hat{x}_1(n) \\ \hat{x}_2(n) \end{bmatrix} \tag{5.140}$$

with

$$\mathbf{T}_2 = \begin{bmatrix} I_{n_1} & \epsilon L \\ -\epsilon H & I_{n_2} - \epsilon^2 HL \end{bmatrix} \tag{5.141}$$

where matrices L and H satisfy equations

$$L\left(a_{22} + \epsilon H a_{12}\right) - \left(a_{11} - \epsilon^2 a_{12}H\right)L + a_{12} = 0 \tag{5.142}$$

$$H a_{11} - a_{22}H + a_{21} - \epsilon^2 H a_{12}H = 0 \tag{5.143}$$

with

$$a_{11} = A_{11} - B_{11}F_{11} - \epsilon^2 B_{12}F_{21}$$
$$a_{12} = A_{12} - B_{11}F_{12} - B_{12}F_{22}$$

149

$$a_{21} = A_{21} - B_{21}F_{11} - B_{22}F_{21}$$
$$a_{22} = A_{22} - B_{22}F_{22} - \epsilon^2 B_{21}F_{12}$$

(5.144)

The optimal feedback control, expressed in the new coordinates, has the form

$$u_1(n) = -f_{11}\hat{\eta}_1(n) - \epsilon f_{12}\hat{\eta}_2(n)$$
$$u_2(n) = -\epsilon f_{21}\hat{\eta}_1(n) - f_{22}\hat{\eta}_2(n)$$

(5.145)

with

$$\hat{\eta}_1(n+1) = \alpha_1\hat{\eta}_1(n) + \beta_{11}v_1(n) + \epsilon\beta_{12}v_2(n)$$
$$\hat{\eta}_2(n+1) = \alpha_2\hat{\eta}_2(n) + \epsilon\beta_{21}v_1(n) + \beta_{22}v_2(n)$$

(5.146)

where

$$f_{11} = F_{11} - \epsilon^2 F_{12}H, \quad f_{12} = F_{12} + (F_{11} - \epsilon^2 F_{12}H)L$$
$$f_{21} = F_{21} - F_{22}H, \quad f_{22} = F_{22} + \epsilon^2(F_{21} - F_{22}H)L$$
$$\alpha_1 = a_{11} - \epsilon^2 a_{12}H, \quad \alpha_2 = a_{22} + \epsilon^2 Ha_{12}$$
$$\beta_{11} = K_{11} - \epsilon^2 L(H + K_{21}), \quad \beta_{12} = K_{12} - LK_{22} - \epsilon^2 LHK_{12}$$
$$\beta_{21} = HK_{11} + K_{21}, \quad \beta_{22} = K_{22} + \epsilon^2 HK_{12}$$

(5.147)

The innovation processes v_1 and v_2 are now given by

$$v_1(n) = y_1(n) - d_{11}\hat{\eta}_1(n) - \epsilon d_{12}\hat{\eta}_2(n)$$
$$v_2(n) = y_2(n) - \epsilon d_{21}\hat{\eta}_1(n) - d_{22}\hat{\eta}_2(n)$$

(5.148)

where

$$d_{11} = C_{11} - \epsilon^2 C_{12}H, \quad d_{12} = C_{11}L + C_{12} - \epsilon^2 C_{12}HL$$
$$d_{21} = C_{21} - C_{22}H, \quad d_{22} = C_{22} + \epsilon^2(C_{21} - C_{22}H)L$$

(5.149)

Approximate control laws are defined by perturbing coefficients $F_{ij}, K_{ij}, i, j = 1, 2$; L and H by $O(\epsilon^k)$, $k = 1, 2, ...$, in other words by using k-th approximations for these coefficients, where k stands for the required order of accuracy, that is

$$u_1^{(k)}(n) = -f_{11}^{(k)}\hat{\eta}_1^{(k)}(n) - \epsilon f_{12}^{(k)}\hat{\eta}_2^{(k)}(n)$$
$$u_2^{(k)}(n) = -\epsilon f_{21}^{(k)}\hat{\eta}_1^{(k)}(n) - f_{22}^{(k)}\hat{\eta}_2^{(k)}(n)$$

(5.150)

with

$$\hat{\eta}_1^{(k)}(n+1) = \alpha_1^{(k)}\hat{\eta}_1^{(k)}(n) + \beta_{11}^{(k)}v_1^{(k)}(n) + \epsilon\beta_{12}^{(k)}v_2^{(k)}(n)$$
$$\hat{\eta}_2^{(k)}(n+1) = \alpha_2^{(k)}\hat{\eta}_2^{(k)}(n) + \epsilon\beta_{21}^{(k)}v_1^{(k)}(n) + \beta_{22}^{(k)}v_2^{(k)}(n)$$

(5.151)

where

$$v_1^{(k)}(n) = y_1(n) - d_{11}^{(k)}\hat{\eta}_1^{(k)}(n) - \epsilon d_{12}^{(k)}\hat{\eta}_2^{(k)}(n)$$
$$v_2^{(k)}(n) = y_2(n) - \epsilon d_{21}^{(k)}\hat{\eta}_1^{(k)}(n) - d_{22}^{(k)}\hat{\eta}_2^{(k)}(n)$$

(5.152)

and

$$f_{ij}^{(k)} = f_{ij} + O\left(\epsilon^k\right), \quad d_{ij}^{(k)} = d_{ij} + O\left(\epsilon^k\right)$$
$$\alpha_{ij}^{(k)} = \alpha_{ij} + O\left(\epsilon^k\right), \quad \beta_{ij}^{(k)} = \beta_{ij} + O\left(\epsilon^k\right)$$
$$i, j = 1, 2$$

(5.153)

The approximate values of $J^{(k)}$ are obtained from the following expression

$$J^{(k)} = \frac{1}{2}E\left\{\sum_{n=0}^{\infty}\left[x^{(k)^T}(n)D^T Dx^{(k)}(n) + u^{(k)^T}(n)Ru^{(k)}(n)\right]\right\}$$
$$= \frac{1}{2}tr\left\{D^T Dq_{11}^{(k)} + f^{(k)^T}Rf^{(k)}q_{22}^{(k)}\right\}$$

(5.154)

where

$$q_{11}^{(k)} = Var\left\{\left(x_1^{(k)} \ x_2^{(k)}\right)^T\right\} \text{ and } q_{22}^{(k)} = Var\left\{\left(\hat{\eta}_1^{(k)} \ \hat{\eta}_2^{(k)}\right)^T\right\}$$
$$u^{(k)} = \begin{bmatrix} u_1^{(k)}(n) \\ u_2^{(k)}(n) \end{bmatrix}, \quad f^{(k)} = \begin{bmatrix} f_{11}^{(k)} & \epsilon f_{12}^{(k)} \\ \epsilon f_{21}^{(k)} & f_{22}^{(k)} \end{bmatrix}$$

(5.155)

The quantities $q_{11}^{(k)}$ and $q_{22}^{(k)}$ can be obtained by studying the variance equation of the following system driven by white noise

$$\begin{bmatrix} x^{(k)}(n+1) \\ \hat{\eta}^{(k)}(n+1) \end{bmatrix} = \begin{bmatrix} A & -Bf^{(k)} \\ \beta^{(k)}C & \alpha^{(k)} - \beta^{(k)}d^{(k)} \end{bmatrix} \begin{bmatrix} x^{(k)}(n) \\ \hat{\eta}^{(k)}(n) \end{bmatrix} \\ + \begin{bmatrix} G & 0 \\ 0 & \beta^{(k)} \end{bmatrix} \begin{bmatrix} w(n) \\ v(n) \end{bmatrix}$$

(5.156)

where

$$\alpha^{(k)} = \begin{bmatrix} \alpha_1^{(k)} & 0 \\ 0 & \alpha_2^{(k)} \end{bmatrix}; \ \beta^{(k)} = \begin{bmatrix} \beta_{11}^{(k)} & \epsilon\beta_{12}^{(k)} \\ \epsilon\beta_{21}^{(k)} & \beta_{22}^{(k)} \end{bmatrix}; \ d^{(k)} = \begin{bmatrix} d_{11}^{(k)} & \epsilon d_{12}^{(k)} \\ \epsilon d_{21}^{(k)} & d_{22}^{(k)} \end{bmatrix}$$

(5.157)

Equation (5.156) can be represented in the composite form

$$\Gamma^{(k)}(n+1) = \Lambda^{(k)}\Gamma^{(k)}(n) + \Pi^{(k)}w(n) \qquad (5.158)$$

with obvious definitions for $\Lambda^{(k)}$, $\Pi^{(k)}$, $\Gamma^{(k)}(n)$, and $w(n)$. The variance of $\Gamma^{(k)}(n)$ at steady state denoted by $q^{(k)}$, is given by the discrete algebraic Lyapunov equation (Kwakernaak and Sivan, 1972)

$$q^{(k)}(n+1) = \Lambda^{(k)}q^{(k)}\Lambda^{(k)^T} + \Pi^{(k)}\overline{W}\Pi^{(k)^T}, \ \overline{W} = diag(W, V)$$

(5.159)

with $q^{(k)}$ partitioned as

$$q^{(k)} = \begin{bmatrix} q_{11}^{(k)} & q_{12}^{(k)} \\ q_{12}^{(k)^T} & q_{22}^{(k)} \end{bmatrix} \qquad (5.160)$$

On the other hand, the optimal value of J has the very well-known form, (Kwakernaak and Sivan, 1972)

$$J^{opt} = \frac{1}{2}tr\left[D^TD\mathbf{Q} + \mathbf{P}K\left(CQC^T + V\right)K^T\right] \qquad (5.161)$$

where $\mathbf{P}$, $\mathbf{Q}$, F, and K are obtained from (5.126)-(5.129).

The near-optimality of the proposed approximate control law (5.150) is established in the following theorem.

Theorem 5.7 *Let x_1 and x_2 be optimal trajectories and J be the optimal value of the performance criterion. Let $x_1^{(k)}$, $x_2^{(k)}$, and $J^{(k)}$ be corresponding quantities under the approximate control law $u^{(k)}$ given by (5.150). Under the condition stated in Assumption 3.2 and the stabilizability-detectability subsystem assumptions, the following hold*

$$J^{opt} - J^{(k)} = O\left(\epsilon^k\right)$$

$$Var\left\{x_i - x_i^{(k)}\right\} = O\left(\epsilon^{2k}\right), \quad i = 1, 2, \quad k = 0, 1, 2, \ldots.$$

(5.162)

◇

The proof of this theorem is rather lengthly and is therefore omitted here. It follows the ideas of Theorems 1 and 2 from (Khalil and Gajic, 1984) obtained for another class of small parameter problems — singularly perturbed systems. These two theorems were proved in the context of weakly coupled linear systems in (Shen and Gajic, 1990a). In addition, due to the discrete nature of the problem, the proof of our Theorem 5.7, utilizes a bilinear transformation from (Power, 1967) which transforms the discrete Lyapunov equation into the continuous one and compares it with the corresponding equation under the optimal control law. More about the proof can be found in (Shen, 1990).

5.8 Case Study: Distillation Column

A real world physical example, a fifth-order distillation column control problem, (Kautsky et al., 1985), demonstrates the efficiency of the proposed method. The problem matrices A and B are

$$A = 10^{-3} \begin{bmatrix} 989.50 & 5.6382 & 0.2589 & 0.0125 & 0.0006 \\ 117.25 & 814.50 & 76.038 & 5.5526 & 0.3700 \\ 8.7680 & 123.87 & 750.20 & 107.96 & 11.245 \\ 0.9108 & 17.991 & 183.81 & 668.34 & 150.78 \\ 0.0179 & 0.3172 & 1.6974 & 13.298 & 985.19 \end{bmatrix}$$

$$B^T = 10^{-3} \begin{bmatrix} 0.0192 & 6.0733 & 8.2911 & 9.1965 & 0.7025 \\ -0.0013 & -0.6192 & -13.339 & -18.442 & -1.4252 \end{bmatrix}$$

These matrices are obtained from (Kautsky et al., 1985) by performing a discretization with the sampling rate $\Delta T = 0.1$.

Remaining matrices are chosen as

$$C = \begin{bmatrix} 1 & 1 & 0 & 0 & 0 \\ 0 & 0 & 1 & 1 & 1 \end{bmatrix}, \quad Q = I_5, \quad R = I_2$$

It is assumed that $G = B$, and that the white noise intensity matrices are given by

$$W_1 = 1, \quad W_2 = 2, \quad V_1 = 0.1, \quad V_2 = 0.1$$

The simulation results are presented in Table 5.4.

In practice, how the problem matrices are partitioned will determine the choice of the coupling parameter which in turn determines the rate of convergence and the domain of attraction of the iterative scheme to the optimal solution. It is desirable to get as small as possible a value of the small coupling parameter. This will speed up the convergence process. However, the small parameter is built into the problem and one can not go beyond the physical limits. The small weak coupling parameter ϵ can be roughly estimated from the strongest coupled matrix — in this case matrix B. Apparently the strongest coupling is in the third row, that is

$$\epsilon = \frac{b_{31}}{b_{32}} = \frac{8.2911}{13.339} \approx 0.62$$

It can be seen that despite the relatively big value of the coupling parameter $\epsilon = 0.62$, we have very rapid convergence to the optimal solution.

k	$J^{(k)}$	$J^{(k)} - J$
0	0.80528×10^{-2}	0.6989×10^{-3}
1	0.75977×10^{-2}	0.2438×10^{-3}
2	0.74277×10^{-2}	0.7380×10^{-4}
4	0.73887×10^{-2}	0.3480×10^{-4}
6	0.73546×10^{-2}	0.5000×10^{-6}
8	0.73539×10^{-2}	$< 1.000 \times 10^{-7}$
optimal	0.73539×10^{-2}	

Table 5.4: Approximate values for criterion

In summary, the near-optimum (up to any desired accuracy) steady state regulators are obtained for the stochastic linear weakly coupled discrete systems. The proposed method reduces considerably the size of required off-line and on-line computations since it introduces full parallelism in the design procedure.

Exercise 5.4: It is well known that the initial condition of the optimal Kalman filter has to be set to the mean value of the system initial state. Derive an expression for the optimal variance of the estimation error in the case when this condition is not satisfied. Consider both the continuous and discrete-time domains. Hint: It is easier to solve this problem in the discrete-time.

$\triangle$

Research Problem 5.1: Extend the presented recursive parallel reduced-order algorithms to the study of Markov chains displaying both slow-fast phenomena and weak coupling (Phillips and Kokotovic, 1981; Delebecque and Quadrant, 1981; Delebecque et al., 1984; Srikant and Basar, 1989; Aldhaheri and Khalil, 1991).

$\triangle$

Appendix 5.1

Consider the continuous time-invariant linear singularly perturbed stochastic system represented in the fast time scale by

$$\dot{x}_1(t) = \epsilon A_1 x_1(t) + \epsilon A_2 x_2(t) + \epsilon B_1 u(t) + \epsilon G_1 w(t)$$
$$\dot{x}_2(t) = A_3 x_1(t) + A_4 x_2(t) + B_2 u(t) + G_2 w(t) \tag{a.1}$$

where $w(t)$ is zero-mean stationary Gaussian white noise.

To obtain the discrete-time description of this system, we write

$$x(t_{n+1}) = \phi(t_{n+1} - t_n) x(t_n) + \left[\int_{t_n}^{t_{n+1}} \phi(t_{n+1} - t) B dt \right] u(t_n)$$

$$+ \int_{t_n}^{t_{n+1}} \phi(t_{n+1} - t) G w(t) dt$$

$$\tag{a.2}$$

where $n = 0, 1, 2, \ldots$, and $\phi(t_{n+1} - t_n)$ is the transition matrix of the system (a.1). Assuming that $t_{n+1} - t_n = \text{constant} = \Delta$ (sampling period), the equation (a.2) can be written in the form

$$x_d(n+1) = A_d x_d(n) + B_d u_d(n) + G_d w_d(n) \tag{a.3}$$

where

$$A_d = e^{A\Delta}, \quad B_d = \int_0^{\Delta} e^{At} B dt \tag{a.4}$$

and

$$A = \begin{bmatrix} \epsilon A_1 & \epsilon A_2 \\ A_3 & A_4 \end{bmatrix}, \quad B = \begin{bmatrix} \epsilon B_1 \\ B_2 \end{bmatrix}, \quad G = \begin{bmatrix} \epsilon G_1 \\ G_2 \end{bmatrix} \tag{a.5}$$

It is easy to see that A_d and B_d have the form

$$A_d = \begin{bmatrix} I + \epsilon A_{11} & \epsilon A_{12} \\ A_{21} & A_{22} \end{bmatrix}, \quad B_d = \begin{bmatrix} \epsilon B_{11} \\ B_{22} \end{bmatrix} \tag{a.6}$$

More analysis is needed about the stochastic nature of the $G_d w_d(n)$ term. Obviously, the mean value of $G_d w_d(n)$ is equal to zero. On the other hand, the corresponding variance of $G_d w_d(n)$ has the order of

$$Var\{G_d w_d(n)\} = \begin{bmatrix} O(\epsilon^2) & O(\epsilon) \\ O(\epsilon) & O(1) \end{bmatrix} \qquad (a.7)$$

which can be interpreted as of having

$$G_d = \begin{bmatrix} O(\epsilon) \\ O(1) \end{bmatrix}, \quad Int\{w_d(n)\} = W_d = O(1) \qquad (a.8)$$

and this justifies the model (5.78) used in that section. Similarly, we can assume the structure of $G_d w_d(n)$ term as

$$G_d = \begin{bmatrix} O(1) \\ O(1) \end{bmatrix}, \quad Int\{w_d(n)\} = W_d = \begin{bmatrix} O(\epsilon^2) & O(\epsilon) \\ O(\epsilon) & O(1) \end{bmatrix} \qquad (a.9)$$

In Section 5.5.1, we adopt the structure given in (a.8).

Chapter 6

Open-Loop Optimal Control Problems

In this chapter, the reduced-order methods with an arbitrary degree of accuracy are presented for solving the linear-quadratic optimal open-loop control problems of singularly perturbed and weakly coupled systems in both continuous and discrete-time domains.

6.1 Open-Loop Singularly Perturbed Control Problem

The optimal open-loop control problem is a two-point boundary value problem with the associated state-costate equations forming the Hamiltonian system. For singularly perturbed system, after modifying some costate variables, the Hamiltonian matrix retains the singularly perturbed form by interchanging some state and costate variables so that it can be block diagonalized via the nonsingular transformation presented in Chapter 3.

The original two-point boundary value problem is transformed into the pure-slow and pure-fast reduced-order completely decoupled initial value problems. By doing this, the stiffness of the singularly perturbed two-point boundary value problem is converted in the problem of an ill-defined linear system of algebraic equations. The proposed method is very suitable for parallel computations since it allows complete parallelism in both slow and fast time scales.

Consider the linear singularly perturbed control system

$$\dot{x}_1 = A_1 x_1 + A_2 x_2 + B_1 u, \qquad x_1(t_0) = x_{10}$$
$$\epsilon \dot{x}_2 = A_3 x_1 + A_4 x_2 + B_2 u, \qquad x_2(t_0) = x_{20}$$

$$(6.1)$$

where $x_i \in \Re^{n_i}$, $i = 1, 2$, $u \in \Re^m$ are state and control variables, respectively, and ϵ is a small positive parameter. As the parameter ϵ tends to zero, the solution of (6.1) behaves nonuniformly, producing a so-called stiff problem (Kokotovic and Khalil, 1986; Kokotovic et al., 1986).

With (6.1), consider the performance criterion

$$J = \frac{1}{2} \int_{t_0}^{T} \left\{ \begin{bmatrix} x_1 \\ x_2 \end{bmatrix}^T Q \begin{bmatrix} x_1 \\ x_2 \end{bmatrix} + u^T R\, u \right\} dt + \frac{1}{2} \begin{bmatrix} x_1(T) \\ x_2(T) \end{bmatrix}^T F \begin{bmatrix} x_1(T) \\ x_2(T) \end{bmatrix}$$

(6.2)

with positive definite R and positive semidefinite Q and F.

The open-loop optimal control problem has the solution given by

$$u(t) = -R^{-1} B^T p(t) \tag{6.3}$$

where $p(t) \in \Re^{n_1 + n_2}$ is a costate variable satisfying (Kwakernaak and Sivan, 1972)

$$\begin{bmatrix} \dot{x}(t) \\ \dot{p}(t) \end{bmatrix} = \begin{bmatrix} A & -S \\ -Q & -A^T \end{bmatrix} \begin{bmatrix} x \\ p \end{bmatrix} \tag{6.4}$$

with boundary conditions expressed in the standard form as

$$M \begin{bmatrix} x(t_0) \\ p(t_0) \end{bmatrix} + N \begin{bmatrix} x(T) \\ p(T) \end{bmatrix} = c \tag{6.5}$$

where

$$M = \begin{bmatrix} I & 0 \\ 0 & 0 \end{bmatrix}, \quad N = \begin{bmatrix} 0 & 0 \\ -F & I \end{bmatrix}, \quad c = \begin{bmatrix} x(t_0) \\ 0 \end{bmatrix} \tag{6.6}$$

for the free endpoint problem, or

$$M = \begin{bmatrix} I & 0 \\ 0 & 0 \end{bmatrix}, \quad N = \begin{bmatrix} 0 & 0 \\ I & 0 \end{bmatrix}, \quad c = \begin{bmatrix} x(t_0) \\ x(T) \end{bmatrix} \tag{6.7}$$

for the fixed endpoint problem. Since condition (6.7) leads to a two-point boundary value problem, causing both the initial and terminal boundary layers, the treatment of this chapter is applicable to the free end problem only.

The matrices $A, Q, B, S,$ and F in the case of singularly perturbed control systems have the forms

$$A = \begin{bmatrix} A_1 & A_2 \\ \frac{A_3}{\epsilon} & \frac{A_4}{\epsilon} \end{bmatrix}, \quad Q = \begin{bmatrix} Q_1 & Q_2 \\ Q_2^T & Q_3 \end{bmatrix}, \quad B = \begin{bmatrix} B_1 \\ \frac{B_2}{\epsilon} \end{bmatrix}$$

$$S = BR^{-1}B^T = \begin{bmatrix} S_1 & \frac{Z}{\epsilon} \\ \frac{Z^T}{\epsilon} & \frac{S_2}{\epsilon^2} \end{bmatrix}, \quad F = \begin{bmatrix} F_1 & \epsilon F_2 \\ \epsilon F_2^T & \epsilon F_3 \end{bmatrix} \qquad (6.8)$$

Exercise 6.1: Consider the optimal open-loop continuous-time control problem (6.1)-(6.4) at steady state ($T \to \infty$, $F = 0$). Derive an expression for the optimal value of the performance criterion at steady state under the open-loop feedback control. Note that the criterion optimal value under the open-loop control must be identical to the criterion optimal value under the closed-loop control. Can you avoid the problem of solving the algebraic Riccati equation?

$$\triangle$$

The approximate optimal solution of the open-loop control for linear singularly perturbed systems has been studied in (Wilde and Kokotovic, 1973), where the problem order was reduced and the stiff problem was avoided successfully by using the classic approach based on the power-series expansions. The theory developed in (Wilde and Kokotovic, 1973) was based on the dichotomy transformation (Wilde and Kokotovic, 1972) which requires the positive definite and negative definite solutions of the corresponding algebraic Riccati equation. It was concluded in (Wilde and Kokotovic, 1973) that the developed method is efficient for an $O(\epsilon)$ accuracy only. In this section, the solution to the optimal open-loop control problem of singularly perturbed systems with an arbitrary order of accuracy is presented.

Partitioning vector p as $p = [p_1^T \quad \epsilon p_2^T]^T$ with $p_1 \in \Re^{n_1}$ and $p_2 \in \Re^{n_2}$, we get

$$\begin{bmatrix} \dot{x}_1 \\ \dot{p}_1 \\ \dot{x}_2 \\ \dot{p}_2 \end{bmatrix} = \begin{bmatrix} T_1 & T_2 \\ \frac{T_3}{\epsilon} & \frac{T_4}{\epsilon} \end{bmatrix} \begin{bmatrix} x_1 \\ p_1 \\ x_2 \\ p_2 \end{bmatrix} \qquad (6.9)$$

where

$$T_1 = \begin{bmatrix} A_1 & -S_1 \\ -Q_1 & -A_1^T \end{bmatrix}, \quad T_2 = \begin{bmatrix} A_2 & -Z \\ -Q_2 & -A_3^T \end{bmatrix}$$

$$\qquad (6.10)$$

$$T_3 = \begin{bmatrix} A_3 & -Z^T \\ -Q_2^T & -A_2^T \end{bmatrix}, \quad T_4 = \begin{bmatrix} A_4 & -S_2 \\ -Q_3 & -A_4^T \end{bmatrix}$$

Note that (6.9) retains the singular perturbation form as (6.1).

Introduce the notation

$$\begin{bmatrix} x_1 \\ p_1 \end{bmatrix} = w, \qquad \begin{bmatrix} x_2 \\ p_2 \end{bmatrix} = \lambda \tag{6.11}$$

and apply the following transformation (Chang, 1972) defined by

$$\mathbf{T}_1 = \begin{bmatrix} I - \epsilon H L & -\epsilon H \\ L & I \end{bmatrix}, \quad \mathbf{T}_1^{-1} = \begin{bmatrix} I & \epsilon H \\ -L & I - \epsilon L H \end{bmatrix} \tag{6.12}$$

where L and H satisfy

$$T_4 L - T_3 - \epsilon L (T_1 - T_2 L) = 0 \tag{6.13}$$

$$-H (T_4 + \epsilon L T_2) + T_2 + \epsilon (T_1 - T_2 L) H = 0 \tag{6.14}$$

The transformation (6.12) applied to (6.9) produces two completely decoupled subsystems

$$\dot{\eta} = (T_1 - T_2 L) \eta \tag{6.15}$$

and

$$\epsilon \dot{\xi} = (T_4 + \epsilon L T_2) \xi \tag{6.16}$$

where

$$\begin{bmatrix} \eta \\ \xi \end{bmatrix} = \mathbf{T}_1 \begin{bmatrix} w \\ \lambda \end{bmatrix} \tag{6.17}$$

The algebraic equations (6.13) and (6.14) can be solved by using any of the recursive algorithms presented in Chapter 8.

The boundary conditions are changed due to an interchange of p_1 and x_2, which modifies matrices in (6.6) as follows

$$M_1 \begin{bmatrix} w(t_0) \\ \lambda(t_0) \end{bmatrix} + N_1 \begin{bmatrix} w(T) \\ \lambda(T) \end{bmatrix} = c_1 \tag{6.18}$$

where

$$M_1 = \begin{bmatrix} I_{n_1} & 0 & 0 & 0 \\ 0 & 0 & 0 & 0 \\ 0 & 0 & I_{n_2} & 0 \\ 0 & 0 & 0 & 0 \end{bmatrix}, \quad c_1 = \begin{bmatrix} x_{10} \\ 0 \\ x_{20} \\ 0 \end{bmatrix}$$

$$N_1 = \begin{bmatrix} 0 & 0 & 0 & 0 \\ -F_1 & I_{n_1} & -\epsilon F_2 & 0 \\ 0 & 0 & 0 & 0 \\ -F_2^T & 0 & -F_3 & I_{n_2} \end{bmatrix} \tag{6.19}$$

The nonsingular transformation (6.12) applied to (6.18) produces

$$M_2 \begin{bmatrix} \eta(t_0) \\ \xi(t_0) \end{bmatrix} + N_2 \begin{bmatrix} \eta(T) \\ \xi(T) \end{bmatrix} = c_1 \tag{6.20}$$

where

$$M_2 = M_1 T_1^{-1}, \quad N_2 = N_1 T_1^{-1} \tag{6.21}$$

Since solutions of (6.15) and (6.16) are given by

$$\eta(t) = e^{(T_1 - T_2 L)(t - t_0)} \eta(t_0) \tag{6.22}$$

$$\xi(t) = e^{\frac{1}{\epsilon}(T_4 + \epsilon L T_2)(t - t_0)} \xi(t_0) \tag{6.23}$$

we can eliminate $\eta(T)$ and $\xi(T)$ from (6.20) such that

$$\left\{ M_2 + N_2 \begin{bmatrix} e^{(T_1 - T_2 L)(T - t_0)} & 0 \\ 0 & e^{\frac{1}{\epsilon}(T_4 + \epsilon L T_2)(T - t_0)} \end{bmatrix} \right\} \begin{bmatrix} \eta(t_0) \\ \xi(t_0) \end{bmatrix} = c_1 \tag{6.24}$$

The system of linear algebraic equations (6.24) can be represented in the form

$$\alpha(\epsilon) \begin{bmatrix} \eta(t_0) \\ \xi(t_0) \end{bmatrix} = c_1 \tag{6.25}$$

It is shown in Lemma 6.1 that $\alpha(\epsilon)$ is invertible, hence $\eta(t_0)$ and $\xi(t_0)$ can be obtained from (6.25).

Lemma 6.1 *The matrix $\alpha(\epsilon)$ is invertible.*

$\diamond$

Proof: Transition matrices of (6.22) and (6.23) can be denoted $\Phi(t - t_0)$ and $\Psi(t - t_0)$, respectively, and partitioned as

$$\Phi(t - t_0) = \begin{bmatrix} \Phi_{11}(t - t_0) & \Phi_{12}(t - t_0) \\ \Phi_{21}(t - t_0) & \Phi_{22}(t - t_0) \end{bmatrix} \tag{6.26}$$

$$\Psi(t - t_0) = \begin{bmatrix} \Psi_{11}(t - t_0) & \Psi_{12}(t - t_0) \\ \Psi_{21}(t - t_0) & \Psi_{22}(t - t_0) \end{bmatrix} \tag{6.27}$$

From (6.24) we have

$$\alpha(\epsilon) = \left(M_2 + N_2 \begin{bmatrix} \Phi(T - t_0) & 0 \\ 0 & \Psi(T - t_0) \end{bmatrix} \right) \tag{6.28}$$

Using expressions for M_2 and N_2 given by (6.18) and (6.20) we get

$$\alpha(\epsilon) = \begin{bmatrix} I_{n_1} & 0 & 0 & 0 \\ * & \Phi_{22} - F_1\Phi_{12} & 0 & 0 \\ * & * & I_{n_2} & 0 \\ * & * & * & \Psi_{22} - F_3\Psi_{12} \end{bmatrix} + O(\epsilon) \tag{6.29}$$

where asterisks denote terms which are not important for the nonsingularity of $\alpha(\epsilon)$. Since matrices $\Phi_{22} - F_1\Phi_{12}$ and $\Psi_{22} - F_3\Psi_{12}$ are invertible (Kalman, 1960), the matrix $\alpha(\epsilon)$ is invertible for sufficiently small values of ϵ. However, in the case of singularly perturbed systems, due to the nature of the fast subsystem transition matrix (6.23), which contains unstable modes, we can observe that $\alpha(0)$ is singular.

Thus, $\alpha(\epsilon)$ is invertible for $0 < \epsilon < \epsilon_1$ and ϵ_1 sufficiently small. In other words, the stiffness of the singularly perturbed system of differential equations is carried over to the stiffness of the linear system of algebraic equations. However, the latter problem is much easier to handle.

Now we are able to find $\eta(t)$ and $\xi(t)$ from (6.15) and (6.16). Using (6.12), we can find $\omega(t)$ and $\lambda(t)$. Partitioning $\omega(t)$ and $\lambda(t)$ according to (6.11), we get values for $p_1(t)$ and $p_2(t)$. The costate variables $p(t)$ and the optimal control law are therefore found.

The only difficulty we have encounted in the procedure is to compute $\alpha(\epsilon)$ in (6.25) where an ill-defined problem occurs when ϵ is extremely small or $(T - t_0)$ is very large because the matrix T_4 contains both stable and unstable modes. In that case we refer to (Wilde and Kokotovic, 1973).

The recursive reduced-order technique for solving the optimal open-loop control problem of linear singularly perturbed systems presented in this section is mostly based on the results obtained in (Su et al., 1992a).

6.2 Case Study: Magnetic Tape Control

In order to illustrate the proposed method, we shall consider a real world problem — a magnetic tape control system (Chow and Kokotovic, 1976). Problem matrices are given in Section 2.2.3. The initial conditions are

$$x^T(t_0) = [-1.3702 \quad 0.10686 \quad -0.53307 \quad 0.83467]$$

and the time interval of interest is specified by $t_0 = 0$ and $T = 1$. Obtained results are presented in Table 6.1.

The approximate control is defined as

$$u^{(k)}(t) = -R^{-1}B^Tp^{(k)}(t) \tag{6.30}$$

where k stands for the number of iterations used to solve recursively equations (6.13)-(6.14). Values for $p^{(k)}(t)$ are obtained by following steps (6.14)-(6.25), with $p^{(k)}(t)$ obtained directly from (6.17) and (6.11). Note that steps (6.13)-(6.25) can be performed by using the method of series expansions, but since it is not recursive in its nature, it can be efficient for an $O(\epsilon)$ accuracy only, as was pointed out in (Wilde and Kokotovic, 1973).

	$t = 0.25$	$t = 0.5$	$t = 1$
$u^{(4)}(t) =$ optimal	3.1719×10^{-1}	3.0299×10^{-1}	-8.2827×10^{-2}
$u^{(3)}(t)$	3.1719×10^{-1}	3.0299×10^{-1}	-8.2827×10^{-2}
$u^{(2)}(t)$	3.1720×10^{-1}	3.0299×10^{-1}	-8.2825×10^{-2}
$u^{(1)}(t)$	3.1712×10^{-1}	3.0287×10^{-1}	-8.2758×10^{-2}
$u^{(0)}(t)$	3.3244×10^{-1}	3.01350×10^{-1}	-7.6749×10^{-2}

Table 6.1: Values of an approximate control at certain time instants

6.3 Open-Loop Weakly Coupled Optimal Control Problem

In this section, we study the open-loop control problem (linear two-point boundary value problem) of weakly coupled systems. Corresponding closed-loop (nonlinear differential Riccati equation) optimal control problems will be studied in Chapter 8.

The recursive reduced-order solution is obtained by exploiting the transformation introduced in Chapter 3. The transformation block diagonalizes the Hamiltonian form of the solution for the optimal linear-quadratic control problem. Completely decoupled sets of reduced-order differential equations are obtained. The convergence to the optimal solution is pretty rapid, due to the fact that the algorithms derived in Chapter 3 have the rate of convergence of at least of $O\left(\epsilon^2\right)$. This produces a lot of savings in the size of computations required. In addition, the proposed method is very suitable for parallel and distributed computations.

It is interesting to point out that the better results are obtained for the open-loop problem since it is less computationally involved than the closed-loop problem (exactly the same sets of differential equations have to be solved, but they differ in the dimensionality).

Consider the linear weakly coupled system

$$
\begin{aligned}
\dot{x}_1 &= A_1 x_1 + \epsilon A_2 x_2 + B_1 u_1 + \epsilon B_2 u_2, & x_1\left(t_0\right) &= x_{10} \\
\dot{x}_2 &= \epsilon A_3 x_1 + A_4 x_2 + \epsilon B_3 u_1 + B_4 u_2, & x_2\left(t_0\right) &= x_{20}
\end{aligned}
\tag{6.31}
$$

with

$$
z = \begin{bmatrix} z_1 \\ z_2 \end{bmatrix} = D \begin{bmatrix} x_1 \\ x_2 \end{bmatrix} = \begin{bmatrix} D_1 & \epsilon D_2 \\ \epsilon D_3 & D_4 \end{bmatrix} \begin{bmatrix} x_1 \\ x_2 \end{bmatrix}
\tag{6.32}
$$

where $x_i \in \Re^{n_i}$, $u_i \in \Re^{m_i}$, $z_i \in \Re^{r_i}$, $i = 1, 2$, are state, control, and output variables, respectively. The system matrices are of appropriate dimensions, and in general, they are bounded functions of a small coupling parameter ϵ. In this section, we will assume that all given matrices are constant.

With (6.31)-(6.32), consider a quadratic performance criterion in the form

$$J = \frac{1}{2} \int_{t_0}^{T} \left\{ \begin{bmatrix} x_1 \\ x_2 \end{bmatrix}^T D^T D \begin{bmatrix} x_1 \\ x_2 \end{bmatrix} + \begin{bmatrix} u_1 \\ u_2 \end{bmatrix}^T R \begin{bmatrix} u_1 \\ u_2 \end{bmatrix} \right\} dt$$

$$+ \frac{1}{2} \begin{bmatrix} x_1(T) \\ x_2(T) \end{bmatrix}^T F \begin{bmatrix} x_1(T) \\ x_2(T) \end{bmatrix} \tag{6.33}$$

with positive definite R and positive semidefinite F, which has to be minimized. It is assumed that matrices F and R have the weakly coupled structure, that is

$$F = \begin{bmatrix} F_1 & \epsilon F_2 \\ \epsilon F_2^T & F_3 \end{bmatrix}, \quad R = \begin{bmatrix} R_1 & 0 \\ 0 & R_2 \end{bmatrix} \tag{6.34}$$

The open-loop optimal control problem of (6.31)-(6.34) has the solution given by

$$u(t) = -R^{-1} B^T p(t) \tag{6.35}$$

where $p(t) \in \Re^{n_1 + n_2}$ is a costate variable satisfying

$$\begin{bmatrix} \dot{p}(t) \\ \dot{x}(t) \end{bmatrix} = \begin{bmatrix} -A^T & -D^T D \\ -S & A \end{bmatrix} \begin{bmatrix} p(t) \\ x(t) \end{bmatrix} \tag{6.36}$$

where

$$A = \begin{bmatrix} A_1 & \epsilon A_2 \\ \epsilon A_3 & A_4 \end{bmatrix}, \quad B = \begin{bmatrix} B_1 & \epsilon B_2 \\ \epsilon B_3 & B_4 \end{bmatrix}$$

$$S = B R^{-1} B^T = \begin{bmatrix} S_1 & \epsilon S_2 \\ \epsilon S_2^T & S_3 \end{bmatrix} \tag{6.37}$$

with boundary conditions expressed in the standard form as

$$W \begin{bmatrix} p(t_0) \\ x(t_0) \end{bmatrix} + G \begin{bmatrix} p(T) \\ x(T) \end{bmatrix} = c \tag{6.38}$$

where

$$W = \begin{bmatrix} 0 & 0 \\ 0 & I \end{bmatrix}, \quad G = \begin{bmatrix} I & -F \\ 0 & 0 \end{bmatrix}, \quad c = \begin{bmatrix} 0 \\ x(t_0) \end{bmatrix} \tag{6.39}$$

Partitioning p into $p_1 \in \Re^{n_1}$ and $p_2 \in \Re^{n_2}$ such that $p = \begin{bmatrix} p_1^T & p_2^T \end{bmatrix}^T$, and rearranging rows in (6.36), we can get

$$\begin{bmatrix} \dot{p}_1 \\ \dot{x}_1 \\ \dot{p}_2 \\ \dot{x}_2 \end{bmatrix} = \begin{bmatrix} T_1 & \epsilon T_2 \\ \epsilon T_3 & T_4 \end{bmatrix} \begin{bmatrix} p_1 \\ x_1 \\ p_2 \\ x_2 \end{bmatrix} \tag{6.40}$$

where $T_i's$, $i = 1,2,3,4$ are given by

$$T_1 = \begin{bmatrix} -A_1^T & -Q_1 \\ -S_1 & A_1 \end{bmatrix}, \quad T_2 = \begin{bmatrix} -A_3^T & -Q_2 \\ -S_2 & A_2 \end{bmatrix}$$

$$T_3 = \begin{bmatrix} -A_2^T & -Q_2^T \\ -S_2^T & A_3 \end{bmatrix}, \quad T_4 = \begin{bmatrix} -A_4^T & -Q_3 \\ -S_3 & A_4 \end{bmatrix} \tag{6.41}$$

with

$$Q_1 = D_1^T D_1 + \epsilon^2 D_3^T D_3, \; Q_2 = D_1^T D_2 + D_3^T D_4, \; Q_3 = D_4^T D_4 + \epsilon^2 D_2^T D_2 \tag{6.42}$$

Introduce the notation

$$\begin{bmatrix} p_1 \\ x_1 \end{bmatrix} = w, \quad \begin{bmatrix} p_2 \\ x_2 \end{bmatrix} = \lambda \tag{6.43}$$

and apply the corresponding weak coupling transformation presented in Chapter 3

$$T_2 = \begin{bmatrix} I & -\epsilon L \\ \epsilon H & I - \epsilon^2 HL \end{bmatrix}, \quad T_2^{-1} = \begin{bmatrix} I - \epsilon^2 LH & \epsilon L \\ -\epsilon H & I \end{bmatrix} \tag{6.44}$$

where L and H satisfy

$$T_1 L + T_2 - LT_4 - \epsilon^2 LT_3 L = 0 \tag{6.45}$$

$$H(T_1 - \epsilon^2 LT_3) - (T_4 + \epsilon^2 T_3 L)H + T_3 = 0 \tag{6.46}$$

This transformation produce a decoupled system of the form

$$\dot{\eta} = (T_1 - \epsilon^2 LT_3)\eta \tag{6.47}$$

$$\dot{\xi} = (T_4 + \epsilon^2 T_3 L)\xi \tag{6.48}$$

with

$$\begin{bmatrix} \eta \\ \xi \end{bmatrix} = \mathbf{T}_2^{-1} \begin{bmatrix} w \\ \lambda \end{bmatrix} \tag{6.49}$$

In order to be able to solve (6.47) and (6.48), we need to find their initial or terminal conditions, which can be obtained as follows. An interchange of rows for p_2 and x_1 in (6.42) will modify matrices defined in (6.38) and (6.39) as follows

$$W_1 \begin{bmatrix} w(t_0) \\ \lambda(t_0) \end{bmatrix} + G_1 \begin{bmatrix} w(T) \\ \lambda(T) \end{bmatrix} = c_1 \tag{6.50}$$

where

$$W_1 = \begin{bmatrix} 0 & 0 & 0 & 0 \\ 0 & I_{n_1} & 0 & 0 \\ 0 & 0 & 0 & 0 \\ 0 & 0 & 0 & I_{n_2} \end{bmatrix}; \quad c_1 = \begin{bmatrix} 0 \\ x_{10} \\ 0 \\ x_{20} \end{bmatrix}$$

$$G_1 = \begin{bmatrix} I_{n_1} & -F_1 & 0 & -\epsilon F_2 \\ 0 & 0 & 0 & 0 \\ 0 & -\epsilon F_2^T & I_{n_2} & -F_3 \\ 0 & 0 & 0 & 0 \end{bmatrix} \tag{6.51}$$

The transformation (6.49) applied to (6.50) produces

$$W_2 \begin{bmatrix} \eta(t_0) \\ \xi(t_0) \end{bmatrix} + G_2 \begin{bmatrix} \eta(T) \\ \xi(T) \end{bmatrix} = c_1 \tag{6.52}$$

where

$$W_2 = W_1 \mathbf{T}_2; \quad G_2 = G_1 \mathbf{T}_2 \tag{6.53}$$

Since solutions of (6.47) and (6.48) are given by

$$\eta(t) = e^{(T_1 - \epsilon^2 L T_3)(t - t_0)} \eta(t_0) \tag{6.54}$$

$$\xi(t) = e^{(T_4 + \epsilon^2 T_3 L)(t - t_0)} \xi(t_0) \tag{6.55}$$

we can eliminate $\eta(T)$ and $\xi(T)$ from (6.52); that is, we have

$$\left(W_2 + G_2 \begin{bmatrix} e^{(T_1 - \epsilon^2 L T_3)(T - t_0)} & 0 \\ 0 & e^{(T_4 + \epsilon^2 T_3 L)(T - t_0)} \end{bmatrix} \right) \begin{bmatrix} \eta(t_0) \\ \xi(t_0) \end{bmatrix} = c_1 \tag{6.56}$$

Equation (6.56) has the form

$$\alpha\left(\epsilon\right) \begin{bmatrix} \eta\left(t_0\right) \\ \xi\left(t_0\right) \end{bmatrix} = c_1 \tag{6.57}$$

with obvious definition for $\alpha\left(\epsilon\right)$. It is shown in the next lemma that this system of linear algebraic equations has unique solution, assuming that a coupling parameter ϵ is sufficiently small.

Lemma 6.2 *The matrix $\alpha\left(\epsilon\right)$ is invertible for sufficiently small values of ϵ.*

$\diamond$

Proof: Let the transition matrices of (6.47) and (6.48) be denoted by $\Phi\left(t - t_0\right)$ and $\Psi\left(t - t_0\right)$, respectively, and partitioned as follows

$$\begin{aligned} \Phi\left(t - t_0\right) &= \begin{bmatrix} \Phi_{11}\left(t - t_0\right) & \Phi_{12}\left(t - t_0\right) \\ \Phi_{21}\left(t - t_0\right) & \Phi_{22}\left(t - t_0\right) \end{bmatrix} \\ \Psi\left(t - t_0\right) &= \begin{bmatrix} \Psi_{11}\left(t - t_0\right) & \Psi_{12}\left(t - t_0\right) \\ \Psi_{21}\left(t - t_0\right) & \Psi_{22}\left(t - t_0\right) \end{bmatrix} \end{aligned} \tag{6.58}$$

From (6.56), we have

$$\alpha\left(\epsilon\right) = \left(W_2 + G_2 \begin{bmatrix} \Phi\left(T - t_0\right) & 0 \\ 0 & \Psi\left(T - t_0\right) \end{bmatrix}\right) \tag{6.59}$$

Using expressions for W_2 and G_2, defined by (6.53) and (6.44), we get

$$\alpha\left(\epsilon\right) = \begin{bmatrix} I_{n_1} & 0 & 0 & 0 \\ \Phi_{21} - F_1\Phi_{11} & \Phi_{22} - F_1\Phi_{12} & 0 & 0 \\ 0 & 0 & I_{n_2} & 0 \\ 0 & 0 & \Psi_{21} - F_3\Psi_{11} & \Psi_{22} - F_3\Psi_{12} \end{bmatrix} + O\left(\epsilon\right) \tag{6.60}$$

Since matrices $\Phi_{22}\left(T - t_0\right) - F_1\Phi_{12}\left(T - t_0\right)$ and $\Psi_{22}\left(T - t_0\right) - F_3\Psi_{12}\left(T - t_0\right)$ are invertible (see (Kirk, 1970), page 211), the matrix $\alpha\left(\epsilon\right)$ is invertible for sufficiently small values of ϵ.

Now we are able to find $\eta\left(t\right)$ and $\xi\left(t\right)$ from (6.53) and (6.54). Using (6.49), we can find $w\left(t\right)$ and $\lambda\left(t\right)$. Partitioning $w\left(t\right)$ and $\lambda\left(t\right)$ according to (6.43) we get values for $p_1\left(t\right)$ and $p_2\left(t\right)$, in other words, one finds the optimal reduced-order open-loop control defined by (6.36).

The transformation matrix T_2 from (6.44) can be easily obtained, with required accuracy, by using numerical algorithms developed in

Section 3.3 for solving (6.45)-(6.46). These algorithms converge with the rate of convergence of at least of $O\left(\epsilon^2\right)$. Thus, after k iterations, one gets the approximation $\mathbf{T}_2^{(k)} = \mathbf{T}_2 + O\left(\epsilon^{2k}\right)$. The use of $\mathbf{T}_2^{(k)}$ in the design procedure instead of $\mathbf{T}_2$ will perturb the coefficients of the corresponding systems of linear differential equations by $O\left(\epsilon^2\right)$, which implies that the approximate solutions of these differential equations are $O\left(\epsilon^2\right)$ close to the exact ones (Kato, 1980). Thus, it is of interest to obtain $\mathbf{T}_2^{(k)}$ with the desired accuracy, which produces the same accuracy in the sought solution.

As a matter of fact, we have obtained the approximate expression for the optimal control in the form

$$u^{(k)}\left(t\right) = -R^{-1}B^T p^{(k)} = u^{opt}\left(t\right) + O\left(\epsilon^{2k}\right) \qquad (6.61)$$

Apparently, as k increases, the approximate control defined in (6.61) converges very rapidly to the optimal solution.

Simulation results for finding the optimal open-loop control in terms of the reduced-order problems are presented in the next section, where a fifth-order distillation column example is solved. It is interesting to point out that the proposed method produces better accuracy for the open-loop control than for the closed-loop control. This can be justified by comparing linear systems of differential equations (6.40) and the corresponding equations for the closed-loop control problem (Section 8.3). The closed loop solution is computationally much more involved since corresponding system of differential equations is of order of $2 \times (2n \times n)$, whereas (6.40) represents the same set of equations of order $2n \times 1$.

Results presented in this section are mostly based on the recent paper by (Su and Gajic, 1991).

6.4 Case Study: Distillation Column

The recursive reduced-order open-loop control problem is demonstrated on a real world problem, a fifth-order distillation column (Petkov et al., 1986). The problem matrices A and B are given by

$$A = \begin{bmatrix} -0.1094 & 0.0628 & 0 & 0 & 0 \\ 1.3060 & -2.1320 & 0.9807 & 0 & 0 \\ 0 & 1.5950 & -3.1490 & 1.5470 & 0 \\ 0 & 0.0355 & 2.6320 & -4.2570 & 1.8550 \\ 0 & 0.00227 & 0 & 0.1636 & -0.1625 \end{bmatrix}$$

$$B = \begin{bmatrix} 0 & 0.0632 & 0.0838 & 0.1004 & 0.0063 \\ 0 & 0 & -0.1396 & -0.2060 & -0.0128 \end{bmatrix}^T$$

Remaining matrices are chosen as

$$D^T D = \begin{bmatrix} 3 & 0 & 0.7 & 0.7 & 0.7 \\ 0 & 3 & 0.7 & 0.7 & 0.7 \\ 0.7 & 0.7 & 3 & 0 & 0 \\ 0.7 & 0.7 & 0 & 3 & 0 \\ 0.7 & 0.7 & 0 & 0 & 3 \end{bmatrix}. \quad R = I_2, \quad F = I_5$$

The initial and final times are selected as $t_0 = 0$ and $T = 1$. The initial conditions are chosen randomly as

$$x_{10} = [-1.259 \quad 1.437]^T, \quad x_{20} = [-0.412 \quad -0.642 \quad 0.877]^T$$

The system is partitioned into two subsystems with $n_1 = 2$, $n_2 = 3$, and $\epsilon = 0.6$. The small parameter ϵ is roughly estimated from the strongest coupled matrix — in this case matrix B — producing $|b_{31}| / |b_{32}| = 0.0838/0.1396 = 0.6$. The open-loop control is obtained with accuracy of 10^{-5} after 6 iterations. Corresponding simulation results for both components of the approximate open-loop control are presented in Table 6.2.

6.5 Open-Loop Discrete Singularly Perturbed Control Problem

A singularly perturbed linear discrete system is represented by (Litkouhi and Khalil, 1984)

$$x_1(k+1) = (I_{n_1} + \epsilon A_1) x_1(k) + \epsilon A_2 x_2(k) + \epsilon B_1 u(k)$$
$$x_2(k+1) = A_3 x_1(k) + A_4 x_2(k) + B_2 u(k) \qquad (6.62)$$
$$x_1(0) = x_{10}, \qquad x_2(0) = x_{20}$$

with slow state variables $x_1 \in \Re^{n_1}$, fast state variables $x_2 \in \Re^{n_2}$, and control inputs $u \in \Re^m$, where ϵ is a small positive parameter. As the

iteration	$u(t = 0)$	$u(t = 0.25)$	$u(t = 0.5)$	$u(t = 0.75)$
6 = optimal control	-0.01033 -0.04169	0.00270 -0.00916	0.01275 0.00411	0.00945 0.02781
5	-0.01033 -0.04169	0.00270 -0.00916	0.01274 0.00411	0.00945 0.02781
4	-0.01039 -0.04167	0.00268 -0.00917	0.01275 0.00407	0.00948 0.02766
3	-0.01073 -0.04153	-0.00262 -0.00938	0.01293 0.00360	0.00980 0.02713
2	0.02831 -0.07095	0.03571 -0.03485	0.03895 -0.00164	0.02694 0.01403
1	0.06505 -0.09423	-0.06219 -0.05219	0.05589 -0.02776	0.03591 0.00699
0	0.69589 -0.34489	0.54689 -0.25504	0.40566 -0.18864	0.25491 -0.11626

Table 6.2: Simulation results for the open-loop control

parameter ϵ tends to zero, the solution behaves nonuniformly, producing a so-called stiff problem (Litkouhi and Khalil, 1985). The performance criterion of the corresponding linear-quadratic control problem is defined by

$$J(k) = \frac{1}{2}x^T(n) Fx(n) + \frac{1}{2}\sum_{k=0}^{n-1}\left[x^T(k) Qx(k) + u^T(k) Ru(k)\right]$$

(6.63)

where

$$x(k) = \begin{bmatrix} x_1(k) \\ x_2(k) \end{bmatrix}, \quad Q = \begin{bmatrix} Q_1 & Q_2 \\ Q_2^T & Q_3 \end{bmatrix} \geq 0$$

$$F = \begin{bmatrix} F_1/\epsilon & F_2 \\ F_2^T & F_3 \end{bmatrix} \geq 0, \quad R > 0$$

(6.64)

The open-loop optimal control problem has the solution given by

$$u(k) = -R^{-1}B^T\lambda(k + 1)$$

(6.65)

where $\lambda(k)$ is a costate variable. The Hamiltonian form of (6.62)-(6.63) can be written as the forward recursion (Lewis, 1986)

$$\begin{bmatrix} x(k+1) \\ \lambda(k+1) \end{bmatrix} = \mathbf{H} \begin{bmatrix} x(k) \\ \lambda(k) \end{bmatrix} \qquad (6.66)$$

where

$$\mathbf{H} = \begin{bmatrix} A + BR^{-1}B^T A^{-T}Q & -BR^{-1}B^T A^{-T} \\ -A^{-T}Q & A^{-T} \end{bmatrix} \qquad (6.67)$$

with boundary conditions expressed in the standard form as

$$M_1 \begin{bmatrix} x(0) \\ \lambda(0) \end{bmatrix} + N_1 \begin{bmatrix} x(n) \\ \lambda(n) \end{bmatrix} = c \qquad (6.68)$$

where $\mathbf{H}$ is the symplectic matrix which has the property that the eigenvalues of $\mathbf{H}$ can be grouped into two disjoint subsets Γ_1 and Γ_2, such that for every $\lambda_c \in \Gamma_1$ there exists $\lambda_d \in \Gamma_2$, which satisfies $\lambda_c \times \lambda_d = 1$, and we can choose either Γ_1 or Γ_2 to contain only the stable eigenvalues (Salgado et al., 1988).

Note that

$$M_1 = \begin{bmatrix} I & 0 \\ 0 & 0 \end{bmatrix}, \quad N_1 = \begin{bmatrix} 0 & 0 \\ -F & I \end{bmatrix}, \quad c = \begin{bmatrix} x(0) \\ 0 \end{bmatrix} \qquad (6.69)$$

for the free ending problem, or

$$M_1 = \begin{bmatrix} I & 0 \\ 0 & 0 \end{bmatrix}, \quad N_1 = \begin{bmatrix} 0 & 0 \\ I & 0 \end{bmatrix}, \quad c = \begin{bmatrix} x(0) \\ x(T) \end{bmatrix} \qquad (6.70)$$

for the fixed endpoint problem.

Exercise 6.2: Consider the general linear-quadratic open-loop discrete-time control problem at steady state. Derive an expression for the optimal value of the performance criterion without introducing the discrete-time algebraic Riccati equation.

$$\triangle$$

For the singularly perturbed discrete system the matrices A and S have the forms

$$A = \begin{bmatrix} I_{n_1} + \epsilon A_1 & \epsilon A_2 \\ A_3 & A_4 \end{bmatrix} \qquad (6.71)$$

$$S = BR^{-1}B^T = \begin{bmatrix} \epsilon^2 S_1 & \epsilon Z \\ \epsilon Z^T & S_2 \end{bmatrix}$$

$$S_1 = B_1 R^{-1} B_1^T, \quad S_2 = B_2 R^{-1} B_2^T, \quad Z = B_1 R^{-1} B_2^T \qquad (6.72)$$

The approximate optimal solution of the open-loop control for linear singularly perturbed systems has been studied in (Naidu, 1988), where the problem order was reduced and the stiff problem was avoided successfully by using the classic approach based on the power-series expansions. The developed method (Naidu, 1988) is efficient for an $O(\epsilon)$ accuracy only. In this section, the results of Section 6.1 and (Su et al., 1992a) are extended to the optimal open-loop control problem of singularly perturbed discrete systems producing the solution with an arbitrary order of accuracy, (Qureshi et al., 1991).

The optimal open-loop control problem is a two-point boundary value problem with the associated state-costate equations forming the Hamiltonian matrix. For singularly perturbed discrete systems, after modifying some costate variables, the Hamiltonian matrix retains the singularly perturbed form by interchanging some state and costate variables so that it can be block diagonalized via the nonsingular transformation introduced in (Chang, 1972). Similar to (Su et al., 1992a), the idea of this section is to exploit the reduced-order subsystems to find the optimal open-loop control in the new coordinates.

Partitioning vector $\lambda(k)$ as $\lambda(k) = [\lambda_1^T(k) \quad \lambda_2^T(k)]^T$ with $\lambda_1(k) \in \Re^{n_1}$ and $\lambda_2(k) \in \Re^{n_2}$, we get

$$\begin{bmatrix} x_1(k+1) \\ x_2(k+1) \\ \lambda_1(k+1) \\ \lambda_2(k+1) \end{bmatrix} = \mathbf{H} \begin{bmatrix} x_1(k) \\ x_2(k) \\ \lambda_1(k) \\ \lambda_2(k) \end{bmatrix} \qquad (6.73)$$

where

$$\mathbf{H} = \begin{bmatrix} I_{n_1} + \epsilon \overline{A_1} & \epsilon \overline{A_2} & \epsilon^2 \overline{S_1} & \epsilon \overline{S_2} \\ \overline{A_3} & \overline{A_4} & \epsilon \overline{S_3} & \overline{S_4} \\ \overline{Q_1} & \overline{Q_2} & I_{n_1} + \epsilon A_{11}^T & A_{21}^T \\ \overline{Q_3} & \overline{Q_4} & \epsilon A_{12}^T & A_{22}^T \end{bmatrix} \qquad (6.74)$$

(see Appendix 6.1).

Interchanging second and third rows in (6.73) produces

OPEN-LOOP CONTROL

$$\begin{bmatrix} x_1(k+1) \\ \epsilon\lambda_1(k+1) \\ x_2(k+1) \\ \lambda_2(k+1) \end{bmatrix} = \begin{bmatrix} I_{n_1} + \epsilon\overline{A_1} & \epsilon\overline{S_1} & \epsilon\overline{A_2} & \epsilon\overline{S_2} \\ \epsilon\overline{Q_1} & I_{n_1} + \epsilon\overline{A_{11}^T} & \epsilon\overline{Q_2} & \epsilon\overline{A_{21}^T} \\ \overline{A_3} & \overline{S_3} & \overline{A_4} & \overline{S_4} \\ \overline{Q_3} & \overline{A_{12}^T} & \overline{Q_4} & \overline{A_{22}^T} \end{bmatrix} \begin{bmatrix} x_1(k) \\ \epsilon\lambda_1(k) \\ x_2(k) \\ \lambda_2(k) \end{bmatrix}$$

$$= \begin{bmatrix} I + \epsilon T_1 & \epsilon T_2 \\ T_3 & T_4 \end{bmatrix} \begin{bmatrix} x_1(k) \\ \epsilon\lambda_1(k) \\ x_2(k) \\ \lambda_2(k) \end{bmatrix}$$

(6.75)

where

$$T_1 = \begin{bmatrix} \overline{A_1} & \overline{S_1} \\ \overline{Q_1} & \overline{A_{11}^T} \end{bmatrix}, \quad T_2 = \begin{bmatrix} \overline{A_2} & \overline{S_2} \\ \overline{Q_2} & \overline{A_{21}^T} \end{bmatrix}$$

$$T_3 = \begin{bmatrix} \overline{A_3} & \overline{S_3} \\ \overline{Q_3} & \overline{A_{12}^T} \end{bmatrix}, \quad T_4 = \begin{bmatrix} \overline{A_4} & \overline{S_4} \\ \overline{Q_4} & \overline{A_{22}^T} \end{bmatrix}$$

(6.76)

Introducing the notation

$$U(k) = \begin{bmatrix} x_1(k) \\ \epsilon\lambda_1(k) \end{bmatrix}, \quad V(k) = \begin{bmatrix} x_2(k) \\ \lambda_2(k) \end{bmatrix}$$

(6.77)

we get the singularly perturbed discrete system under new notation

$$U(k+1) = (I + \epsilon T_1) U(k) + \epsilon T_2 V(k)$$
$$V(k+1) = T_3 U(k) + T_4 V(k)$$

(6.78)

Applying Chang's transformation (Chang, 1972)

$$\mathbf{T_1} = \begin{bmatrix} I - \epsilon HL & -\epsilon H \\ L & I \end{bmatrix}, \quad \mathbf{T_1^{-1}} = \begin{bmatrix} I & \epsilon H \\ -L & I - \epsilon LH \end{bmatrix}$$

(6.79)

$$\begin{bmatrix} U(k) \\ V(k) \end{bmatrix} = \mathbf{T_1} \begin{bmatrix} \overline{U}(k) \\ \overline{V}(k) \end{bmatrix}$$

to (6.78) produces two completely decoupled subsystems

$$\overline{U}(k+1) = (I + \epsilon T_1 - \epsilon T_2 L)\overline{U}(k),$$
$$\overline{V}(k+1) = (T_4 + \epsilon LT_2)\overline{V}(k)$$

(6.80)

where L and H satisfy

$$0 = H + T_2 - HT_4 + \epsilon (T_1 - T_2 L) H - \epsilon H L T_2$$
$$0 = -L + T_4 L - T_3 - \epsilon L (T_1 - T_2 L) \tag{6.81}$$

Expanding (6.67) by using the partitioned matrices given by (6.64) and (6.71)-(6.72), and identifying the terms for the matrix T_4, we obtain

$$T_4 = \begin{bmatrix} A_4 + S_2 A_4^{-T} Q_3 & -S_2 A_4^{-T} \\ -A_4^{-T} Q_3 & A_4^{-T} \end{bmatrix} + O(\epsilon) \tag{6.82}$$

which is an $O(\epsilon)$ perturbation of the Hamiltonian matrix of the fast subsystem. Thus, the matrix T_4 has no eigenvalues on the unit circle, so that $(T_4 - I)$ is a nonsingular matrix, which implies the existence of the unique solutions for L and H in (6.81). Matrices L and H can be obtained by using the Newton recursive algorithm from (Gajic et al., 1990) with the rate of convergence is $O\left(\epsilon^{2^j}\right)$, where j is the number of iterations used to solve L in (6.81).

The boundary conditions are changed due to an interchange of $\lambda_1(k)$ and $x_2(k)$, which modifies matrices in (6.67) as follows

$$M_2 \begin{bmatrix} U(0) \\ V(0) \end{bmatrix} + N_2 \begin{bmatrix} U(n) \\ V(n) \end{bmatrix} = c_1 \tag{6.83}$$

where

$$M_2 = \begin{bmatrix} I_{n_1} & 0 & 0 & 0 \\ 0 & 0 & 0 & 0 \\ 0 & 0 & I_{n_2} & 0 \\ 0 & 0 & 0 & 0 \end{bmatrix}, \quad c_1 = \begin{bmatrix} x_1(0) \\ 0 \\ x_2(0) \\ 0 \end{bmatrix}$$

$$N_2 = \begin{bmatrix} 0 & 0 & 0 & 0 \\ -F_1 & I_{n_1} & -\epsilon F_2 & 0 \\ 0 & 0 & 0 & 0 \\ -F_2^T & 0 & -F_3 & I_{n_2} \end{bmatrix} \tag{6.84}$$

The nonsingular transformation (6.79) applied to (6.83) produces

$$M_3 \begin{bmatrix} \overline{U}(0) \\ \overline{V}(0) \end{bmatrix} + N_3 \begin{bmatrix} \overline{U}(n) \\ \overline{V}(n) \end{bmatrix} = c_1 \tag{6.85}$$

where

$$M_3 = M_2 \mathbf{T_1}, \quad N_3 = N_2 \mathbf{T_1} \tag{6.86}$$

Solutions of (6.80) are then given by

$$\overline{U}(k) = (I + \epsilon T_1 - \epsilon T_2 L)^k \overline{U}(0)$$
$$\overline{V}(k) = (T_4 + \epsilon L T_2)^k \overline{V}(0) \tag{6.87}$$

We can eliminate $\overline{U}(n)$ and $\overline{V}(n)$ from (6.85) such that

$$\alpha(\epsilon) \begin{bmatrix} \overline{U}(0) \\ \overline{V}(0) \end{bmatrix} = c_1 \tag{6.88}$$

where

$$\alpha(\epsilon) = \left\{ M_3 + N_3 \begin{bmatrix} (I + \epsilon T_1 - \epsilon T_2 L)^n & 0 \\ 0 & (T_4 + \epsilon L T_2)^n \end{bmatrix} \right\} \tag{6.89}$$

$\overline{U}(0)$ and $\overline{V}(0)$ can be obtained from (6.88) provided the matrix $\alpha(\epsilon)$ is invertible. It is shown in Appendix 6.2 that the matrix $\alpha(\epsilon)$ is invertible for sufficiently small values of ϵ. Thus, we are able to find $\overline{U}(k)$ and $\overline{V}(k)$ from (6.87). Using (6.79), we can find $U(k)$ and $V(k)$.

After getting the solutions of $U(k)$ and $V(k)$, we can use the following relations to get the values for $\lambda_1(k)$ and $\lambda_2(k)$.

$$\begin{bmatrix} x_1(k) \\ \epsilon\lambda_1(k) \end{bmatrix} = \begin{bmatrix} U_1(k) \\ U_2(k) \end{bmatrix} = U(k), \quad \begin{bmatrix} x_2(k) \\ \lambda_2(k) \end{bmatrix} = \begin{bmatrix} V_1(k) \\ V_2(k) \end{bmatrix} = V(k) \tag{6.90}$$

The only difficulty we may encounter in the procedure to compute $\alpha(\epsilon)$ in (6.89) is when an ill-defined problem occurs due to presence of unstable modes in T_4 giving rise to large value of $(T_4 + \epsilon L T_2)^n$ for large values of n. In such a case we refer to the $O(\epsilon)$ solution as given in (Naidu, 1988).

6.6 Case Study: F-8 Aircraft Control Problem

In order to demonstrate the proposed method, we study the linearized model of F-8 aircraft from Section 2.6. The problem matrices $A, B, Q,$ and R are given in Section 2.6. The system initial condition is chosen as

$$x^T(0) = [1 \quad 1 \quad 1 \quad 1]$$

and the terminal penalty matrix is assumed to be

$$F = diag[0.5 \quad 0.5 \quad 0.01 \quad 0.01]$$

Number of iterations j	$J^{(j)}$	$J_{opt} - J^{(j)}$
1	2.6070	0.2787
2	2.3292	0.0009
3	2.3285	0.0002
4	2.3283	< 0.000001

Table 6.3: Values of the performance criterion

The small perturbation parameter ϵ equals 1/30 and the terminal time is $n = 9$.

With the proposed method, simulation results for an approximate open-loop optimal control (6.65) are obtained by using the package MATLAB (Hill, 1988). The approximate and optimal values of the performance criterion are presented in Table 6.3.

Table 6.4 shows the approximate and optimal values of the control input $u(k)$. The approximate control is defined as

$$u^{(j)}(k) = -R^{-1}B^T\lambda^{(j)}(k+1)$$

where j stands for the number of iterations used to solve recursively equation (6.81).

We can see that the control $u(k)$ and the performance criterion converge very rapidly to the optimum values. It can be seen that the error of the performance criterion reduces with the rate of $O(\epsilon^2)$, which is consistent with the obtained analytical results.

6.7 Open-Loop Discrete Weakly Coupled Control Problem

In this section, we will study the open-loop optimal control problem of discrete weakly coupled systems in terms of the reduced-order difference equations. A weakly coupled linear discrete system is represented by (Gajic et al., 1990)

$$x_1(k+1) = A_1x_1(k) + \epsilon A_2x_2(k) + B_1u_1(k) + \epsilon B_2u_2(k)$$
$$x_2(k+1) = \epsilon A_3x_1(k) + A_4x_2(k) + \epsilon B_3u_1(k) + B_4u_2(k)$$
$$x_1(0) = x_{10}, \quad x_2(0) = x_{20}$$

$$(6.91)$$

179

k	$u^{(0)}(k)$	$u^{(1)}(k)$	$u^{(2)}(k)$	$u^{(3)}(k) = u_{opt}(k)$
0	0.3838 -0.0063	0.3950 -0.0063	0.3948 -0.0063	0.3947 -0.0063
1	0.4876 -0.0060	0.4977 -0.0060	0.4973 -0.0060	0.4973 -0.0060
2	0.5120 -0.0056	0.5217 -0.0056	0.5214 -0.0056	0.5214 -0.0056
3	0.5495 -0.0052	0.5601 -0.0052	0.5599 -0.0053	0.5599 -0.0053
4	0.5664 -0.0048	0.5769 -0.0048	0.5767 -0.0048	0.5767 -0.0048
5	0.6250 -0.0044	0.6358 -0.0044	0.6357 -0.0044	0.6357 -0.0044
6	0.6713 -0.0039	0.6825 -0.0039	0.6825 -0.0039	0.6825 -0.0039
7	0.5265 -0.0035	0.5359 -0.0034	0.5359 -0.0034	0.5359 -0.0034
8	0.8580 -0.0029	0.8695 -0.0029	0.8694 -0.0029	0.8694 -0.0029
9	0.8929 -0.0021	0.9055 -0.0021	0.9060 -0.0021	0.9059 -0.0021

Table 6.4: Approximate and optimal values of $u(k)$

with state variables $x_i \in \Re^{n_i}$ and control inputs $u_i \in \Re^{m_i}$, $i = 1, 2$, respectively, where ϵ is a small coupling parameter. We will assume that all given matrices are constant. The performance criterion of the corresponding linear-quadratic control problem is defined by (6.63) with

$$x(k) = \begin{bmatrix} x_1(k) \\ x_2(k) \end{bmatrix}, \quad Q = \begin{bmatrix} Q_1 & \epsilon Q_2 \\ \epsilon Q_2^T & Q_3 \end{bmatrix} \geq 0$$

$$F = \begin{bmatrix} F_1 & \epsilon F_2 \\ \epsilon F_2^T & F_3 \end{bmatrix} \geq 0, \quad R = \begin{bmatrix} R_1 & 0 \\ 0 & R_2 \end{bmatrix} > 0$$

(6.92)

The open-loop optimal control problem has the solution given by (6.65).

The Hamiltonian form of this optimal control problem is given by (6.66)-(6.67), that is

$$\begin{bmatrix} x(k+1) \\ \lambda(k+1) \end{bmatrix} = \mathbf{H} \begin{bmatrix} x(k) \\ \lambda(k) \end{bmatrix}$$

where

$$\mathbf{H} = \begin{bmatrix} A + BR^{-1}B^T A^{-T} Q & -BR^{-1}B^T A^{-T} \\ -A^{-T} Q & A^{-T} \end{bmatrix}$$

with boundary conditions expressed in the standard form as

$$M \begin{bmatrix} x(0) \\ \lambda(0) \end{bmatrix} + N \begin{bmatrix} x(n) \\ \lambda(n) \end{bmatrix} = c \qquad (6.93)$$

Note that

$$M = \begin{bmatrix} I & 0 \\ 0 & 0 \end{bmatrix}, \quad N = \begin{bmatrix} 0 & 0 \\ -F & I \end{bmatrix}, \quad c = \begin{bmatrix} x(0) \\ 0 \end{bmatrix} \qquad (6.94)$$

Matrices A and S have the forms

$$A = \begin{bmatrix} A_1 & \epsilon A_2 \\ \epsilon A_3 & A_4 \end{bmatrix}, \quad S = BR^{-1}B^T = \begin{bmatrix} S_1 & \epsilon Z \\ \epsilon Z^T & S_2 \end{bmatrix} \qquad (6.95)$$

Similar to Section 6.3, for discrete weakly coupled systems, the Hamiltonian matrix retains the weakly coupled form by interchanging some state and costate variables so that it can be block diagonalized via the nonsingular transformations presented in Sections 3.3 and 3.4.

Partitioning vector $\lambda(k)$ as $\lambda(k) = [\lambda_1^T(k) \quad \lambda_2^T(k)]^T$ with $\lambda_1(k) \in \Re^{n_1}$ and $\lambda_2(k) \in \Re^{n_2}$, we get

$$\begin{bmatrix} x_1(k+1) \\ x_2(k+1) \\ \lambda_1(k+1) \\ \lambda_2(k+1) \end{bmatrix} = \begin{bmatrix} \overline{A_1} & \epsilon \overline{A_2} & \overline{S_1} & \epsilon \overline{S_2} \\ \epsilon \overline{A_3} & \overline{A_4} & \epsilon \overline{S_3} & \overline{S_4} \\ \overline{Q_1} & \epsilon \overline{Q_2} & A_{11}^T & \epsilon A_{21}^T \\ \epsilon \overline{Q_3} & \overline{Q_4} & \epsilon A_{12}^T & A_{22}^T \end{bmatrix} \begin{bmatrix} x_1(k) \\ x_2(k) \\ \lambda_1(k) \\ \lambda_2(k) \end{bmatrix} = \mathbf{H} \begin{bmatrix} x_1(k) \\ x_2(k) \\ \lambda_1(k) \\ \lambda_2(k) \end{bmatrix} \qquad (6.96)$$

(see Appendix 6.3).

Interchanging the second and third rows in (6.96) produces

$$\begin{bmatrix} x_1(k+1) \\ \lambda_1(k+1) \\ x_2(k+1) \\ \lambda_2(k+1) \end{bmatrix} = \begin{bmatrix} \overline{A_1} & \overline{S_1} & \epsilon\overline{A_2} & \epsilon\overline{S_2} \\ \overline{Q_1} & A_{11}^T & \epsilon Q_2 & \epsilon A_{21}^T \\ \epsilon\overline{A_3} & \epsilon\overline{S_3} & \overline{A_4} & \overline{S_4} \\ \epsilon\overline{Q_3} & \epsilon A_{12}^T & \overline{Q_4} & A_{22}^T \end{bmatrix} \begin{bmatrix} x_1(k) \\ \lambda_1(k) \\ x_2(k) \\ \lambda_2(k) \end{bmatrix}$$

$$= \begin{bmatrix} T_1 & \epsilon T_2 \\ \epsilon T_3 & T_4 \end{bmatrix} \begin{bmatrix} x_1(k) \\ \lambda_1(k) \\ x_2(k) \\ \lambda_2(k) \end{bmatrix}$$

(6.97)

where

$$T_1 = \begin{bmatrix} \overline{A_1} & \overline{S_1} \\ \overline{Q_1} & A_{11}^T \end{bmatrix}, \quad T_2 = \begin{bmatrix} \overline{A_2} & \overline{S_2} \\ \overline{Q_2} & A_{21}^T \end{bmatrix}$$

(6.98)

$$T_3 = \begin{bmatrix} \overline{A_3} & \overline{S_3} \\ \overline{Q_3} & A_{12}^T \end{bmatrix}, \quad T_4 = \begin{bmatrix} \overline{A_4} & \overline{S_4} \\ \overline{Q_4} & A_{22}^T \end{bmatrix}$$

Introducing the notation

$$U(k) = \begin{bmatrix} x_1(k) \\ \lambda_1(k) \end{bmatrix}, \quad V(k) = \begin{bmatrix} x_2(k) \\ \lambda_2(k) \end{bmatrix}$$

(6.99)

we get the weakly coupled discrete system in the new coordinates

$$U(k+1) = T_1 U(k) + \epsilon T_2 V(k)$$
$$V(k+1) = \epsilon T_3 U(k) + T_4 V(k)$$

(6.100)

Applying a nonsingular transformation of the form

$$\mathbf{T_2} = \begin{bmatrix} I & -\epsilon L \\ \epsilon H & I - \epsilon^2 HL \end{bmatrix}, \quad \mathbf{T_2^{-1}} = \begin{bmatrix} I - \epsilon^2 LH & \epsilon L \\ -\epsilon H & I \end{bmatrix}$$

(6.101)

$$\begin{bmatrix} U(k) \\ V(k) \end{bmatrix} = \mathbf{T_2} \begin{bmatrix} \overline{U}(k) \\ \overline{V}(k) \end{bmatrix}$$

to (6.100) is producing two completely decoupled subsystems

$$\overline{U}(k+1) = \left(T_1 - \epsilon^2 LT_3\right) \overline{U}(k)$$
$$\overline{V}(k+1) = \left(T_4 + \epsilon^2 T_3 L\right) \overline{V}(k)$$

(6.102)

where L and H satisfy

$$H \left(T_1 - \epsilon^2 LT_3\right) - \left(T_4 + \epsilon^2 T_3 L\right) H + T_3 = 0$$
$$T_1 L + T_2 - LT_4 - \epsilon^2 LT_3 L = 0 \tag{6.103}$$

Matrices L and H can be obtained with required accuracy, by using numerical techniques developed in Chapter 3 for solving (6.103), where the rate of convergence is $O\left(\epsilon^2\right)$. Thus, after j iterations, one gets the approximations $L^{(j)} = L + O\left(\epsilon^{2j}\right)$ and $H^{(j)} = H + O\left(\epsilon^{2j}\right)$. Using $L^{(j)}$, $H^{(j)}$ instead of L and H, will perturb the coefficients of the corresponding systems of linear differential equations by $O\left(\epsilon^2\right)$, which implies that the same accuracy of the system solutions is obtained.

The boundary conditions are changed due to an interchange of $\lambda_1\left(k\right)$ and $x_2\left(k\right)$ which modifies matrices in (6.93) as follows

$$M_1 \begin{bmatrix} U\left(0\right) \\ V\left(0\right) \end{bmatrix} + N_1 \begin{bmatrix} U\left(n\right) \\ V\left(n\right) \end{bmatrix} = c_1 \tag{6.104}$$

where

$$M_1 = \begin{bmatrix} I_{n_1} & 0 & 0 & 0 \\ 0 & 0 & 0 & 0 \\ 0 & 0 & I_{n_2} & 0 \\ 0 & 0 & 0 & 0 \end{bmatrix}, \quad c_1 = \begin{bmatrix} x_1\left(0\right) \\ 0 \\ x_2\left(0\right) \\ 0 \end{bmatrix}$$

$$N_1 = \begin{bmatrix} 0 & 0 & 0 & 0 \\ -F_1 & I_{n_1} & -\epsilon F_2 & 0 \\ 0 & 0 & 0 & 0 \\ -\epsilon F_2^T & 0 & -F_3 & I_{n_2} \end{bmatrix} \tag{6.105}$$

The nonsingular transformation (6.101) applied to (6.104) produces

$$M_2 \begin{bmatrix} \overline{U}\left(0\right) \\ \overline{V}\left(0\right) \end{bmatrix} + N_2 \begin{bmatrix} \overline{U}\left(n\right) \\ \overline{V}\left(n\right) \end{bmatrix} = c_1 \tag{6.106}$$

where

$$M_2 = M_1 T_2, \quad N_2 = N_1 T_2 \tag{6.107}$$

Solutions of (6.102) are then given by

$$\overline{U}\left(k\right) = \left(T_1 - \epsilon^2 LT_3\right)^k \overline{U}\left(0\right)$$
$$\overline{V}\left(k\right) = \left(T_4 + \epsilon^2 T_3 L\right)^k \overline{V}\left(0\right) \tag{6.108}$$

We can eliminate $\overline{U}\left(n\right)$ and $\overline{V}\left(n\right)$ from (6.106) such that

183

$$\beta(\epsilon)\begin{bmatrix} \overline{U}(0) \\ \overline{V}(0) \end{bmatrix} = c_1 \qquad (6.109)$$

where

$$\beta(\epsilon) = \left\{ M_2 + N_2 \begin{bmatrix} (T_1 - \epsilon^2 L T_3)^n & 0 \\ 0 & (T_4 + \epsilon^2 T_3 L)^n \end{bmatrix} \right\} \qquad (6.110)$$

It is shown in Appendix 6.4 that $\beta(\epsilon)$ is nonsingular, that is, this system of linear algebraic equations has a unique solution, assuming that a coupling parameter ϵ is sufficiently small. Since $\beta(\epsilon)$ is invertible, hence $\overline{U}(0)$ and $\overline{V}(0)$ can be obtained.

Now we are able to find $\overline{U}(k)$ and $\overline{V}(k)$ from (6.108). Using (6.101), we can find $U(k)$ and $V(k)$.

After getting the solutions of $U(k)$ and $V(k)$, we can use the following relations to get the values for $\lambda_1(k)$ and $\lambda_2(k)$.

$$\begin{aligned} \begin{bmatrix} x_1(k) \\ \lambda_1(k) \end{bmatrix} &= \begin{bmatrix} U_1(k) \\ U_2(k) \end{bmatrix} = U(k) \\ \begin{bmatrix} x_2(k) \\ \lambda_2(k) \end{bmatrix} &= \begin{bmatrix} V_1(k) \\ V_2(k) \end{bmatrix} = V(k) \end{aligned} \qquad (6.111)$$

The costate variable $\lambda(k)$ and the optimal control law are therefore found such that

$$u(k) = -R^{-1}B^T \lambda^{(j)}(k+1) + O(\epsilon^{2j}) \qquad (6.112)$$

Apparently, as j increases, the control defined in (6.112) converges very rapidly to the optimal solution.

6.8 Numerical Example

In order to demonstrate the proposed method, a discrete system (Katzberg, 1977) is studied. The system matrices are

$$A = \begin{bmatrix} 0.964 & 0.18 & 0.017 & 0.019 \\ -0.342 & 0.802 & 0.162 & 0.179 \\ 0.016 & 0.019 & 0.983 & 0.181 \\ 0.144 & 0.179 & -0.163 & 0.82 \end{bmatrix}$$

$$B^T = \begin{bmatrix} 0.019 & 0.180 & 0.005 & -0.054 \\ 0.001 & 0.019 & 0.019 & 0.181 \end{bmatrix}$$

with initial conditions

$$x^T(0) = \begin{bmatrix} 1 & 1 & 1 & 1 \end{bmatrix}$$

The weighting matrices are chosen as $R = I_2$, $Q = 0.1I_4$ with the terminal penalty matrix $F = 0.5I_4$. The small coupling parameter ϵ is 0.329, and the final time is $n = 5$.

Table 6.5 shows the approximate and optimal values of control $u(k)$. The approximate control is defined as

$$u^{(j)}(k) = -R^{-1}B^T\lambda^{(j)}(k+1)$$

where j stands for the number of iterations used to solve L recursively in equation (6.103). The recursive solution for L after six iterations is given in (6.113). For $j = 0$

$$L^{(j+1)} - L^{(j)} = \begin{bmatrix} -1.1436 & -1.3305 & 0.0562 & -0.2300 \\ 1.8946 & 0.1277 & 0.3022 & -0.2789 \\ 0.8015 & -1.7028 & -2.5535 & 4.2259 \\ 2.4466 & 0.0174 & -2.787 & 1.5939 \end{bmatrix}$$

and for $j = 6$

$$L^{(j+1)} - L^{(j)} = 10^{-5}\begin{bmatrix} -0.0009 & -0.0013 & -0.0049 & 0.0021 \\ 0.0001 & -0.0079 & -0.0091 & -0.0007 \\ -0.0391 & 0.1256 & 0.112 & -0.1037 \\ -0.0499 & 0.0656 & 0.0934 & -0.059 \end{bmatrix}$$

$$(6.113)$$

In order to indicate the relative differences among optimal and approximate control strategies, they are also presented in Figures 6.1-6.2. The optimal (solid lines) and approximate system trajectories are presented in Figure 6.3-6.4.

Exercise 6.3: Use the MATLAB package to discretize a magnetic tape control system from Section 2.2.3 with the sampling rate $T = 1$. Then write a program to find the discrete-time optimal open-loop control of the corresponding singularly perturbed system in terms of the reduced-order slow and fast subsystems. Repeat the procedure with the sampling rate $T = 0.1$.

$\triangle$

k	$u^{(0)}(k)$	$u^{(1)}(k)$	$u^{(2)}(k)$	$u^{(3)}(k)$	$u^{(4)}(k)$	$u^{(5)}(k) =$ optimal
1	-0.0745 -0.3616	-0.1228 -0.1738	-0.0612 -0.0606	0.0397 -0.1632	0.0147 -0.1333	0.0131 -0.1323
2	0.0709 -0.3208	-0.0674 -0.1354	-0.0214 -0.0430	0.0695 -0.1171	0.0465 -0.0926	0.0447 -0.0912
3	0.1532 -0.3278	-0.0255 -0.0898	0.0123 -0.0259	0.0883 -0.0759	0.0685 -0.0569	0.0668 -0.0556
4	0.1955 -0.3835	0.0074 -0.0409	0.0429 -0.0047	0.0994 -0.0390	0.0834 -0.0251	0.0819 -0.0239
5	0.2178 -0.4884	0.0359 0.0083	0.0744 0.0257	0.1068 -0.0056	0.0946 0.0042	0.0934 0.0053

Table 6.5: Approximate and optimal values of $u(k)$

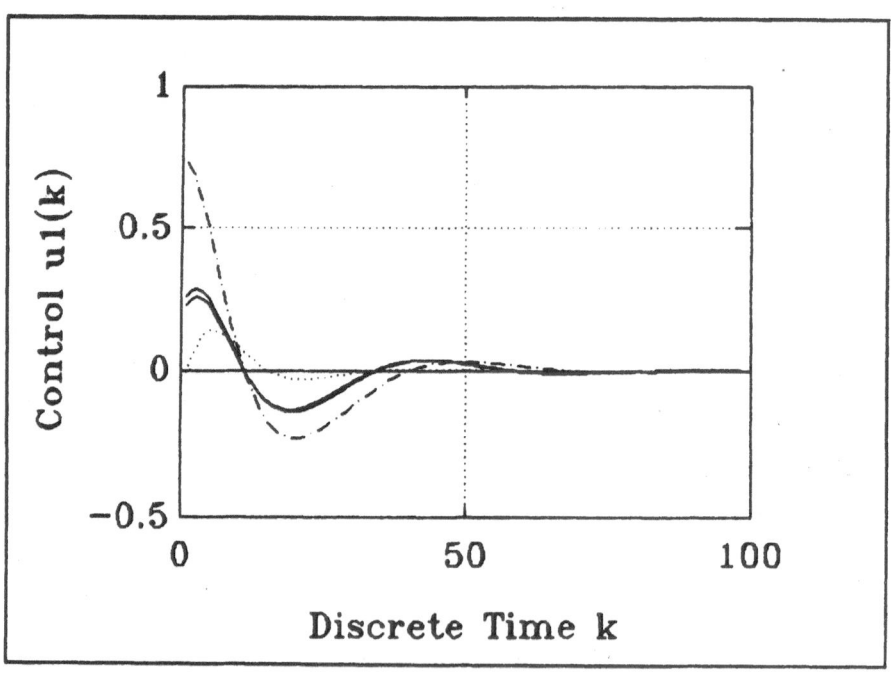

Figure 6.1: Optimal and approximate control strategies $u_1(k)$

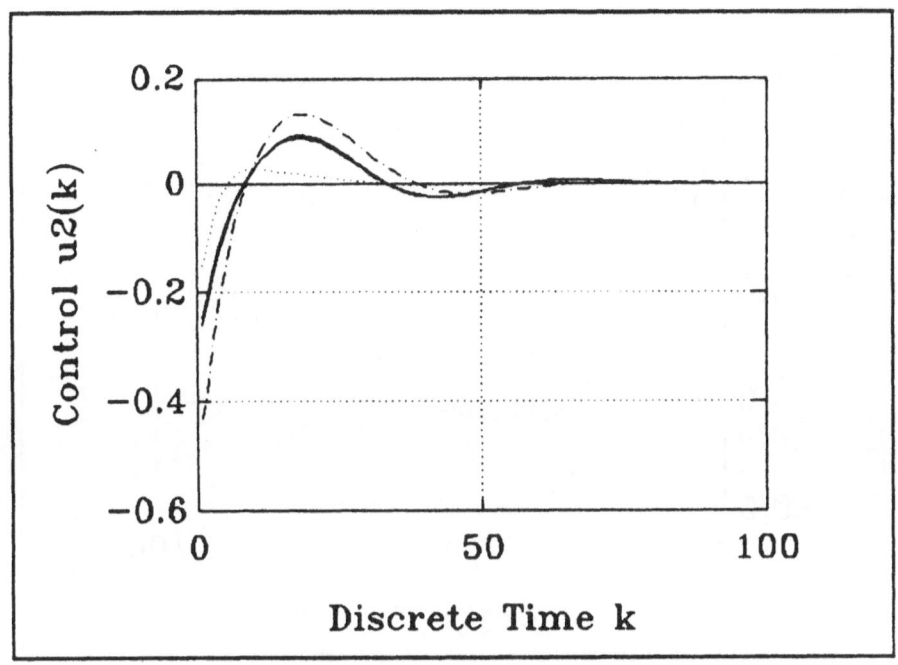

Figure 6.2: Optimal and approximate control strategies $u_2(k)$

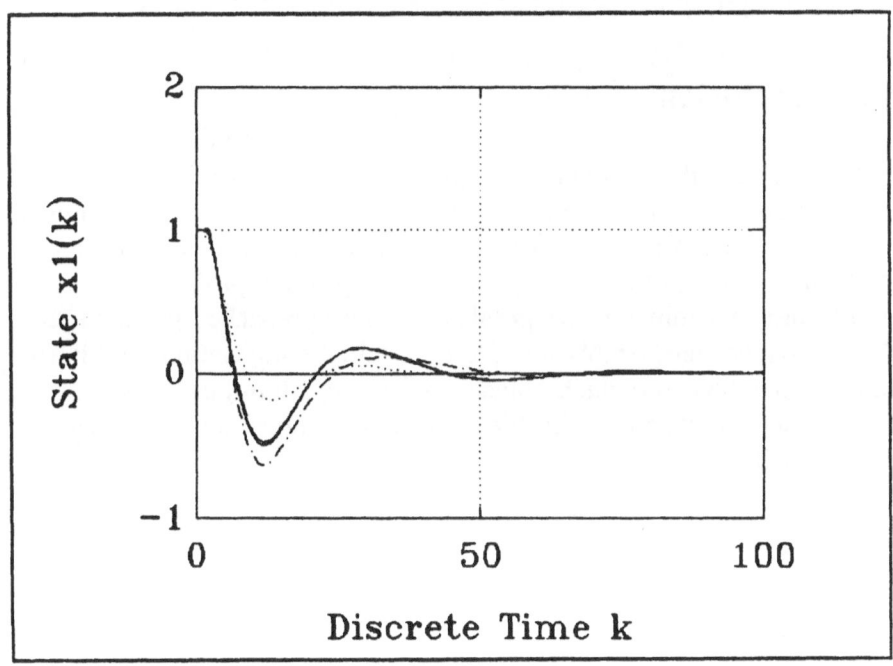

Figure 6.3: Optimal and approximate system trajectories $x_1(k)$

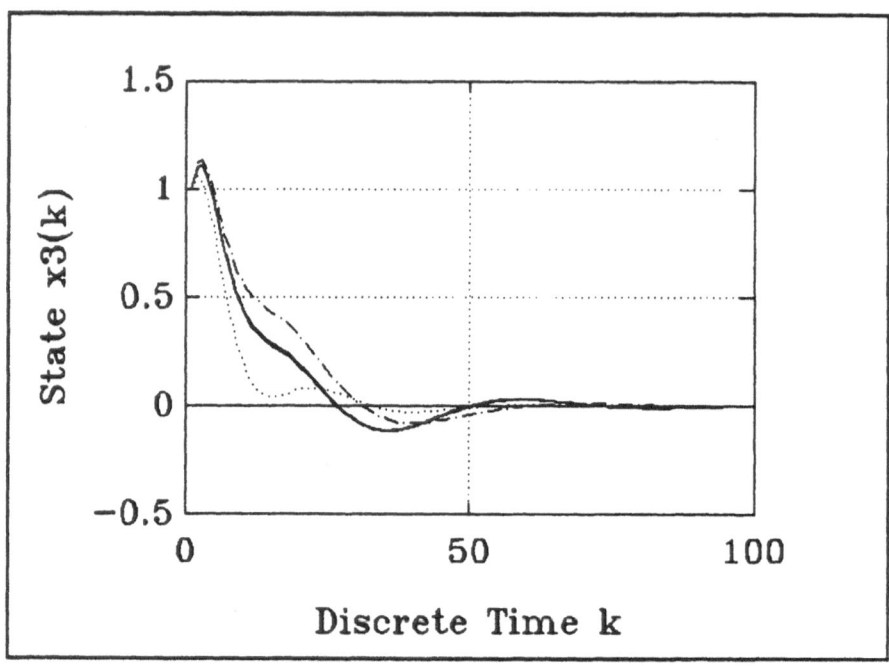

Figure 6.4: Optimal and approximate system trajectories $x_3(k)$

6.9 Conclusion

In this chapter, the optimal finite time open-loop continuous and discrete control problems for both singularly perturbed and weakly coupled systems are solved with any desired accuracy in terms of reduced-order problems. Corresponding two-point boundary value problems are converted into two initial value problems. The approaches given in this chapter reduce considerably the size of required computations and introduce full parallelism in the problems under study. It has been shown that the obtained results are applicable to both continuous-time and discrete-time domains.

Appendix 6.1

From (6.67)

$$H = \begin{bmatrix} A + BR^{-1}B^T A^{-T}Q & -BR^{-1}B^T A^{-T} \\ -A^{-T}Q & A^{-T} \end{bmatrix} \qquad (a.1)$$

Since A^{-T} has the same structure as A^T, that is

$$A^{-T} = \begin{bmatrix} I_{n_1} + O(\epsilon) & O(1) \\ O(\epsilon) & O(1) \end{bmatrix} \qquad (a.2)$$

then

$$A^{-T}Q = \begin{bmatrix} I_{n_1} + O(\epsilon) & O(1) \\ O(\epsilon) & O(1) \end{bmatrix} \begin{bmatrix} O(1) & O(1) \\ O(1) & O(1) \end{bmatrix} = \begin{bmatrix} O(1) & O(1) \\ O(1) & O(1) \end{bmatrix}$$

$$BR^{-1}B^T A^{-T} = \begin{bmatrix} O(\epsilon^2) & O(\epsilon) \\ O(\epsilon) & O(1) \end{bmatrix} \begin{bmatrix} I_{n_1} + O(\epsilon) & O(1) \\ O(\epsilon) & O(1) \end{bmatrix}$$

$$= \begin{bmatrix} O(\epsilon^2) & O(\epsilon) \\ O(\epsilon) & O(1) \end{bmatrix}$$

$$A + BR^{-1}B^T A^{-T}Q = \begin{bmatrix} I + O(\epsilon) & O(\epsilon) \\ O(1) & O(1) \end{bmatrix}$$

$$H = \begin{bmatrix} I_{n_1} + \epsilon \overline{A_1} & \epsilon \overline{A_2} & \epsilon^2 \overline{S_1} & \epsilon \overline{S_2} \\ \overline{A_3} & \overline{A_4} & \epsilon \overline{S_3} & \overline{S_4} \\ \overline{Q_1} & \overline{Q_2} & I_{n_1} + \epsilon A_{11}^T & A_{21}^T \\ \overline{Q_3} & \overline{Q_4} & \epsilon A_{12}^T & A_{22}^T \end{bmatrix}$$

$$(a.3)$$

Note that it is easy to obtain overlined matrices in the process of programming and it is of no interest to get corresponding analytical expressions.

Appendix 6.2

Lemma 6.3 *If the optimal control problem defined in (6.62)-(6.63) satisfies the stabilizability-detectability assumption, the matrix $\alpha(\epsilon)$ in (6.89) is invertible.*

◇

189

Proof: The matrix $\alpha(\epsilon)$ can be written as

$$\alpha(\epsilon) = M_3 + N_3 \begin{bmatrix} I & 0 \\ 0 & T_4^n \end{bmatrix} + O(\epsilon) \qquad \text{(b.1)}$$

Let

$$T_4^n = \begin{bmatrix} \phi_{11} & \phi_{12} \\ \phi_{21} & \phi_{22} \end{bmatrix} \qquad \text{(b.2)}$$

then by using expressions for M_3 and N_3 given by (6.86), we obtain

$$\alpha(\epsilon) = \begin{bmatrix} I_{n_1} & 0 & 0 & 0 \\ * & I_{n_1} & 0 & 0 \\ * & * & I_{n_2} & 0 \\ * & * & * & \phi_{22} - F_3\phi_{12} \end{bmatrix} + O(\epsilon) \qquad \text{(b.3)}$$

where asterisks denote terms which are not important for the non-singularity of $\alpha(\epsilon)$.

From (b.3), note that if the matrix $\phi_{22} - F_3\phi_{12}$ is invertible, the matrix $\alpha(\epsilon)$ will be invertible for sufficiently small value of ϵ. It will be shown in the following that the invertibility of $\phi_{22} - F_3\phi_{12}$ follows from the assumption that the system is stabilizable-detectable. From (6.79) and (6.87), we can write

$$\begin{bmatrix} U(n) \\ V(n) \end{bmatrix} = T_1^{-1} \begin{bmatrix} I + O(\epsilon) & 0 \\ 0 & T_4^n + O(\epsilon) \end{bmatrix} T_1 \begin{bmatrix} U(0) \\ V(0) \end{bmatrix} \qquad \text{(b.4)}$$

By using the values of T_1 and T_1^{-1} from (6.79), we obtain

$$\begin{aligned} U(n) &= U(0) + O(\epsilon) \\ V(n) &= (T_4^n L - L)U(0) + T_4^n V(0) + O(\epsilon) \end{aligned} \qquad \text{(b.5)}$$

Let

$$T_4^n L - L = \begin{bmatrix} \psi_{11} & \psi_{12} \\ \psi_{21} & \psi_{22} \end{bmatrix} \qquad \text{(b.6)}$$

then by using (6.77) and (b.6) in (b.5) yields

$$\begin{aligned} x_2(n) &= \psi_{11}x_1(0) + \phi_{11}x_2(0) + \phi_{12}\lambda_2(0) + O(\epsilon) \\ \lambda_2(n) &= \psi_{21}x_1(0) + \phi_{21}x_2(0) + \phi_{22}\lambda_2(0) + O(\epsilon) \end{aligned} \qquad \text{(b.7)}$$

From the boundary condition $\lambda(n) = Fx(n)$, we have

$$\lambda_2(n) = F_2^T x_1(n) + F_3 x_2(n) \qquad \text{(b.8)}$$

Since $U(n) = U(0) + O(\epsilon)$, therefore, $x_1(n) = x_1(0) + O(\epsilon)$. Using this fact and substituting the value of $x_2(n)$ from (b.7) into (b.8), we get

$$
\begin{aligned}
(\phi_{22} - F_3\phi_{12})\lambda_2(0) &= (F_3\phi_{11} - \phi_{21})x_2(0) \\
&+ (F_2^T - \phi_{21} + F_3\phi_{11})x_1(0) + O(\epsilon)
\end{aligned}
\tag{b.9}
$$

Since the system is stabilizable-detectable, the control $u(0)$, and hence $\lambda_2(0)$ exist, which concludes that $(\phi_{22} - F_3\phi_{12})$ must be invertible. Thus, for sufficiently small ϵ, the matrix $\alpha(\epsilon)$ is invertible.

Appendix 6.3

From (6.67)

$$
H = \begin{bmatrix} A + BR^{-1}B^T A^{-T}Q & -BR^{-1}B^T A^{-T} \\ -A^{-T}Q & A^{-T} \end{bmatrix}
\tag{c.1}
$$

Since A^{-T} has same structure as A^T, that is

$$
A^{-T} = \begin{bmatrix} O(1) & O(\varepsilon) \\ O(\varepsilon) & O(1) \end{bmatrix}
\tag{c.2}
$$

then

$$
A^{-T}Q = \begin{bmatrix} O(1) & O(\epsilon) \\ O(\epsilon) & O(1) \end{bmatrix}\begin{bmatrix} O(1) & O(\epsilon) \\ O(\epsilon) & O(1) \end{bmatrix} = \begin{bmatrix} O(1) & O(\epsilon) \\ O(\epsilon) & O(1) \end{bmatrix}
$$

$$
\begin{aligned}
BR^{-1}B^T A^{-T} &= \begin{bmatrix} O(1) & O(\epsilon) \\ O(\epsilon) & O(1) \end{bmatrix}\begin{bmatrix} O(1) & O(\epsilon) \\ O(\epsilon) & O(1) \end{bmatrix} \\
&= \begin{bmatrix} O(1) & O(\epsilon) \\ O(\epsilon) & O(1) \end{bmatrix}
\end{aligned}
$$

$$
A + BR^{-1}B^T A^{-T}Q = \begin{bmatrix} O(1) & O(\epsilon) \\ O(\epsilon) & O(1) \end{bmatrix}
$$

$$
H = \begin{bmatrix} \overline{A_1} & \epsilon\overline{A_2} & \overline{S_1} & \epsilon\overline{S_2} \\ \epsilon\overline{A_3} & \overline{A_4} & \epsilon\overline{S_3} & \overline{S_4} \\ \overline{Q_1} & \epsilon\overline{Q_2} & A_{11}^T & \epsilon A_{21}^T \\ \epsilon\overline{Q_3} & \overline{Q_4} & \epsilon A_{12}^T & A_{22}^T \end{bmatrix}
\tag{c.3}
$$

Note again that it is easy to obtain the overlined matrices in the process of programming and it is of no interest to obtain corresponding analytical expressions.

Appendix 6.4

Let the transition matrices of the difference equations given in (6.102) be denoted as $\phi(k)$ and $\psi(k)$ respectively, and let partition them as follows

$$
\begin{aligned}
\phi(k) &= \begin{bmatrix} \phi_1(k) & \phi_2(k) \\ \phi_3(k) & \phi_4(k) \end{bmatrix} \\
\psi(k) &= \begin{bmatrix} \psi_1(k) & \psi_2(k) \\ \psi_3(k) & \psi_4(k) \end{bmatrix}
\end{aligned}
\tag{d.1}
$$

From (6.109) we have

$$
\beta(\epsilon) = \left\{ M_2 + N_2 \begin{bmatrix} \phi(k) & 0 \\ 0 & \psi(k) \end{bmatrix} \right\}
\tag{d.2}
$$

Using expressions for M_2 and N_2 defined by (6.107) we get

$$
\beta(\epsilon) = \begin{bmatrix} I_{n_1} & 0 & 0 & 0 \\ \phi_3(k) - F_1\phi_1(k) & \phi_4(k) - F_1\phi_2(k) & 0 & 0 \\ 0 & 0 & I_{n_2} & 0 \\ 0 & 0 & \psi_3(k) - F_3\psi_1(k) & \psi_4(k) - F_3\psi_2(k) \end{bmatrix}
$$

$$
+ O(\epsilon)
$$

$$
\tag{d.3}
$$

It is left as an exercise to the reader to show that under stabilizability-detectability conditions imposed on the subsystems, the matrices $\phi_4(n) - F_1\phi_2(n)$ and $\psi_4(n) - F_3\psi_2(n)$ are invertible. Thus, the matrix $\beta(\epsilon)$ is invertible for sufficiently small values of ϵ.

Exercise 6.4: Following the arguments of Appendix 6.3 show that the matrices $\phi_4(n) - F_1\phi_2(n)$ and $\psi_4(n) - F_3\psi_2(n)$ are invertible under stabilizability-detectability conditions imposed on the subsystems.

$\triangle$

Chapter 7

Exact Decompositions of Algebraic Riccati Equations

In this chapter, the algebraic Riccati equations of both singularly perturbed and weakly coupled control systems are completely and exactly decomposed into two reduced-order algebraic Riccati equations. The decomposed algebraic Riccati equations are nonsymmetric ones. It is shown that the Newton method is very efficient for solving the obtained nonsymmetric algebraic Riccati equations. Due to complete and exact decomposition of the Riccati equations, we have obtained the parallel algorithms for solving these equations. The presented procedure might produce a new insight in the singularly perturbed and weakly coupled optimal filtering and control problems since the corresponding reduced-order optimal filters and controllers are completely decoupled. The decompositions of the algebraic Lyapunov equations for both singularly perturbed and weakly coupled systems are presented in Section 3.5 in the context of the complete decomposition of the differential Lyapunov equations.

7.1 The Exact Decomposition of the Singularly Perturbed Algebraic Riccati Equation

A linear singularly perturbed control system is given by

$$\begin{aligned}
\dot{x}_1 &= A_1 x_1 + A_2 x_2 + B_1 u, & x_1(t_0) &= x_{10} \\
\epsilon \dot{x}_2 &= A_3 x_1 + A_4 x_2 + B_2 u, & x_2(t_0) &= x_{20}
\end{aligned} \tag{7.1}$$

where $x_i \in \Re^{n_i}$, $i = 1, 2$, $u \in \Re^m$ are state and control variables, respectively, and ϵ is a small positive parameter. As a parameter ϵ tends to zero, the solution behaves nonuniformly, producing a so-called stiff problem (Kokotovic et al., 1986).

The main equation of the linear optimal control theory — the Riccati equation, can be obtained from the Hamiltonian matrix. For singularly perturbed systems, after modifying some costate variables, the Hamiltonian matrix retains the singularly perturbed form by interchanging some state and costate variables so that it can be block diagonalized via the nonsingular transformation studied in Chapter 3.

The task of this section is to exploit the reduced-order subsystems to find the exact solution of the global algebraic Riccati equation in terms of the reduced-order problems — both leading to the nonsymmetric algebraic Riccati equations: pure-slow and pure-fast. It is shown that the $O(\epsilon)$ perturbations of these nonsymmetric algebraic Riccati equations are symmetric ones and equal to the well-known first-order approximations of the slow and fast algebraic Riccati equations. The solutions of the symmetric reduced-order algebraic Riccati equations play the role of the initial guesses for the Newton method which is very efficient for solving the obtained nonsymmetric Riccati equations. Furthermore, the proposed method is very suitable for parallel computations since it allows complete parallelism. In addition, due to complete and exact decomposition of the algebraic Riccati equation, the optimal filtering and control at steady state might be performed independently and in parallel in slow and fast time scales.

With (7.1), consider the performance criterion

$$J = \frac{1}{2} \int_{t_0}^{\infty} \left\{ \begin{bmatrix} x_1 \\ x_2 \end{bmatrix}^T Q \begin{bmatrix} x_1 \\ x_2 \end{bmatrix} + u^T R\, u \right\} dt \qquad (7.2)$$

with positive definite R and positive semidefinite Q.

The open-loop optimal control problem of (7.1)-(7.2) has the solution

$$u = -R^{-1} B^T p \qquad (7.3)$$

where $p \in \Re^{n_1+n_2}$ is a costate variable satisfying (Kwakernaak and Sivan, 1972)

$$\begin{bmatrix} \dot{x} \\ \dot{p} \end{bmatrix} = \begin{bmatrix} A & -S \\ -Q & -A^T \end{bmatrix} \begin{bmatrix} x \\ p \end{bmatrix} \qquad (7.4)$$

with

$$A = \begin{bmatrix} A_1 & A_2 \\ \frac{A_3}{\epsilon} & \frac{A_4}{\epsilon} \end{bmatrix}, \quad Q = \begin{bmatrix} Q_1 & Q_2 \\ Q_2^T & Q_3 \end{bmatrix}, \quad x = \begin{bmatrix} x_1 \\ x_2 \end{bmatrix} \quad (7.5)$$

$$B = \begin{bmatrix} B_1 \\ \frac{B_2}{\epsilon} \end{bmatrix}, \quad S = BR^{-1}B^T = \begin{bmatrix} S_1 & \frac{Z}{\epsilon} \\ \frac{Z^T}{\epsilon} & \frac{S_2}{\epsilon^2} \end{bmatrix}$$

The optimal closed-loop control law has the very-well known form

$$u = -R^{-1}B^T p = -R^{-1}B^T Px \quad (7.6)$$

where P satisfies the algebraic Riccati equation given by

$$0 = PA + A^T P + Q - PSP \quad (7.7)$$

Our main goal is to find the solution of (7.7) in terms of the solutions of the reduced-order pure-slow and pure-fast algebraic Riccati equations.

Partitioning p such that $p = [p_1^T \quad \epsilon p_2^T]^T$ with $p_1 \in \Re^{n_1}$ and $p_2 \in \Re^{n_2}$ and interchanging second and third rows in (7.4), we get

$$\begin{bmatrix} \dot{x}_1 \\ \dot{p}_1 \\ \dot{x}_2 \\ \dot{p}_2 \end{bmatrix} = \begin{bmatrix} T_1 & T_2 \\ \frac{T_3}{\epsilon} & \frac{T_4}{\epsilon} \end{bmatrix} \begin{bmatrix} x_1 \\ p_1 \\ x_2 \\ p_2 \end{bmatrix} \quad (7.8)$$

where

$$T_1 = \begin{bmatrix} A_1 & -S_1 \\ -Q_1 & -A_1^T \end{bmatrix}, \quad T_2 = \begin{bmatrix} A_2 & -Z \\ -Q_2 & -A_3^T \end{bmatrix}$$

$$\quad (7.9)$$

$$T_3 = \begin{bmatrix} A_3 & -Z^T \\ -Q_2^T & -A_2^T \end{bmatrix}, \quad T_4 = \begin{bmatrix} A_4 & -S_2 \\ -Q_3 & -A_4^T \end{bmatrix}$$

It is important to notice that (7.8) retains the singular perturbation form. Also, the matrix T_4 is the Hamiltonian matrix of the fast subsystem, and it is nonsingular under stabilizability-detectability conditions imposed on the fast subsystem (Kwakernaak and Sivan, 1972).

Introduce the notation

$$\begin{bmatrix} x_1 \\ p_1 \end{bmatrix} = w, \quad \begin{bmatrix} x_2 \\ p_2 \end{bmatrix} = \lambda \quad (7.10)$$

and the transformation (Chang, 1972) defined by

$$\mathbf{T}_1 = \begin{bmatrix} I - \epsilon H L & -\epsilon H \\ L & I \end{bmatrix}, \quad \mathbf{T}_1^{-1} = \begin{bmatrix} I & \epsilon H \\ -L & I - \epsilon L H \end{bmatrix} \tag{7.11}$$

where L and H satisfy

$$T_4 L - T_3 - \epsilon L (T_1 - T_2 L) = 0 \tag{7.12}$$

$$-H (T_4 + \epsilon L T_2) + T_2 + \epsilon (T_1 - T_2 L) H = 0 \tag{7.13}$$

The unique solutions of (7.12) and (7.13) exist under condition that T_4 is nonsingular.

The transformation (7.11) applied to (7.8) produces two completely decoupled subsystems

$$\dot{\eta} = (T_1 - T_2 L) \eta \tag{7.14}$$

and

$$\epsilon \dot{\xi} = (T_4 + \epsilon L T_2) \xi \tag{7.15}$$

where

$$\begin{bmatrix} \eta \\ \xi \end{bmatrix} = \mathbf{T}_1 \begin{bmatrix} w \\ \lambda \end{bmatrix} \tag{7.16}$$

The algebraic equations (7.12) and (7.13) can be solved by using any of the recursive algorithms developed in (Grodt and Gajic, 1988; Kokotovic et al., 1980).

The rearrangement and modification of variables in (7.8) is done by using the permutation matrix E_1 of the form

$$\begin{bmatrix} x_1 \\ p_1 \\ x_2 \\ p_2 \end{bmatrix} = \begin{bmatrix} I_{n_1} & 0 & 0 & 0 \\ 0 & 0 & I_{n_1} & 0 \\ 0 & I_{n_2} & 0 & 0 \\ 0 & 0 & 0 & \frac{I_{n_2}}{\epsilon} \end{bmatrix} \begin{bmatrix} x_1 \\ x_2 \\ p_1 \\ \epsilon p_2 \end{bmatrix} = E_1 \begin{bmatrix} x \\ p \end{bmatrix} \tag{7.17}$$

Combining (7.16) and (7.17), we obtain the relationship between the original coordinates and the new ones

$$\begin{bmatrix} \eta_1 \\ \xi_1 \\ \eta_2 \\ \xi_2 \end{bmatrix} = E_2^T T_1 E_1 \begin{bmatrix} x \\ p \end{bmatrix} = \Pi \begin{bmatrix} x \\ p \end{bmatrix} = \begin{bmatrix} \Pi_1 & \Pi_2 \\ \Pi_3 & \Pi_4 \end{bmatrix} \begin{bmatrix} x \\ p \end{bmatrix} \qquad (7.18)$$

where E_2 is a permutation matrix in the form

$$E_2 = \begin{bmatrix} I_{n_1} & 0 & 0 & 0 \\ 0 & 0 & I_{n_1} & 0 \\ 0 & I_{n_2} & 0 & 0 \\ 0 & 0 & 0 & I_{n_2} \end{bmatrix} \qquad (7.19)$$

Since $p = Px$, where P satisfies the algebraic Riccati equation (7.7), it follows that

$$\begin{bmatrix} \eta_1 \\ \xi_1 \end{bmatrix} = (\Pi_1 + \Pi_2 P)\, x, \quad \begin{bmatrix} \eta_2 \\ \xi_2 \end{bmatrix} = (\Pi_3 + \Pi_4 P)\, x \qquad (7.20)$$

In the original coordinates, the required optimal solution has a closed-loop nature. We have the same attribute for the new systems (7.14) and (7.15); that is

$$\begin{bmatrix} \eta_2 \\ \xi_2 \end{bmatrix} = \begin{bmatrix} P_1 & 0 \\ 0 & P_2 \end{bmatrix} \begin{bmatrix} \eta_1 \\ \xi_1 \end{bmatrix} \qquad (7.21)$$

Then, (7.20) and (7.21) yield

$$\begin{bmatrix} P_1 & 0 \\ 0 & P_2 \end{bmatrix} = (\Pi_3 + \Pi_4 P)(\Pi_1 + \Pi_2 P)^{-1} \qquad (7.22)$$

Following the same logic, we can find P reversely by introducing

$$E_1^{-1} T_1^{-1} E_2 = \Omega = \begin{bmatrix} \Omega_1 & \Omega_2 \\ \Omega_3 & \Omega_4 \end{bmatrix} \qquad (7.23)$$

where

$$E_1^{-1} = \begin{bmatrix} I_{n_1} & 0 & 0 & 0 \\ 0 & 0 & I_{n_2} & 0 \\ 0 & I_{n_1} & 0 & 0 \\ 0 & 0 & 0 & \epsilon I_{n_2} \end{bmatrix} \qquad (7.24)$$

and it yields

$$P = \left(\Omega_3 + \Omega_4 \begin{bmatrix} P_1 & 0 \\ 0 & P_2 \end{bmatrix}\right)\left(\Omega_1 + \Omega_2 \begin{bmatrix} P_1 & 0 \\ 0 & P_2 \end{bmatrix}\right)^{-1} \qquad (7.25)$$

It is shown in Appendix 7.1 that the required matrices in (7.22) and (7.25) are invertible. Partitioning (7.14) and (7.15) as

$$\begin{bmatrix} \dot{\eta}_1 \\ \dot{\eta}_2 \end{bmatrix} = \begin{bmatrix} a_1 & a_2 \\ a_3 & a_4 \end{bmatrix}\begin{bmatrix} \eta_1 \\ \eta_2 \end{bmatrix} = (T_1 - T_2 L)\begin{bmatrix} \eta_1 \\ \eta_2 \end{bmatrix} \qquad (7.26)$$

$$\epsilon \begin{bmatrix} \dot{\xi}_1 \\ \dot{\xi}_2 \end{bmatrix} = \begin{bmatrix} b_1 & b_2 \\ b_3 & b_4 \end{bmatrix}\begin{bmatrix} \xi_1 \\ \xi_2 \end{bmatrix} = (T_4 + \epsilon L T_2)\begin{bmatrix} \xi_1 \\ \xi_2 \end{bmatrix} \qquad (7.27)$$

and using (7.21) yield to two reduced-order nonsymmetric algebraic Riccati equations

$$0 = P_1 a_1 - a_4 P_1 - a_3 + P_1 a_2 P_1 \qquad (7.28)$$

$$0 = P_2 b_1 - b_4 P_2 - b_3 + P_2 b_2 P_2 \qquad (7.29)$$

where

$$\begin{bmatrix} a_1 & a_2 \\ a_3 & a_4 \end{bmatrix} = \begin{bmatrix} A_1 - A_2 L_1 + Z L_3 & -S_1 - A_2 L_2 + Z L_4 \\ -Q_1 + Q_2 L_1 + A_3^T L_3 & -A_1^T + Q_2 L_2 + A_3^T L_4 \end{bmatrix}$$

$$\begin{bmatrix} b_1 & b_2 \\ b_3 & b_4 \end{bmatrix} = \begin{bmatrix} A_4 + \epsilon(L_1 A_2 - L_2 Q_2) & -S_2 - \epsilon(L_1 Z + L_2 A_3^T) \\ -Q_3 + \epsilon(L_3 A_2 - L_4 Q_2) & -A_4^T - \epsilon(L_3 Z + L_4 A_3^T) \end{bmatrix}$$

$$\qquad (7.30)$$

with

$$L = \begin{bmatrix} L_1 & L_2 \\ L_3 & L_4 \end{bmatrix}$$

The pure-slow algebraic Riccati equation (7.28) is nonsymmetric and it is given by

$$P_1 (A_1 - A_2 L_1 + Z L_3) + (A_1 - L_2^T Q_2^T - L_4^T A_3)^T P_1$$
$$+ (Q_1 - Q_2 L_1 - A_3^T L_3) - P_1 (S_1 + A_2 L_2 - Z L_4) P_1 = 0$$
$$\qquad (7.31)$$

The pure-fast algebraic Riccati equation (7.29) is also nonsymmetric

$$P_2 (A_4 + \epsilon(L_1 A_2 - L_2 Q_2)) + (A_4^T + \epsilon(L_3 Z + L_4 A_3^T)) P_2$$
$$+ (Q_3 - \epsilon(L_3 A_2 - L_4 Q_2)) - P_2 (S_2 + \epsilon(L_1 Z + L_2 A_3^T)) P_2 = 0$$
$$\qquad (7.32)$$

but its $O(\epsilon)$ approximation is a symmetric one, that is

$$P_2 A_4 + A_4^T P_2 + Q_3 - P_2 S_2 P_2 + O(\epsilon) = 0 \qquad (7.33)$$

In addition, it can be shown (see Appendix 7.2) that (7.31) is an $O(\epsilon)$ perturbation of the first-order approximate slow algebraic Riccati equation obtained in (Chow and Kokotovic, 1976)

$$P_s A_s + A_s P_s + Q_s - P_s S_s P_s = 0 \qquad (7.34)$$

where A_s, Q_s, and S_s can be found in Section 2.2.2.

The nonsymmetric algebraic Riccati equation was studied in (Medanic, 1982). An algorithm for solving general nonsymmetric algebraic Riccati equation was derived in (Avramovic, 1979) — see also (Avramovic et al., 1980).

Using (7.33)-(7.34) and the implicit function theorem (Ortega and Rheinboldt, 1970), the existence of the unique solutions of (7.31) and (7.32) are guaranteed by the following lemma.

Lemma 7.1 *If the triples $\left(A_4, B_2, \sqrt{Q_3}\right)$ and $\left(A_s, \sqrt{S_s}, \sqrt{Q_s}\right)$ are stabilizable-detectable, then $\exists \epsilon_0 > 0$ such that $\forall \epsilon \leq \epsilon_0$ unique solutions of (7.31) and (7.32) exist.*

$\diamond$

From (7.33) one can obtain an $O(\epsilon)$ approximation for P_2 as

$$P_2^{(0)} A_4 + A_4^T P_2^{(0)} + Q_3 - P_2^{(0)} S_2 P_2^{(0)} = 0 \qquad (7.35)$$

Having obtained a good initial guess, the Newton type algorithm can be used very efficiently for solving (7.33). The Newton algorithm is given by

$$P_2^{(i+1)} \left(b_1 + b_2 P_2^{(i)}\right) - \left(b_4 - P_2^{(i)} b_2\right) P_2^{(i+1)} = b_3 + P_2^{(i)} b_2 P_2^{(i)}$$
$$i = 0, 1, 2, \dots$$
$$(7.36)$$

with an initial guess obtained from (7.35).

The pure-slow equation (7.31) can be solved by using the Newton algorithm also, with an initial guess obtained from (7.34). The Newton algorithm for (7.31) is given by

$$P_1^{(i+1)} \left(a_1 + a_2 P_1^{(i)}\right) - \left(a_4 - P_1^{(i)} a_2\right) P_1^{(i+1)} = a_3 + P_1^{(i)} a_2 P_1^{(i)}$$
$$P_1^{(0)} = P_s, \qquad i = 0, 1, 2, \dots$$
$$(7.37)$$

It is important to notice that the total number of scalar quadratic algebraic equations in (7.31) and (7.32) is $n_1^2 + n_2^2$. On the other hand, the global algebraic Riccati equation (7.7) contains $\frac{1}{2}(n_1 + n_2)(n_1 + n_2 + 1)$ scalar algebraic equations. Thus, the proposed method can reduce the number of equations if

$$n_1^2 + n_2^2 < \frac{1}{2}(n_1 + n_2)(n_1 + n_2 + 1) \tag{7.38}$$

or

$$(n_1 - n_2)^2 < n_1 + n_2 \tag{7.39}$$

It is interesting to point out that we were not able to extend these results to the differential Riccati equation of singularly perturbed systems. The problem has arisen in finding the terminal conditions for the reduced-order differential Riccati equations. Namely, by interchanging columns in the Hamiltonian matrix and performing corresponding transformations the analytical expressions that relate state and costate variables in new coordinates are quite complicated. In other words, there are not exist any more simple linear relations in the forms $\eta_2(t_f) = P_1(t_f)\eta_1(t_f)$ and $\xi_2(t_f) = P_1(t_f)\xi_1(t_f)$, where t_f stands for the finite final time.

Using solutions of both pure-slow and pure-fast Riccati equations and formulas (7.21) and (7.26), we can get completely decoupled slow and fast subsystems in the form

$$\begin{aligned}
\dot{\eta}_1 &= (a_1 + a_2 P_1)\,\eta_1 \\
\epsilon\dot{\xi}_1 &= (b_1 + b_2 P_2)\,\xi_1
\end{aligned} \tag{7.40}$$

The interpretation of the result presented by (7.40) is that the optimal processing (filtering or control) can be completely performed at the local levels (slow and fast subsystems). The global solution in the original coordinates is then obtained at any time instant by using formula (7.20), that is

$$x = (\Pi_1 + \Pi_2 P)^{-1}\begin{bmatrix} \eta_1 \\ \xi_1 \end{bmatrix} \tag{7.41}$$

where P is given by (7.25).

Note that completely decoupled slow and fast Kalman filters for singularly perturbed systems were obtained in (Khalil and Gajic, 1984) by applying the decoupling transformation (Chang, 1972) to the feedback form (after the global algebraic Riccati equation is solved) of the full-order global Kalman filter. In this section, the transformation (Chang, 1972) is applied to the Hamiltonian matrix (which contains the open-loop information only), leading to the decoupled closed-loop reduced-order optimal filtering (and/or control). However, the use of the formula (7.40) in optimal filtering (first of all) and control of singularly perturbed linear systems should be much more clarified and can be the subject for future research.

In summary, we have obtained the solution of the global (full-order) algebraic Riccati equation of singularly perturbed systems in terms of pure-slow and pure-fast reduced-order algebraic Riccati equations. Instead of solving $(n_1 + n_2)(n_1 + n_2 + 1)/2$ equations for symmetric P in (7.7), we solve $(n_1^2 + n_2^2)$ equations in (7.31) and (7.32). This is more efficient if n_1 and n_2 are selected to be close to each other. Furthermore, due to the split into two independent subsystems, the advantage of parallel computations becomes significant in this case.

The importance of the presented results is in the fact that the optimal control and filtering can be completely and exactly decomposed into slow and fast time scales.

7.2 Numerical Example

In order to illustrate the proposed method, we consider a real world problem — a magnetic tape control system (Chow and Kokotovic, 1976). The problem matrices are given in Section 2.2.3.

The optimal global solution from (7.7) is

$$
P_{exact} = \begin{bmatrix} 7.5400 & 6.1704 & 0.40534 & 0.10000 \\ 6.1704 & 7.4673 & 0.39510 & 0.089202 \\ 0.40534 & 0.39510 & 0.13044 & 0.024396 \\ 0.10000 & 0.089202 & 0.024396 & 0.006200 \end{bmatrix}
$$

Solutions of pure-slow and pure-fast algebraic Riccati equations obtained from (7.36) and (7.37) are

$$
P_1 = \begin{bmatrix} 7.2437 & 5.5037 \\ 5.8884 & 6.8214 \end{bmatrix}, \quad P_2 = \begin{bmatrix} 1.0411 & 0.18501 \\ 0.17853 & 0.047413 \end{bmatrix}
$$

Using (7.25), the obtained solution for P is found to be identical to P_{exact}.

Research Problem 7.1: Develop the parallel algorithm for complete decomposition of the singularly perturbed discrete-time algebraic Riccati equation into pure-slow and pure-fast reduced-order Riccati equations.

$$\triangle$$

Research Problem 7.2: Apply the methodology presented in Section 7.1 to the optimal control and filtering of linear singularly perturbed systems attempting to achieve the complete decomposition of the control and filtering tasks into independent slow and fast time scale problems. Study this problem in both continuous and discrete-time domains.

$$\triangle$$

7.3 The Exact Decomposition of the Weakly Coupled Algebraic Riccati Equation

Consider the linear weakly coupled system

$$\begin{aligned} \dot{x}_1 &= A_1 x_1 + \epsilon A_2 x_2 + B_1 u_1 + \epsilon B_2 u_2, & x_1(t_0) &= x_{10} \\ \dot{x}_2 &= \epsilon A_3 x_1 + A_4 x_2 + \epsilon B_3 u_1 + B_4 u_2, & x_2(t_0) &= x_{20} \end{aligned} \tag{7.42}$$

with

$$z = \begin{bmatrix} z_1 \\ z_2 \end{bmatrix} = D \begin{bmatrix} x_1 \\ x_2 \end{bmatrix} = \begin{bmatrix} D_1 & \epsilon D_2 \\ \epsilon D_3 & D_4 \end{bmatrix} \begin{bmatrix} x_1 \\ x_2 \end{bmatrix} \tag{7.43}$$

where $x_i \in \Re^{n_i}$, $u_i \in \Re^{m_i}$, $z_i \in \Re^{r_i}$, $i = 1, 2$, are state, control, and output variables, respectively. The system matrices are of appropriate dimensions and, in general, they are bounded functions of a small coupling parameter ϵ (Gajic, et al., 1990). We will assume that all given matrices are constant.

With (7.42)-(7.43), consider the performance criterion

$$J = \frac{1}{2} \int\limits_{t_0}^{\infty} \left\{ \begin{bmatrix} x_1 \\ x_2 \end{bmatrix}^T D^T D \begin{bmatrix} x_1 \\ x_2 \end{bmatrix} + \begin{bmatrix} u_1 \\ u_2 \end{bmatrix}^T R \begin{bmatrix} u_1 \\ u_2 \end{bmatrix} \right\} dt \tag{7.44}$$

with positive definite R, which has to be minimized. It is assumed that the matrix R has the weakly coupled structure, that is

$$R = \begin{bmatrix} R_1 & 0 \\ 0 & R_2 \end{bmatrix} \tag{7.45}$$

The optimal closed-loop control law has the very well-known form

$$u = \begin{bmatrix} u_1 \\ u_2 \end{bmatrix} = -R^{-1} \begin{bmatrix} B_1 & \epsilon B_2 \\ \epsilon B_3 & B_4 \end{bmatrix}^T P \begin{bmatrix} x_1 \\ x_2 \end{bmatrix} = -R^{-1} B^T P x \quad (7.46)$$

where P satisfies the algebraic Riccati equation given by

$$0 = PA + A^T P + Q - PSP \quad (7.47)$$

with

$$A = \begin{bmatrix} A_1 & \epsilon A_2 \\ \epsilon A_3 & A_4 \end{bmatrix}, \qquad S = BR^{-1}B^T = \begin{bmatrix} S_1 & \epsilon S_2 \\ \epsilon S_2^T & S_3 \end{bmatrix}, \quad (7.48)$$

and

$$Q = D^T D = \begin{bmatrix} Q_1 & \epsilon Q_2 \\ \epsilon Q_2^T & Q_3 \end{bmatrix} \quad (7.49)$$

The open-loop optimal control problem of (7.42)-(7.45) has the solution given by (7.3) where $p \in \Re^{n_1+n_2}$ is a costate variable satisfying (7.4). Partitioning p into $p_1 \in \Re^{n_1}$ and $p_2 \in \Re^{n_2}$ and rearranging rows in (7.4), we can get

$$\begin{bmatrix} \dot{x}_1 \\ \dot{p}_1 \\ \dot{x}_2 \\ \dot{p}_2 \end{bmatrix} = \begin{bmatrix} T_1 & \epsilon T_2 \\ \epsilon T_3 & T_4 \end{bmatrix} \begin{bmatrix} x_1 \\ p_1 \\ x_2 \\ p_2 \end{bmatrix} \quad (7.50)$$

where $T_i's$, $i = 1,2,3,4$, are given by

$$T_1 = \begin{bmatrix} A_1 & -S_1 \\ -Q_1 & -A_1^T \end{bmatrix}, \qquad T_2 = \begin{bmatrix} A_2 & -S_2 \\ -Q_2 & -A_3^T \end{bmatrix}$$

$$T_3 = \begin{bmatrix} A_3 & -S_2^T \\ -Q_2^T & -A_2^T \end{bmatrix}, \qquad T_4 = \begin{bmatrix} A_4 & -S_3 \\ -Q_3 & -A_4^T \end{bmatrix} \quad (7.51)$$

Introducing the notation

$$\begin{bmatrix} x_1 \\ p_1 \end{bmatrix} = w, \qquad \begin{bmatrix} x_2 \\ p_2 \end{bmatrix} = \lambda \quad (7.52)$$

and applying the transformation presented in Section 3.3

$$\begin{bmatrix} \eta \\ \xi \end{bmatrix} = T_2^{-1} \begin{bmatrix} w \\ \lambda \end{bmatrix} \tag{7.53}$$

$$T_2 = \begin{bmatrix} I & -\epsilon L \\ \epsilon H & I - \epsilon^2 H L \end{bmatrix}, \quad T_2^{-1} = \begin{bmatrix} I - \epsilon^2 L H & \epsilon L \\ -\epsilon H & I \end{bmatrix} \tag{7.54}$$

where L and H satisfy

$$T_1 L + T_2 - L T_4 - \epsilon^2 L T_3 L = 0 \tag{7.55}$$

$$H \left(T_1 - \epsilon^2 L T_3 \right) - \left(T_4 + \epsilon^2 T_3 L \right) H + T_3 = 0 \tag{7.56}$$

will produce a decoupled form

$$\dot{\eta} = \left(T_1 - \epsilon^2 L T_3 \right) \eta \tag{7.57}$$

$$\dot{\xi} = \left(T_4 + \epsilon^2 T_3 L \right) \xi \tag{7.58}$$

The rearrangement of states in (7.50) is done by using a permutation matrix E of the form

$$\begin{bmatrix} x_1 \\ p_1 \\ x_2 \\ p_2 \end{bmatrix} = \begin{bmatrix} I_{n_1} & 0 & 0 & 0 \\ 0 & 0 & I_{n_1} & 0 \\ 0 & I_{n_2} & 0 & 0 \\ 0 & 0 & 0 & I_{n_2} \end{bmatrix} \begin{bmatrix} x_1 \\ x_2 \\ p_1 \\ p_2 \end{bmatrix} = E \begin{bmatrix} x \\ p \end{bmatrix} \tag{7.59}$$

Combining (7.53) and (7.59), we obtain the relationship between the original coordinates and the new ones

$$\begin{bmatrix} \eta_1 \\ \xi_1 \\ \eta_2 \\ \xi_2 \end{bmatrix} = E^T T_2^{-1} E \begin{bmatrix} x \\ p \end{bmatrix} = \Pi \begin{bmatrix} x \\ p \end{bmatrix} = \begin{bmatrix} \Pi_1 & \Pi_2 \\ \Pi_3 & \Pi_4 \end{bmatrix} \begin{bmatrix} x \\ p \end{bmatrix} \tag{7.60}$$

Since $p = Px$, where P satisfies the algebraic Riccati equation (7.47), it follows that

$$\begin{bmatrix} \eta_1 \\ \xi_1 \end{bmatrix} = \left(\Pi_1 + \Pi_2 P \right) x, \quad \begin{bmatrix} \eta_2 \\ \xi_2 \end{bmatrix} = \left(\Pi_3 + \Pi_4 P \right) x \tag{7.61}$$

In the original coordinates, the required optimal solution has a closed-loop nature. We have the same attribute for the new systems (7.57) and (7.58), that is

$$\begin{bmatrix} \eta_2 \\ \xi_2 \end{bmatrix} = \begin{bmatrix} P_1 & 0 \\ 0 & P_2 \end{bmatrix} \begin{bmatrix} \eta_1 \\ \xi_1 \end{bmatrix} \qquad (7.62)$$

Then (7.61) and (7.62) yield

$$\begin{bmatrix} P_1 & 0 \\ 0 & P_2 \end{bmatrix} = (\Pi_3 + \Pi_4 P)(\Pi_1 + \Pi_2 P)^{-1} \qquad (7.63)$$

Following the same logic, we can find P reversely by introducing

$$E^T T_2 E = \Omega = \begin{bmatrix} \Omega_1 & \Omega_2 \\ \Omega_3 & \Omega_4 \end{bmatrix} \qquad (7.64)$$

and it yields

$$P = \left(\Omega_3 + \Omega_4 \begin{bmatrix} P_1 & 0 \\ 0 & P_2 \end{bmatrix} \right) \left(\Omega_1 + \Omega_2 \begin{bmatrix} P_1 & 0 \\ 0 & P_2 \end{bmatrix} \right)^{-1} \qquad (7.65)$$

The invertibility of the matrices defined in (7.63) and (7.65) is proved in Appendix 7.3.

Partitioning (7.57) and 7.58) as

$$\begin{bmatrix} \dot{\eta}_1 \\ \dot{\eta}_2 \end{bmatrix} = \begin{bmatrix} a_1 & a_2 \\ a_3 & a_4 \end{bmatrix} \begin{bmatrix} \eta_1 \\ \eta_2 \end{bmatrix} \qquad (7.66)$$

$$\begin{bmatrix} \dot{\xi}_1 \\ \dot{\xi}_2 \end{bmatrix} = \begin{bmatrix} b_1 & b_2 \\ b_3 & b_4 \end{bmatrix} \begin{bmatrix} \xi_1 \\ \xi_2 \end{bmatrix} \qquad (7.67)$$

where

$$\begin{aligned} a_1 &= A_1 + O\left(\epsilon^2\right), & a_2 &= -S_1 + O\left(\epsilon^2\right) \\ a_3 &= -Q_1 + O\left(\epsilon^2\right), & a_4 &= -A^T + O\left(\epsilon^2\right) \end{aligned} \qquad (7.68)$$

$$\begin{aligned} b_1 &= A_4 + O\left(\epsilon^2\right), & b_2 &= -S_3 + O\left(\epsilon^2\right) \\ b_3 &= -Q_3 + O\left(\epsilon^2\right), & b_4 &= -A^T + O\left(\epsilon^2\right) \end{aligned} \qquad (7.69)$$

and using (7.62) yield two reduced-order nonsymmetric algebraic Riccati equations

$$0 = P_1 a_1 - a_4 P_1 - a_3 + P_1 a_2 P_1 \qquad (7.70)$$

$$0 = P_2 b_1 - b_4 P_2 - b_3 + P_2 b_2 P_2 \qquad (7.71)$$

From (7.68) and (7.69), it follows that the $O\left(\epsilon^2\right)$ perturbations of the nonsymmetric algebraic Riccati equations (7.70) and (7.71) are symmetric, namely,

$$P_1 A_1 + A_1^T P_1 + D_1^T D_1 - P_1 B_1 R_1^{-1} B_1^T P_1 + O\left(\epsilon^2\right) = 0 \qquad (7.72)$$

$$P_2 A_4 + A_4^T P_2 + D_4^T D_4 - P_2 B_4 R_2^{-1} B_4^T P_2 + O\left(\epsilon^2\right) = 0 \qquad (7.73)$$

Using these facts and the implicit function theorem (Ortega and Rheinboldt, 1970), the existence of the unique solutions of (7.70) and (7.71) is guaranteed under the following lemma.

Lemma 7.2 *If both the triples* (A_1, B_1, D_1) *and* (A_4, B_4, D_4) *are stablizable-detectable, then* $\exists \epsilon_0 > 0$ *such that* $\forall \epsilon \leq \epsilon_0$ *the solutions of (7.70) and (7.71) exist.*

$\diamond$

Two numerical methods can be proposed for solving (7.70) and (7.71), namely, the fixed point and Newton methods similar to those developed in (Gajic and Shen, 1989). The Newton method leads to the following recursive scheme

$$P_1^{(i+1)} \left(a_1 + a_2 P_1^{(i)}\right) - \left(a_4 - P_1^{(i)} a_2\right) P_1^{(i+1)} = a_3 + P_1^{(i)} a_2 P^{(i)} \qquad (7.74)$$

where the initial condition is obtained from

$$P_1^{(0)} A_1 + A_1^T P_1^{(0)} + Q_1 - P_1^{(0)} S_1 P_1^{(0)} = 0 \qquad (7.75)$$

Similar formulas hold for (7.71).

It is interesting to point out the the proposed method is not applicable for the differential Riccati equation of weakly coupled systems, because there is no way to find the terminal conditions for the reduced-order nonsymmetric differential Riccati equations.

7.4 Case Study: A Satellite Control Problem

To demonstrate the presented method, we have solved a fourth-order example, a satellite control problem considered in (Ackerson and Fu, 1970). Problem matrices are given by

$$A = \begin{bmatrix} 0 & 0.667 & 0 & 0 \\ -0.667 & 0 & 0 & 0 \\ 0 & 0 & 0 & 1.53 \\ 0 & 0 & 1.53 & 0 \end{bmatrix}, \quad B = \begin{bmatrix} 0 & 0.2 \\ 1 & 0 \\ 0.4 & 0 \\ 0 & 1 \end{bmatrix}$$

Penalty matrices Q and R are chosen as identities.

Results obtained from (7.70) and (7.71) are given by

$$P_1 = \begin{bmatrix} 2.2201 & 0.45889 \\ 0.4410 & 1.2749 \end{bmatrix}, \quad P_2 = \begin{bmatrix} 1.5056 & 0.1947 \\ 0.22817 & 1.2782 \end{bmatrix}$$

which by the use of the formula of (7.65) produce

$$P = \begin{bmatrix} 2.2437 & 0.46218 & 0.13613 & -0.10735 \\ 0.46218 & 1.3456 & -0.2091 & -0.24753 \\ 0.13613 & -0.2091 & 1.5375 & 0.24817 \\ -0.10735 & -0.24753 & 0.24817 & 1.3396 \end{bmatrix}$$

Exactly the same result has been obtained by using the classical global method for solving the algebraic Riccati equation.

Research Problem 7.3: Decompose the discrete-time algebraic Riccati equation of weakly coupled systems into two reduced-order completely independent nonsymmetric algebraic Riccati equations. Generalize obtained result to the problem of N weakly coupled subsystems.

$\triangle$

Research Problem 7.4: Decompose the optimal control and filtering tasks of linear weakly coupled systems into completely independent reduced-order optimal control and filtering subproblems. Generalize obtained results to the weakly coupled linear control systems composed of N subsystems.

$\triangle$

7.5 Conclusion

In this chapter, the optimal steady-state, closed-loop control problems of singularly perturbed and weakly coupled systems are solved by way of the reduced-order nonsymmetric algebraic Riccati equations. Since the decomposed Riccati equations are completely independent, the processing time for the optimal control and filtering problems is reduced. The results presented in this chapter are based on the work by (Su and Gajic, 1992; Su et al., 1992b).

Appendix 7.1

It is easy to show that

$$\begin{bmatrix} \Omega_1 & \Omega_2 \\ \Omega_3 & \Omega_4 \end{bmatrix} = E_1^{-1} T_1^{-1} E_2 = \begin{bmatrix} I_{n_1} & 0 & 0 & 0 \\ -L_1 & I_{n_2} & -L_2 & 0 \\ 0 & 0 & I_{n_1} & 0 \\ 0 & 0 & 0 & 0 \end{bmatrix} + O(\epsilon) \quad \text{(a.1)}$$

which implies

$$\Omega_1 = \begin{bmatrix} I_{n_1} & 0 \\ -L_1 & I_{n_2} \end{bmatrix} + O(\epsilon), \qquad \Omega_2 = \begin{bmatrix} 0 & 0 \\ -L_2 & 0 \end{bmatrix} + O(\epsilon) \quad \text{(a.2)}$$

Then, the matrix

$$\Omega_1 + \Omega_2 \begin{bmatrix} P_1 & 0 \\ 0 & P_2 \end{bmatrix} = \begin{bmatrix} I_{n_1} & 0 \\ -L_1 - L_2 P_1 & I_{n_2} \end{bmatrix} + O(\epsilon) \quad \text{(a.3)}$$

is invertible for sufficiently small values of ϵ.

Similarly

$$\begin{bmatrix} \Pi_1 & \Pi_2 \\ \Pi_3 & \Pi_4 \end{bmatrix} = E_2^T T_1 E_1 = \begin{bmatrix} I_{n_1} & 0 & 0 & -H_2 \\ L_1 & I_{n_2} & 0 & 0 \\ 0 & 0 & I_{n_1} & -H_4 \\ L_3 & 0 & 0 & \frac{I_{n_2}}{\epsilon} \end{bmatrix} + O(\epsilon) \quad \text{(a.4)}$$

with

$$\Pi_1 = \begin{bmatrix} I_{n_1} & 0 \\ L_1 & I_{n_2} \end{bmatrix} + O(\epsilon), \qquad \Pi_2 = \begin{bmatrix} 0 & -H_2 \\ 0 & 0 \end{bmatrix} + O(\epsilon) \quad \text{(a.5)}$$

imply that the matrix

$$\Pi_1 + \Pi_2 P = \begin{bmatrix} I_{n_1} & 0 \\ L_1 & I_{n_2} \end{bmatrix} + O(\epsilon) \quad \text{(a.6)}$$

is invertible for sufficiently small values of ϵ. In this appendix, we have used the following notation for the partitioned matrix H

$$H = \begin{bmatrix} H_1 & H_2 \\ H_3 & H_4 \end{bmatrix} \quad \text{(a.7)}$$

Appendix 7.2

From (a.1) we have

$$\Omega_3 + \Omega_4 \begin{bmatrix} P_1 & 0 \\ 0 & P_2 \end{bmatrix} = \begin{bmatrix} P_1 & 0 \\ 0 & 0 \end{bmatrix} + O(\epsilon) \qquad (b.1)$$

Using (b.1) and (a.3) in formula (7.25) produces

$$P = \begin{bmatrix} P_1 & 0 \\ 0 & 0 \end{bmatrix} \begin{bmatrix} I_{n_1} & 0 \\ -L_1 - L_2 P_1 & I_{n_2} \end{bmatrix}^{-1} + O(\epsilon) \qquad (b.2)$$

or

$$P = \begin{bmatrix} P_1 & 0 \\ 0 & 0 \end{bmatrix} + O(\epsilon) \qquad (b.3)$$

It is very well known that the structure of the solution for P is given by (Kokotovic, et al., 1986)

$$P = \begin{bmatrix} P_0 & \epsilon P_m \\ \epsilon P_m^T & \epsilon P_f \end{bmatrix} \qquad (b.4)$$

which implies

$$P_1 = P_0 + O(\epsilon) \qquad (b.5)$$

On the other hand, P_0 is $O(\epsilon)$ close to the solution of (7.34), that is to P_s (Chow and Kokotovic, 1976), so that

$$P_1 = P_s + O(\epsilon) \qquad (b.6)$$

Appendix 7.3

According to (7.60) and (7.64), it can be seen that

$$\Pi = \begin{bmatrix} I_{n_1} + O(\epsilon^2) & O(\epsilon) & O(\epsilon^2) & O(\epsilon) \\ O(\epsilon) & I_{n_2} & O(\epsilon) & 0 \\ O(\epsilon^2) & O(\epsilon) & I_{n_1} + O(\epsilon^2) & O(\epsilon) \\ O(\epsilon) & 0 & O(\epsilon) & I_{n_2} \end{bmatrix} \qquad (c.1)$$

and

$$\Omega = \begin{bmatrix} I_{n_1} & O\left(\epsilon\right) & 0 & O\left(\epsilon\right) \\ O\left(\epsilon\right) & I_{n_2} + O\left(\epsilon^2\right) & O\left(\epsilon\right) & O\left(\epsilon^2\right) \\ 0 & O\left(\epsilon\right) & I_{n_1} & O\left(\epsilon\right) \\ O\left(\epsilon\right) & O\left(\epsilon^2\right) & O\left(\epsilon\right) & I_{n_2} + O\left(\epsilon^2\right) \end{bmatrix} \qquad \text{(c.2)}$$

Therefore,

$$(\Pi_1 + \Pi_2 P) = I_{n_1+n_2} + O\left(\epsilon\right) \qquad \text{(c.3)}$$

$$\left(\Omega_1 + \Omega_2 \begin{bmatrix} P_1 & 0 \\ 0 & P_2 \end{bmatrix}\right) = I_{n_1+n_2} + O\left(\epsilon\right) \qquad \text{(c.4)}$$

There exists $\epsilon_1 > 0$ such that $\forall \epsilon \leq \epsilon_1$ the required matrices are invertible.

Chapter 8

Differential and Difference Riccati Equations

In this chapter, we study the main equations of the finite time optimal closed-loop linear-quadratic control problems, namely, the differential and difference Riccati equations, for both singularly perturbed and weakly coupled systems. A unique approach to the solutions of these Riccati equations is developed by performing the block diagonalization of the corresponding Hamiltonian matrices.

8.1 Recursive Solution of the Singularly Perturbed Differential Riccati Equation

A differential Riccati equation of a singularly perturbed system (Kokotovic and Khalil, 1986) is given by

$$-\dot{P}(t) = P(t)A + A^T P(t) + Q - P(t)SP(t), \quad P(T) = F \quad (8.1)$$

where

$$A = \begin{bmatrix} A_1 & A_2 \\ \frac{A_3}{\epsilon} & \frac{A_4}{\epsilon} \end{bmatrix}, \quad Q = \begin{bmatrix} Q_1 & Q_2 \\ Q_2^T & Q_3 \end{bmatrix}, \quad Q \geq 0$$

$$S = BR^{-1}B^T = \begin{bmatrix} S_1 & \frac{Z}{\epsilon} \\ \frac{Z^T}{\epsilon} & \frac{S_2}{\epsilon^2} \end{bmatrix}, \quad F = \begin{bmatrix} F_1 & \epsilon F_2 \\ \epsilon F_2^T & \epsilon F_3 \end{bmatrix}, \quad R > 0 \quad (8.2)$$

are $n \times n$ constant matrices and ϵ is a small positive parameter. The presence of a small parameter ϵ makes this problem numerically ill-defined, producing a so-called stiff numerical problem (huge slope at terminal time), (Miranker, 1981). In order to overcome this difficulty, the Taylor series expansion approach, with respect to a small parameter ϵ has been taken in (Yackel and Kokotovic, 1973) leading to a family of well-defined reduced-order problems. However, the Taylor series expansion method is not recursive in its application. When one is interested in a high degree of accuracy, or when ϵ is not very small, the size of computations required can be considerable. In such cases, the advantage of using the series expansion method (the important theoretical tool) is questionable from the numerical point of view, and sometimes (see example in Section 8.2) that method is almost not applicable.

In this section, we will exploit the known Hamiltonian form of the solution of the Riccati equation (Kwakernaak and Sivan, 1972), and a nonsingular transformation (Chang, 1972) in order to obtain an efficient recursive numerical method for solving (8.1). The Chang transformation is used to block diagonalize the Hamiltonian, so that the required solution is obtained in terms of reduced-order problems. In addition, an efficient Newton-type algorithm (with the quadratic rate of convergence, that is, $O\left(\epsilon^{2^k}\right)$ — where k is a number of iterations) is developed for solving algebraic equations comprising the Chang transformation.

The solution of (8.1) can be sought in the form

$$P(t) = M(t) N^{-1}(t) \qquad (8.3)$$

where matrices $M(t)$ and $N(t)$ satisfy a system of linear equations (Kwakernaak and Sivan, 1972)

$$\begin{aligned} \dot{M}(t) &= -A^T M(t) - QN(t), & M(T) &= F \\ \dot{N}(t) &= -SM(t) + AN(t), & N(T) &= I \end{aligned} \qquad (8.4)$$

and $N(t)$ is assumed to be nonsingular for $\forall t, t < T$. This approach is considered as the most efficient numerical method for the solution of the differential Riccati equation (Kenney and Leipnik, 1985), where the invertibility problem of $N(t)$ is solved by performing a reinitialization along the path $t_0 < t < T$ whenever $N(t)$ is close to being singular.

Knowing the nature of the solution of (8.1), which is properly scaled as (Kokotovic and Khalil, 1986; Yackel and Kokotovic, 1973)

$$P(t) = \begin{bmatrix} P_1(t) & \epsilon P_2(t) \\ \epsilon P_2^T(t) & \epsilon P_3(t) \end{bmatrix}, \quad P(T) = F = \begin{bmatrix} F_1 & \epsilon F_2 \\ \epsilon F_2^T & \epsilon F_3 \end{bmatrix} \qquad (8.5)$$

where $\dim P_1 = n_1 \times n_1$, $\dim P_3 = n_2 \times n_2$, $n_1 + n_2 = n$ (n_1-slow variables, n_2-fast variables), we introduce compatible partitions of $M(t)$

and $N(t)$ matrices

$$M(t) = \begin{bmatrix} M_1(t) & M_2(t) \\ M_3(t) & M_4(t) \end{bmatrix}, \quad N(t) = \begin{bmatrix} N_1(t) & N_2(t) \\ N_3(t) & N_4(t) \end{bmatrix} \quad (8.6)$$

The invertibility of $N(t)$ for every t, $t_0 \leq t < T$, plays an important role in the proposed method. The condition under which $N(t)$ is an invertible matrix is stated in the following lemma.

Lemma 8.1 *If the triple* $(A, B, \sqrt{Q})$ *is stabilizable-observable, then the matrix* $N(t)$, *with* $N(T) = I$ *is invertible for any* $t \in (t_0, T)$.

◇

Proof: By using a dichotomy transformation introduced in (Wilde and Kokotovic, 1972),

$$\begin{bmatrix} M \\ N \end{bmatrix} = \begin{bmatrix} K & P \\ I & I \end{bmatrix} \begin{bmatrix} \widehat{M} \\ \widehat{N} \end{bmatrix} \quad (8.7)$$

$$\begin{bmatrix} \widehat{M} \\ \widehat{N} \end{bmatrix} = \begin{bmatrix} (\underline{K} - \underline{P})^{-1} & -(\underline{K} - \underline{P})^{-1}\underline{P} \\ -(\underline{K} - \underline{P})^{-1} & I + (\underline{K} - \underline{P})^{-1}\underline{P} \end{bmatrix} \begin{bmatrix} M \\ N \end{bmatrix} \quad (8.8)$$

where $\underline{P}$ and $\underline{K}$ are unique positive definite and negative definite solutions of the algebraic Riccati equation corresponding to (8.1), the system (8.4) can be transformed in

$$\begin{bmatrix} \dot{\widehat{M}} \\ \dot{\widehat{N}} \end{bmatrix} = \begin{bmatrix} A - S\underline{K} & 0 \\ 0 & A - S\underline{P} \end{bmatrix} \begin{bmatrix} \widehat{M} \\ \widehat{N} \end{bmatrix} \quad (8.9)$$

with terminal conditions

$$\widehat{M}(T) = (\underline{K} - \underline{P})^{-1}(F - \underline{P})$$
$$\widehat{N}(T) = I + (\underline{K} - \underline{P})^{-1}(F - \underline{P}) = I + \widehat{M}(T)$$

It is known that $(A - S\underline{K})$ is an unstable matrix and that matrix $(A - S\underline{P})$ is stable (Wilde and Kokotovic, 1972). The solution of (8.9) is given by

$$\widehat{M}(t) = e^{(A - S\underline{K})(t - T)}\widehat{M}(T)$$
$$\widehat{N}(t) = e^{(A - S\underline{P})(t - T)}\widehat{N}(T) \quad (8.10)$$

Using (8.7)-(8.10) it can be easily shown that

$$N(t) = e^{(A-S\underline{K})(t-T)}$$

$$\times \left[I + \left(I - e^{S(\underline{P}-\underline{K})(t-T)}(\underline{K} - \underline{P})^{-1}(\underline{P} - F)\right)\right] N(T)$$

that is

$$N(t) = \phi(t - T) N(T) \tag{8.11}$$

with obvious definition of $\phi(t - T)$. Since $\phi(t - T)$ plays the role of the transition matrix of $N(t)$ and by very well-known facts is nonsingular, the regularity of $N(t)$ is determined by $N(T)$ only. Thus, having chosen $N(T)$ as an identity will assure the nonsingularity of $N(t)$ for any $t < T$, and prove the given lemma.

Partitioning (8.4), according to (8.6), will reveal a decoupled structure, that is, equations for M_1, M_3, N_1, and N_3 are independent of equations for M_2, M_4, N_2, and N_4 and vice versa. Introducing the notation

$$U = \begin{bmatrix} M_1 \\ N_1 \end{bmatrix}, \quad \epsilon V = \begin{bmatrix} M_3 \\ \epsilon N_3 \end{bmatrix}, \quad X = \begin{bmatrix} M_2 \\ N_2 \end{bmatrix}, \quad \epsilon Y = \begin{bmatrix} M_4 \\ \epsilon N_4 \end{bmatrix} \tag{8.12}$$

$$T_1 = \begin{bmatrix} -A_1^T & -Q_1 \\ -S_1 & A_1 \end{bmatrix}, \quad T_2 = \begin{bmatrix} -A_3^T & -Q_2 \\ -Z & A_2 \end{bmatrix}$$

$$\tag{8.13}$$

$$T_3 = \begin{bmatrix} -A_2^T & -Q_2^T \\ -Z^T & A_3 \end{bmatrix}, \quad T_2 = \begin{bmatrix} -A_4^T & -Q_3 \\ -S_2 & A_4 \end{bmatrix}$$

and after doing some algebra, we get two systems of singularly perturbed matrix differential equations

$$\dot{U} = T_1 U + T_2 V, \quad U(T) = \begin{bmatrix} F_1 \\ I \end{bmatrix}$$

$$\epsilon \dot{V} = T_3 U + T_4 V, \quad V(T) = \begin{bmatrix} F_2 \\ 0 \end{bmatrix} \tag{8.14}$$

$$\dot{X} = T_1 X + T_2 Y, \quad X(T) = \begin{bmatrix} \epsilon F_2 \\ 0 \end{bmatrix}$$

$$\epsilon \dot{Y} = T_3 X + T_4 Y, \quad Y(T) = \begin{bmatrix} F_3 \\ I \end{bmatrix} \tag{8.15}$$

Note that these two systems have exactly the same form and they differ in terminal conditions only. From this point we will proceed by applying the Chang transformation to (8.14) and (8.15). This transformation is defined by (Chang, 1972)

$$T_1 = \begin{bmatrix} I - \epsilon H L & -\epsilon H \\ L & I \end{bmatrix} \tag{8.16}$$

and

$$T_1^{-1} = \begin{bmatrix} I & \epsilon H \\ -L & I - \epsilon L H \end{bmatrix} \tag{8.17}$$

where L and H satisfy

$$T_4 L - T_3 - \epsilon L (T_1 - T_2 L) = 0 \tag{8.18}$$

$$-H (T_4 + \epsilon L T_2) + T_2 + \epsilon (T_1 - T_2 L) H = 0 \tag{8.19}$$

Applying this transformation to (8.14) and (8.15) we get

$$\dot{\hat{U}} (t) = (T_1 - T_2 L) \hat{U} (t), \quad \hat{U} (T) = (I - \epsilon H L) U (T) - \epsilon H V (T) \tag{8.20}$$

$$\epsilon \dot{\hat{V}} (t) = (T_4 + \epsilon L T_2) \hat{V} (t), \quad \hat{V} (T) = L U (T) + V (T) \tag{8.21}$$

$$\dot{\hat{X}} (t) = (T_1 - T_2 L) \hat{X} (t), \quad \hat{X} (T) = (I - \epsilon H L) X (T) - \epsilon H Y (T) \tag{8.22}$$

$$\epsilon \dot{\hat{Y}} (t) = (T_4 + \epsilon L T_2) \hat{Y} (t), \quad \hat{Y} (T) = L X (T) + Y (T) \tag{8.23}$$

Solutions of (8.20)-(8.23) are given by

$$\hat{U} (t) = e^{(T_1 - T_2 L)(t-T)} \hat{U} (T) \tag{8.24}$$

$$\hat{V} (t) = e^{\frac{1}{\epsilon}(T_4 + \epsilon L T_2)(t-T)} \hat{V} (T) \tag{8.25}$$

$$\hat{X} (t) = e^{(T_1 - T_2 L)(t-T)} \hat{X} (T) \tag{8.26}$$

$$\hat{Y} (t) = e^{\frac{1}{\epsilon}(T_4 + \epsilon L T_2)(t-T)} \hat{Y} (T) \tag{8.27}$$

so that in the original coordinates we have

217

$$U(t) = e^{(T_1 - T_2 L)(t-T)} \widehat{U}(T) + \epsilon H e^{\frac{1}{\epsilon}(T_4 + \epsilon L T_2)(t-T)} \widehat{V}(T) \qquad (8.28)$$

$$V(t) = -L e^{(T_1 - T_2 L)(t-T)} \widehat{U}(T) + (I - \epsilon L H) e^{\frac{1}{\epsilon}(T_4 + \epsilon L T_2)(t-T)} \widehat{V}(T)$$

$$(8.29)$$

$$X(t) = e^{(T_1 - T_2 L)(t-T)} \widehat{X}(T) + \epsilon H e^{\frac{1}{\epsilon}(T_4 + \epsilon L T_2)(t-T)} \widehat{Y}(T) \qquad (8.30)$$

$$Y(t) = -L e^{(T_1 - T_2 L)(t-T)} \widehat{X}(T) + (I - \epsilon L H) e^{\frac{1}{\epsilon}(T_4 + \epsilon L T_2)(t-T)} \widehat{Y}(T)$$

$$(8.31)$$

Partitioning (8.28)-(8.31) according to (8.12) will produce all components of matrices $M(t)$ and $N(t)$, that is,

$$\begin{bmatrix} M_1(t) \\ N_1(t) \end{bmatrix} = \begin{bmatrix} U_1(t) \\ U_2(t) \end{bmatrix} = U(t), \qquad \begin{bmatrix} M_2(t) \\ N_2(t) \end{bmatrix} = \begin{bmatrix} X_1(t) \\ X_2(t) \end{bmatrix} = X(t)$$

$$\begin{bmatrix} \frac{1}{\epsilon} M_3(t) \\ N_3(t) \end{bmatrix} = \begin{bmatrix} V_1(t) \\ V_2(t) \end{bmatrix} = V(t), \qquad \begin{bmatrix} \frac{1}{\epsilon} M_4(t) \\ N_4(t) \end{bmatrix} = \begin{bmatrix} Y_1(t) \\ Y_2(t) \end{bmatrix} = Y(t)$$

so that the required solution of (8.1) is given by

$$P(t) = \begin{bmatrix} U_1(t) & X_1(t) \\ \epsilon V_1(t) & \epsilon Y_1(t) \end{bmatrix} \begin{bmatrix} U_2(t) & X_2(t) \\ V_2(t) & Y_2(t) \end{bmatrix}^{-1} \qquad (8.32)$$

Thus, in order to get the numerical solution of (8.1), that is $P(t)$, which has dimensions $n \times n = (n_1 + n_2) \times (n_1 + n_2)$, we have to solve two simple algebraic equations (8.18) and (8.19) of dimensions of $(2n_2 \times 2n_1)$ and $(2n_1 \times 2n_2)$, respectively. The existing numerical algorithms for solving (8.18) and (8.19) can be found in (Gajic, 1986; Kokotovic et al., 1980). Then, two exponential forms $exp[(T_1 - T_2 L)(t - T)]$ and $exp[\frac{1}{\epsilon}(T_4 + \epsilon L T_2)(t - T)]$ have to be transformed in the matrix forms by using some of the well-known approaches (Molen and Von Loan, 1978). Finally, the inversion of the matrix $N(t)$ has to be performed.

The algebraic equations (8.18), which are weakly nonlinear equations and (8.19), a linear Lyapunov-type equation, play the crucial role in the developed method and a very important role in the linear theory of singular perturbations (Kokotovic and Khalil, 1986). The existing methods for solving (8.18) and (8.19) are recursive type algorithms with a rate of convergence of $O\left(\epsilon^k\right)$, where k represents the number of iterations (Gajic, 1986; Kokotovic et al., 1980). In this section, a new method for solving (8.18) and (8.19) with a quadratic rate of convergence, that is, $O\left(\epsilon^{2^k}\right)$, will be presented (Grodt and Gajic, 1988). This method is based on the Newton recursive scheme. It is the very well-known fact that the Newton method converges quadratically in the neighborhood of the sought solution and that its main problem is in the choice of the initial guess. For the algebraic equation (8.18) the initial guess is easily obtained with the accuracy of $O\left(\epsilon\right)$, by setting $\epsilon = 0$ in that equation, that is

$$L^{(0)} = T_4^{-1}T_3 = L + O\left(\epsilon\right) \tag{8.33}$$

Thus, the Newton sequence will be $O\left(\epsilon^2\right), O\left(\epsilon^4\right), O\left(\epsilon^8\right), ..., O\left(\epsilon^{2k}\right)$ close to the exact solution, respectively, in each iteration.

The Newton-type algorithm of (8.18), can be constructed by setting $L^{(i+1)} = L^{(i)} + \Delta L^{(i)}$ and neglecting $O\left(\Delta L\right)^2$ terms. This will produce a Lyapunov-type equation of the form

$$D_1^{(i)} L^{(i+1)} + L^{(i+1)} D_2^{(i)} = Q^{(i)} \tag{8.34}$$

where

$$D_1^{(i)} = T_4 + \epsilon L^{(i)}T_2, \quad D_2^{(i)} = -\epsilon\left(T_1 - T_2 L^{(i)}\right)$$
$$Q^{(i)} = T_3 + \epsilon L^{(i)}T_2 L^{(i)} \quad i = 0,1,2,...$$

with the initial condition given by (8.33).

Having found the solution of (8.18), up to the required degree of accuracy, one can get the solution of (8.19) by solving directly the algebraic Lyapunov equation of the form

$$M^{(i)}D_1^{(i)} + D_2^{(i)}M^{(i)} = T_2 \tag{8.35}$$

which implies $M^{(i)} = M + O\left(\epsilon^{2^i}\right)$.

Note that the existence of the solutions of (8.18) and (8.19) are guaranteed by the nonsingularity of T_4. The sufficient condition for the convergence of the algorithm (8.34) is given by (Belanger and McGillivray, 1976)

$$\| \Delta L^{(i)} \| \leq \| Q^{(i)} \| = \| T_3 + \epsilon L^{(i)}T_2 L^{(i)} \| \tag{8.36}$$

which is almost always satisfied, except for some special cases, for example, $T_3 \approx 0$ and $T_2 \approx 0$, which corresponds to a system already in a block diagonal form.

One has to point out, that contrary to previously used algorithms for solving (8.18)-(8.19) (Gajic, 1986; Kokotovic et al., 1980), which require recursive solution of linear equations, in the proposed method one is faced with the recursive solution of the Lyapunov equations. Thus, for the price of speeding up the convergence from $O\left(\epsilon^k\right)$ to $O\left(\epsilon^{2^k}\right)$ slightly more computations have to be performed per iteration. However, the size of computations required is of the same order, that is of $O\left(n^3\right)$ for both the solutions of the Lyapunov and solution of linear equations, so that the comparison of the rate of convergence of these two algorithms plays the dominant role. In order to demonstrate the efficiency of the proposed algorithm, we have run a fifth-order example.

Example 8.1

Matrices T_1, T_2, T_3, and T_4 are chosen randomly (standard deviation equal to 1, and mean value equal to zero) such that T_4 is the invertible matrix. The simulation results for different values of a small parameter are given in Table 8.1. It can be seen that the Newton method is much more powerful than the successive approximation recursive scheme (Gajic, 1986; Kokotovic et al., 1980). In Table 8.2 we have shown the propagation of the error per iteration when $\epsilon = 0.2$ for the Newton method.

$$T_1 = \begin{bmatrix} -2.014 & -0.058 & 0.499 & 0.585 & 1.372 \\ 1.366 & -0.805 & 0.320 & 0.548 & 0.950 \\ -0.952 & 0.747 & 0.984 & -1.816 & -1.563 \\ -1.241 & 0.758 & -1.126 & 0.497 & -0.131 \\ 0.663 & -0.021 & -0.640 & -0.296 & 1.375 \end{bmatrix}$$

$$T_2 = \begin{bmatrix} -1.796 & -0.009 & -0.840 & 1.819 & 0.794 \\ 0.158 & 0.467 & 1.324 & -0.123 & 0.629 \\ -0.433 & 0.248 & -1.181 & -1.426 & 0.297 \\ -1.599 & 0.269 & -0.133 & -0.845 & -0.769 \\ 1.967 & -0.565 & 0.776 & 1.419 & -0.450 \end{bmatrix}$$

$$T_3 = \begin{bmatrix} -1.496 & -0.666 & 0.699 & 1.262 & -0.731 \\ 1.43 & 0.563 & 0.812 & -1.300 & -0.616 \\ -0.521 & -0.962 & -0.141 & -1.159 & 0.939 \\ 1.071 & -0.943 & 0.017 & 0.696 & 1.295 \\ 1.397 & -1.436 & 0.843 & -1.488 & 0.524 \end{bmatrix}$$

$$T_4 = \begin{bmatrix} -1.367 & -0.885 & -0.506 & -1.174 & 1.435 \\ 0.133 & 1.319 & 1.244 & 0.892 & -1.221 \\ -0.296 & 1.333 & 1.002 & -0.927 & -0.794 \\ 0.780 & 1.358 & 0.607 & -0.511 & 0.671 \\ -0.999 & 0.914 & -1.320 & -0.556 & -1.135 \end{bmatrix}$$

ϵ	Number of required iteration such that	$\|L^{(i+1)} - L^{(i)}\|_\infty < 10^{-7}$
	Newton method	Successive approximations
0.3	6	*
0.2	5	*
0.1	4	*
0.04	4	19
0.02	4	11
0.01	3	7
0.001	2	4

Table 8.1: Dependence of the number of iterations on ϵ (* = no convergence)

$\epsilon = 0.2$ i	$\|L^{(i+1)} - L^{(i)}\|_\infty < 10^{-7}$
1	2.40745×10^0
2	7.80653×10^{-1}
3	4.21800×10^{-2}
4	0.88748×10^{-4}
5	0.17808×10^{-8}

Table 8.2: Propagation of the error per iteration for a constant value of ϵ for the Newton method

The Hamiltonian method developed in this section will be used for the numerical solution of the singularly perturbed matrix differential Riccati equation. Since the matrices $M(t)$ and $N(t)$ contain unstable modes of the Hamiltonian also (Kwakernaak and Sivan, 1972), then even though a product $M(t)$ and $N^{-1}(t)$ tends to a constant as $t \to \infty$, the inversion of the nonsingular matrix $N(t)$, which contains huge elements, will hurt the accuracy.

The reinitialization version of the Hamiltonian approach, which leads to the known Kalman-Englar method (Kwakernaak and Sivan, 1972), is considered as the most efficient numerical method for the solution of the general matrix differential Riccati equation. The reinitialization technique applied to the previously obtained formulas will modify (8.4), (8.14)-(8.15), respectively, in

$$M(k\Delta t) = P(k\Delta t) \qquad (8.37)$$

$$U(k\Delta t) = \begin{bmatrix} P_1(k\Delta t) \\ I \end{bmatrix}, \quad V(k\Delta t) = \begin{bmatrix} P_2^T(k\Delta t) \\ 0 \end{bmatrix} \qquad (8.38)$$

$$X(k\Delta t) = \begin{bmatrix} \epsilon P_2(k\Delta t) \\ 0 \end{bmatrix}, \quad Y(k\Delta t) = \begin{bmatrix} P_3(k\Delta t) \\ I \end{bmatrix} \qquad (8.39)$$

where k represents the number of steps and Δt is an integration step. This will introduce slight modifications in formulas (8.20)-(8.31), namely, instead of the final time T a discrete time $k\Delta t$ has to be used. These changes can be implemented very easily from the programming point of view.

The recursive method for the numerical solution for the singularly perturbed Riccati differential equation proposed in this section is very important in two cases: a) ϵ is not very small; b) high order of accuracy is required. The first case represents one of the main problems in the modern numerical analysis of the singularly perturbed problems. It was pointed by (Hemker, 1983) that "numerical analysis of singular perturbation problems mainly concentrates on the following question: how to find a numerical approximation to the solution for small as well as intermediate values of ϵ, where no short asymptotic expansion is available. Or, more general, how to construct a single numerical method that can be applied both in the case of extremely small ϵ and for larger values of ϵ, where one wouldn't consider the problem as singularly perturbed any longer". Results reported in this section resolve that problem in the case of the singularly perturbed matrix Riccati differential equation.

8.2 Case Study: A Synchronous Machine Connected to an Infinite Bus

The recursive solution of the differential matrix Riccati equation of singularly perturbed systems is demonstrated on a seventh-order model of a synchronous machine connected to an infinite bus (Kokotovic et al., 1980). The system matrix A is given by

$$A = \begin{bmatrix} -0.58 & 0 & 0 & -0.27 & 0 & 0.2 & 0 \\ 0 & -1 & 0 & 0 & 0 & 1 & 0 \\ 0 & 0 & -5 & 2.1 & 0 & 0 & 0 \\ 0 & 0 & 0 & 0 & 337 & 0 & 0 \\ -0.14 & 0 & 0.14 & -0.2 & -0.28 & 0 & 0 \\ 0 & 0 & 0 & 0 & 0 & 0.08 & 2 \\ -173 & 66.7 & -116 & 40.9 & 0 & -66.7 & -16.7 \end{bmatrix}$$

Remaining matrices are chosen as $Q = I$, $F = 0$; S_1, S_2, and Z have all entries equal to 1. The eigenvalues of A are $-8.53 \pm j8.22$, -3.93, $-0.326 \pm j0.56$, $-0.86 \pm j8.37$. Two fast and five slow variables are separated by the choice of the small singular perturbation parameter $\epsilon = 0.4$ (roughly the ratio of 3.93 and 8.53). Simulation results for the element $P_{11}(t)$ are given in Table 8.3.

It can be seen that in order to get the accuracy of four decimal digits, it takes 12 iterations (the successive approximation method was used for solving algebraic equations composing the Chang transformation — in order to be able to compare the proposed recursive scheme to the power-series expansion method, since both methods are producing the same order of accuracy). This result is expected since $O\left(0.4^{12}\right) \approx 10^{-5}$. That means if the power-series expansion method had been used, in order to get the same accuracy, it would have required 12 terms, that is (Yackel and Kokotovic, 1973)

$$P(t, \epsilon) = \sum_{m=0}^{11} \frac{\epsilon^m}{m!} \left\{ P_s^{(m)}(t) + P_f^{(m)}(\tau) \right\} + O\left(\epsilon^{12}\right), \quad \tau = \frac{t - T}{\epsilon}$$

where

$$P_s^{(m)}(t) = \begin{bmatrix} P_{1s}^{(m)}(t) & \epsilon P_{2s}^{(m)}(t) \\ \epsilon P_{2s}^{(m)^T}(t) & \epsilon P_{3s}^{(m)}(t) \end{bmatrix}$$

$$P_f^{(m)}(\tau) = \begin{bmatrix} P_{1f}^{(m)}(\tau) & \epsilon P_{2f}^{(m)}(\tau) \\ \epsilon P_{2f}^{(m)T}(\tau) & \epsilon P_{3f}^{(m)}(\tau) \end{bmatrix}$$

It is shown in (Yackel and Kokotovic, 1973), (pp. 21, formula 32) that the right-hand sides of differential equations for $P_{1f}^{(1)}(\tau)$, $P_{2f}^{(1)}(\tau)$, and $P_{3f}^{(1)}(\tau)$ contain respectively 7, 23, and 22 terms, each consisting of a product of two or three matrices. Thus, the size of computations required for only an $O(\epsilon^2)$ accuracy is already enormous. The complexity of the right-hand side of differential equations for $P_f^{(m)}(\tau)$ grows extremely quickly with the increase of m so that this nice theoretical method is not convenient for the practical computations. For an $O(\epsilon^{12})$ accuracy, the right-hand sides of the differential equations for the power-series expansion method will contain hundreds or even thousands of terms, and this example can not be efficiently solved by using the power-series expansion method.

8.3 Recursive Solution of the Differential Riccati Equation of Weakly Coupled Systems

In this section, we study the finite time closed-loop optimal control problem of weakly coupled systems. The recursive reduced-order solution will be obtained by exploiting the transformation introduced in (Gajic and Shen, 1989) which will block diagonalize the Hamiltonian form of the solution for the optimal linear-quadratic control problem. Completely decoupled sets of reduced-order differential equations are obtained. The convergence to the optimal solution is pretty rapid, due to the fact that the algorithms derived in (Gajic and Shen, 1989) have the rate of convergence of at least of $O(\epsilon^2)$. This produces a lot of savings in the size of computations required.

Consider the linear weakly coupled system

$$\begin{aligned} \dot{x}_1 &= A_1 x_1 + \epsilon A_2 x_2 + B_1 u_1 + \epsilon B_2 u_2, & x_1(t_0) &= x_{10} \\ \dot{x}_2 &= \epsilon A_3 x_1 + A_4 x_2 + \epsilon B_3 u_1 + B_4 u_2, & x_2(t_0) &= x_{20} \end{aligned} \quad (8.40)$$

with

$$z = \begin{bmatrix} z_1 \\ z_2 \end{bmatrix} = D \begin{bmatrix} x_1 \\ x_2 \end{bmatrix} = \begin{bmatrix} D_1 & \epsilon D_2 \\ \epsilon D_3 & D_4 \end{bmatrix} \begin{bmatrix} x_1 \\ x_2 \end{bmatrix} \quad (8.41)$$

where $x_i \in R^{n_i}$, $u_i \in R^{m_i}$, $z_i \in R^{r_i}$, $i = 1, 2$, are state, control, and output variables, respectively. The system matrices are of appropriate

time = t	0.1	0.5	1.0
$P_{11} = P_{exact}$	1.9699	6.6483	9.6600
$P_{11}^{(12)}$	"	"	"
$P_{11}^{(11)}$	"	"	9.6599
$P_{11}^{(10)}$	1.9698	"	9.6601
$P_{11}^{(9)}$	1.9700	6.6484	9.6598
$P_{11}^{(8)}$	1.9696	6.6482	9.6602
$P_{11}^{(7)}$	1.9703	6.6487	9.6603
$P_{11}^{(6)}$	1.9694	6.6471	9.6572
$P_{11}^{(5)}$	1.9703	6.6500	9.6671
$P_{11}^{(4)}$	1.9720	6.6496	9.6477
$P_{11}^{(3)}$	1.9537	6.6488	9.6991
$P_{11}^{(2)}$	2.0603	6.6520	9.5417
$P_{11}^{(1)}$	1.9847	6.7926	9.8624
$P_{11}^{(0)}$	1.9742	7.0256	10.4610

Table 8.3: Simulation results for the element $P_{11}(t)$

dimensions and, in general, they are bounded functions of a small coupling parameter ϵ (Gajic et al., 1990; Harkara et al., 1989; Petrovic and Gajic, 1988). In this section, we will assume that all given matrices are constant.

With (8.40)-(8.41), consider the performance criterion

$$J = \frac{1}{2} \int_{t_0}^{T} \left\{ \begin{bmatrix} x_1 \\ x_2 \end{bmatrix}^T D^T D \begin{bmatrix} x_1 \\ x_2 \end{bmatrix} + \begin{bmatrix} u_1 \\ u_2 \end{bmatrix}^T R \begin{bmatrix} u_1 \\ u_2 \end{bmatrix} \right\} dt$$

$$+ \frac{1}{2} \begin{bmatrix} x_1(T) \\ x_2(T) \end{bmatrix}^T F \begin{bmatrix} x_1(T) \\ x_2(T) \end{bmatrix} \tag{8.42}$$

with positive definite R and positive semidefinite F, which has to be minimized. It is assumed that matrices F and R have the weakly coupled structures, that is

$$F = \begin{bmatrix} F_1 & \epsilon F_2 \\ \epsilon F_2^T & F_3 \end{bmatrix}, \qquad R = \begin{bmatrix} R_1 & 0 \\ 0 & R_2 \end{bmatrix} \tag{8.43}$$

The optimal closed-loop control law has the very well-known form (Kwakernaak and Sivan, 1972)

$$u = \begin{bmatrix} u_1 \\ u_2 \end{bmatrix} = -R^{-1} \begin{bmatrix} B_1 & \epsilon B_2 \\ \epsilon B_3 & B_4 \end{bmatrix}^T P \begin{bmatrix} x_1 \\ x_2 \end{bmatrix} = -R^{-1} B^T P x \tag{8.44}$$

where P satisfies the differential Riccati equation given by

$$-\dot{P} = PA + A^T P + D^T D - PSP, \qquad P(T) = F \tag{8.45}$$

with

$$A = \begin{bmatrix} A_1 & \epsilon A_2 \\ \epsilon A_3 & A_4 \end{bmatrix}, \qquad S = BR^{-1}B^T = \begin{bmatrix} S_1 & \epsilon S_2 \\ \epsilon S_2^T & S_3 \end{bmatrix} \tag{8.46}$$

Due to weakly coupled structure of all coefficients in (8.45), the solution of that equation has the form

$$P = \begin{bmatrix} P_1 & \epsilon P_2 \\ \epsilon P_2^T & P_3 \end{bmatrix} \tag{8.47}$$

In this section, we will exploit the Hamiltonian form of the solution of the Riccati differential equation and a nonsingular transformation introduced in (Gajic and Shen, 1989) in order to obtain an efficient recursive method for solving (8.45).

The solution of (8.45) can be sought in the form

$$P(t) = M(t) N^{-1}(t) \qquad (8.48)$$

where matrices $M(t)$ and $N(t)$ satisfy a system of linear equations (Kwakernaak and Sivan, 1972)

$$\dot{M} = -A^T M(t) - D^T D N(t), \qquad M(T) = F \qquad (8.49)$$

$$\dot{N}(t) = -S M(t) + A N(t), \qquad N(T) = I \qquad (8.50)$$

Lemma 8.1, proved in Section 8.1, guarantees the existence of the invertible solution for $N(t)$ for all t.

Knowing the nature of the solution of (8.45), we introduce compatible partitions of $M(t)$ and $N(t)$ matrices as

$$M(t) = \begin{bmatrix} M_1(t) & \epsilon M_2(t) \\ \epsilon M_3(t) & M_4(t) \end{bmatrix}, \quad N(t) = \begin{bmatrix} N_1(t) & \epsilon N_2(t) \\ \epsilon N_3(t) & N_4(t) \end{bmatrix} \qquad (8.51)$$

Partitioning (8.49) and (8.50), according to (8.46) and (8.51), will reveal a decoupled structure, that is, M_1, M_3, N_1, and N_3 are independent of equations for M_2, M_4, N_2, and N_4 and vice versa. Introducing the notation

$$U = \begin{bmatrix} M_1 \\ N_1 \end{bmatrix}, \quad V = \begin{bmatrix} \epsilon M_3 \\ \epsilon N_3 \end{bmatrix}, \quad X = \begin{bmatrix} \epsilon M_2 \\ \epsilon N_2 \end{bmatrix}, \quad Y = \begin{bmatrix} M_4 \\ N_4 \end{bmatrix} \qquad (8.52)$$

and

$$T_1 = \begin{bmatrix} -A_1^T & -Q_1 \\ -S_1 & A_1 \end{bmatrix}, \quad T_2 = \begin{bmatrix} -A_3^T & -Q_2 \\ -S_2 & A_2 \end{bmatrix}$$

$$T_3 = \begin{bmatrix} -A_2^T & -Q_2^T \\ -S_2^T & A_3 \end{bmatrix}, \quad T_4 = \begin{bmatrix} -A_4^T & -Q_3 \\ -S_3 & A_4 \end{bmatrix} \qquad (8.53)$$

where

$$Q_1 = D_1^T D_1 + \epsilon^2 D_3^T D_3, \; Q_2 = D_1^T D_2 + D_3^T D_4$$
$$Q_3 = D_4^T D_4 + \epsilon^2 D_2^T D_2$$

and after doing some algebra, we get two independent systems of weakly coupled matrix differential equations

$$\dot{U} = T_1 U + \epsilon T_2 V$$
$$\dot{V} = \epsilon T_3 U + T_4 V \qquad (8.54)$$

with terminal conditions

$$U(T) = \begin{bmatrix} F_1 \\ I \end{bmatrix}, \quad V(T) = \begin{bmatrix} \epsilon F_2^T \\ 0 \end{bmatrix} \qquad (8.55)$$

and

$$\dot{X} = T_1 X + \epsilon T_2 Y$$
$$\dot{Y} = \epsilon T_3 X + T_4 Y \qquad (8.56)$$

with terminal conditions

$$X(T) = \begin{bmatrix} \epsilon F_2 \\ 0 \end{bmatrix}, \quad Y(T) = \begin{bmatrix} F_3 \\ I \end{bmatrix} \qquad (8.57)$$

Note that these two systems have exactly the same form and they differ in terminal conditions only. From this point, we will proceed by applying the decoupling transformation introduced in (Gajic and Shen, 1989). This transformation is defined by

$$T_2 = \begin{bmatrix} I & -\epsilon L \\ \epsilon H & I - \epsilon^2 H L \end{bmatrix}, \quad T_2^{-1} = \begin{bmatrix} I - \epsilon^2 LH & \epsilon L \\ -\epsilon H & I \end{bmatrix} \qquad (8.58)$$

where L and H satisfy

$$T_1 L + T_2 - L T_4 - \epsilon^2 L T_3 L = 0 \qquad (8.59)$$

$$H(T_1 - \epsilon^2 L T_3) - (T_4 + \epsilon^2 T_3 L) H + T_3 = 0 \qquad (8.60)$$

Applied to (8.54)-(8.57), it will produce

$$\dot{\hat{U}} = (T_1 - \epsilon^2 L T_3)\hat{U}, \qquad \hat{U}(T) = U(T) - \epsilon L V(T) \qquad (8.61)$$

$$\dot{\hat{V}} = (T_4 + \epsilon^2 T_3 L)\hat{V}, \qquad \hat{V}(T) = \epsilon H U(T) + (I - \epsilon^2 H L) V(T) \qquad (8.62)$$

and

$$\dot{\hat{X}} = (T_1 - \epsilon^2 L T_3)\hat{X}, \qquad \hat{X}(T) = X(T) - \epsilon L Y(T) \qquad (8.63)$$

$$\dot{\hat{Y}} = (T_4 + \epsilon^2 T_3 L)\hat{Y}, \qquad \hat{Y}(T) = \epsilon H X(T) + (I - \epsilon^2 H L) Y(T) \qquad (8.64)$$

Solutions of (8.61)-(8.64) are given by

$$\widehat{U}(t) = e^{(T_1 - \epsilon^2 LT_3)(t-T)}\widehat{U}(T) \tag{8.65}$$

$$\widehat{V}(t) = e^{(T_4 + \epsilon^2 T_3 L)(t-T)}\widehat{V}(T) \tag{8.66}$$

$$\widehat{X}(t) = e^{(T_1 - \epsilon^2 LT_3)(t-T)}\widehat{X}(T) \tag{8.67}$$

$$\widehat{Y}(t) = e^{(T_4 + \epsilon^2 T_3 L)(t-T)}\widehat{Y}(T) \tag{8.68}$$

so that in the original coordinates we have

$$U(t) = \left(I - \epsilon^2 LH\right) e^{(T_1 - \epsilon^2 LT_3)(t-T)}\widehat{U}(T) + \epsilon L e^{(T_4 + \epsilon^2 T_3 L)(t-T)}\widehat{V}(T) \tag{8.69}$$

$$V(t) = -\epsilon H e^{(T_1 - \epsilon^2 LT_3)(t-T)}\widehat{U}(T) + e^{(T_4 + \epsilon^2 T_3 L)(t-T)}\widehat{V}(T) \tag{8.70}$$

$$X(t) = \left(I - \epsilon^2 LH\right) e^{(T_1 - \epsilon^2 LT_3)(t-T)}\widehat{X}(T) + \epsilon L e^{(T_4 + \epsilon^2 T_3 L)(t-T)}\widehat{Y}(T) \tag{8.71}$$

$$Y(t) = -\epsilon H e^{(T_1 - \epsilon^2 LT_3)(t-T)}\widehat{X}(T) + e^{(T_4 + \epsilon^2 T_3 L)(t-T)}\widehat{Y}(T) \tag{8.72}$$

Partitioning $U(t), V(t), X(t)$, and $Y(t)$ according to (8.52) will produce all components of the matrices $M(t)$ and $N(t)$; that is

$$\begin{bmatrix} M_1(t) \\ N_1(t) \end{bmatrix} = \begin{bmatrix} U_1(t) \\ U_2(t) \end{bmatrix} = U(t), \quad \begin{bmatrix} \epsilon M_2(t) \\ \epsilon N_2(t) \end{bmatrix} = \begin{bmatrix} X_1(t) \\ X_2(t) \end{bmatrix} = X(t)$$

$$\begin{bmatrix} \epsilon M_3(t) \\ \epsilon N_3(t) \end{bmatrix} = \begin{bmatrix} V_1(t) \\ V_2(t) \end{bmatrix} = V(t), \quad \begin{bmatrix} M_4(t) \\ N_4(t) \end{bmatrix} = \begin{bmatrix} Y_1(t) \\ Y_2(t) \end{bmatrix} = Y(t)$$

so that the required solution of (8.45) is given by

$$P(t) = \begin{bmatrix} U_1(t) & X_1(t) \\ V_1(t) & Y_1(t) \end{bmatrix} \begin{bmatrix} U_2(t) & X_2(t) \\ V_2(t) & Y_2(t) \end{bmatrix}^{-1} \tag{8.74}$$

Thus, in order to get the solution of (8.45), $P(t)$, which has dimensions $n \times n = (n_1 + n_2) \times (n_1 + n_2)$, we have to solve two simple algebraic equations (8.59) and (8.60) of dimensions $(2n_2 \times 2n_1)$ and $(2n_1 \times 2n_2)$, respectively. The efficient numerical algorithm based on the fixed point iterations and the Newton method for solving (8.59) and (8.60) can be found in Section 3.3. Then, two exponential forms $exp\left[(T_1 - \epsilon^2 LT_3)(t - T)\right]$ and $exp\left[(T_4 + \epsilon^2 T_3 L)(t - T)\right]$ have to be transformed in the matrix forms by using some of the well-known approaches (Molen and Van Loan, 1978). Finally, the inversion of the matrix $N(t)$ has to be performed.

As discussed in Section 8.1, the matrices $M(t)$ and $N(t)$ contain unstable models of the Hamiltonian, and the reinitialization version of the Hamiltonian approach avoids that problem. The reinitialization technique applied to the problem under consideration will modify only terminal conditions in formulas (8.49), (8.55), and (8.57), respectively,

$$M(k\Delta t) = P(k\Delta t) \qquad (8.75)$$

$$U(k\Delta t) = \begin{bmatrix} P_1(k\Delta t) \\ I \end{bmatrix}, \quad V(k\Delta t) = \begin{bmatrix} \epsilon P_2^T(k\Delta t) \\ 0 \end{bmatrix} \qquad (8.76)$$

$$X(k\Delta t) = \begin{bmatrix} \epsilon P_2(k\Delta t) \\ 0 \end{bmatrix}, \quad Y(k\Delta t) = \begin{bmatrix} P_3(k\Delta t) \\ I \end{bmatrix} \qquad (8.77)$$

where k represents the number of steps and Δt is an integration step.

The transformation matrix T_2 from (8.58) can be easily obtained, with required accuracy, by using numerical techniques developed in (Gajic and Shen, 1989) for solving (8.59)-(8.60). They converge with the rate of convergence of at least of $O(\epsilon^2)$. Thus, after k iterations, one gets the approximation $T_2^{(k)} = T_2 + O(\epsilon^{2k})$. The use of $T_2^{(k)}$ in (8.61)-(8.64) instead of T_2, will perturb the coefficients of the corresponding systems of linear differential equations by $O(\epsilon^{2k})$, which implies that the approximate solutions of these differential equations are $O(\epsilon^{2k})$ close to the exact ones (Kato, 1980). Thus, it is of interest to obtain $T_2^{(k)}$ with the desired accuracy, which produces the same accuracy in the sought solution.

The recursive reduced-order solution of the differential Riccati equation of weakly coupled systems is demonstrated in the next section where a real world example is considered.

8.4 Case Study: Gas Absorber

A real world example, a six-plate gas absorber (De Vlieger et al., 1982) is considered to demonstrate the proposed method.

The problem matrices A and B are given by

$$A = \begin{bmatrix} -1.173 & 0.6341 & 0 & 0 & 0 & 0 \\ 0.5390 & -1.173 & 0.6341 & 0 & 0 & 0 \\ 0 & 0.5390 & -1.173 & 0.6341 & 0 & 0 \\ 0 & 0 & 0.5390 & -1.173 & 0.6341 & 0 \\ 0 & 0 & 0 & 0.5390 & -1.173 & 0.6341 \\ 0 & 0 & 0 & 0 & 0.5390 & -1.173 \end{bmatrix}$$

$$B^T = \begin{bmatrix} 0.5390 & 0 & 0 & 0 & 0 & 0 \\ 0 & 0 & 0 & 0 & 0 & 0.6341 \end{bmatrix}$$

Remaining matrices are chosen as

$$D^T D = \begin{bmatrix} 1 & 0 & 0 & 0 & 0 & 0 \\ 0 & 1 & 0 & 0 & 0 & 0 \\ 0 & 0 & 1 & 0 & 0 & 0 \\ 0 & 0 & 0 & 2 & 0 & 0 \\ 0 & 0 & 0 & 0 & 2 & 0 \\ 0 & 0 & 0 & 0 & 0 & 2 \end{bmatrix}, \quad R = 0.1 I_2, \quad F = I_6$$

The initial and final times are selected as $t_0 = 0$ and $T = 1$. The initial condition are

$$x_{10} = [-0.0306 \ -0.0568 \ -0.0788]^T$$
$$x_{20} = [-0.0977 \ -0.1138 \ -0.1273]^T$$

The system is partitioned into two subsystems with $n_1 = 3$, $n_2 = 3$, and $\epsilon = 0.37$. The small parameter ϵ is built into the problem. It can be roughly estimated from the strongest coupled matrix — in this case matrix A — producing $|a_{34}| / (|a_{32}| + |a_{33}|) = 0.6341/1.7120 = 0.37$. The simulation results for the differential Riccati equation are presented in Table 8.4. After performing 4 iterations, we have obtained the accuracy of 10^{-5}.

We have solved also the open-loop optimal control for the same example by using the corresponding recursive reduced-order method presented in Chapter 6. Corresponding simulation results for both components of the approximate open-loop control are presented in Table 8.5.

iteration	$t = 0.25$	$t = 0.5$	$t = 1$
4 = optimal	0.51024	0.39942	0.35868
3	0.51024	0.39942	0.35867
2	0.51031	0.40003	0.36180
1	0.51022	0.39922	0.35808
0	0.51023	0.39939	0.36066

Table 8.4: Simulation result for the element
$P_{11}(t)$ of the Riccati differential equation

iteration	$u(t = 0)$	$u(t = 0.25)$	$u(t = 0.5)$	$u(t = 0.75)$
optimal 5	0.17112 0.64956	0.12257 0.33987	0.08678 0.18081	0.05618 0.08738
4	0.17106 0.64956	0.12252 0.33987	0.08674 0.18081	0.05615 0.08738
3	0.16264 0.64959	0.11538 0.33988	0.08088 0.18081	0.05123 0.08738
2	0.30392 0.65187	0.23349 0.34082	0.17643 0.18105	0.12555 0.08723
1	0.19365 0.66051	0.12912 0.34421	0.08322 0.18181	0.04314 0.08667
0	0.56931 0.70882	0.37896 0.36203	0.24325 0.18507	0.12551 0.08298

Table 8.5: Simulation results for the open-loop control

By comparing linear systems of differential equations (8.54)-(8.57) and (6.40), apparently the closed-loop solution is computationally much more involved since (8.54) and (8.56) are of the order of $2(2n \times n)$, whereas (6.40) represents the same set of equations of order $2n$.

8.5 Reduced-Order Solution of the Singularly Perturbed Matrix Difference Riccati Equation

In this section, we study the solution of the singularly perturbed matrix difference Riccati equation using the approach presented in Section 8.1. The order reduction is achieved via the use of the Chang transformation applied to the Hamiltonian matrix of a singularly perturbed linear-quadratic control problem. Since the decoupling transformation can be obtained, up to an arbitrary degree of accuracy at very low cost, this approach produces an efficient numerical method for solving singularly perturbed difference Riccati equation. The results are demonstrated on a real world example.

A singularly perturbed linear discrete system is represented by (Litkouhi and Khalil, 1985)

$$
\begin{aligned}
x_1\,(k+1) &= (I_{n_1} + \epsilon A_1)\,x_1\,(k) + \epsilon A_2 x_2\,(k) + \epsilon B_1 u\,(k) \\
x_2\,(k+1) &= A_3 x_1\,(k) + A_4 x_2\,(k) + B_2 u\,(k)
\end{aligned}
\tag{8.78}
$$

with slow state variables $x_1 \in \Re^{n_1}$, fast state variables $x_2 \in \Re^{n_2}$, and control inputs $u \in \Re^m$. The performance criterion of the corresponding linear-quadratic control problem is defined by

$$
J\,(k) = \frac{1}{2} x^T\,(n)\,Fx\,(n) + \frac{1}{2} \sum_{k=0}^{n-1} \left[x^T\,(k)\,Qx\,(k) + u^T\,(k)\,Ru\,(k) \right]
\tag{8.79}
$$

where

$$
x\,(k) = \begin{bmatrix} x_1\,(k) \\ x_2\,(k) \end{bmatrix}, \quad Q = Q^T = \begin{bmatrix} Q_1 & Q_2 \\ Q_2^T & Q_3 \end{bmatrix} \geq 0
\tag{8.80}
$$

$$
F = \begin{bmatrix} F_1/\epsilon & F_2 \\ F_2^T & F_3 \end{bmatrix} \geq 0, \qquad R = R^T > 0
$$

The Hamiltonian form of (8.78)-(8.79) can be written as the back recursion (Lewis, 1986)

$$
\begin{aligned}
\begin{bmatrix} x\,(k) \\ \lambda\,(k) \end{bmatrix} &= \begin{bmatrix} A^{-1} & A^{-1}BR^{-1}B^T \\ QA^{-1} & A^T + QA^{-1}BR^{-1}B^T \end{bmatrix} \begin{bmatrix} x\,(k+1) \\ \lambda\,(k+1) \end{bmatrix} \\
&= \mathbf{H} \begin{bmatrix} x\,(k+1) \\ \lambda\,(k+1) \end{bmatrix}
\end{aligned}
\tag{8.81}
$$

where $\mathbf{H}$ is the symplectic matrix which has the property that the eigenvalues of $\mathbf{H}$ can be grouped into two disjoint subsets Γ_1 and Γ_2, such that for every $\lambda_c \in \Gamma_1$ there exists $\lambda_d \in \Gamma_2$, which satisfies $\lambda_c \times \lambda_d = 1$, and we can choose either Γ_1 or Γ_2 to contain only the stable eigenvalues (Salgado et al., 1988).

The optimal feedback control law has the very well-known form

$$u(k) = -R^{-1}B^T\lambda(k+1) = -R^{-1}B^TP(k+1)x(k+1) \quad (8.82)$$

where $P(k)$ satisfies the difference Riccati equation given by

$$\begin{aligned} P(k) &= Q + A^TP(k+1)[I + SP(k+1)]^{-1}A \\ &= Q + A^TP(k+1)A \\ -A^TP(k+1)B&[R + B^TP(k+1)B]^{-1}B^TP(k+1)A \end{aligned} \quad (8.83)$$

with

$$A = \begin{bmatrix} I_{n_1} + \epsilon A_1 & \epsilon A_2 \\ A_3 & A_4 \end{bmatrix} \quad (8.84)$$

$$S = BR^{-1}B^T = \begin{bmatrix} \epsilon^2 S_1 & \epsilon Z \\ \epsilon Z^T & S_2 \end{bmatrix}$$

$$S_1 = B_1 R^{-1} B_1^T, \quad S_2 = B_2 R^{-1} B_2^T, \quad Z = B_1 R^{-1} B_2^T \quad (8.85)$$

The presence of a small parameter ϵ makes this problem numerically ill-defined (Litkouhi and Khalil, 1985). In order to overcome this difficulty and obtain an efficient numerical method for solving (8.83), we will utilize the known Hamiltonian form (8.81) of the solution of the difference Riccati equation and the nonsingular Chang transformation (Chang, 1972). The Hamiltonian form can "linearize" the difference Riccati equation and the Chang transformation is used to block diagonalize the Hamiltonian, so that the required solution of the Riccati equation is obtained in terms of reduced-order problems. An efficient Newton-type algorithm (Gajic et al., 1990), (with the quadratic rate of convergence, that is, $O\left(\epsilon^{2^i}\right)$, where i is a number of iterations) is used for solving algebraic equations, which results in forming the Chang transformation.

The solution of (8.83) can be sought in the form (see Appendix 8.1)

$$P(k) = M(k)N^{-1}(k) \quad (8.86)$$

where matrices $M(k)$ and $N(k)$ satisfy a system of linear equations

$$\begin{aligned} N(k) &= A^{-1}N(k+1) + A^{-1}BR^{-1}B^TM(k+1) \\ M(k) &= QA^{-1}N(k+1) + (A^T + QA^{-1}BR^{-1}B^T)M(k+1) \end{aligned}$$
$$(8.87)$$

with $M(n) = F$, $N(n) = I$.

The following lemma guarantees the existence of the invertible solution for $N(k)$.

Lemma 8.2 *If the triple $\left(A, B, \sqrt{Q}\right)$ is stabilizable-observable then the matrix $N(k)$, with $N(n) = I$ is invertible for any $k = 0, 1, 2, ...n$.*

◇

Proof: See Appendix 8.2.

The solution of (8.83) is properly scaled as (Litkouhi and Khalil, 1984)

$$P(k) = \begin{bmatrix} P_1(k)/\epsilon & P_2(k) \\ P_2^T(k) & P_3(k) \end{bmatrix}, \ P(n) = F = \begin{bmatrix} F_1/\epsilon & F_2 \\ F_2^T & F_3 \end{bmatrix} \quad (8.88)$$

where $dim P_1 = n_1 \times n_1$, $dim P_3 = n_2 \times n_2$.

Let compatible partitions of matrices $M(k)$ and $N(k)$ be

$$M(k) = \begin{bmatrix} M_1(k) & M_2(k) \\ M_3(k) & M_4(k) \end{bmatrix}, \ N(k) = \begin{bmatrix} N_1(k) & N_2(k) \\ N_3(k) & N_4(k) \end{bmatrix} \quad (8.89)$$

Partitioning (8.87), according to (8.89), will reveal a decoupled structure, that is, equations for $M_1(k)$, $M_3(k)$, $N_1(k)$, and $N_3(k)$ are independent of equations for $M_2(k)$, $M_4(k)$, $N_2(k)$, and $N_4(k)$ and vice versa.

$$\begin{bmatrix} N_1(k) \\ N_3(k) \\ M_1(k) \\ M_3(k) \end{bmatrix} = \begin{bmatrix} I_{n_1} + \epsilon\overline{A_1} & \epsilon\overline{A_2} & \epsilon^2\overline{S_1} & \epsilon\overline{S_2} \\ \overline{A_3} & \overline{A_4} & \epsilon\overline{S_3} & \overline{S_4} \\ \overline{Q_1} & \overline{Q_2} & I_{n_1} + \epsilon A_{11}^T & A_{21}^T \\ \overline{Q_3} & \overline{Q_4} & \epsilon A_{12}^T & A_{22}^T \end{bmatrix} \begin{bmatrix} N_1(k+1) \\ N_3(k+1) \\ M_1(k+1) \\ M_3(k+1) \end{bmatrix}$$

$$= \mathbf{H} \begin{bmatrix} N_1(k+1) \\ N_3(k+1) \\ M_1(k+1) \\ M_3(k+1) \end{bmatrix}$$

$$(8.90)$$

$$\begin{bmatrix} N_2(k) \\ N_4(k) \\ M_2(k) \\ M_4(k) \end{bmatrix} = \begin{bmatrix} I_{n_1} + \epsilon\overline{A_1} & \epsilon\overline{A_2} & \epsilon^2\overline{S_1} & \epsilon\overline{S_2} \\ \overline{A_3} & \overline{A_4} & \epsilon\overline{S_3} & \overline{S_4} \\ \overline{Q_1} & \overline{Q_2} & I_{n_2} + \epsilon A_{11}^T & A_{21}^T \\ \overline{Q_3} & \overline{Q_4} & \epsilon A_{12}^T & A_{22}^T \end{bmatrix} \begin{bmatrix} N_2(k+1) \\ N_4(k+1) \\ M_2(k+1) \\ M_4(k+1) \end{bmatrix}$$

$$= \mathbf{H} \begin{bmatrix} N_2(k+1) \\ N_4(k+1) \\ M_2(k+1) \\ M_4(k+1) \end{bmatrix}$$

$$(8.91)$$

For details of these calculations see Appendix 8.3. Interchanging the second and third rows in (8.90) and (8.91), respectively, produces

$$
\begin{bmatrix} N_1(k) \\ \epsilon M_1(k) \\ N_3(k) \\ M_3(k) \end{bmatrix} = \begin{bmatrix} I_{n_1} + \epsilon \overline{A_1} & \epsilon \overline{S_1} & \epsilon \overline{A_2} & \epsilon \overline{S_2} \\ \epsilon \overline{Q_1} & I_{n_1} + \epsilon \overline{A_{11}^T} & \epsilon \overline{Q_2} & \epsilon \overline{A_{21}^T} \\ \overline{A_3} & \overline{S_3} & \overline{A_4} & \overline{S_4} \\ \overline{Q_3} & \overline{A_{12}^T} & \overline{Q_4} & \overline{A_{22}^T} \end{bmatrix}
$$

$$
\times \begin{bmatrix} N_1(k+1) \\ \epsilon M_1(k+1) \\ N_3(k+1) \\ M_3(k+1) \end{bmatrix} = \begin{bmatrix} I + \epsilon T_1 & \epsilon T_2 \\ T_3 & T_4 \end{bmatrix} \begin{bmatrix} N_1(k+1) \\ \epsilon M_1(k+1) \\ N_3(k+1) \\ M_3(k+1) \end{bmatrix} \tag{8.92}
$$

$$
\begin{bmatrix} N_2(k) \\ \epsilon M_2(k) \\ N_4(k) \\ M_4(k) \end{bmatrix} = \begin{bmatrix} I_{n_1} + \epsilon \overline{A_1} & \epsilon \overline{S_1} & \epsilon \overline{A_2} & \epsilon \overline{S_2} \\ \epsilon \overline{Q_1} & I_{n_1} + \epsilon \overline{A_{11}^T} & \epsilon \overline{Q_2} & \epsilon \overline{A_{21}^T} \\ \overline{A_3} & \overline{S_3} & \overline{A_4} & \overline{S_4} \\ \overline{Q_3} & \overline{A_{12}^T} & \overline{Q_4} & \overline{A_{22}^T} \end{bmatrix}
$$

$$
\times \begin{bmatrix} N_2(k+1) \\ \epsilon M_2(k+1) \\ N_4(k+1) \\ M_4(k+1) \end{bmatrix} = \begin{bmatrix} I + \epsilon T_1 & \epsilon T_2 \\ T_3 & T_4 \end{bmatrix} \begin{bmatrix} N_2(k+1) \\ \epsilon M_2(k+1) \\ N_4(k+1) \\ M_4(k+1) \end{bmatrix} \tag{8.93}
$$

where

$$
T_1 = \begin{bmatrix} \overline{A_1} & \overline{S_1} \\ \overline{Q_1} & \overline{A_{11}^T} \end{bmatrix}, \quad T_2 = \begin{bmatrix} \overline{A_2} & \overline{S_2} \\ \overline{Q_2} & \overline{A_{21}^T} \end{bmatrix}
$$

$$
T_3 = \begin{bmatrix} \overline{A_3} & \overline{S_3} \\ \overline{Q_3} & \overline{A_{12}^T} \end{bmatrix}, \quad T_4 = \begin{bmatrix} \overline{A_4} & \overline{S_4} \\ \overline{Q_4} & \overline{A_{22}^T} \end{bmatrix} \tag{8.94}
$$

Introducing the notation

$$
U(k) = \begin{bmatrix} N_1(k) \\ \epsilon M_1(k) \end{bmatrix}, \quad V(k) = \begin{bmatrix} N_3(k) \\ M_3(k) \end{bmatrix}
$$

$$
X(k) = \begin{bmatrix} N_2(k) \\ \epsilon M_2(k) \end{bmatrix}, \quad Y(k) = \begin{bmatrix} N_4(k) \\ M_4(k) \end{bmatrix} \tag{8.95}
$$

we get two systems of singularly perturbed difference equations

$$
U(k) = (I + \epsilon T_1) U(k+1) + \epsilon T_2 V(k+1)
$$
$$
V(k) = T_3 U(k+1) + T_4 V(k+1) \tag{8.96}
$$

$$X(k) = (I + \epsilon T_1) X(k+1) + \epsilon T_2 Y(k+1)$$
$$Y(k) = T_3 X(k+1) + T_4 Y(k+1) \tag{8.97}$$

with terminal conditions

$$U(n) = \begin{bmatrix} I \\ F_1 \end{bmatrix}, \quad V(n) = \begin{bmatrix} 0 \\ F_2^T \end{bmatrix}$$
$$X(n) = \begin{bmatrix} 0 \\ \epsilon F_2 \end{bmatrix}, \quad Y(n) = \begin{bmatrix} I \\ F_3 \end{bmatrix} \tag{8.98}$$

Note that systems (8.96) and (8.97) have exactly the same form and the only difference is in the terminal conditions.

Applying the Chang transformation given by

$$\mathbf{T}_1 = \begin{bmatrix} I - \epsilon ML & -\epsilon M \\ L & I \end{bmatrix}, \quad \mathbf{T}_1^{-1} = \begin{bmatrix} I & \epsilon M \\ -L & I - \epsilon LM \end{bmatrix} \tag{8.99}$$

where L and M satisfy

$$0 = M + T_2 - MT_4 + \epsilon(T_1 - T_2L)M - \epsilon MLT_2$$
$$0 = -L + T_4L - T_3 - \epsilon L(T_1 - T_2L) \tag{8.100}$$

to (8.96) and (8.97) produces

$$\overline{U}(k) = (I + \epsilon T_1 - \epsilon T_2 L)\overline{U}(k+1)$$
$$\overline{V}(k) = (T_4 + \epsilon LT_2)\overline{V}(k+1) \tag{8.101}$$

$$\overline{X}(k) = (I + \epsilon T_1 - \epsilon T_2 L)\overline{X}(k+1)$$
$$\overline{Y}(k) = (T_4 + \epsilon LT_2)\overline{Y}(k+1) \tag{8.102}$$

with the terminal conditions

$$\overline{U}(n) = (I - \epsilon ML)U(n) - \epsilon MV(n)$$
$$\overline{V}(n) = LU(n) + V(n)$$
$$\overline{X}(n) = (I - \epsilon ML)X(n) - \epsilon MY(n)$$
$$\overline{Y}(n) = LX(n) + Y(n) \tag{8.103}$$

Matrices L an M can be obtained by using the recursive algorithm from (Gajic et al., 1990). Solutions of (8.101)-(8.102) are given by

$$\overline{U}(k) = (I + \epsilon T_1 - \epsilon T_2 L)^{n-k}\overline{U}(n)$$
$$\overline{V}(k) = (T_4 + \epsilon LT_2)^{n-k}\overline{V}(n)$$

$$\overline{X}(k) = (I + \epsilon T_1 - \epsilon T_2 L)^{n-k} \overline{X}(n)$$
$$\overline{Y}(k) = (T_4 + \epsilon L T_2)^{n-k} \overline{Y}(n)$$

(8.104)

The solutions in the original coordinates are

$$U(k) = (I + \epsilon T_1 - \epsilon T_2 L)^{n-k} \overline{U}(n) + \epsilon M (T_4 + \epsilon L T_2)^{n-k} \overline{V}(n)$$

$$V(k) = -L (I + \epsilon T_1 - \epsilon T_2 L)^{n-k} \overline{U}(n)$$
$$+ (I - \epsilon L M) (T_4 + \epsilon L T_2)^{n-k} \overline{V}(n)$$

$$X(k) = (I + \epsilon T_1 - \epsilon T_2 L)^{n-k} \overline{X}(n) + \epsilon M (T_4 + \epsilon L T_2)^{n-k} \overline{Y}(n)$$

$$Y(k) = -L (I + \epsilon T_1 - \epsilon T_2 L)^{n-k} \overline{X}(n)$$
$$+ (I - \epsilon L M) (T_4 + \epsilon L T_2)^{n-k} \overline{Y}(n)$$

(8.105)

Partitioning (8.105), according to (8.95), will produce all components of matrices $M(k)$ and $N(k)$, that is

$$\begin{bmatrix} N_1(k) \\ \epsilon M_1(k) \end{bmatrix} = \begin{bmatrix} U_1(k) \\ U_2(k) \end{bmatrix} = U(k)$$

$$\begin{bmatrix} N_3(k) \\ M_3(k) \end{bmatrix} = \begin{bmatrix} V_1(k) \\ V_2(k) \end{bmatrix} = V(k)$$

(8.106)

$$\begin{bmatrix} N_2(k) \\ \epsilon M_2(k) \end{bmatrix} = \begin{bmatrix} X_1(k) \\ X_2(k) \end{bmatrix} = X(k)$$

$$\begin{bmatrix} N_4(k) \\ M_4(k) \end{bmatrix} = \begin{bmatrix} Y_1(k) \\ Y_2(k) \end{bmatrix} = Y(k)$$

Then the required solution of (8.83) is given by

$$P(k) = \begin{bmatrix} U_2(k)/\epsilon & X_2(k)/\epsilon \\ V_2(k) & Y_2(k) \end{bmatrix} \begin{bmatrix} U_1(k) & X_1(k) \\ V_1(k) & Y_1(k) \end{bmatrix}^{-1}$$

(8.107)

Thus, in order to get the solutions of (8.83), that is $P(k)$, which has $dim\ n \times n = (n_1 + n_2) \times (n_1 + n_2)$, we only solve two simple algebraic equations (8.100) of dimensions of $(2n_2 \times 2n_1)$ and $(2n_1 \times 2n_2)$, respectively. The existing numerical algorithms for solving (8.100) can be found in (Gajic et al., 1990) where the rate of convergence is $O\left(\epsilon^{2^i}\right)$, i is a number of iterations.

Results presented in this section follows very closely the derivations done in (Shen, 1992).

8.6 Case Study: Linearized Model of an F-8 Aircraft

In order to demonstrate the proposed method, a linearized model of an F-8 aircraft (Litkouhi, 1983) in the singularly perturbed continuous-time form (fast time version) is studied, with the system matrix A and the control matrix B given in Section 2.6.1. The small perturbation parameter ϵ is chosen as 1/30. Remaining matrices are chosen as $R = I_2, Q = 10^{-2}I_4$, and the terminal condition

$$P\left(n\right) = F = diag\left[0.5, 0.5, 0.01, 0.01\right]$$

With the proposed method, the simulation results for the L equation (8.100), presented in Table 8.6, and the solution of the singularly perturbed matrix difference Riccati equation (8.83) are obtained by using the package L-A-S for the computer aided control system design (West et al., 1985).

The terminal time is selected as $n = 8$ with k equals 4. The obtained solution P_{app}, is identical to the solution of the global Riccati difference equation (8.83) obtained by using any standard method (Pappas et al., 1980). However, in our method we have been using the reduced-order algorithm and the problem of ill-conditioning due to the singularly perturbed structure is eliminated.

$$P_{app} = \begin{bmatrix} 0.88976 & -0.077469 & -0.015048 & 0.00014416 \\ -0.077469 & 0.55719 & -0.016686 & 0.0047727 \\ -0.015048 & -0.016686 & 0.019299 & -0.0029608 \\ 0.00014416 & 0.0047727 & -0.0029608 & 0.011119 \end{bmatrix}$$

8.7 Reduced-Order Solution of the Weakly Coupled Matrix Difference Riccati Equation

The solution of the algebraic Riccati equation for weakly coupled discrete systems has been obtained in terms of the reduced-order continuous-time

i	error
0	1.00436 x 10^{-1}
1	2.00665 x 10^{-2}
2	1.44568 x 10^{-3}
3	1.09863 x 10^{-4}
4	7.74528 x 10^{-6}
5	4.91643 x 10^{-7}

Table 8.6: Approximate solution of L equation
where error is defined as $\|L^{(i+1)} - L^{(i)}\|_\infty$

algebraic Riccati equations via the use of a bilinear transformation (Shen and Gajic, 1990b). In this section, we use the approach developed in Section 8.3 to get the solution of the weakly coupled difference Riccati equation, up to any order of accuracy, by solving the reduced-order linear difference equations.

The weakly coupled linear discrete system is represented by (Shen and Gajic, 1990b)

$$
\begin{aligned}
x_1\,(k+1) &= A_1 x_1\,(k) + \epsilon A_2 x_2\,(k) + B_1 u_1\,(k) + \epsilon B_2 u_2\,(k) \\
x_2\,(k+1) &= \epsilon A_3 x_1\,(k) + A_4 x_2\,(k) + \epsilon B_3 u_1\,(k) + B_4 u_2\,(k)
\end{aligned}
\tag{8.108}
$$

with states $x_i \in \Re^{n_i}$, and control inputs $u_i \in \Re^{m_i}$, $i = 1, 2$, where ϵ is a small coupling parameter. The performance criterion of the corresponding linear-quadratic discrete control problem is defined as in (8.79), taking into account the presence of two control agents, that is

$$
u\,(k) = \begin{bmatrix} u_1\,(k) \\ u_2\,(k) \end{bmatrix}, \quad R = R^T = \begin{bmatrix} R_1 & 0 \\ 0 & R_2 \end{bmatrix} > 0
\tag{8.109}
$$

Introducing the notation

$$
A = \begin{bmatrix} A_1 & \epsilon A_2 \\ \epsilon A_3 & A_4 \end{bmatrix}, \quad B = \begin{bmatrix} B_1 & \epsilon B_2 \\ \epsilon B_3 & B_4 \end{bmatrix}
\tag{8.110}
$$

the Hamiltonian form of this optimal control problem can be written as the back recursion identical to (8.81)

$$\begin{bmatrix} x(k) \\ \lambda(k) \end{bmatrix} = \begin{bmatrix} A^{-1} & A^{-1}BR^{-1}B^T \\ QA^{-1} & A^T + QA^{-1}BR^{-1}B^T \end{bmatrix} \begin{bmatrix} x(k+1) \\ \lambda(k+1) \end{bmatrix}$$
$$= H \begin{bmatrix} x(k+1) \\ \lambda(k+1) \end{bmatrix}$$

The optimal control law has the very well-known form given by (8.82), that is,

$$u(k) = -R^{-1}B^T\lambda(k+1) = -R^{-1}B^T P(k+1) x(k+1)$$

where $P(k)$ satisfies the difference Riccati equation

$$\begin{aligned} P(k) &= Q + A^T P(k+1)[I + SP(k+1)]^{-1} A \\ &= Q + A^T P(k+1) A \end{aligned} \qquad (8.111)$$
$$- A^T P(k+1) B \left[R + B^T P(k+1) B \right]^{-1} B^T P(k+1) A$$

with

$$S = BR^{-1}B^T = \begin{bmatrix} S_1 & \epsilon S_2 \\ \epsilon S_2^T & S_3 \end{bmatrix}$$

$$S_1 = B_1 R_1^{-1} B_1^T + \epsilon^2 B_2 R_2^{-1} B_2^T, \quad S_2 = \left(B_1 R_1^{-1} B_3^T + B_2 R_2^{-1} B_4^T \right),$$
$$S_3 = B_4 R_2^{-1} B_2^T + \epsilon^2 B_3 R_1^{-1} B_3^T$$
$$(8.112)$$

In order to obtain an efficient numerical method for solving (8.111) in terms of the reduced-order problem, we will utilize the known Hamiltonian form of the solution of the difference Riccati equation and a nonsingular decoupling transformation from (Gajic and Shen, 1989).

The presented method for solving difference Riccati equation of weakly coupled discrete systems is dual to the one developed in Section 8.3 for the reduced-order solution of the differential Riccati equation of weakly coupled continuous systems.

The solution of (8.111) can be sought in the form

$$P(k) = M(k) N^{-1}(k) \qquad (8.113)$$

where matrices $M(k)$ and $N(k)$ satisfy a system of linear equations

$$N(k) = A^{-1}N(k+1) + A^{-1}BR^{-1}B^T M(k+1), \qquad N(n) = I$$
$$M(k) = QA^{-1}N(k+1) + \left(A^T + QA^{-1}BR^{-1}B^T \right) M(k+1),$$
$$M(n) = F$$
$$(8.114)$$

Lemma 8.2 guarantees the existence of the invertible solution for $N(k)$, for all values of k.

Due to the weakly coupled structure of all coefficients in (8.111), the solution of that equation has the form

$$P(k) = \begin{bmatrix} P_1(k) & \epsilon P_2(k) \\ \epsilon P_2^T(k) & P_3(k) \end{bmatrix}, \quad P(n) = F = \begin{bmatrix} F_1 & \epsilon F_2 \\ \epsilon F_2^T & F_3 \end{bmatrix} \quad (8.115)$$

where $dim P_1 = n_1 \times n_1$, $dim P_3 = n_2 \times n_2$.

Let compatible partitions of matrices $M(k)$ and $N(k)$ be

$$M(k) = \begin{bmatrix} M_1(k) & \epsilon M_2(k) \\ \epsilon M_3(k) & M_4(k) \end{bmatrix}, \quad N(k) = \begin{bmatrix} N_1(k) & \epsilon N_2(k) \\ \epsilon N_3(k) & N_4(k) \end{bmatrix} \quad (8.116)$$

Partitioning (8.114), according to (8.116), will reveal a decoupled structure, that is, equations for M_1, M_3, N_1, and N_3 are independent of equations for M_2, M_4, N_2, and N_4 and vice versa

$$\begin{bmatrix} N_1(k) \\ \epsilon N_3(k) \\ M_1(k) \\ \epsilon M_3(k) \end{bmatrix} = \begin{bmatrix} \overline{A_1} & \epsilon \overline{A_2} & \overline{S_1} & \epsilon \overline{S_2} \\ \epsilon \overline{A_3} & \overline{A_4} & \epsilon \overline{S_3} & \overline{S_4} \\ \overline{Q_1} & \epsilon \overline{Q_2} & A_{11}^T & \epsilon A_{21}^T \\ \epsilon \overline{Q_3} & \overline{Q_4} & \epsilon A_{12}^T & A_{22}^T \end{bmatrix} \begin{bmatrix} N_1(k+1) \\ \epsilon N_3(k+1) \\ M_1(k+1) \\ \epsilon M_3(k+1) \end{bmatrix}$$

$$= H \begin{bmatrix} N_1(k+1) \\ \epsilon N_3(k+1) \\ M_1(k+1) \\ \epsilon M_3(k+1) \end{bmatrix} \quad (8.117)$$

$$\begin{bmatrix} \epsilon N_2(k) \\ N_4(k) \\ \epsilon M_2(k) \\ M_4(k) \end{bmatrix} = \begin{bmatrix} \overline{A_1} & \epsilon \overline{A_2} & \overline{S_1} & \epsilon \overline{S_2} \\ \epsilon \overline{A_3} & \overline{A_4} & \epsilon \overline{S_3} & \overline{S_4} \\ \overline{Q_1} & \epsilon \overline{Q_2} & A_{11}^T & \epsilon A_{21}^T \\ \epsilon \overline{Q_3} & \overline{Q_4} & \epsilon A_{12}^T & A_{22}^T \end{bmatrix} \begin{bmatrix} \epsilon N_2(k+1) \\ N_4(k+1) \\ \epsilon M_2(k+1) \\ M_4(k+1) \end{bmatrix}$$

$$= H \begin{bmatrix} \epsilon N_2(k+1) \\ N_4(k+1) \\ \epsilon M_2(k+1) \\ M_4(k+1) \end{bmatrix} \quad (8.118)$$

where newly defined quantities are given in Appendix 8.4.

Interchanging the second and third rows in (8.117) and (8.118), respectively, produces

$$\begin{bmatrix} N_1(k) \\ M_1(k) \\ \epsilon N_3(k) \\ \epsilon M_3(k) \end{bmatrix} = \begin{bmatrix} \overline{A_1} & \overline{S_1} & \epsilon \overline{A_2} & \epsilon \overline{S_2} \\ \overline{Q_1} & A_{11}^T & \epsilon \overline{Q_2} & \epsilon A_{21}^T \\ \epsilon \overline{A_3} & \epsilon \overline{S_3} & \overline{A_4} & \overline{S_4} \\ \epsilon \overline{Q_3} & \epsilon A_{12}^T & \overline{Q_4} & A_{22}^T \end{bmatrix} \begin{bmatrix} N_1(k+1) \\ M_1(k+1) \\ \epsilon N_3(k+1) \\ \epsilon M_3(k+1) \end{bmatrix}$$

$$= \begin{bmatrix} T_1 & \epsilon T_2 \\ \epsilon T_3 & T_4 \end{bmatrix} \begin{bmatrix} N_1(k+1) \\ M_1(k+1) \\ \epsilon N_3(k+1) \\ \epsilon M_3(k+1) \end{bmatrix} \qquad (8.119)$$

$$\begin{bmatrix} \epsilon N_2(k) \\ \epsilon M_2(k) \\ N_4(k) \\ M_4(k) \end{bmatrix} = \begin{bmatrix} \overline{A_1} & \overline{S_1} & \epsilon \overline{A_2} & \epsilon \overline{S_2} \\ \overline{Q_1} & A_{11}^T & \epsilon \overline{Q_2} & \epsilon A_{21}^T \\ \epsilon \overline{A_3} & \epsilon \overline{S_3} & \overline{A_4} & \overline{S_4} \\ \epsilon \overline{Q_3} & \epsilon A_{12}^T & \overline{Q_4} & A_{22}^T \end{bmatrix} \begin{bmatrix} \epsilon N_2(k+1) \\ \epsilon M_2(k+1) \\ N_4(k+1) \\ M_4(k+1) \end{bmatrix}$$

$$= \begin{bmatrix} T_1 & \epsilon T_2 \\ \epsilon T_3 & T_4 \end{bmatrix} \begin{bmatrix} \epsilon N_2(k+1) \\ \epsilon M_2(k+1) \\ N_4(k+1) \\ M_4(k+1) \end{bmatrix} \qquad (8.120)$$

where

$$T_1 = \begin{bmatrix} \overline{A_1} & \overline{S_1} \\ \overline{Q_1} & A_{11}^T \end{bmatrix}, \quad T_2 = \begin{bmatrix} \overline{A_2} & \overline{S_2} \\ \overline{Q_2} & A_{21}^T \end{bmatrix}$$

$$\qquad (8.121)$$

$$T_3 = \begin{bmatrix} \overline{A_3} & \overline{S_3} \\ \overline{Q_3} & A_{12}^T \end{bmatrix}, \quad T_4 = \begin{bmatrix} \overline{A_4} & \overline{S_4} \\ \overline{Q_4} & A_{22}^T \end{bmatrix}$$

Introducing the notation

$$U(k) = \begin{bmatrix} N_{1k} \\ M_{1k} \end{bmatrix}, \quad V(k) = \begin{bmatrix} \epsilon N_3(k) \\ \epsilon M_3(k) \end{bmatrix}$$

$$X(k) = \begin{bmatrix} \epsilon N_2(k) \\ \epsilon M_2(k) \end{bmatrix}, \quad Y(k) = \begin{bmatrix} N_4(k) \\ M_4(k) \end{bmatrix} \qquad (8.122)$$

we get two independent systems of weakly coupled difference equations

$$U(k) = T_1 U(k+1) + \epsilon T_2 V(k+1)$$
$$V(k) = \epsilon T_3 U(k+1) + T_4 V(k+1) \qquad (8.123)$$

$$X(k) = T_1 X(k+1) + \epsilon T_2 Y(k+1)$$
$$Y(k) = \epsilon T_3 X(k+1) + T_4 Y(k+1) \qquad (8.124)$$

with terminal conditions

$$U(n) = \begin{bmatrix} I \\ F_1 \end{bmatrix}, \quad V(n) = \begin{bmatrix} 0 \\ \epsilon F_2^T \end{bmatrix}, \quad X(n) = \begin{bmatrix} 0 \\ \epsilon F_2 \end{bmatrix}, \quad Y(n) = \begin{bmatrix} I \\ F_3 \end{bmatrix}$$

$$(8.125)$$

Note that systems (8.123) and (8.124) have exactly the same form and the only difference is in the terminal conditions.

Applying the decoupling transformation (Gajic and Shen, 1989)

$$\mathbf{T_2} = \begin{bmatrix} I & -\epsilon L \\ \epsilon H & I - \epsilon^2 H L \end{bmatrix}, \quad \mathbf{T_2^{-1}} = \begin{bmatrix} I - \epsilon^2 L H & \epsilon L \\ -\epsilon H & I \end{bmatrix} \qquad (8.126)$$

where L and H satisfy

$$\begin{aligned} T_1 L + T_2 - L T_4 - \epsilon^2 L T_3 L &= 0 \\ H \left(T_1 - \epsilon^2 L T_3 \right) - \left(T_4 + \epsilon^2 T_3 L \right) H + T_3 &= 0 \end{aligned} \qquad (8.127)$$

to (8.123) and (8.124) produces

$$\begin{aligned} \overline{U}(k) &= \left(T_1 - \epsilon^2 L T_3 \right) \overline{U}(k+1) \\ \overline{V}(k) &= \left(T_4 + \epsilon^2 T_3 L \right) \overline{V}(k+1) \end{aligned} \qquad (8.128)$$

$$\begin{aligned} \overline{X}(k) &= \left(T_1 - \epsilon^2 L T_3 \right) \overline{X}(k+1) \\ \overline{Y}(k) &= \left(T_4 + \epsilon^2 T_3 L \right) \overline{Y}(k+1) \end{aligned} \qquad (8.129)$$

with terminal conditions

$$\overline{U}(n) = U(n) - \epsilon L V(n), \quad \overline{V}(n) = \epsilon H U(n) + \left(I - \epsilon^2 H L \right) V(n)$$

$$\overline{X}(n) = X(n) - \epsilon L Y(n), \quad \overline{Y}(n) = \epsilon H X(n) + \left(I - \epsilon^2 H L \right) Y(n) \qquad (8.130)$$

Matrices L an H can be easily obtained, at the very low cost, by using the recursive algorithm from (Gajic and Shen, 1989). Solutions of (8.128)-(8.129) are given by

$$\overline{U}(k) = \left(T_1 - \epsilon^2 L T_3 \right)^{n-k} \overline{U}(n)$$

$$\overline{V}(k) = \left(T_4 + \epsilon^2 T_3 L \right)^{n-k} \overline{V}(n)$$

$$\qquad (8.131)$$

$$\overline{X}(k) = \left(T_1 - \epsilon^2 L T_3 \right)^{n-k} \overline{X}(n)$$

$$\overline{Y}(k) = \left(T_4 + \epsilon^2 T_3 L \right)^{n-k} \overline{Y}(n)$$

Corresponding solutions in the original coordinates are

$$U(k) = (I - \epsilon^2 LH)(T_1 - \epsilon^2 LT_3)^{n-k} \overline{U}(n)$$
$$+ \epsilon L (T_4 + \epsilon^2 T_3 L)^{n-k} \overline{V}(n)$$

$$V(k) = -\epsilon K (T_1 - \epsilon^2 LT_3)^{n-k} \overline{U}(n) + (T_4 + \epsilon^2 T_3 L)^{n-k} \overline{V}(n)$$

$$X(k) = (I - \epsilon^2 LH)(T_1 - \epsilon^2 LT_3)^{n-k} \overline{X}(n)$$
$$+ \epsilon L (T_4 + \epsilon^2 T_3 L)^{n-k} \overline{Y}(n)$$

$$Y(k) = -\epsilon H (T_1 - \epsilon^2 LT_3)^{n-k} \overline{X}(n) + (T_4 + \epsilon^2 T_3 L)^{n-k} \overline{Y}(n) \tag{8.132}$$

Partitioning (8.132), according to (8.122), will produce all components of matrices $M(k)$ and $N(k)$, that is

$$\begin{bmatrix} N_1(k) \\ M_1(k) \end{bmatrix} = \begin{bmatrix} U_1(k) \\ U_2(k) \end{bmatrix} = U(k), \qquad \begin{bmatrix} \epsilon N_3(k) \\ \epsilon M_3(k) \end{bmatrix} = \begin{bmatrix} V_1(k) \\ V_2(k) \end{bmatrix} = V(k)$$

$$\begin{bmatrix} \epsilon N_2(k) \\ \epsilon M_2(k) \end{bmatrix} = \begin{bmatrix} X_1(k) \\ X_2(k) \end{bmatrix} = X(k), \qquad \begin{bmatrix} N_4(k) \\ M_4(k) \end{bmatrix} = \begin{bmatrix} Y_1(k) \\ Y_2(k) \end{bmatrix} = Y(k) \tag{8.133}$$

Then the required solution of (8.111) is given by

$$P(k) = \begin{bmatrix} U_2(k) & X_2(k) \\ V_2(k) & Y_2(k) \end{bmatrix} \begin{bmatrix} U_1(k) & X_1(k) \\ V_1(k) & Y_1(k) \end{bmatrix}^{-1} \tag{8.134}$$

Thus, in order to get the solution of (8.111), $P(k)$, which has $dim P(k) = n \times n = (n_1 + n_2) \times (n_1 + n_2)$, we solve two simple algebraic equations (8.127) of dimensions of $(2n_2 \times 2n_1)$ and $(2n_1 \times 2n_2)$, respectively. The existing numerical algorithms based on the fixed point iterations and the Newton method for solving (8.127) can be found in (Gajic and Shen, 1989). In addition, the $(n - k)$-th powers of the matrices $T_1 - \epsilon^2 LT$ and $T_4 + \epsilon^2 T_3 L$ have to be found.

8.8 Numerical Example

In order to demonstrate the proposed method, a discrete system from (Katzberg, 1977) is studied. The problem matrices A and B are given in Section 6.8. Remaining matrices are chosen as $R = 0.5I_2, Q = 0.1I_4$, and the terminal condition is given by

$$P(n) = F = \begin{bmatrix} 0.9 & 0 & 0.3 & 0 \\ 0 & 0.9 & 0 & 0.3 \\ 0.3 & 0 & 0.9 & 0 \\ 0 & 0.3 & 0 & 0.9 \end{bmatrix}$$

The small weak coupling parameter ϵ is built in the problem and can be roughly estimated from the strongest coupled matrix (matrix A). The strongest coupling is in the fourth row, where

$$\epsilon = \frac{0.323}{0.983} \approx 0.329$$

With the proposed method, the simulation results for (8.133) and the solution of the weakly coupled matrix difference Riccati equation (8.134) are obtained by using the package L-A-S for the computer aided control system design (West et al., 1985)

$$\begin{bmatrix} U_1(k) & X_1(k) \\ V_1(k) & Y_1(k) \end{bmatrix} = \begin{bmatrix} 0.063 & -1.182 & 0.497 & 0.394 \\ 2.698 & 1.828 & -1.475 & -1.049 \\ 0.632 & 0.388 & 0.446 & -1.326 \\ -2.002 & -0.973 & 1.724 & 2.447 \end{bmatrix}$$

$$\begin{bmatrix} U_2(k) & X_2(k) \\ V_2(k) & Y_2(k) \end{bmatrix} = \begin{bmatrix} 0.569 & -1.190 & 0.495 & 0.077 \\ 1.050 & 0.809 & 0.163 & 0.430 \\ 0.639 & 0.349 & 0.996 & -0.651 \\ 0.125 & 0.477 & 0.933 & 1.086 \end{bmatrix}$$

$$P_{app} = \begin{bmatrix} 1.273 & 0.121 & 0.181 & -0.023 \\ 0.121 & 0.814 & 0.314 & 0.675 \\ 0.181 & 0.314 & 1.192 & 0.485 \\ -0.023 & 0.675 & 0.485 & 1.000 \end{bmatrix}$$

The terminal time is selected as $n = 8$ and $k = 4$. The obtained solution P_{app}, is identical to the solution of the global Riccati difference equation (8.111).

Appendix 8.1

Rewrite (8.87) (Pappas et al., 1980) as

$$\begin{bmatrix} I & BR^{-1}B^T \\ 0 & A^T \end{bmatrix} \begin{bmatrix} N_{k+1} \\ M_{k+1} \end{bmatrix} = \begin{bmatrix} A & 0 \\ -Q & I \end{bmatrix} \begin{bmatrix} N_k \\ M_k \end{bmatrix} \tag{a.1}$$

which implies the following two relations

$$AN_k = N_{k+1} + BR^{-1}B^T M_{k+1} \quad -QN_k + M_k = A^T M_{k+1} \tag{a.2}$$

Then

$$A = N_{k+1}N_k^{-1} + BR^{-1}B^T M_{k+1}N_k^{-1}$$
$$A^T = N_k^{-T}N_{k+1}^T + N_k^{-T}M_{k+1}^T BR^{-1}B^T \tag{a.3}$$
$$M_k N_k^{-1} = A^T M_{k+1}N_k^{-1} + Q$$

Assuming that N_k is invertible, substitute (a.3) in (8.83). We obtain

$$A^T P_{k+1} A - A^T P_{k+1} B \left(R + B^T P_{k+1} B \right)^{-1} B^T P_{k+1} A + Q$$

$$= A^T M_{k+1} N_{k+1}^{-1} \left(N_{k+1} N_k^{-1} + BR^{-1}B^T M_{k+1} N_k^{-1} \right)$$
$$+ M_k N_k^{-1} - A^T M_{k+1} N_k^{-1}$$
$$- A^T M_{k+1} N_{k+1}^{-1} B \left(R + B^T M_{k+1} N_{k+1}^{-1} B \right)^{-1} B^T M_{k+1} N_{k+1}^{-1} A$$

$$= A^T M_{k+1} N_{k+1}^{-1} BR^{-1}B^T M_{k+1} N_k^{-1}$$
$$- A^T M_{k+1} N_{k+1}^{-1} B \left(R + B^T M_{k+1} N_{k+1}^{-1} B \right)^{-1} B^T M_{k+1} N_k^{-1} + M_k N_k^{-1}$$

$$= A^T M_{k+1} N_{k+1}^{-1} BR^{-1}B^T M_{k+1} N_k^{-1}$$
$$- A^T M_{k+1} N_{k+1}^{-1} B \left(R + B^T M_{k+1} N_{k+1}^{-1} B \right)^{-1} B^T M_{k+1}$$
$$\times \left(N_k^{-1} + N_{k+1}^{-1} BR^{-1}B^T M_{k+1} N_k^{-1} \right) + M_k N_k^{-1}$$

$$= A^T M_{k+1} N_{k+1}^{-1} BR^{-1}B^T M_{k+1} N_k^{-1}$$
$$- A^T M_{k+1} N_{k+1}^{-1} B \left(R + B^T M_{k+1} N_{k+1}^{-1} B \right)^{-1} B^T M_{k+1} N_k^{-1}$$
$$- A^T M_{k+1} N_{k+1}^{-1} B \left(R + B^T M_{k+1} N_{k+1}^{-1} B \right)^{-1}$$
$$\times B^T M_{k+1} N_{k+1}^{-1} BR^{-1}B^T M_{k+1} N_k^{-1} + M_k N_k^{-1}$$

$$= A^T M_{k+1} N_{k+1}^{-1} B \left(R + B^T M_{k+1} N_{k+1}^{-1} B \right)^{-1}$$
$$\times \left[\left(R + B^T M_{k+1} N_{k+1}^{-1} B \right) R^{-1} - I - B^T M_{k+1} N_{k+1}^{-1} B R^{-1} \right] \quad \text{(a.4)}$$
$$\times B^T M_{k+1} N_k^{-1} + M_k N_k^{-1} = M_k N_k^{-1} = P_k$$

Appendix 8.2

Using the discrete version of the dichotomy transformation (Wilde and Kokotovic, 1972) we have

$$\begin{bmatrix} N(k) \\ M(k) \end{bmatrix} = \begin{bmatrix} I & I \\ P & K \end{bmatrix} \begin{bmatrix} \widehat{N}(k) \\ \widehat{M}(k) \end{bmatrix}, \quad N(n) = I, \; M(n) = F \quad \text{(b.1)}$$

and

$$\begin{bmatrix} \widehat{N}(k) \\ \widehat{M}(k) \end{bmatrix} = \begin{bmatrix} I + (K - P)^{-1} P & -(K - P)^{-1} \\ -(K - P)^{-1} P & (K - P)^{-1} \end{bmatrix} \begin{bmatrix} N(k) \\ M(k) \end{bmatrix} \quad \text{(b.2)}$$

where P and K are unique positive definite and negative definite solutions of the discrete-time algebraic Riccati equation corresponding to (8.83). These two solutions exist under the conditions stated in Lemma 8.2.

The system (8.87) can be transformed in

$$\begin{bmatrix} \widehat{N}(k) \\ \widehat{M}(k) \end{bmatrix} =$$
$$\begin{bmatrix} A^{-1} \left(I + B R^{-1} B^T P \right) & 0 \\ 0 & A^{-1} \left(I + B R^{-1} B^T K \right) \end{bmatrix} \begin{bmatrix} \widehat{N}(k+1) \\ \widehat{M}(k+1) \end{bmatrix}$$
$$\text{(b.3)}$$

with terminal conditions

$$\widehat{N}(n) = I + (K - P)^{-1} (P - F)$$
$$\widehat{M}(n) = (K - P)^{-1} (F - P) \quad \text{(b.4)}$$

The solution of (b.3) is given by

$$\hat{N}(k) = \left[A^{-1}\left(I + BR^{-1}B^T P\right)\right]^{n-k} \hat{N}(n)$$
$$\widehat{M}(k) = \left[A^{-1}\left(I + BR^{-1}B^T K\right)\right]^{n-k} \widehat{M}(n) \qquad \text{(b.5)}$$

Using (b.4)-(b.5) it can be shown that

$$N(k) = \left[A^{-1}\left(I + BR^{-1}B^T P\right)\right]^{n-k} \left[I + (K-P)^{-1}(P-F)\right]$$
$$+ \left[A^{-1}\left(I + BR^{-1}B^T K\right)\right]^{n-k} (K-P)^{-1}(F-P) \qquad \text{(b.6)}$$

that is

$$N(k) = \phi(n-k)N(n), \quad N(n) = I \qquad \text{(b.7)}$$

with obvious definition of $\phi(n-k)$. Since $\phi(n-k)$ plays the role of the transition matrix, it is nonsingular by the fact that the matrix A is nonsingular. Note that the matrix $(I + BR^{-1}B^T P)$, defined in (b.5) is also nonsingular. The regularity of $N(k)$ is determined by $N(n)$ only. Thus, having chosen $N(n)$ as an identity will assure the nonsingularity of $N(k)$ for any $k < n$, and prove the given Lemma.

Appendix 8.3

From (8.81)

$$\mathbf{H} = \begin{bmatrix} A^{-1} & A^{-1}BR^{-1}B^T \\ QA^{-1} & A^T + QA^{-1}BR^{-1}B^T \end{bmatrix} \qquad \text{(c.1)}$$

Since A^{-1} has the same structure as A, that is

$$A^{-1} = \begin{bmatrix} I + O(\epsilon) & O(\epsilon) \\ O(1) & O(1) \end{bmatrix} \qquad \text{(c.2)}$$

then

$$QA^{-1} = \begin{bmatrix} O(1) & O(1) \\ O(1) & O(1) \end{bmatrix} \begin{bmatrix} I+O(\epsilon) & O(\epsilon) \\ O(1) & O(1) \end{bmatrix} = \begin{bmatrix} O(1) & O(1) \\ O(1) & O(1) \end{bmatrix}$$

$$\begin{aligned} A^{-1}BR^{-1}B^T &= \begin{bmatrix} I+O(\epsilon) & O(\epsilon) \\ O(1) & O(1) \end{bmatrix} \begin{bmatrix} O(\epsilon^2) & O(\epsilon) \\ O(\epsilon) & O(1) \end{bmatrix} \\ &= \begin{bmatrix} O(\epsilon^2) & O(\epsilon) \\ O(\epsilon) & O(1) \end{bmatrix} \end{aligned}$$

$$A^T + QA^{-1}BR^{-1}B^T = \begin{bmatrix} I+O(\epsilon) & O(1) \\ O(\epsilon) & O(1) \end{bmatrix}$$

$$H = \begin{bmatrix} I+\epsilon\overline{A_1} & \epsilon\overline{A_2} & \epsilon^2\overline{S_1} & \epsilon\overline{S_2} \\ \overline{A_3} & \overline{A_4} & \epsilon\overline{S_3} & \overline{S_4} \\ \overline{Q_1} & \overline{Q_2} & I+\epsilon A_{11}^T & A_{21}^T \\ \overline{Q_3} & \overline{Q_4} & \epsilon A_{12}^T & A_{22}^T \end{bmatrix}$$

(c.3)

Appendix 8.4

From (8.81) we have

$$H = \begin{bmatrix} A^{-1} & A^{-1}BR^{-1}B^T \\ QA^{-1} & A^T+QA^{-1}BR^{-1}B^T \end{bmatrix} \tag{d.1}$$

Since A^{-1} has the same structure as A, that is

$$A^{-1} = \begin{bmatrix} O(1) & O(\epsilon) \\ O(\epsilon) & O(1) \end{bmatrix} \tag{d.2}$$

then

$$QA^{-1} = \begin{bmatrix} O(1) & O(\epsilon) \\ O(\epsilon) & O(1) \end{bmatrix} \begin{bmatrix} O(1) & O(\epsilon) \\ O(\epsilon) & O(1) \end{bmatrix} = \begin{bmatrix} O(1) & O(\epsilon) \\ O(\epsilon) & O(1) \end{bmatrix}$$

$$A^{-1}BR^{-1}B^T = \begin{bmatrix} O(1) & O(\epsilon) \\ O(\epsilon) & O(1) \end{bmatrix} \begin{bmatrix} O(1) & O(\epsilon) \\ O(\epsilon) & O(1) \end{bmatrix} = \begin{bmatrix} O(1) & O(\epsilon) \\ O(\epsilon) & O(1) \end{bmatrix}$$

$$A^T + QA^{-1}BR^{-1}B^T = \begin{bmatrix} O(1) & O(\epsilon) \\ O(\epsilon) & O(1) \end{bmatrix}$$

$$= H \begin{bmatrix} \overline{A_1} & \epsilon\overline{A_2} & \overline{S_1} & \epsilon\overline{S_2} \\ \epsilon\overline{A_3} & \overline{A_4} & \epsilon\overline{S_3} & \overline{S_4} \\ \overline{Q_1} & \epsilon\overline{Q_2} & A_{11}^T & \epsilon A_{21}^T \\ \epsilon\overline{Q_3} & \overline{Q_4} & \epsilon A_{12}^T & A_{22}^T \end{bmatrix}$$

(d.3)

PART TWO — Applications

Quasi Singularly Perturbed and Quasi Weakly Coupled Systems

Singularly Perturbed Weakly Coupled Systems

Stochastic Output Feedback of Discrete Systems

Differential Games

High Gain and Cheap Control Problems

Linear Approach to Bilinear Control Systems

Chapter 9

Quasi Singularly Perturbed and Weakly Coupled Linear Control Systems

Several structures of the linear-quadratic control problems containing small parameters can be studied efficiently by using the methodology similar to the one presented in the previous chapters. We call these structures quasi singularly perturbed and quasi weakly coupled (quasi SP&WC systems). Namely, the quasi singularly perturbed and quasi weakly coupled linear-quadratic control problems are very closely related to the standard singularly perturbed and weakly coupled control problems. However, these similarities are not obvious, and very often, in many applications, the quasi singularly perturbed and quasi weakly coupled structures are producing the parallel reduced-order algorithms of the simpler structures and under milder conditions than the standard singularly perturbed and standard weakly coupled linear-quadratic control problems.

9.1 Linear Control of Quasi Singularly Perturbed Hydro Power Plants

In this section, we consider a special class of linear control systems represented by the standard singularly perturbed system matrix and with the control input matrix having three different nonstandard forms. Many real systems (such as hydro power plants, systems with only few actuators) possess the control structure studied in this section. The obtained results are quite simplified, (comparing to the standard singularly perturbed control systems), and in one case the optimal solution of the algebraic Riccati equation is completely determined in terms of the reduced-order algebraic Lyapunov equations. The proposed method is successfully applied to the reduced-order design of optimal controllers for the real hydro power plant of the Serbian power system.

In this section, we study structures corresponding to the real hydro power plants. We call them quasi (nearly) singularly perturbed systems since they contain the singularly perturbed system matrix (like the standard singularly perturbed linear systems), but they have different structures for the control matrix. The control matrix of the standard singularly perturbed system is given by

$$B = \begin{bmatrix} B_1 \\ \frac{B_2}{\epsilon} \end{bmatrix}, \quad \epsilon - small \ positive \ parameter \qquad (9.1)$$

Three different structures for the control matrix will be studied in this section since they bring different and interesting solutions

$$1) \quad B = \begin{bmatrix} 0 \\ B_2 \end{bmatrix}, \qquad 2) \quad B = \begin{bmatrix} B_1 \\ 0 \end{bmatrix}, \qquad 3) \quad B = \begin{bmatrix} B_1 \\ B_2 \end{bmatrix} \qquad (9.2)$$

In the first structure, the system is weakly controlled through the fast modes only; in the second one, it is strongly controlled through the slow modes; and the third one contains both strongly controlled slow modes and weakly controlled fast modes. All three structures can be encountered in the hydro power plant controllers design.

The optimal solution to the first structure ("weakly controlled fast mode structure") is obtained under the strongest assumptions (both slow and fast open-loop system matrices have to be stable), but the solution of the global algebraic Riccati equation is completely given in terms of the reduced-order algebraic Lyapunov equations. The second case ("strongly controlled slow modes structure") and the third one ("strongly controlled slow modes and weakly controlled fast modes") demand the solution of one reduced-order local Riccati equation corresponding to the slow subsystem. It is important to point out that the solution to the real 14th-order hydro power control system, corresponding to the second case, and 11th-order hydro power plant corresponding to the third case, are obtained by the presented reduced-order recursive method, but the global method fails to produce an answer in both cases. We have used a very reliable package L-A-S (West et al., 1985) for computer aided control system design, and its eigenvector approach for solving the algebraic Riccati equation (it happens that the transformation matrices are close to being singular in both cases).

9.2 Case Study: Hydro Power Plant

The presented approach will be demonstrated for three different structures by using an example of the Serbian hydro power plant.

9.2.1 Weakly Controlled Fast Modes Structure

Consider a linear dynamical system in the form

$$\begin{bmatrix} \dot{x}_1 \\ \dot{x}_2 \end{bmatrix} = \begin{bmatrix} A_1 & A_2 \\ \frac{A_3}{\epsilon} & \frac{A_4}{\epsilon} \end{bmatrix} \begin{bmatrix} x_1 \\ x_2 \end{bmatrix} + \begin{bmatrix} 0 \\ B_2 \end{bmatrix} u \qquad (9.3)$$

where $x_i \in \Re^{n_i}$, $i = 1, 2$, are state vectors, $u \in \Re^m$ is a control input and ϵ is a small positive parameter. This is a special class of linear dynamical systems represented, in general, by

$$\dot{x} = Ax + Bu \qquad (9.4)$$

with

$$x = \begin{bmatrix} x_1 \\ x_2 \end{bmatrix}, \quad A = \begin{bmatrix} A_1 & A_2 \\ \frac{A_3}{\epsilon} & \frac{A_4}{\epsilon} \end{bmatrix}, \quad B = \begin{bmatrix} 0 \\ B_2 \end{bmatrix} \qquad (9.5)$$

A quadratic type cost functional to be minimized is associated with (9.3) in the form

$$J = \frac{1}{2} \int_0^\infty (x^T Q x + u^T R u) \, dt, \quad Q \geq 0, \quad R > 0 \qquad (9.6)$$

For the purpose of this section, we assume that the structure of the matrix Q is

$$Q = \begin{bmatrix} Q_1 & Q_2 \\ Q_2^T & Q_3 \end{bmatrix} \qquad (9.7)$$

All problem matrices defined in (9.3)-(9.7) are constant and of appropriate dimensions.

The structure defined in (9.3) corresponds to the singularly perturbed systems (Kokotovic et al., 1986) with the control input weakly influencing fast modes only. This structure has not been studied in the literature from the order reduction point of view. Motivated by results reported in (Gajic et al., 1990), we will show that in this case one is able to design the optimal controller by using the reduced-order parallel recursive algorithm and that the solution is extremely simplified. Namely, the algebraic Riccati equation is solved completely in terms of the reduced-order algebraic Lyapunov equations.

The optimal problem of minimizing (9.6) along trajectories of (9.3) has the very well-known solution given by

$$u_{opt} = -F_{opt}x = -R^{-1}B^T Px \qquad (9.8)$$

where P is the positive semidefinite stabilizing solution of the algebraic Riccati equation

$$A^T P + PA + Q - PSP = 0, \qquad S = BR^{-1}B^T \qquad (9.9)$$

It can be shown that the nature of the solution of (9.9) is

$$P = \begin{bmatrix} P_1 & \epsilon P_2 \\ \epsilon P_2^T & \epsilon P_3 \end{bmatrix} \qquad (9.10)$$

Partitioning (9.9) compatible to (9.5), (9.7) and (9.10), we get three matrix algebraic equations

$$P_1 A_1 + A_1^T P_1 + P_2 A_3 + A_3^T P_2^T + Q_1 - \epsilon^2 P_2 S_3 P_2^T = 0 \qquad (9.11)$$

$$P_2 A_4 + \epsilon A_1^T P_2 + Q_2 + A_3^T P_3 + P_1 A_2 - \epsilon^2 P_2 S_3 P_3 = 0 \qquad (9.12)$$

$$P_3 A_4 + A_4^T P_3 + Q_3 + \epsilon \left(P_2^T A_2 + A_2^T P_2 \right) - \epsilon^2 P_3 S_3 P_3 = 0 \qquad (9.13)$$

where $S_3 = B_2 R^{-1} B_2^T$.

Since ϵ is a small parameter, we can define $O(\epsilon)$ approximation of (9.11)-(9.13) as follows

$$P_1^{(0)} A_0 + A_0^T P_1^{(0)} + Q_0 = 0 \qquad (9.14)$$

$$P_2^{(0)} A_4 + Q_2 + A_3^T P_3^{(0)} + P_1^{(0)} A_2 = 0 \qquad (9.15)$$

$$P_3^{(0)} A_4 + A_4^T P_3^{(0)} + Q_3 = 0 \qquad (9.16)$$

where

$$A_0 = A_1 - A_2 A_4^{-1} A_3$$
$$Q_0 = Q_1 - A_3^T A_4^{-T} Q_2^T - Q_2 A_4^{-1} A_3 + A_3^T A_4^{-T} Q_3 A_4^{-1} A_3 \qquad (9.17)$$

The corresponding approximate solution of (9.10) is now given by

$$P^{(0)} = \begin{bmatrix} P_1^{(0)} & \epsilon P_2^{(0)} \\ \epsilon P_2^{(0)T} & \epsilon P_3^{(0)} \end{bmatrix} = P + O(\epsilon) \qquad (9.18)$$

On the contrary to the standard singularly perturbed systems (Koko-tovic et al., 1986), where the zeroth-approximation is given in terms of two reduced-order Riccati equations, in this case we need to solve only two reduced-order algebraic Lyapunov equations.

The unique positive semidefinite stabilizing solution $P^{(0)}$, obtained from (9.14)-(9.16), exists under the following assumption.

Assumption 9.1 Slow and fast subsystem matrices, respectively, A_0 and A_4 are stable.

$\triangle$

Defining the approximation errors as

$$P_j = P_j^{(0)} + \epsilon E_j, \qquad j = 1, 2, 3 \qquad (9.19)$$

and using (9.18) in (9.11)-(9.13) and (9.14)-(9.16), we get the following error equations

$$E_1 A_0 + A_0^T E_1 = -A_0^T P_2 A_4^{-1} A_3 - A_3^T A_4^{-T} P_2^T A_0$$
$$+\epsilon^2 \left(P_2 - A_3^T A_4^{-T} P_3 \right) S_3 \left(P_2 - A_3^T A_4^{-T} P_3 \right)^T \qquad (9.20)$$

$$E_2 A_4 = \epsilon P_2 S_2 P_3 - E_1 A_2 - A_3^T E_3 - A_1^T P_2 \qquad (9.21)$$

$$E_3 A_4 + A_4^T E_3 = -P_2^T A_2 - A_2^T P_2 + \epsilon P_3^T S_3 P_3 \qquad (9.22)$$

Let us propose the following reduced-order parallel algorithm (in the spirit of those developed in (Gajic et al., 1990) for solving (9.20)-(9.22).

Algorithm 9.1:

$$E_1^{(i+1)} A_0 + A_0^T E_1^{(i+1)} = -A_0^T P_2^{(i)} A_4^{-1} A_3 - A_3^T A_4^{-T} P_2^{(i)^T} A_0$$
$$+\epsilon^2 \left(P_2^{(i)} - A_3^T A_4^{-T} P_3^{(i)} \right) S_3 \left(P_2^{(i)} - A_3^T A_4^{-T} P_3^{(i)} \right)^T \quad (9.23)$$

$$E_2^{(i+1)} A_4 = \epsilon P_2^{(i)} S_2 P_3^{(i)} - E_1^{(i+1)} A_2 - A_3^T E_3^{(i+1)} - A_1^T P_2^{(i)} \quad (9.24)$$

$$E_3^{(i+1)} A_4 + A_4^T E_3^{(i+1)} = -P_2^{(i)^T} A_2 - A_2^T P_2^{(i)} + \epsilon P_3^{(i)} S_3 P_3^{(i)} \quad (9.25)$$

where

$$P_j^{(i)} = P_j^{(0)} + \epsilon E_j^{(i)}, \qquad j = 1, 2, 3 \qquad i = 0, 1, 2, 3... \qquad (9.26)$$

with initial conditions

$$E_1^{(0)} = 0, \ E_2^{(0)} = 0, \ E_3^{(0)} = 0 \qquad (9.27)$$

$\triangle$

The following theorem indicates the features of the proposed algorithm (9.23)-(9.27).

Theorem 9.1 *Under conditions stated in Assumption 9.1, the algorithm (9.23)-(9.27) converges to the exact solution of the error term, and thus to the required solution P, with the rate of convergence of $O(\epsilon)$, that is*

$$\left\| E_j - E_j^{(i+1)} \right\| = O(\epsilon) \left\| E_j - E_j^{(i)} \right\| \qquad (9.28)$$

or equivalently

$$\left\| E_j - E_j^{(i)} \right\| = O\left(\epsilon^{(i+1)} \right), \quad i = 0, 1, 2, ...; \quad j = 1, 2, 3 \qquad (9.29)$$

$\diamond$

The proof of this theorem follows the ideas presented in Chapter 2 and thus, it is omitted. Instead, we will demonstrate the results of Theorem 9.1 on a real hydro power plant.

In summary, the solution of the global algebraic Riccati equation (9.9) is obtained up to any arbitrary order of accuracy from the reduced-order algebraic Lyapunov equations (9.14) and (9.16), and by performing iterations on the algebraic Lyapunov equations (9.23) and (9.25).

The approximate optimal feedback gain $F^{(i)}$ is now defined by

$$F^{(i)} = -R^{-1} B^T P^{(i)} \qquad (9.30)$$

where

$$P^{(i)} = \begin{bmatrix} P_1^{(0)} + \epsilon E_1^{(i)} & \epsilon \left(P_2^{(0)} + \epsilon E_2^{(i)} \right) \\ \epsilon \left(P_2^{(0)} + \epsilon E_2^{(i)} \right)^T & \epsilon \left(P_3^{(0)} + \epsilon E_3^{(i)} \right) \end{bmatrix} \qquad (9.31)$$

The approximate criterion is obtained from

$$J_{app}^{(i)} = tr \left\{ V^{(i)} \right\} \qquad (9.32)$$

where $V^{(i)}$ satisfies

$$\left(A - SP^{(i)} \right)^T V^{(i)} + V^{(i)} \left(A - SP^{(i)} \right) + Q + P^{(i)} SP^{(i)} = 0 \quad (9.33)$$

Case Study 1

The proposed methodology is applied to the design of the optimal voltage controller of the real hydro power plant of the Serbian power system. The hydro power plant is treated as one-unit synchronous generator connected to an infinite bus system through the transmission line (Skataric, 1989). Linearized mathematical model of the synchronous generator in dq reference frame is obtained under the assumption that both transient effects in stator windings and in damper windings are not negligible (Anderson and Fouad, 1984). Also, the synchronous generator is assumed to be equipped with the first order exciter. The state space is given by

$$x^T = [\Delta\theta \quad \Delta\omega \quad \Delta u_f \quad \Delta\psi_d \quad \Delta\psi_q \quad \Delta\psi_f \quad \Delta\psi_D \quad \Delta\psi_Q]$$

where

$\Delta\theta$ — torque angle in *rad*.

$\Delta\omega$ — rotation speed in *p.u.*

Δu_f — excitation voltage in *p.u.*

$\Delta\psi_d$ — d-axis stator windings flux linkage

$\Delta\psi_q$ — q-axis stator windings flux linkage

$\Delta\psi_f$ — excitation flux linkage

$\Delta\psi_D$ — d-axis damper windings flux linkage

$\Delta\psi_Q$ — q-axis damper windings flux linkage.

261

The control input given by $u = \Delta u_{vr}$ represents the control signal to voltage regulation system.

The system matrices for the considered nominal point are given by

$$
A = \begin{bmatrix}
0 & 314.16 & 0 & 0 & 0 & 0 & 0 & 0 \\
0 & -0.286 & 0 & 0.147 & 0.528 & -0.134 & -0.04 & -0.276 \\
0 & 0 & -100 & 0 & 0 & 0 & 0 & 0 \\
255.38 & -152.49 & 0 & -13.72 & 511.14 & 8.51 & 2.556 & -135.04 \\
-182.84 & -319.5 & 0 & -534.9 & -12.24 & 137 & 41.136 & 8.389 \\
0 & 0 & 314.16 & 0.446 & 0 & -0.523 & 0.0375 & 0 \\
0 & 0 & 0 & 21.646 & 0 & 6.094 & -29.6 & 0 \\
0 & 0 & 0 & 0 & 87.236 & 0 & 0 & -97.74
\end{bmatrix}
$$

$$
B^T = [0 \quad 0 \quad 0.184 \quad 0 \quad 0 \quad 0 \quad 0 \quad 0]
$$

Weighting matrices Q and R are chosen as identity matrices.

The eigenvalues of the matrix A are given by -100, -74.8, $-25.167 \pm j520.79$, -27.906, $-0.391 \pm j8.51$, -0.295. Apparently this system has the singularly perturbed structure with five fast and three slow variables. By interchanging third and sixth rows in matrices A and B we get the required structure studied in this section. It is easy to check that Assumption 9.1 is satisfied.

Due to the special structure of the matrix B, the feedback control will not affect very small slow eigenvalues so that the system will remain almost marginally stable under the feedback control. In addition, in order to have only two time scales we need that $det\{A_4\} = O(1)$ and $det\{A_0\} = O(1)$, (Chow and Kokotovic, 1983). These two problems can be facilitated by choosing the performance criterion of the form (Singh et al., 1987)

$$
J = \frac{1}{2} \int_0^\infty e^{2\alpha t} \left(x^T Q x + u^T R u \right) dt \tag{9.37}
$$

Parameters α and ϵ are chosen as $\alpha = 1$, and $\epsilon = 0.1$.

Simulation results are obtained by using the L-A-S package (West et al., 1985) for computer aided control system design. Obtained results are presented in Table 9.1.

From Table 9.1 we can notice very good numerical behavior of the proposed algorithm consistent with the statement of Theorem 9.1.

i	$J_{app} - J_{opt}$
0	0.161 x 10^0
1	0.209 x 10^{-2}
2	0.156 x 10^{-4}
3	0.224 x 10^{-5}
4	0.119 x 10^{-6}
5	0.259 x 10^{-7}
6	0.662 x 10^{-9}
7	0.727 x 10^{-10}

Table 9.1: Errors in the criterion approximation per iteration

9.2.2 Strongly Controlled Slow Modes Structure

In this case the partition form of the algebraic Riccati equation (9.9) is given by

$$P_1 A_1 + A_1^T P_1 + P_2 A_3 + A_3^T P_2^T + Q_1 - P_1 S_1 P_1 = 0 \qquad (9.38)$$

$$P_2 A_4 + \epsilon A_1^T P_2 + Q_2 - \epsilon P_1 S_1 P_2 + A_3^T P_3 + P_1 A_2 = 0 \qquad (9.39)$$

$$P_3 A_4 + A_4^T P_3 + Q_3 + \epsilon P_2^T A_2 + \epsilon A_2^T P_2 - \epsilon^2 P_2^T S_1 P_2 = 0 \qquad (9.40)$$

where $S_1 = B_1 R^{-1} B_1^T$.

Following similar arguments as in Section 9.2.1, we get the following expressions for the zeroth-order approximation

$$P_1^{(0)} A_0 + A_0^T P_1^{(0)} + Q_0 - P_1^{(0)} S_1 P_1^{(0)} = 0 \qquad (9.41)$$

$$P_2^{(0)} A_4 + Q_2 + A_3^T P_3^{(0)} + P_1^{(0)} A_2 = 0 \qquad (9.42)$$

$$P_3^{(0)} A_4 + A_4^T P_3^{(0)} + Q_3 = 0 \qquad (9.43)$$

Thus, the zeroth-order approximation of P can be obtained in terms of one reduced-order Riccati and one reduced-order Lyapunov equations.

The unique positive semidefinite stabilizing solution $P^{(0)}$, obtained from (9.41)-(9.43) exists under the following assumption.

Assumption 9.2 The triple $\left(A_0, B_1, \sqrt{Q_0}\right)$ is stabilizable-detectable and the matrix A_4 is stable.

$$\triangle$$

The approximate solution is given by (9.26) with the approximation errors obtained from the following reduced-order parallel algorithm (derived similarly to Section 9.2.1).

Algorithm 9.2:

$$
\begin{aligned}
E_1^{(i+1)} A_s + A_s^T E_1^{(i+1)} &= A_3^T A_4^{-T} P_2^{(i)^T} A_s + A_s^T P_2^{(i)} A_4^{-1} A_3 \\
&+ \epsilon \left(E_1^{(i)} S_1 E_1^{(i)} + A_3^T A_4^{-T} P_2^{(i)^T} S_1 P_2^{(i)} A_4^{-1} A_3 \right)
\end{aligned}
\tag{9.44}
$$

$$
E_3^{(i+1)} A_4 + A_4^T E_3^{(i+1)} = -P_2^{(i)^T} A_2 - A_2^T P_2^{(i)} + \epsilon P_2^{(i)^T} S_1 P_2^{(i)} \tag{9.45}
$$

$$
E_2^{(i+1)} A_4 = P_1^{(i+1)} S_1 P_2^{(i)} - E_1^{(i+1)} A_2 - A_3^T E_3^{(i+1)} - A_1^T P_2^{(i)} \tag{9.46}
$$

where $A_s = A_0 - S_1 P_1^{(0)}$ is a stable, slow subsystem, feedback matrix. The initial conditions for this algorithm are set to zero.

$$\triangle$$

The following theorem indicates the features of the proposed algorithm (9.44)-(9.46).

Theorem 9.2 *Under conditions stated in Assumption 9.2, the algorithm (9.44)-(9.46) converges to the exact solution of the error term with the rate of convergence of $O(\epsilon)$, that is*

$$
\left\| E_j - E_j^{(i+1)} \right\| = O(\epsilon) \left\| E_j - E_j^{(i)} \right\| \tag{9.47}
$$

or equivalently

$$
\left\| E_j - E_j^{(i)} \right\| = O\left(\epsilon^{(i+1)}\right), \quad i = 0, 1, 2, \ldots; \quad j = 1, 2, 3 \tag{9.48}
$$

$$\diamond$$

The proof of this theorem is omitted for the reason explained in Section 9.2.1.

Case Study 2

Strongly controlled slow modes structure and the presented recursive approach are encountered in the optimal turbine controller design of the same low head hydro power plant with the static excitation system of power system stabilizer type (Skataric, 1989). For this type of of hydro power plants assumption of rigid water hammer holds. Also, subtransient effects in stator winding are neglected. The linearized state space is given by

$$x^T = [\Delta\theta \quad \Delta\omega \quad \Delta\psi_f \quad \Delta u_f \quad \Delta x_{1u} \quad \Delta x_{2u} \quad \Delta x_{2u}$$
$$\Delta x_{1f} \quad \Delta x_{2f} \quad \Delta x_{3f} \quad \Delta x_{4f} \quad \Delta q \quad \Delta a \quad \Delta\varphi^o]$$

Newly defined variables represent

x_{1u}, x_{2u}, x_{3u} — voltage input variables in $p.u.$

$x_{1f}, x_{2f}, x_{3f}, x_{4f}$ — frequency input variables in $p.u.$

q — water flow through the turbine in $p.u.$

Q — gate opening in $p.u.$

φ^o — runner blade position in $p.u.$

The control variable $u = \Delta u_{tr}$ represents the input to the turbine governing system. The nonzero entries in the matrices $A^{14\times14}$ and $B^{14\times1}$ are given by

$a_{12} = a_{34} = 314$, $a_{21} = -0.228$, $a_{22} = -0.565$, $a_{23} = -0.113$, $a_{2,12} = 0.595$,

$a_{2,13} = -0.371$, $a_{2,14} = -0.182$, $a_{31} = -0.168$, $a_{32} = -0.305$, $a_{33} = -0.308$,

$a_{44} = -100$, $a_{45} = -21.36$, $a_{46} = a_{47} = a_{4,10} = a_{4,11} = 0.184$, $a_{49} = 427.4$,

$a_{51} = -1.13$, $9a_{52} = 9.02$, $a_{53} = 6.38$, $a_{55} = -21.28$, $a_{65} = 3746$,

$a_{66} = a_{77} = a_{10,10} = a_{11,11} = -32.26$, $a_{75} = -1613$, $a_{82} = 33.3$, $a_{88} = -33.3$,

$a_{98} = 21.27$, $a_{99} = -21.27$, $a_{10,9} = -74921$, $a_{11,10} = 18548$, $a_{12,2} = 0.566$,

$a_{12,12} = -1.463$, $a_{12,13} = 1.208$, $a_{12,14} = 0.614$, $a_{13,13} = -2$, $a_{14,14} = -0.714$

$$b_{13,1} = 2, \quad b_{14,1} = 0.714$$

The eigenvalues of the matrix A are -106.15, -32.26, -32.26, -32.26, -27.94 $\pm$ j14.11, -21.29, -11.302 $\pm$ j1.56, -2, -1.52 $\pm$ j 10.646, -1.47, -0.714.

Weighting matrices are chosen as $Q = 10^{-3}I$, $R = I$.

By interchanging 4th, 5th, and 6th rows with 12th, 13th, and 14th rows in matrices A and B, the structure considered in this section is obtained, such that Assumption 9.2 is satisfied. We have studied this problem with 6 slow and 8 fast variables and with $\epsilon = 0.66$. It can be seen that the matrix A contains huge elements so that the numerics of the problem is very ill-defined. In order to improve the numerical behavior of this algorithm few different scalings of state variables were

k	J_{app}
1	42.762
5	42.415
10	42.312
15	42.286
20	42.283
25	42.281
30	42.281
35	42.280
40	42.280 = J_{opt}

Table 9.2: Approximate values of the performance criterion

used. We have obtained the best results by using the following scalings $x_{12} = 100x_{12}$, $x_2 = 20x_2$, $x_7 = 20x_7$. Even with this scaling the eigenvector method for the solution of the global algebraic Riccati equations (9.9) failed to produce the answer (West et al., 1985). However, the proposed reduced-order parallel algorithm (9.44)-(9.46) has produced the correct result despite of the relatively big value for the small parameter ϵ. Simulation results are presented in Table 9.2.

9.2.3 Weakly Controlled Fast Modes and Strongly Controlled Slow Modes Structure

In this case the partitioned solution of the algebraic Riccati equation (9.9) is given by

$$P_1 A_1 + A_1^T P_1 + P_2 A_3 + A_3^T P_2^T + Q_1 - P_1 S_1 P_1$$
$$-\epsilon \left(P_2 S_2^T P_1 + P_1 S_2 P_2^T \right) - \epsilon^2 P_2 S_3 P_2^T = 0 \tag{9.52}$$

$$P_2 A_4 + \epsilon A_1^T P_2 + Q_2 - \epsilon P_1 S_1 P_2 + A_3^T P_3 + P_1 A_2$$
$$-\epsilon^2 P_2 S_2^T P_2 - \epsilon P_1 S_2 P_3 - \epsilon^2 P_2 S_3 P_3 = 0 \tag{9.53}$$

$$P_3 A_4 + A_4^T P_3 + Q_3 + \epsilon \left(P_2^T A_2 + A_2^T P_2 \right)$$
$$-\epsilon^2 \left(P_2^T S_1 P_2 + P_2^T S_2 P_3 + P_3 S_3 P_3 + P_3 S_2^T P_2 \right) = 0 \tag{9.54}$$

where $S_2 = B_1 R^{-1} B_2^T$.

266

Following similar arguments as in Section 9.2.1, we get exactly the same expressions for the zeroth-order approximation as in Section 9.2.2, namely, equations (9.41)-(9.43). Thus, the unique positive semidefinite stabilizing solution $P^{(0)}$, in this case, exists under Assumption 9.2.

Defining the approximation errors like in (9.19) we get the following error equations

$$
\begin{aligned}
& E_1 D_1 + D_1^T E_1 + E_2 A_3 + A_3^T E_2 \\
& = -P_2 S_2^T P_1 - P_1 S_2 P_2^T + \epsilon E_1 S_1 E_1 + \epsilon P_2 S_3 P_2^T
\end{aligned}
\tag{9.55}
$$

$$
\begin{aligned}
E_2 A_4 + A_3^T E_3 + E_1 A_2 &= \epsilon \left(P_2 S_2^T P_2 + P_2 S_3 P_3 \right) \\
&\quad + P_1 S_1 P_2 + P_1 S_2 P_3 - A_1^T P_2
\end{aligned}
\tag{9.56}
$$

$$
\begin{aligned}
E_3 A_4 + A_4^T E_3 + P_2^T A_2 + A_2^T P_2 \\
= \epsilon \left(P_2^T S_1 P_2 + P_2^T S_2 P_3 + P_3 S_3 P_3 + P_3 S_2^T P_2 \right)
\end{aligned}
\tag{9.57}
$$

where $D_1 = A_1 - S_1 P_1^{(0)}$.

Let us propose the following reduced-order parallel algorithm for solving (9.55)-(9.57).

Algorithm 9.3:

$$
\begin{aligned}
& E_1^{(i+1)} A_s + A_s^T E_1^{(i+1)} = H_1^{(i)} + H_1^{(i)^T} - H_2^{(i)} \\
& + \epsilon \left(H_3^{(i)} + A_3^T A_4^{-T} H_4^{(i)} A_4^{-1} A_3 - H_5^{(i)} A_4^{-1} A_3 - A_3^T A_4^{-T} H_5^{(i)^T} \right)
\end{aligned}
\tag{9.58}
$$

$$
E_3^{(i+1)} A_4 + A_4^T E_3^{(i+1)} = -P_2^{(i)^T} A_2 - A_2^T P_2^{(i)} + \epsilon H_4^{(i)}
\tag{9.59}
$$

$$
E_2^{(i+1)} = \left[\epsilon H_5^{(i,i+1)} - A_3^T E_3^{(i+1)} - E_1^{(i+1)} A_2 - H_6^{(i,i+1)} \right] A_4^{-1}
\tag{9.60}
$$

where

$$
H_1^{(i)} = \left(A_0^T P_2^{(i)} - P_1^{(i)} S_1 P_2^{(i)} - P_1^{(i)} S_2 P_3^{(i)} \right) A_4^{-1} A_3
$$
$$
H_2^{(i)} = P_2^{(i)} S_2^T P_1^{(i)} + P_1^{(i)} S_2 P_2^{(i)^T}
$$
$$
H_3^{(i)} = P_2^{(i)} S_3 P_2^{(i)^T} + E_1^{(i)} S_1 E_1^{(i)}
$$
$$
H_4^{(i)} = P_2^{(i)^T} S_1 P_2^{(i)} + P_2^{(i)^T} S_2 P_3^{(i)} + P_3^{(i)} S_3 P_3^{(i)} + P_3^{(i)} S_2^T P_2^{(i)}
$$
$$
H_5^{(i)} = P_2^{(i)} S_2^T P_2^{(i)} + P_2^{(i)} S_3 P_3^{(i)}
$$
$$
H_5^{(i,i+1)} = P_2^{(i)} S_2^T P_2^{(i)} + P_2^{(i)} S_3 P_3^{(i+1)}
$$
$$
H_6^{(i,i+1)} = P_2^{(i)} S_2^T P_2^{(i)} + P_2^{(i)} S_3 P_3^{(i+1)}
$$

with $P_j^{(i)}$ satisfying (9.19) and initial conditions given by (9.27).

$$\triangle$$

The following theorem indicates the features of the proposed algorithm (9.58)-(9.60).

Theorem 9.3 *Under conditions stated in Assumption 9.2, the algorithm (9.58)-(9.60) converges to the exact solution of the error term with the rate of convergence of $O(\epsilon)$, that is*

$$\left\| E_j - E_j^{(i+1)} \right\| = O(\epsilon) \left\| E_j - E_j^{(i)} \right\| \tag{9.62}$$

or equivalently

$$\left\| E_j - E_j^{(i)} \right\| = O\left(\epsilon^{(i+1)}\right), \quad i = 0, 1, 2, ...; \; j = 1, 2, 3 \tag{9.63}$$

$$\diamond$$

Case Study 3

The developed procedure is applied to the synthesis of the optimal hydro power plant control by an overall optimal regulator commonly designed for both active power-frequency and reactive power-voltage control loops (Skataric, 1989). Assumptions made in the case studies 1 and 2 hold in this case also. The state space model is of order eleven, and the state variables are ordered as

$$x^T = [\Delta\theta \quad \Delta\omega \quad \Delta u_f \quad \Delta\psi_d \quad \Delta\psi_q \quad \Delta\psi_f \quad \Delta\psi_D \quad \Delta\psi_Q \quad \Delta q \quad \Delta a \quad \Delta\varphi^\circ]$$

The control vector is given by $u = [\Delta u_{vr} \quad \Delta u_{tr}]$.
The nonzero entries in the matrices $A^{11\times11}$ and $B^{11\times2}$ are

$$a_{12} = a_{63} = 314.16, \; a_{21} = -0.595, \; a_{24} = 0.147, \; a_{25} = 0.528, \; a_{26} = -0.134,$$
$$a_{27} = -0.04, \; a_{28} = -0.276, \; a_{29} = -0.371, \; a_{2,10} = -0.182, \; a_{2,11} = 0.594,$$
$$a_{31} = -100, \; a_{41} = 255.38, \; a_{42} = -152.49, \; a_{44} = -13.72, \; a_{45} = 511.14,$$
$$a_{46} = 8.51, \; a_{47} = 2.556, \; a_{48} = -135.04, \; a_{51} = -182.82, \; a_{52} = -319.15,$$
$$a_{54} = -534.99, \; a_{55} = -12.54, \; a_{56} = 137, \; a_{57} = 41.136, \; a_{58} = 8.389,$$
$$a_{64} = 0.446, \; a_{66} = -0.523, \; a_{67} = 0.0375, \; a_{74} = 21.646, \; a_{76} = 6.094,$$
$$a_{77} = -29.609, \; a_{85} = 87.236, \; a_{88} = -97.747, \; a_{99} = -2, \; a_{10,10} = -0.714,$$
$$a_{11,2} = 0.566, \; a_{11,9} = 1.208, \; a_{11,10} = 0.614, \; a_{11,11} = -1.463$$

$$b_{31} = 0.184, \; b_{92} = 2, \; b_{10,2} = 0.714$$

Weighting matrices are chosen as $Q = 0.1 \times I_{11}$, and $R = I_2$.

The required structure studied in this section is obtained by interchanging 3th, 4th, and 5th rows in matrices A and B by 9th, 10th, and 11th rows.

The eigenvalues of the matrix A are given by -100, -74.806, -27.895, $-25.167 \pm j520.79$, -2, -1.469, -0.714, $-0.544 \pm j8.483$, -0.295. We have studied this singularly perturbed problem with 6 slow and fast variables and with $\epsilon = 0.1$. For this ordering of the state variables Assumption 9.2 is satisfied.

Simulation results, representing the absolute error between the approximate and optimal values of the performance criterion, are presented in Table 9.3.

i	$J_{app} - J_{opt}$
0	$0.127 \times 10^{+1}$
1	0.740×10^{-1}
2	0.292×10^{-3}
3	0.674×10^{-4}
4	0.364×10^{-5}
5	0.250×10^{-6}
6	0.379×10^{-7}
7	0.215×10^{-8}
8	0.109×10^{-10}

Table 9.3: Errors in the performance criterion per iteration

It is important to point out that in this case the eigenvector approach for solving the global algebraic Riccati equation (9.9) failed to produce the answer also. On the other hand, it can be seen from Table 9.3 that the method proposed in this section is numerically very efficient.

9.3 Reduced-Order Design of Optimal Controller for Quasi Weakly Coupled Linear Systems

In this section, we consider a special class of linear systems having block diagonally dominant system matrix and with the control input influencing only one of subsystems. The optimal reduced-order controllers are designed through the recursive reduced-order algorithm which converges quickly to the required optimal solution. Many real world systems (such as power systems, chemical reactors, flexible structures, and, in general, systems with only few actuators) possess the control structure studied in this section. We call these structures quasi (nearly) weakly coupled since they contain the diagonally block dominant system matrix (like the standard weakly coupled systems), but they have only one decision maker (weakly coupled systems require at least two decision makers).

Consider a linear dynamical system composed of two subsystems in the form

$$\begin{bmatrix} \dot{x}_1 \\ \dot{x}_2 \end{bmatrix} = \begin{bmatrix} A_1 & \epsilon A_2 \\ \epsilon A_3 & A_4 \end{bmatrix} \begin{bmatrix} x_1 \\ x_2 \end{bmatrix} + \begin{bmatrix} B_1 \\ 0 \end{bmatrix} u \qquad (9.67)$$

where $x_i \in \Re^{n_i}$, $i = 1, 2$, are state vectors, $u \in \Re^m$ is a control input vector, and ϵ is a small parameter. This is a special class of linear dynamical systems represented, in general, by

$$\dot{x} = Ax + Bu \qquad (9.68)$$

with

$$x = \begin{bmatrix} x_1 \\ x_2 \end{bmatrix}, \qquad A = \begin{bmatrix} A_1 & \epsilon A_2 \\ \epsilon A_3 & A_4 \end{bmatrix}, \qquad B = \begin{bmatrix} B_1 \\ 0 \end{bmatrix} \qquad (9.69)$$

A quadratic type functional to be minimized is associated with (9.67) in the form

$$J = \frac{1}{2} \int_0^\infty (x^T Q x + u^T R u)\, dt, \qquad Q \geq 0, \quad R > 0 \qquad (9.70)$$

For the purpose of this section we assume that the structure of the matrix Q is consistent with the system matrix A, that is

$$Q = \begin{bmatrix} Q_1 & \epsilon Q_2 \\ \epsilon Q_2^T & Q_3 \end{bmatrix} \tag{9.71}$$

All problem matrices defined in (9.67)-(9.71) are constant and of appropriate dimensions.

The structure defined in (9.67) corresponds to the weakly interconnected subsystems (Kokotovic et al., 1969) with the control input influencing only one of them. This structure has not been studied in the literature from the order reduction point of view. The purpose of this section is not to derive new theoretical concepts. Instead, its main goal is to show that certain classes of linear optimal control problems can be studied by using the developed reduced-order recursive theory for the weakly coupled linear control systems.

The optimal problem of minimizing (9.70) along trajectories of (9.67) has the very well-known solution given by

$$u_{opt} = -F_{opt}x = -R^{-1}B^T Px \tag{9.72}$$

where P is the positive semidefinite stabilizing solution of the algebraic Riccati equation

$$A^T P + PA + Q - PSP = 0, \quad S = BR^{-1}B^T \tag{9.73}$$

It can be shown that the nature of the solution of (9.73) is

$$P = \begin{bmatrix} P_1 & \epsilon P_2 \\ \epsilon P_2^T & P_3 \end{bmatrix} \tag{9.74}$$

Partitioning (9.73) compatible to (9.69), (9.71), and (9.74), we get three matrix algebraic equations

$$P_1 A_1 + A_1^T P_1 + Q_1 - P_1 S_1 P_1 + \epsilon^2 \left(P_2 A_3 + A_3^T P_2 \right) = 0 \tag{9.75}$$

$$P_2 A_4 + A_1^T P_2 + Q_2 - P_1 S_1 P_2 + A_3^T P_3 + P_1 A_2 = 0 \tag{9.76}$$

$$P_3 A_4 + A_4^T P_3 + Q_3 + \epsilon^2 \left(P_2^T A_2 + A_2^T P_2 - P_2^T S_1 P_2 \right) = 0 \tag{9.77}$$

where $S_1 = B_1 R^{-1} B_1^T$.

Since ϵ is a small parameter, we can define $O\left(\epsilon^2\right)$ approximation of (9.75)-(9.77) as follows

$$P_1^{(0)} A_1 + A_1^T P_1^{(0)} + Q_1 - P_1^{(0)} S_1 P_1^{(0)} = 0 \tag{9.78}$$

$$P_2^{(0)} A_4 + \left(A_1 - S_1 P_1^{(0)}\right)^T P_2^{(0)} + Q_2 + A_3^T P_3^{(0)} + P_1^{(0)} A_2 = 0 \tag{9.79}$$

$$P_3^{(0)} A_4 + A_4^T P_3^{(0)} + Q_3 = 0 \tag{9.80}$$

so that the required solution (9.74) satisfies

$$P^{(0)} = \begin{bmatrix} P_1^{(0)} & \epsilon P_2^{(0)} \\ \epsilon P_2^{(0)T} & P_3^{(0)} \end{bmatrix} = P + O\left(\epsilon^2\right) \tag{9.81}$$

On the contrary to the standard weakly coupled systems (Kokotovic et al., 1969), where the zeroth-approximation is given in terms of two reduced-order Riccati equations, for the quasi weakly coupled systems we need to solve only one reduced-order Riccati equation (9.78).

The unique positive semidefinite stabilizing solution $P^{(0)}$, obtained from (9.78)-(9.80), exists under the following assumption.

Assumption 9.3 The triple $\left(A_1, B_1, \sqrt{Q_1}\right)$ is stabilizable-detectable and the matrix A_4 is stable.

$\triangle$

Defining the approximation errors as

$$P_j = P_j^{(0)} + \epsilon^2 E_j, \qquad j = 1, 2, 3 \tag{9.82}$$

and using (9.82) in (9.75)-(9.77) and (9.78)-(9.80), we get the following error equations

$$E_1 D_1 + D_1^T E_1 = -P_2 A_3 - A_3^T P_2^T + \epsilon^2 E_1 S_1 E_1 \tag{9.83}$$

$$E_2 A_4 + D_1^T E_2 = \epsilon^2 E_1 S_1 E_2 - E_1 A_2 - A_3^T E_3 - E_1 S_1 P_1^{(0)} \tag{9.84}$$

$$E_3 A_4 + A_4^T E_3 = P_2^T A_2 + A_2^T P_2 - P_2^T S_1 P_2 \tag{9.85}$$

where $D_1 = A_1 - S_1 P_1^{(0)}$, is a stable matrix (Gajic et al., 1990).

Let us propose the following reduced-order parallel algorithm for solving (9.83)-(9.85).

Algorithm 9.4:

$$E_1^{(i+1)} D_1 + D_1^T E_1^{(i+1)} = -P_2^{(i)} A_3 - A_3^T P_2^{(i)^T} + \epsilon^2 E_1^{(i)} S_1 E_1^{(i)} \quad (9.86)$$

$$E_3^{(i+1)} A_4 + A_4^T E_3^{(i+1)} = P_2^{(i)^T} A_2 + A_2^T P_2^{(i)} - P_2^{(i)^T} S_1 P_2^{(i)} \quad (9.87)$$

$$E_2^{(i+1)} A_4 + D_1^T E_2^{(i+1)} = \epsilon^2 E_1^{(i+1)} S_1 E_2^{(i)} - E_1^{(i+1)} D_{12} - A_3^T E_3^{(i+1)} \quad (9.88)$$

where

$$P_j^{(i)} = P_j^{(0)} + \epsilon^2 E_j^{(i)}, \qquad j = 1, 2, 3, \qquad i = 0, 1, 2, 3... \quad (9.89)$$

and $D_{12} = A_2 - S_1 P_2^{(0)}$ with initial conditions

$$E_1^{(0)} = 0, \ E_2^{(0)} = 0, \ E_3^{(0)} = 0 \quad (9.90)$$

$$\triangle$$

The following theorem indicates the features of the proposed algorithm (9.86)-(9.89).

Theorem 9.4 *Under conditions stated in Assumption 9.3, the algorithm (9.86)-(9.90) converges to the exact solution of the error term with the rate of convergence of $O\left(\epsilon^2\right)$, that is*

$$\left\| E_j - E_j^{(i+1)} \right\| = O\left(\epsilon^2\right) \left\| E_j - E_j^{(i)} \right\| \quad (9.91)$$

or equivalently

$$\left\| E_j - E_j^{(i)} \right\| = O\left(\epsilon^{2(i+1)}\right), \quad i = 0, 1, 2, ... \quad (9.92)$$

$$\diamond$$

The proof of this theorem follows ideas reported in (Gajic et al., 1990), and thus, is omitted. In the first step of the proof, the nonsingularity of the Jacobian of (9.83)-(9.85) at $\epsilon = 0$ has to be established. In the second step, the estimates of the errors given in (9.91)-(9.92) are obtained from (9.83)-(9.85) and (9.86)-(9.88). We will justify results stated in Theorem 9.4 on several real control system examples (Section 9.4).

273

Notice that from (9.91) and (9.92) we have

$$\left\| P - P^{(i)} \right\| = O\left(\epsilon^{2(i+1)} \right) \tag{9.93}$$

where

$$P^{(i)} = \begin{bmatrix} P_1^{(0)} + \epsilon^2 E_1^{(i)} & \epsilon \left(P_2 + \epsilon^2 E_2^{(i)} \right) \\ \epsilon \left(P_2 + \epsilon^2 E_2^{(i)} \right)^T & P_3^{(0)} + \epsilon^2 E_3^{(i)} \end{bmatrix} \tag{9.94}$$

The approximate optimal gain $F^{(i)}$ is now defined by

$$F^{(i)} = -R^{-1} B^T P^{(i)} \tag{9.95}$$

and the approximate criterion is obtained from

$$J_{app}^{(i)} = tr\left\{ V^{(i)} \right\} \tag{9.96}$$

where $V^{(i)}$ satisfies

$$\left(A - SP^{(i)} \right)^T V^{(i)} + V^{(i)} \left(A - SP^{(i)} \right) + Q + P^{(i)} SP^{(i)} = 0 \tag{9.97}$$

Using the criterion approximation theorem (Kokotovic and Cruz, 1969), we have that (9.93) implies

$$\left| J_{opt} - J_{app}^{(i)} \right| = O\left(\epsilon^{4(i+1)} \right), \quad i = 0, 1, 2, \dots \tag{9.98}$$

In some applications, like power systems, the open-loop system matrix A is stable and the elements in the matrix B_1 are all of $O(\epsilon)$. In such cases the presented algorithm can be even more simplified under the following assumption.

Assumption 9.4 The stability of the matrix A implies stability of the partitioned matrices A_1 and A_4.

$$\triangle$$

The zeroth-order approximations in (9.78) and (9.79) are now defined by

$$P_1^{(0)} A_1 + A_1^T P_1^{(0)} + Q_1 = 0 \tag{9.99}$$

$$P_2^{(0)} A_4 + A_1^T P_2^{(0)} + Q_2 + A_3^T P_3^{(0)} + P_1^{(0)} A_2 = 0 \qquad (9.100)$$

Introducing the notation

$$B_1 = \epsilon B_{1p}, \quad S_{1p} = B_{1p} R^{-1} B_{1p}^T \qquad (9.101)$$

the modified algorithm (9.86)-(9.88) gets the form.
Algorithm 9.5:

$$E_1^{(i+1)} A_1 + A_1^T E_1^{(i+1)} = -P_2^{(i)} A_3 - A_3^T P_2^{(i)^T} + P_1^{(i)} S_{1p} P_1^{(i)} \quad (9.102)$$

$$E_3^{(i+1)} A_4 + A_4^T E_3^{(i+1)} = -P_2^{(i)^T} A_2 - A_2^T P_2^{(i)} + \epsilon^2 P_2^{(i)^T} S_{1p} P_2^{(i)} \quad (9.103)$$

$$E_2^{(i+1)} A_4 + A_1^T E_2^{(i+1)} = P_1^{(i+1)} S_{1p} P_1^{(i+1)} - E_1^{(i+1)} A_2 - A_3^T E_3^{(i+1)}$$
$$(9.104)$$
$$\triangle$$

Thus, the complete solution, in this case, is obtained in terms of the Lyapunov equations only.

9.4 Case Studies

In the previous section, we have shown how to generate the solution of the algebraic Riccati equation in terms of the reduced-order subsystems. Having obtained this solution, (9.94), allows us to construct an approximation to the optimal control

$$u^{(i)}(t) = F^{(i)} x^{(i)}(t)$$

where $F^{(i)}$ is given by (9.95) and $x^{(i)}(t)$ satisfies

$$\dot{x}^{(i)}(t) = \left(A - BF^{(i)} \right) x^{(i)}(t)$$

Using (9.93) and (9.95), it follows that the control law $u^{(i)}(t)$ and the approximate trajectories $x^{(i)}(t)$ are suboptimal in the sense

$$x^{(i)}(t) = x_{opt}(t) + O\left(\epsilon^{2(i+1)} \right), \quad u^{(i)}(t) = u_{opt}(t) + O\left(\epsilon^{2(i+1)} \right)$$

The approximate feedback control $u^{(i)}(t)$ applied to the system produces the approximate performance index (9.96) with its property established in (9.98).

In this section, we consider three real physical system control problems: chemical reactor, F-4 fighter aircraft, and multimachine power system and demonstrate the near-optimality with respect to the performance criterion.

9.4.1 Chemical Reactor

The model of a chemical reactor has been studied in (Patnaik et al., 1980). The system and input matrices are given by

$$A = \begin{bmatrix} -4.019 & 5.12 & 0 & 0 & -2.082 & 0 & 0 & 0 & 0.87 \\ -0.346 & 0.986 & 0 & 0 & -2.34 & 0 & 0 & 0 & 0.97 \\ -7.909 & 15.407 & -4.069 & 0 & -6.45 & 0 & 0 & 0 & 2.68 \\ -21.816 & 35.606 & -0.339 & -3.87 & -17.8 & 0 & 0 & 0 & 7.39 \\ -60.196 & 98.188 & -7.907 & 0.34 & -53.008 & 0 & 0 & 0 & 20.4 \\ 0 & 0 & 0 & 0 & 94 & -147.2 & 0 & 53.2 & 0 \\ 0 & 0 & 0 & 0 & 0 & 94 & -147.2 & 0 & 0 \\ 0 & 0 & 0 & 0 & 0 & 12.8 & 0 & -31.6 & 0 \\ 0 & 0 & 0 & 0 & 12.8 & 0 & 0 & 18.8 & -31.6 \end{bmatrix}$$

$$B^T = \begin{bmatrix} 0.010 & 0.003 & 0.009 & 0.024 & 0.068 & 0 & 0 & 0 & 0 \\ -0.011 & -0.021 & -0.059 & -0.162 & -0.445 & 0 & 0 & 0 & 0 \\ -0.151 & 0 & 0 & 0 & 0 & 0 & 0 & 0 & 0 \end{bmatrix}$$

Weighting matrices Q and R are chosen as identities.

This control system problem can be decoupled according to Section 9.3 with $n_1 = 5$ and $n_2 = 4$, where the first five state variables comprise the first subsystem. Using the formula for an estimate of a small coupling parameter ϵ suggested by (Shen and Gajic, 1990a), we have obtained ϵ = 0.47 = 94/200.4.

Simulation results are presented in Table 9.4. Obtained results reveal that the accuracy of $O(10^{-10})$ is obtained after only 7 iterations despite relatively big value of the coupling parameter ϵ. This is consistent with the results given in Theorem 9.4 and formula (9.98) since $(0.47)^{32}$ = 0.32146×10^{-10}. Thus, the presented method is very efficient even in the case when ϵ is not "small enough" — the standard assumption for all small parameter theories. Even more, by using the presented method the accuracy of an arbitrary order is easily achieved. All simulation results in this chapter are obtained by using the L-A-S package for computer aided control systems design (West et al., 1985).

9.4.2 F-4 Fighter Aircraft

An F-4 fighter aircraft (the actuator case) is considered in (Harvey and Stein, 1978). This model is described by the following system and control input matrices

i	$J_{app} - J_{opt}$
0	0.13910×10^{-2}
1	0.15714×10^{-3}
2	0.14805×10^{-4}
3	0.13045×10^{-5}
4	0.10936×10^{-6}
5	0.81286×10^{-8}
6	0.39972×10^{-9}
7	0.33651×10^{-10}

Table 9.4: Errors in the performance criterion per iteration

$$ B^T = \begin{bmatrix} 0 & 0 & 0 & 0 & 20 & 0 \\ 0 & 0 & 0 & 0 & 0 & 10 \end{bmatrix} $$

$$ A = \begin{bmatrix} -0.746 & 0.387 & -12.9 & 0 & 0.952 & 6.05 \\ 0.024 & -0.174 & 4.31 & 0 & -1.76 & -0.416 \\ 0.006 & -0.999 & -0.0578 & 0.0369 & 0.0092 & -0.0012 \\ 1 & 0 & 0 & 0 & 0 & 0 \\ 0 & 0 & 0 & 0 & -10 & 0 \\ 0 & 0 & 0 & 0 & 0 & -5 \end{bmatrix} $$

Weighting matrices Q and R are chosen as $Q = I_6$, $R = I_2$.

Even though the aircraft is not inherently weakly coupled system, we will show that the presented algorithm can be applied to the reduced-order controller design of this aircraft with a prescribed degree of stability. The system is decomposed with $n_1 = 4$ and $n_2 = 2$, where the first four state variables comprise the first subsystem. The eigenvalues of the matrix A are given by -0.006, -0.765, $-0.103 \pm j2.093$, -5, -10. In order to have the weakly coupled structure for the matrix A we need that $det\{A_1\} = O(1)$ and $det\{A_4\} = O(1)$, (Chow and Kokotovic, 1983). However, in this example $det\{A_1(0)\} = 0.021278$. This can be facilitated by choosing the performance criterion in the form (which assures a prescribed degree of stability)

$$J = \int_0^\infty e^{2\alpha t} \left(x^T Q x + u^T R u \right) \ dt \qquad (9.105)$$

The consequence of this is that the actual system matrix that we are working with is $A + \alpha I$. For this modified system matrix and for α = -10 the strongest coupling is in the first row, so that an estimate of ϵ (Shen and Gajic, 1990a) is given by 7.002 / 24.033 = 0.291349. Since $A + \alpha I$ is a stable matrix the required conditions from Assumption 9.3 are satisfied. Simulation results for the performance criterion are presented in Table 9.5.

i	$J_{app} - J_{opt}$
0	0.73118×10^{-3}
1	0.97281×10^{-5}
2	0.16248×10^{-6}
3	0.25505×10^{-8}
4	0.39449×10^{-10}
5	0.34106×10^{-12}

Table 9.5: Errors in the performance criterion per iteration

Note that $(0.291349)^{24}$ = 0.1399×10^{-12} so that this example perfectly matches the results established in Theorem 9.4 and formula (9.98).

Since $A + \alpha I$ is diagonally dominant for α large enough, and thus, weakly coupled, the presented method is more general and is applicable to the systems which are not inherently weakly coupled.

The importance of the higher order approximations for weakly coupled systems is demonstrated in (Shen and Gajic, 1990a), where the $O\left(\epsilon^6\right)$ accuracy was required in order to stabilize the closed-loop system.

9.4.3 Multimachine Power System

The nearly weakly coupled structure studied in this section can be found in power systems. The efficiency of the proposed reduced-order algorithm (9.86)-(9.88) is demonstrated on the design example for the decentralized multivariable excitation controllers in a multimachine power system.

We consider a complex multimachine power system, composed of N synchronous machine-regulator units and connected to the network which includes transformers, lines and load. In these studies it is customary to treat the synchronous generators in plant as one equivalent machine and to use the assumption that the turbine torques are constant, as the changes in these torques are slow in comparison with phenomena of significance in voltage regulation. Furthermore, the electromagnetic transient processes in armature windings of the machines and the elements of the network are usually neglected as well as transient processes in the damping winding as less significant in the problem under consideration.

The linearized model obtained under these assumptions will be used in this section. Each of the synchronous machines is described by Park's equations with a field circuit in the direct axis. The synchronous generators are assumed to be equipped with first order exciters. The network is represented by constant admitances and reduced by eliminating nongenerator basis. Loads are represented by constant admitances and are included in the network admitance matrix.

The linearized equations of the considered multimachine power system are written in the state space form as

$$\dot{x} = Ax + \sum_{i=1}^{N} B_i u_i, \qquad i = 1, 2, 3, ...N \qquad (9.106)$$

where $x^T = [\, x_1^T \quad x_2^T \quad .... \quad x_N^T \,]$, $x_i^T = [\, \delta_{iN} \quad \omega_i \quad \psi_{fi} \quad E_{fdi} \,]$, $i = 1, 2, ..., N$, with δ_{iN} being the load angle with respect to the angle of the reference machine, ω_i the rotor angular velocity, ψ_{fi} the field flux linkage, and E_{fdi} the exiter state variable of the i-th machine. All variables represent small deviations from the operating point.

In this section, we study the real example that represents the portion of the Serbian grid in isolated operation composed of two hydro power plants (Arnautovic, 1988; Arnautovic and Medanic, 1990). Each machine is equipped with the fast exciter whose parameters and operating points are given in Appendix 9.1.

Matrices A and B of the corresponding linearized model are given by

$$A = \begin{bmatrix} 0 & -314.159 & 314.159 & 0 & 0 & 0 & 0 \\ 0.003 & -0.131 & -0.012 & -0.141 & -0.006 & 0 & 0 \\ -0.271 & -0.352 & -2.763 & -0.182 & -0.371 & 0 & 0 \\ 0.005 & -0.290 & -0.008 & -0.373 & 0.005 & 314.159 & 0 \\ -0.290 & -0.127 & -0.724 & 0.025 & -1.261 & 0 & 314.159 \\ 0 & 0 & 0 & 0 & 0 & -33.333 & 0 \\ 0 & 0 & 0 & 0 & 0 & 0 & -33.333 \end{bmatrix}$$

$$B^T = \begin{bmatrix} 0 & 0 & 0 & 0 & 0 & 0.062 & 0 \\ 0 & 0 & 0 & 0 & 0 & 0 & 0.201 \end{bmatrix}$$

Weighting matrices Q and R are chosen as $Q = I_7, R = I_2$.

Apparently, the matrix A has the weakly coupled structure. By interchanging rows in matrices A and B, we can get the nearly weakly coupled structure defined in (9.69). The eigenvalues of the matrix A are given by -0.048, -0.549, -0.822, -1.555 $\pm$ j 9.164, -33.33, -33.33. It can be seen that both conditions of Assumption 9.4 are satisfied for this power system example. Due to the special structure of the matrix B, the feedback control will affect only slightly some of the very small eigenvalues so that the system will remain almost marginally stable under the feedback control. In addition, in order to have the weakly coupled structure for the matrix A we need that $det\{A_1\} = O(1)$ and $det\{A_4\} = O(1)$, (Chow and Kokotovic, 1983). These two problems can be facilitated by choosing the performance criterion in the form

$$J = \int_0^\infty e^{2\alpha t} \left(x^T Q x + u^T R u \right) \, dt$$

In order to improve the numerical behavior of the proposed algorithm it is advisable to balance the elements in the matrix A (some of them are very large) by introducing simple scalings in the form $\bar{x}_i = k x_i$, with $k = 0.1$, $i = 5, 6$, and $k = 0.3$ for $i = 1, 2$. Parameters α and ϵ are chosen as $\alpha = 1$ and $\epsilon = 0.4$.

Simulation results are obtained by using the L-A-S package for computer-aided control system design. Obtained results are presented in Table 9.6.

i	J_{app}	$J_{opt} - J_{app}$
0	122.144	26.198
1	96.285	0.339
2	96.008	0.062
3	95.954	0.008
4	95.948	1.5×10^{-3}
5	95.946	2.0×10^{-4}
6	95.946	8.0×10^{-5}
optimal	95.946	

Table 9.6: Approximate values for the performance index

9.5 Reduced-Order Solution for a Class of Linear-Quadratic Optimal Control Problems

The reduced-order solution is obtained for a class of linear-quadratic optimal control problems having weakly interconnected system matrix, strongly connected control matrix, and with a special structure for the state penalty matrix. An example demonstrates the effectiveness of the proposed reduced-order algorithm. The presented method is very well suited for parallel implementation.

Consider a linear dynamical system given by

$$\dot{x} = Ax + Bu \qquad (9.107)$$

with

$$x = \begin{bmatrix} x_1 \\ x_2 \end{bmatrix}, \quad A = \begin{bmatrix} A_1 & \epsilon A_2 \\ \epsilon A_3 & A_4 \end{bmatrix}, \quad B = \begin{bmatrix} B_1 \\ B_2 \end{bmatrix} \qquad (9.108)$$

where $x_i \in \Re^{n_i}$, $i = 1, 2$, are state vectors, $u \in \Re^m$ is a control input and ϵ is a small parameter. A quadratic type functional to be minimized is associated with (9.107) in the form

$$J = \frac{1}{2} \int_0^\infty [x^T Q x + u^T R u]\, dt, \quad Q \geq 0, \quad R > 0 \qquad (9.109)$$

All matrices defined in (9.107)-(9.109) are constant and of appropriate dimensions.

The system matrix A, defined in (9.108) has the structure of the weakly coupled systems (Kokotovic et al., 1969; Gajic et al., 1990). However, due to strongly coupled control matrix B this system does not belong to the class of weakly coupled linear control systems. In this section, we will show that despite strong coupling coming from the input matrix, the order-reduction can be achieved, like in the case of purely weakly coupled systems, by using the specific structure for the state penalty matrix.

Up to authors best knowledge the problem order-reduction through the choice of the state penalty matrix Q has not been studied in the control literature. Thus, the engineering relevance of this section is to study the linear-quadratic optimal control problem of (9.107)-(9.109), in the spirit of parallel and distributed reduced-order algorithms (Bertsekas and Tsitsiklis, 1991), under the following assumption.

Assumption 9.5 The state penalty matrix Q has the structure

$$Q = \begin{bmatrix} Q_1 & \epsilon Q_2 \\ \epsilon Q_2 & \epsilon Q_3 \end{bmatrix} \qquad (9.110)$$

$\triangle$

This choice of the matrix Q is quite common in engineering practice since the control engineers hardly penalize all state variables by weighting factors of the same magnitude, especially for large scale systems.

The optimal problem of minimizing (9.109) along trajectories of (9.107) has the very well-known solution given by

$$u_{opt} = -F_{opt} x = -R^{-1} B^T P x \qquad (9.111)$$

where P is the positive semidefinite stabilizing solution of the algebraic Riccati equation

$$A^T P + P A + Q - P S P = 0, \quad S = B R^{-1} B^T \qquad (9.112)$$

In the following, we will show that the nature of the solution of (9.112) subject to the partition of the problem matrices defined in (9.108)-(9.110) is given by

$$P = \begin{bmatrix} P_1 & \epsilon P_2 \\ \epsilon P_2^T & \epsilon P_3 \end{bmatrix} \qquad (9.113)$$

and then, we derive the reduced-order algorithm for finding P.

Partitioning (9.112) compatible to (9.108), (9.110), and (9.111), we get three algebraic equations

$$P_1 A_1 + A_1^T P_1 + Q_1 - P_1 S_1 P_1 + \epsilon^2 \left(P_2 A_3 + A_3^T P_2^T \right)$$
$$-\epsilon \left(P_1 S_2 P_2^T + P_2 S_2^T P_1 + \epsilon P_2 S_3 P_2^T \right) = 0 \qquad (9.114)$$

$$P_1 A_2 + P_2 A_4 + A_1^T P_2 - P_1 S_1 P_2 - P_1 S_2 P_3 + Q_2$$
$$+\epsilon \left(A_3^T P_3 - P_2 S_2^T P_2 - P_2 S_3 P_3 \right) = 0 \qquad (9.115)$$

$$P_3 A_4 + A_4^T P_3 + Q_3$$
$$+\epsilon \left(P_2^T A_2 + A_2^T P_2 - P_2^T S_1 P_2 - P_2^T S_2 P_3 - P_3 S_2^T P_2 - P_3 S_3 P_3 \right) = 0 \qquad (9.116)$$

where

$$S_1 = B_1 R^{-1} B_1^T, \quad S_2 = B_1 R^{-1} B_2^T, \quad S_3 = B_2 R^{-1} B_2^T \qquad (9.117)$$

Since ϵ is a small parameter we can define an $O(\epsilon)$ approximation of (9.114)-(9.116) as follows

$$P_1^{(0)} A_1 + A_1^T P_1^{(0)} + Q_1 - P_1^{(0)} S_1 P_1^{(0)} = 0 \qquad (9.118)$$

$$P_1^{(0)} A_2 + P_2^{(0)} A_4 + A_1^T P_2^{(0)} - P_1^{(0)} S_1 P_2^{(0)} - P_1^{(0)} S_2 P_3^{(0)} + Q_2 = 0 \qquad (9.119)$$

$$P_3^{(0)} A_4 + A_4^T P_3^{(0)} + Q_3 = 0 \qquad (9.120)$$

The unique positive semidefinite stabilizing solution for $P^{(0)}$, obtained from (9.118)-(9.120), and defined by

$$P^{(0)} = \begin{bmatrix} P_1^{(0)} & \epsilon P_2^{(0)} \\ \epsilon P_2^{(0)T} & \epsilon P_3^{(0)} \end{bmatrix} \qquad (9.121)$$

exists under the following assumption.

Assumption 9.6 The triple $\left(A_1, B_1, \sqrt{Q_1} \right)$ is stabilizable-detectable and the matrix A_4 is stable.

$\triangle$

Since all solutions obtained from (9.118)-(9.120) are O(1), it can be concluded that our staring assumption (9.113) about the nature of the solution of (9.112) is correct.

Remark 9.1 The nature of the solution of the algebraic Riccati equation for a linear-quadratic optimal control problem defined in this section is exactly the same as the nature of the solution of the algebraic Riccati equation of a singularly perturbed linear-quadratic control problem.

$$\triangle$$

From (9.118)-(9.120) we have obtained the first-order approximation of the required solution in terms of the completely decomposed reduced-order algebraic equations.

In the next step, we will derive the reduced-order parallel algorithm, based on the fixed point iterations, for obtaining the solution of P up to any arbitrary degree of accuracy.

Defining the approximation errors as

$$P_j = P_j^{(0)} + \epsilon E_j, \quad j = 1, 2, 3 \tag{9.122}$$

and using (9.121) in (9.114)-(9.116) and (9.118)-(9.120), we get the following error equations

$$E_1 D_1 + D_1^T E_1 = \epsilon E_1 S_1 E_1 - \epsilon \left(P_2 A_3 + A_3^T P_2^T \right) \\ + P_1 S_2 P_2^T + P_2 S_2^T P_1 + \epsilon P_2 S_3 P_2^T \tag{9.123}$$

$$E_2 A_4 + D_1^T E_2 = \epsilon \left(E_1 S_1 E_2 + E_1 S_2 E_3 \right) - E_1 D_2 \\ + P_1^{(0)} S_2 E_3 - A_3^T P_3 + P_2 S_2^T P_2 + P_2 S_3 P_3 \tag{9.124}$$

$$E_3 A_4 + A_4^T E_3 \\ = P_2^T S_1 P_2 + P_2^T S_2 P_3 + P_3 S_2^T P_2 + P_3 S_3 P_3 - P_2^T A_2 - A_2^T P_2 \tag{9.125}$$

where

$$D_1 = A_1 - S_1 P_1^{(0)}, \quad D_2 = A_2 - S_1 P_2^{(0)} - S_2 P_3^{(0)} \tag{9.126}$$

Let us propose the following reduced-order parallel algorithm for solving (9.123)-(9.125).

Algorithm 9.6:

$$E_1^{(i+1)} D_1 + D_1^T E_1^{(i+1)} = \epsilon E_1^{(i)} S_1 E_1^{(i)} - \epsilon \left(P_2^{(i)} A_3 + A_3^T P_2^{(i)^T} \right) \\ + P_1^{(i)} S_2 P_2^{(i)^T} + P_2^{(i)} S_2^T P_1^{(i)} + \epsilon P_2^{(i)} S_3 P_2^{(i)^T} \tag{9.127}$$

$$E_2^{(i+1)} A_4 + D_1^T E_2^{(i+1)} = \epsilon \left(E_1^{(i+1)} S_1 E_2^{(i)} + E_1^{(i+1)} S_2 E_3^{(i+1)} \right)$$
$$-E_1^{(i+1)} D_2 + P_1^{(0)} S_2 E_3^{(i+1)} - A_3^T P_3^{(i+1)} + P_2^{(i)} S_2^T P_2^{(i)} + P_2^{(i)} S_3 P_3^{(i+1)}$$

$$(9.128)$$

$$E_3^{(i+1)} A_4 + A_4^T E_3^{(i+1)} = P_2^{(i)^T} S_1 P_2^{(i)} + P_2^{(i)^T} S_2 P_3^{(i)} + P_3^{(i)} S_2^T P_2^{(i)}$$
$$+ P_3^{(i)} S_3 P_3^{(i)} - P_2^{(i)^T} A_2 - A_2^T P_2^{(i)}$$

$$(9.129)$$

with initial conditions

$$E_1^{(0)} = 0, \quad E_2^{(0)} = 0, \quad E_3^{(0)} = 0 \qquad (9.130)$$

$$\triangle$$

The following theorem indicates the features of the proposed algorithm (9.127)-(9.129).

Theorem 9.5 *Under conditions stated in Assumptions 9.5 and 9.6, the algorithm (9.127)-(9.130) converges to the exact solution of the error term, and thus to the required solution P, with the rate of convergence of $O(\epsilon)$, that is*

$$\left\| E_j - E_j^{(i+1)} \right\| = O(\epsilon) \left\| E_j - E_j^{(i)} \right\| \qquad (9.131)$$

or equivalently

$$\left\| E_j - E_j^{(i)} \right\| = O\left(\epsilon^{(i+1)} \right), \quad i = 0, 1, 2, ..; \quad j = 1, 2, 3 \qquad (9.132)$$

$$\diamond$$

Proof: As a starting point, we need to show the existence of a bounded solution of (9.123)-(9.125) in the neighborhood of $\epsilon = 0$. By the implicit function theorem (Ortega and Rheinboldt, 1970), it is enough to show that the corresponding Jacobian is nonsingular at $\epsilon = 0$. The Jacobian at $\epsilon = 0$ is given by

$$J(\epsilon)_{|\epsilon=0} = \begin{bmatrix} J_{11} & 0 & 0 \\ J_{21} & J_{22} & J_{23} \\ 0 & 0 & J_{33} \end{bmatrix} \qquad (9.133)$$

with

$$J_{11} = I_{n_1} \otimes D_1^T + D_1^T \otimes I_{n_1}$$
$$J_{22} = I_{n_1} \otimes D_1^T + A_4^T \otimes I_{n_2} \qquad (9.134)$$
$$J_{33} = I_{n_2} \otimes A_4^T + A_4^T \otimes I_{n_2}$$

where $\otimes$ stands for the Kronecker product representation. For the Jacobian to be nonsingular the block diagonal elements J_{ii}, $i = 1, 2, 3$, have to be nonsingular. The matrix D_1 is a closed-loop matrix, and thus stable by the well-known property of the algebraic Riccati equation and by Assumption 9.5. By the same assumption, the matrix A_4 is stable, so that by the property of the Kronecker product (Lancaster and Tismenetsky, 1985), matrices J_{ii} are nonsingular. Thus, for ϵ small enough, the Jacobian is nonsingular.

In the next step, we have to show the convergence of the algorithm (9.127)-(9.130) and give an estimate of the rate of convergence. For $i = 0$, (9.123) and (9.127) imply

$$\left(E_1 - E_1^{(0)} \right) D_1 + D_1^T \left(E_1 - E_1^{(0)} \right) = \epsilon f_1 \left(E_1, E_2; \epsilon \right) \qquad (9.135)$$

Since D_1 is stable and E_1 and E_2 are bounded it follows that

$$\left\| E_1 - E_1^{(0)} \right\| = O\left(\epsilon \right) \qquad (9.136)$$

Similarly from (9.125) and (9.129) we have

$$\left(E_3 - E_3^{(0)} \right) A_4 + A_4^T \left(E_3 - E_3^{(0)} \right) = \epsilon f_3 \left(E_2, E_3; \epsilon \right) \qquad (9.137)$$

so that

$$\left\| E_3 - E_3^{(0)} \right\| = O\left(\epsilon \right) \qquad (9.138)$$

The use of the same arguments in (9.124) and (9.128) produces

$$\left\| E_2 - E_2^{(0)} \right\| = O\left(\epsilon \right) \qquad (9.139)$$

Continuing the same procedure and by induction, we conclude that

$$\left\| E_1 - E_1^{(i)} \right\| = O\left(\epsilon^{(i+1)} \right) \qquad (9.140)$$

$$\left\| E_2 - E_2^{(i)} \right\| = O\left(\epsilon^{(i+1)} \right) \qquad (9.141)$$

$$\left\| E_3 - E_3^{(i)} \right\| = O\left(\epsilon^{(i+1)} \right) \qquad (9.142)$$

with $i = 1, 2, \ldots$, which completes the proof of Theorem 9.5.

•

It is obvious that the proposed algorithm (9.127)-(9.130) can be implemented as a synchronous one (Bertsekas and Tsitsiklis, 1991). The study is underway to prove the convergence of the corresponding asynchronous algorithm.

9.5.1 Numerical Example

The following fourth-order linear-quadratic control problem example demonstrates the efficiency of the proposed method. Problem matrices are taken from (Shien and Tsay, 1982)

$$A = \begin{bmatrix} -0.75 & 0.28125 & 0.15 & 0.31875 \\ -0.25 & -1.15625 & -0.35 & -0.24375 \\ -0.75 & 0.28125 & -3.25 & -0.28125 \\ -0.25 & -1.15625 & 1.25 & -1.84375 \end{bmatrix}, \quad B = \begin{bmatrix} 1 & 0 \\ 0 & 1 \\ 1 & 0 \\ 0 & 1 \end{bmatrix}$$

$$y = Cx = \begin{bmatrix} 1.5 & 1.9375 & 0.5 & 0.0625 \\ -0.25 & 0.71875 & 0.25 & 0.28125 \end{bmatrix} x$$

$$Q = C^T C, \quad R = I_2, \quad \epsilon = 0.1$$

Simulation results are presented in Table 9.7. The optimal value for the criterion is $J_{opt} = 1.8222$. From Table 9.7 we can notice very good numerical behavior of the proposed algorithm consistent with the statement of Theorem 9.5.

Other examples of weakly coupled systems having strong coupling through the input matrix are: binary distillation column considered in (Bhattacharyya et al., 1983) and L-1011 fighter aircraft (Beale and Shafai, 1989). These control systems can be numerically decomposed and solved in terms of the reduced-order problems by choosing the state penalty matrix according to Assumption 9.5 — see next section.

The presented method is applicable to almost any linear control system with a prescribed degree of stability (Anderson and Moore, 1990), since in that case we are working with $A + \alpha I$ which is block diagonally dominant for α large enough. In some cases the overlapping idea of (Siljak, 1991) can be used to achieve the desired structure.

i	$J^{(i)} - J_{opt}$
1	0.16855 x 10^{-2}
2	0.66638 x 10^{-4}
3	0.28597 x 10^{-5}
4	0.22424 x 10^{-6}
5	0.14966 x 10^{-7}
6	0.10897 x 10^{-8}
7	0.47606 x 10^{-10}
8	0.33538 x 10^{-11}
9	0.33538 x 10^{-12}

Table 9.7: Errors in the criterion approximation per iteration

9.6 Case Studies

We demonstrate results of the previous section on two real world examples: L-1011 fighter aircraft and distillation column linear-quadratic optimal control problems.

9.6.1 Case Study 1: L-1011 Fighter Aircraft

A mathematical model of L-1011 fighter aircraft can be found in (Beale and Shafai, 1989). The problem matrices are given by

$$
A = \begin{bmatrix} 0 & 1 & 0 & 0 \\ 0 & -1.89 & 0.39 & -5.53 \\ 0 & -0.034 & -2.98 & 2.43 \\ 0.034 & -0.0011 & -0.99 & -0.21 \end{bmatrix}, \quad B = \begin{bmatrix} 0 & 0 \\ 0.36 & -1.6 \\ -0.95 & -0.032 \\ 0.03 & 0 \end{bmatrix}
$$

$$
Q = \begin{bmatrix} 2.313 & 2.727 & 0.688 & 0.023 \\ 2.727 & 4.271 & 1.148 & 0.323 \\ 0.688 & 1.148 & 0.313 & 0.102 \\ 0.023 & 0.323 & 0.102 & 0.083 \end{bmatrix}, \quad R = I_2
$$

This control system is decomposed into two subsystems, each of order two. Small parameter ϵ is chosen as $\epsilon = 0.3$. The optimal performance index is J_{opt} = 7.239. Simulation results for the performance criterion are presented in Table 9.8.

It can be seen that the obtained numerical results are consistent with the established analytical relationship.

i	$J^{(i)} - J_{opt}$
1	$0.1114 \times 10^{+1}$
2	0.8048×10^{-1}
3	0.1529×10^{-1}
4	0.3569×10^{-2}
5	0.4790×10^{-3}
6	0.6193×10^{-4}
7	0.1513×10^{-4}
8	0.2542×10^{-5}
9	0.2832×10^{-6}
10	0.5678×10^{-7}
11	0.1066×10^{-7}
12	0.1183×10^{-8}

Table 9.8: Difference between approximate and optimal criteria

9.6.2 Case Study 2: Distillation Column

Mathematical model of a binary distillation column with condenser, reboiler, and nine plates is given by (Bhattacharyya et al., 1983)

$$
A = \begin{bmatrix}
-0.991 & 0.529 & 0 & 0 & 0 & 0 & 0 & 0 \\
0.522 & -1.051 & 0.596 & 0 & 0 & 0 & 0 & 0 \\
0 & 0.522 & -1.118 & 0.596 & 0 & 0 & 0 & 0 \\
0 & 0 & 0.522 & -1.548 & 0.718 & 0 & 0 & 0 \\
0 & 0 & 0 & 0.922 & -1.640 & 0.799 & 0 & 0 \\
0 & 0 & 0 & 0 & 0.922 & -1.721 & 0.901 & 0 \\
0 & 0 & 0 & 0 & 0 & 0.922 & -1.823 & 1.021 \\
0 & 0 & 0 & 0 & 0 & 0 & 0.922 & -1.943
\end{bmatrix}
$$

$$B^T = 10^{-3} \begin{bmatrix} 3.84 & 4 & 37.6 & 3.08 & 2.36 & 2.88 & 3.08 & 3 \\ -2.88 & -3.04 & -2.80 & -2.32 & -3.32 & -3.82 & -4.12 & -3.96 \end{bmatrix}$$

$$Q = \begin{bmatrix} 1 & 0 & 0 & 0 & 0.5 & 0 & 0 & 0.1 \\ 0 & 1 & 0 & 0 & 0.1 & 0 & 0 & 0 \\ 0 & 0 & 1 & 0 & 0 & 0.5 & 0 & 0 \\ 0 & 0 & 0 & 1 & 0 & 0 & 0 & 0 \\ 0.5 & 0.1 & 0 & 0 & 0.1 & 0 & 0 & 0 \\ 0 & 0 & 0.5 & 0 & 0 & 0.1 & 0 & 0 \\ 0 & 0 & 0 & 0 & 0 & 0 & 0.1 & 0 \\ 0.1 & 0 & 0 & 0 & 0 & 0 & 0 & 0.1 \end{bmatrix}, \quad R = I_2, \quad \epsilon = 0.2$$

The optimal value of the performance criterion is $J_{opt} = 6.1656$. Obtained simulation results are presented in Table 9.9. The results are consistent with the statement of the corresponding theorem.

9.7 Notes

Results presented in this chapter are mostly based on the work of D. Skataric (Skataric et al., 1990, 1991; Skataric and Gajic, 1992; Skataric, 1992). The study of the quasi singularly perturbed and quasi weakly coupled systems is not complete. There are many other classes of the linear-quadratic optimal control problems with small parameters that can be decomposed into the reduced-order subproblems. The presented results, obtained in the continuous-time domain, can serve as a guideline. Their extension to the discrete-time domain is also an interesting research area.

i	$J_{app}^{(i)} - J_{opt}^{(i)}$
0	0..0186
1	0.0060
2	0.0020
3	7.2365×10^{-4}
4	2.6123×10^{-4}
5	9.5219×10^{-5}
6	3.4870×10^{-5}
7	1.2799×10^{-5}
8	4.7030×10^{-6}
9	1.7291×10^{-6}
10	6.3585×10^{-7}
11	2.3386×10^{-7}
12	8.6016×10^{-8}
13	3.1639×10^{-8}
14	1.1638×10^{-8}
15	4.2807×10^{-9}
16	1.5746×10^{-9}
17	5.7919×10^{-10}
18	2.1305×10^{-10}
19	7.8363×10^{-11}
20	2.8817×10^{-11}

Table 9.9: Difference between approximate and optimal criteria

Appendix 9.1

The operating points for these two machines are given as follows.

Machine No.1: P_1 = 170MW, Q_1 = 82 MWAr, V_1 = 15.75 kV.

Machine No.2: P_2 = 24.5 MW, Q_2 = $-$ 6MVAr, V_2 = 6.5 kV, $V_2 = 6.5/\underline{-4.5}°$. Machine and exciter data are presented in Table 9.10.

Unit no.	No.1	No.2
Synchronous machine		
Rated MVA	190	28
Rated kV	15.75	6.3
Xd(p.u.)	1.245	1
Xq(p.u.)	0.925	0.7
Xd'(p.u.)	0.373	0.42
Xad(p.u.)	1.145	0.85
Tdo'(s)	6.5	1.65
ra(p.u.)	0.00285	0.0107
Ta(s)	11.06	2.45
D(p.u.)	1	1
Exciter		
Te(s)	0.03	0.03
Ke(p.u.)	0.00185	0.00604

Table 9.10: Synchronous machine and fast exciter data

Chapter 10

Singularly Perturbed Weakly Coupled Linear Control Systems

10.1 Introduction

In mathematical models of many real physical systems small parameters appear. Two large classes of small parameter problems have been studied so far, extensively in the context of control theory: 1) singularly perturbed systems (Kokotovic et al., 1986; Gajic et al., 1990) and 2) weakly coupled systems (Gajic et al., 1990). However, motivated by the models of the real physical systems, we have found that many of them have both singularly perturbed and weakly coupled structures. Even more, the structure of many systems with slow-fast phenomena and weak coupling can not be put either in the standard singularly perturbed or standard weakly coupled forms. In this chapter, we study systems that are at the same time both singularly perturbed and weakly coupled.

A special class of singularly perturbed weakly coupled linear systems has been studied in the concept of multimodeling (Khalil and Kokotovic, 1978; Khalil, 1980b; Saksena and Cruz, 1981a, 1981b; Saksena and Basar, 1982; Saksena et al., 1983; Gajic and Khalil, 1986; Gajic, 1988; Zhuang and Gajic, 1991), where the weak coupling is allowed between fast variables only (see also, Ozguner, 1979). In this chapter, we will study the effect of weak coupling between slow and fast variables. The obtained solution will be given in terms of a ratio of two small parameters. Let ϵ_1 and ϵ_2 represent a small positive weak coupling and small positive singular perturbation parameters, respectively, then one can study any of the following three cases

$$1) \quad 0 < \; m \leq \frac{\epsilon_1}{\epsilon_2} \leq M < \infty,$$

$$2) \quad \frac{\epsilon_1}{\epsilon_2} \rightarrow 0, \quad 3) \; \frac{\epsilon_2}{\epsilon_1} \rightarrow 0 \qquad (10.1)$$

In the first structure, which is the subject of this chapter, the system is both singularly perturbed and weakly coupled. In the second structure, it is predominantly weakly coupled, and in the third one it is predominantly singularly perturbed; so that they can be studied by using the corresponding techniques derived in (Kokotovic et al., 1986; Gajic et al., 1990). Note that pure singularly perturbed systems involving many small parameters of the same magnitude have been studied under the name of "multiparameter singular perturbations" (for example, Khalil and Kokotovic 1979a, 1979b).

The approach taken in this chapter is in the spirit of the reduced-order fixed point iterations (Gajic et al., 1990), and the parallel synchronous algorithms (Bertsekas and Tsitsiklis, 1989, 1991).

The study of this chapter reveals one very important feature of this kind of systems displaying slow-fast phenomena. Namely, the stabilizability-detectability condition is imposed directly on the given subsystems, on the contrary to pure singularly perturbed systems where this condition has to be imposed on the slow subsystem matrices, which depend on the given problem matrices on a quite complicated manner.

The obtained results for singularly perturbed weakly coupled linear systems are extended in Section 10.4 to the so-called quasi singularly perturbed weakly coupled linear systems (Skataric, 1992). Several real world control problems are solved in order to demonstrate the efficiency of the proposed synchronous reduced-order parallel algorithms.

10.2 Singularly Perturbed Weakly Coupled Linear Control Systems

The singularly perturbed weakly coupled linear dynamical control system has the form consistent with both the singularly perturbed and weakly coupled systems

$$\begin{bmatrix} \dot{x}_1 \\ \dot{x}_2 \end{bmatrix} = \begin{bmatrix} A_1 & \epsilon_1 A_2 \\ \frac{\epsilon_1 A_3}{\epsilon_2} & \frac{A_4}{\epsilon_2} \end{bmatrix} \begin{bmatrix} x_1 \\ x_2 \end{bmatrix} + \begin{bmatrix} B_1 & \epsilon_1 B_2 \\ \frac{\epsilon_1 B_3}{\epsilon_2} & \frac{B_4}{\epsilon_2} \end{bmatrix} \begin{bmatrix} u_1 \\ u_2 \end{bmatrix} \qquad (10.2)$$

where $x_i \in \Re^{n_i}$, $i = 1, 2$, are state vectors, $u \in \Re^{m_i}$, $i = 1, 2$, are control inputs. This is a special class of linear dynamical control systems represented, in general, by

$$\dot{x} = Ax + Bu \qquad (10.3)$$

$$x = \begin{bmatrix} x_1 \\ x_2 \end{bmatrix}, \quad u = \begin{bmatrix} u_1 \\ u_2 \end{bmatrix}$$

$$A = \begin{bmatrix} A_1 & \epsilon_1 A_2 \\ \frac{\epsilon_1 A_3}{\epsilon_2} & \frac{A_4}{\epsilon_2} \end{bmatrix}, \quad B = \begin{bmatrix} B_1 & \epsilon_1 B_2 \\ \frac{\epsilon_1 B_3}{\epsilon_2} & \frac{B_4}{\epsilon_2} \end{bmatrix} \qquad (10.4)$$

A quadratic type functional to be minimized is associated with (10.2) in the form

$$J = \frac{1}{2} \int_0^\infty \left(x^T Q x + u^T R u \right) dt, \quad Q \geq 0, \quad R > 0 \qquad (10.5)$$

For the purpose of this chapter, we assume that the structures of the matrices Q and R are

$$Q = \begin{bmatrix} Q_1 & \epsilon_1 Q_2 \\ \epsilon_1 Q_2^T & Q_3 \end{bmatrix}, \quad R = \begin{bmatrix} R_1 & 0 \\ 0 & R_2 \end{bmatrix} \qquad (10.6)$$

which is consistent with both the singularly perturbed and weakly coupled penalty matrices used in (Kokotovic et al., 1986; Gajic et al., 1990). All problem matrices defined in (10.2)-(10.6) are constant and of appropriate dimensions.

The control structure defined by (10.2)-(10.6) has not been studied in the literature so far. Motivated by the results reported in (Gajic et al., 1990), and by the importance of the parallel and distributed computations (Bertsekas and Tsitsiklis, 1989, 1991), we will show that in this case one is able to design the optimal controllers by using the reduced-order parallel synchronous algorithms. Even more, the obtained results are applicable under milder conditions than for pure singularly perturbed linear-quadratic control problems. Namely, the stabilizability-detectability conditions are imposed directly on the subsystem matrices A_1, B_1, Q_1 and A_4, B_4, Q_3. For pure singularly perturbed systems the stabilizability-detectability condition is imposed on A_s, B_s, Q_s (see Section 2.2.2), which depend on a quite complicated manner on the original problem matrices. Thus, for pure singularly perturbed systems one is not able to test directly the required stabilizability-detectability conditions.

The optimal problem of minimizing (10.5) along trajectories of (10.2) has the very well-known solution given by

$$u_{opt} = -F_{opt}x = -R^{-1}B^T P x \qquad (10.7)$$

where P is the positive semidefinite stabilizing solution of the algebraic Riccati equation

$$A^T P + PA + Q - PSP = 0, \quad S = BR^{-1}B^T \tag{10.8}$$

For the development of the parallel reduced-order algorithms for solving (10.8), it is very important to discover the proper nature of the solution of (10.8) in terms of small parameters ϵ_1 and ϵ_2. By studying partitioned equations of (10.8), it is shown (Skataric, 1992) that the solution of (10.8) is properly scaled as

$$P = \begin{bmatrix} P_1 & \epsilon_1 \epsilon_2 P_2 \\ \epsilon_1 \epsilon_2 P_2^T & \sqrt{\epsilon_1 \epsilon_2} P_3 \end{bmatrix} \tag{10.9}$$

Partitioning (10.8) compatible to (10.4), (10.6), and (10.9), we get three matrix algebraic equations

$$0 = P_1 A_1 + A_1^T P_1 + Q_1 - P_1 S_1 P_1 + \epsilon_1^2 \left(A_3^T P_2^T + P_2 A_3 \right)$$
$$-\epsilon_1^4 P_2 S_3 P_2^T - \epsilon_1^2 \left(P_1 S_2 P_1 + P_2 Z^T P_1 + P_1 Z P_2^T + P_2 S_4 P_2^T \right) \tag{10.10}$$

$$P_2 A_4 + Q_2 + P_1 A_2 + \alpha \left(A_3^T P_3 - P_2 S_4 P_3 - P_1 Z P_3 \right) - \epsilon_1^2 \alpha P_2 S_3 P_3$$
$$+ \epsilon_2 \left(A_1^T P_2 - P_1 S_1 P_2 \right) - \epsilon_1^2 \epsilon_2 \left(P_1 S_2 P_2 + P_2 Z^T P_2 \right) = 0 \tag{10.11}$$

$$\alpha P_3 A_4 + \alpha A_4^T P_3 + Q_3 - \alpha^2 P_3 S_4 P_3$$
$$+ \epsilon_1^2 \epsilon_2 \left(P_2^T A_2 + A_2^T P_2 \right) - \epsilon_1^2 \epsilon_2 \alpha \left(P_3 Z^T P_2 + P_2^T Z P_3 \right) \tag{10.12}$$
$$- \epsilon_1^2 \epsilon_2^2 \left(P_2^T S_1 P_2 + \epsilon_1^2 P_2^T S_2 P_2 \right) - \epsilon_1^2 \alpha^2 P_3 S_3 P_3 = 0$$

where

$$S_1 = B_1 R_1^{-1} B_1^T, \quad S_2 = B_2 R_2^{-1} B_2^T, \quad S_3 = B_3 R_1^{-1} B_3^T$$
$$S_4 = B_4 R_2^{-1} B_4^T, \quad Z = B_1 R_1^{-1} B_3^T + B_2 R_2^{-1} B_4^T \tag{10.13}$$

with

$$\alpha = \left| \sqrt{\frac{\epsilon_1}{\epsilon_2}} \right| \tag{10.14}$$

Since ϵ_1 and ϵ_2 are small parameters, we can define $O(\epsilon)^1$ approximation of (10.10)-(10.12) as follows

$$P_1^{(0)} A_1 + A_1^T P_1^{(0)} + Q_1 - P_1^{(0)} S_1 P_1^{(0)} = 0 \tag{10.15}$$

---

[1]   In the case of two parameters, $O\left(\epsilon^k\right)$ stands for $C\epsilon^k$, where C is a bounded constant, k is any arbitrary constant and ϵ is any norm of a two-dimensional vector composed of ϵ_1 and ϵ_2.

$$P_2^{(0)}\left(A_4 - \alpha S_4 P_3^{(0)}\right) + Q_2 + P_1^{(0)} A_2 + \alpha \left(A_3^T P_3^{(0)} - P_1^{(0)} Z P_3^{(0)}\right) = 0 \tag{10.16}$$

$$\alpha P_3^{(0)} A_4 + \alpha A_4^T P_3^{(0)} + Q_3 - \alpha^2 P_3^{(0)} S_4 P_3^{(0)} = 0 \tag{10.17}$$

Corresponding solution of (10.8) is now given by

$$P^{(0)} = \begin{bmatrix} P_1^{(0)} & \epsilon_1 \epsilon_2 P_2^{(0)} \\ \epsilon_1 \epsilon_2 P_2^{(0)T} & \sqrt{\epsilon_1 \epsilon_2} P_3^{(0)} \end{bmatrix} = P + O(\epsilon) \tag{10.18}$$

The unique positive semidefinite stabilizing solution $P^{(0)}$, obtained from (10.15)-(10.17), exists under the assumption that the triples $\left(A_1, B_1, \sqrt{Q_1}\right)$ and $\left(A_4, \sqrt{\alpha} B_4, \sqrt{\frac{Q_3}{\alpha}}\right)$ are stabilizable-detectable. However, due to the structure of the controllability-observability matrices, we can eliminate the α-dependence so that we need the following assumption.

Assumption 10.1 The triples $\left(A_1, B_1, \sqrt{Q_1}\right)$ and $\left(A_4, B_4, \sqrt{Q_3}\right)$ are stabilizable-detectable.

$\triangle$

Defining the approximation errors as

$$P_j = P_j^{(0)} + \epsilon E_j, \quad j = 1, 2, 3; \quad \epsilon = |\sqrt{\epsilon_1 \epsilon_2}| \tag{10.19}$$

and using (10.19) in (10.10)-(10.12) and (10.15)-(10.17), we get the following error equations

$$E_1 D_1 + D_1^T E_1 = \epsilon E_1 S_1 E_1 - \epsilon_1 \alpha \left(P_2 A_3 + A_3^T P_2^T\right)$$
$$+ \epsilon_1 \alpha \left(P_1 S_2 P_1 + P_2 Z^T P_1 + P_1 Z P_2^T + P_2 S_4 P_2^T\right) + \epsilon_1^3 \alpha P_2 S_3 P_2^T \tag{10.20}$$

$$E_2 D_3 + E_1 D_{23} + \alpha D_{21}^T E_3 = \epsilon_1 \left(E_2 S_4 E_3 + E_1 Z E_3\right)$$
$$- \frac{1}{\alpha} \left(A_1 - S_1 P_1\right)^T P_2 + \epsilon_1 \epsilon \left(P_1 S_2 P_2 + P_2 Z^T P_2\right) + \epsilon_1 \alpha^2 P_2 S_3 P_3 \tag{10.21}$$

$$E_3 D_3 + D_3^T E_3 = \epsilon_1 E_3 S_4 E_3 + \epsilon_1 \alpha^2 P_3 S_3 P_3$$
$$-\epsilon_1 \epsilon_2 \left(A_2^T P_2 + P_2^T A_2 \right) + \epsilon_1 \epsilon_2^2 \left(P_2^T S_1 P_2 + \epsilon_1^2 P_2^T S_2 P_2 \right) \qquad (10.22)$$
$$+\epsilon_1 \epsilon_2 \alpha \left(P_3 Z^T P_2 + P_2^T Z P_3 \right)$$

where

$$D_1 = A_1 - S_1 P_1^{(0)}, \quad D_3 = A_4 - \alpha S_4 P_3^{(0)},$$
$$D_{23} = A_2 - \alpha Z P_3^{(0)}, \quad D_{21} = A_3 - S_4 P_2^{(0)^T} - Z^T P_1^{(0)} \qquad (10.23)$$

Note that all nonlinear terms and all cross coupling terms in (10.20)-(10.22) are multiplied by small parameters. This fact suggests the following reduced-order parallel synchronous algorithm for solving (10.20)-(10.22), (Gajic and Skataric, 1991).
Algorithm 10.1:

$$E_1^{(i+1)} D_1 + D_1^T E_1^{(i+1)} = \varepsilon E_1^{(i)} S_1 E_1^{(i)} + \epsilon_1 \alpha P_2^{(i)} S_4 P_2^{(i)^T}$$
$$-\epsilon_1 \alpha \left(P_2^{(i)} A_3 + A_3^T P_2^{(i)^T} \right) + \epsilon_1^3 \alpha P_2^{(i)} S_3 P_2^{(i)^T} \qquad (10.24)$$
$$+\epsilon_1 \alpha \left(P_1^{(i)} S_2 P_1^{(i)} + P_2^{(i)} Z^T P_1^{(i)} + P_1^{(i)} Z P_2^{(i)^T} \right)$$

$$E_2^{(i+1)} D_3 + E_1^{(i+1)} D_{23} + \alpha D_{21}^T E_3^{(i+1)}$$
$$= \epsilon_1 \left(E_2^{(i)} S_4 E_3^{(i+1)} + E_1^{(i+1)} Z E_3^{(i+1)} \right) + \epsilon_1 \alpha^2 P_2^{(i)} S_3 P_3^{(i+1)}$$
$$+\epsilon_1 \varepsilon \left(P_1^{(i+1)} S_2 P_2^{(i)} + P_2^{(i)} Z^T P_2^{(i)} \right) - \frac{1}{\alpha} \left(A_1 - S_1 P_1^{(i+1)} \right)^T P_2^{(i)} \qquad (10.25)$$

$$E_3^{(i+1)} D_3 + D_3^T E_3^{(i+1)} = \epsilon_1 E_3^{(i)} S_4 E_3^{(i)} + \epsilon_1 \alpha^2 P_3^{(i)} S_3 P_3^{(i)}$$
$$-\epsilon_1 \epsilon_2 \left(A_2^T P_2^{(i)} + P_2^{(i)^T} A_2 \right) + \epsilon_1 \epsilon_2 \alpha \left(P_3^{(i)} Z^T P_2^{(i)} + P_2^{(i)^T} Z P_3^{(i)} \right)$$
$$+\epsilon_1 \epsilon_2^2 \left(P_2^{(i)^T} S_1 P_2^{(i)} + \epsilon_1^2 P_2^{(i)^T} S_2 P_2^{(i)} \right) \qquad (10.26)$$

where

$$P_j^{(i)} = P_j^{(0)} + \varepsilon E_j^{(i)}, \quad j = 1, 2, 3, \quad i = 0, 1, 2, 3 \ldots \qquad (10.27)$$

with initial conditions

$$E_1^{(0)} = 0, \quad E_2^{(0)} = 0, \quad E_3^{(0)} = 0 \qquad (10.28)$$

$\triangle$

Note that we have parallelized our algorithm by using the Gauss-Seidel iterations. The similar algorithm could have been derived by using the Jacobi-type iterations, but that algorithm would be slower (Bertsekas and Tsitsiklis, 1989, 1991). Since equations (10.24)-(10.26) are completely decoupled, the solution of (10.20)-(10.22) can be obtained by using three processors working in parallel and exchanging intermediate results after each iteration. The work of these three processors has to be synchronized by a global clock.

The following theorem indicates features of the proposed algorithm (10.24)-(10.28).

Theorem 10.1 *Under conditions stated in Assumption 10.1, the algorithm (10.24)-(10.28) converges to the exact solution of the error term, and thus to the required solution P, with the rate of convergence of $O(\epsilon)$, that is*

$$\left\| E_j - E_j^{(i+1)} \right\| = O(\epsilon) \left\| E_j - E_j^{(i)} \right\| \tag{10.29}$$

or equivalently

$$\left\| E_j - E_j^{(i)} \right\| = O\left(\epsilon^{(i+1)}\right), \quad i = 0, 1, 2, ...; \quad j = 1, 2, 3 \tag{10.30}$$

◇

Proof: The proof of this theorem can be found in (Skataric, 1992).

●

The approximate optimal feedback gain $F^{(i)}$ for the problem under consideration is given by

$$F^{(i)} = -R^{-1} B^T P^{(i)} \tag{10.31}$$

where

$$P^{(i)} = \begin{bmatrix} P_1^{(i)} & \epsilon_1 \epsilon_2 P_2^{(i)} \\ \epsilon_1 \epsilon_2 P_2^{(i)T} & \sqrt{\epsilon_1 \epsilon_2} P_3^{(i)} \end{bmatrix} \tag{10.32}$$

The approximate criterion is obtained from

$$J_{app}^{(i)} = tr\left\{ V^{(i)} \right\} \tag{10.33}$$

where $V^{(i)}$ satisfies

$$\left(A - SP^{(i)}\right)^T V^{(i)} + V^{(i)} \left(A - SP^{(i)}\right) + Q + P^{(i)} S P^{(i)} = 0 \tag{10.34}$$

10.3 Case Studies

Many real control systems possess at the same time both the singularly perturbed and the weakly coupled forms. In this section, we present two of them. In Section 10.5, we will study additional two real control systems having the quasi singularly perturbed weakly coupled forms.

10.3.1 Case Study 1: A Model of Supported Beam

The mathematical model of a supported beam in the state space form is given by (Hsieh et al., 1989)

$$A = \begin{bmatrix} 0 & 1 & 0 & 0 \\ -1 & -0.01 & 0 & 0 \\ 0 & 0 & 0 & 1 \\ 0 & 0 & -16 & -0.04 \end{bmatrix}, \quad B = \begin{bmatrix} 0 & 0 \\ 0.5878 & -1 \\ 0 & 0 \\ 0.9511 & 2 \end{bmatrix}$$

Weighting matrices Q and R are chosen as identities. We have solved this problem for $n_1 = 2$ and $\epsilon_1 = \epsilon_2 = 0.1$. Simulation results for the performance criterion are presented in Table 10.1. It is intersting to point out that in this example the proposed algorithm converges despite the fact that in iterations 1, 2, and 3 the approximate solution for the algebraic Riccati equation has lost its positive semidefiniteness.

The corresponding MATLAB program is given in Appendix 10.1.

10.3.2 Case Study 2: A Satellite Control Problem

We demonstrate the result of Theorem 10.1 on a satellite control example from Section 7.4. All of the problem matrices and the required parameters are given in Section 7.4. Simulation results for the approximate criterion are presented in Table 10.2. I can be seen that the simulation results are consistent with the statement of Theorem 10.1.

10.4 Quasi Singularly Perturbed Weakly Coupled Linear Control Systems

The quasi singularly perturbed weakly coupled structures are induced by the system matrix having singularly perturbed weakly coupled form as in (10.4) and by the control input matrix having one of the nonstandard structures, namely

$$\dot{x} = Ax + Bu \tag{10.35}$$

i	J_{app}
0	12.0616
1	*
2	*
3	*
4	11.3799
5	11.3372
6	11.3249
7	11.3248
9	11.3247
10	11.3246
13	11.3245
16	11.3244
20	11.3243
29	11.3242 = J_{opt}

Table 10.1: Approximate values for criterion
* = solution of the Riccati equation indefinite

with

$$i) \ B = \begin{bmatrix} B_1 \\ 0 \end{bmatrix}, \quad ii) \ B = \begin{bmatrix} 0 \\ B_2 \end{bmatrix}, \quad iii) \ B = \begin{bmatrix} B_1 \\ B_2 \end{bmatrix} \quad (10.36)$$

This implies the existence of only one control agent. All of these three structures appear in the real control systems (see case studies in Section 10.5).

Case i) It can be shown that the matrix P preserves the structure given by (10.9). With the system matrix A given by (10.4) and the newly defined matrix S as

$$S = \begin{bmatrix} Z_1 & 0 \\ 0 & 0 \end{bmatrix}, \quad Z_1 = B_1 R^{-1} B_1^T \quad (10.37)$$

the algebraic Riccati equation (10.8) is partitioned according to

$$P_1 A_1 + A_1^T P_1 + Q_1 - P_1 Z_1 P_1 + \epsilon_1^2 \left(A_3^T P_2^T + P_2 A_3 \right) = 0 \quad (10.38)$$

301

i	J_{app}
0	13.8580
1	12.0499
2	11.7434
3	11.4806
4	11.4683
5	11.4640
6	11.4557
7	11.4533
8	11.4528
9	11.4529
10	11.4528
11	11.4527
	11.4527 = J_{opt}
i	$J_{app} - J_{opt}$
12	5.7729 x 10^{-6}
16	3.3605 x 10^{-7}
20	1.0437 x 10^{-8}
25	1.0225 x 10^{-10}
30	1.0072 x 10^{-12}

Table 10.2: Approximate and optimal values for criterion

$$P_2 A_4 + Q_2 + P_1 A_2 + \alpha A_3^T P_3 + \epsilon_2 \left(A_1^T P_2 - P_1 Z_1 P_2 \right) = 0 \quad (10.39)$$

$$\alpha P_3 A_4 + \alpha A_4^T P_3 + Q_3 + \epsilon_1^2 \epsilon_2 \left(P_2^T A_2 + A_2^T P_2 \right) - \varepsilon^4 P_2^T Z_1 P_2 = 0 \quad (10.40)$$

Following the same arguments as in Section 10.2, the reduced-order solution is obtained as

$$P_1^{(0)} A_1 + A_1^T P_1^{(0)} + Q_1 - P_1^{(0)} Z_1 P_1^{(0)} = 0 \qquad (10.41)$$

$$P_2^{(0)} A_4 + Q_2 + P_1^{(0)} A_2 + \alpha A_3^T P_3^{(0)} = 0 \qquad (10.42)$$

$$\alpha P_3^{(0)} A_4 + \alpha A_4^T P_3^{(0)} + Q_3 = 0 \qquad (10.43)$$

The unique solutions of (10.43)-(10.45) exist under the following assumption.

Assumption 10.2 The triple $(A_1, B_1, \sqrt{Q_1})$ is stabilizable-detectable and the matrix A_4 is stable.

$\triangle$

Defining the approximation errors as in (10.19), we get the following expressions for the error equations.

$$E_1 D_1 + D_1^T E_1 = \varepsilon E_1 Z_1 E_1 - \epsilon_1 \alpha \left(P_2 A_3 + A_3^T P_2^T \right) \qquad (10.44)$$

$$E_2 A_4 + E_1 A_2 + \alpha A_3^T E_3 = -\frac{1}{\alpha} (A_1 - Z_1 P_1)^T P_2 \qquad (10.45)$$

$$E_3 A_4 + A_4^T E_3 = -\varepsilon^2 \left(A_2^T P_2 + P_2^T A_2 \right) + \epsilon_1 \epsilon_2^2 P_2^T Z_1 P_2 \qquad (10.46)$$

where $D_1 = A_1 - Z_1 P_1^{(0)}$.

The following parallel synchronous algorithm is proposed for solving the error equations (10.44)-(10.46).

Algorithm 10.2:

$$E_1^{(i+1)} D_1 + D_1^T E_1^{(i+1)} = \epsilon_1 E_1^{(i)} Z_1 E_1^{(i)} - \epsilon_1 \alpha \left(P_2^{(i)} A_3 + A_3^T P_2^{(i)^T} \right) \qquad (10.47)$$

$$E_2^{(i+1)} A_4 + E_1^{(i+1)} A_2 + \alpha A_3^T E_3^{(i+1)} = -\frac{1}{\alpha} \left(A_1 - Z_1 P_1^{(i+1)} \right)^T P_2^{(i)} \qquad (10.48)$$

$$E_3^{(i+1)} A_4 + A_4^T E_3^{(i+1)}$$
$$= -\varepsilon^2 \left(A_2^T P_2^{(i)} + P_2^{(i)^T} A_2 \right) + \epsilon_1 \epsilon_2^2 P_2^{(i)^T} Z_1 P_2^{(i)} \qquad (10.49)$$

with $P_j^{(i)}$ and the initial conditions given in (10.27) and (10.28).

$\triangle$

The following theorem summarizes features of the algorithm (10.47)-(10.49).

Theorem 10.2 *Under conditions stated in Assumption 10.2, the algorithm (10.47)-(10.49) converges to the exact solution of the error term, and thus to the required solution P, with the rate of convergence of $O(\epsilon)$, that is*

$$\left\| E_j - E_j^{(i+1)} \right\| = O(\epsilon) \left\| E_j - E_j^{(i)} \right\| \tag{10.50}$$

or equivalently

$$\left\| E_j - E_j^{(i)} \right\| = O\left(\epsilon^{(i+1)}\right), \quad i = 0, 1, 2, \ldots; \quad j = 1, 2, 3 \tag{10.51}$$

◇

The proof of Theorem 10.2 follows the ideas of the proof of Theorem 10.1, and thus is omitted.

Case ii) It can be shown that in this case the matrix P also preserves the structure given by (10.9). With the system matrix A given by (10.4) and the newly defined matrix S given by

$$S = \begin{bmatrix} 0 & 0 \\ 0 & Z_4 \end{bmatrix}, \quad Z_4 = B_2 R^{-1} B_2^T \tag{10.52}$$

the algebraic Riccati equation (10.8) is partitioned as

$$P_1 A_1 + A_1^T P_1 + Q_1 - \epsilon^4 P_2 Z_4 P_2^T + \epsilon_1^2 \left(A_3^T P_2^T + P_2 A_3 \right) = 0 \tag{10.53}$$

$$P_2 A_4 + Q_2 + P_1 A_2 + \alpha A_3^T P_3 - \epsilon_2 \epsilon P_2 Z_4 P_3 + \epsilon_2 A_1^T P_2 = 0 \tag{10.54}$$

$$\alpha P_3 A_4 + \alpha A_4^T P_3 + Q_3 - \epsilon^2 P_3 Z_4 P_3 + \epsilon_1^2 \epsilon_2 \left(P_2^T A_2 + A_2^T P_2 \right) = 0 \tag{10.55}$$

The reduced-order solution is obtained as

$$P_1^{(0)} A_1 + A_1^T P_1^{(0)} + Q_1 = 0 \tag{10.56}$$

$$P_2^{(0)} A_4 + Q_2 + P_1^{(0)} A_2 + \alpha A_3^T P_3^{(0)} = 0 \tag{10.57}$$

$$\alpha P_3^{(0)} A_4 + \alpha A_4^T P_3^{(0)} + Q_3 = 0 \tag{10.58}$$

The unique solutions of (10.56)-(10.58) exist under the following assumption.

SINGULARLY PERTURBED WEAKLY COUPLED SYSTEMS

Assumption 10.3 The matrices A_1 and A_4 are stable.

△

Defining the approximation errors as in (10.19), we get the following expressions for the error equations

$$E_1 A_1 + A_1^T E_1 = -\epsilon_1 \alpha \left(P_2 A_3 + A_3^T P_2^T \right) + \varepsilon^3 P_2 Z_4 P_2^T \qquad (10.59)$$

$$E_2 A_4 + E_1 A_2 + \alpha A_3^T E_3 = \epsilon_2 P_2 Z_4 P_3 - \frac{1}{\alpha} A_1^T P_2 \qquad (10.60)$$

$$E_3 A_4 + A_4^T E_3 = -\varepsilon^2 \left(A_2^T P_2 + P_2^T A_2 \right) + \epsilon_2 E_3 Z_4 E_3 \qquad (10.61)$$

The following parallel synchronous algorithm is proposed for solving the error equations.
Algorithm 10.3:

$$E_1^{(i+1)} A_1 + A_1^T E_1^{(i+1)}$$
$$= \varepsilon^3 P_2^{(i)} Z_4 P_2^{(i)^T} - \epsilon_1 \alpha \left(P_2^{(i)} A_3 + A_3^T P_2^{(i)^T} \right) \qquad (10.62)$$

$$E_2^{(i+1)} A_4 + E_1^{(i+1)} A_2 + \alpha A_3 E_3^{(i+1)}$$
$$= \epsilon_2 P_2^{(i)} Z_4 P_3^{(i+1)} - \frac{1}{\alpha} A_1^T P_2^{(i)} \qquad (10.63)$$

$$E_3^{(i+1)} A_4 + A_4^T E_3^{(i+1)} = -\varepsilon^2 \left(A_2^T P_2^{(i)} + P_2^{(i)^T} A_2 \right) + \epsilon_2 P_3^{(i)} Z_4 P_3^{(i)} \qquad (10.64)$$

with $P_j^{(i)}$ and the initial conditions given in (10.27) and (10.28).

△

The following theorem summarizes the features of the algorithm (10.62)-(10.64).
Theorem 10.3 *Under conditions stated in Assumption 10.3, the algorithm (10.62)-(10.64) converges to the exact solution of the error term, and thus to the required solution P, with the rate of convergence of $O(\epsilon)$, that is*

$$\left\| E_j - E_j^{(i+1)} \right\| = O(\epsilon) \left\| E_j - E_j^{(i)} \right\| \qquad (10.65)$$

or equivalently

$$\left\| E_j - E_j^{(i)} \right\| = O\left(\epsilon^{(i+1)} \right), \quad i = 0,1,2,...; \quad j = 1,2,3 \qquad (10.66)$$

◇

The proof of this theorem is omitted for the reason explained before.

305

Case iii) It can be shown that the matrix P preserves the structure given by (10.9). With the system matrix A given by (10.4) and the newly defined matrix S as

$$S = \begin{bmatrix} Z_1 & Z_2 \\ Z_2^T & Z_4 \end{bmatrix}, \quad Z_2 = B_1 R^{-1} B_2^T \tag{10.67}$$

the algebraic Riccati equation (10.8) is partitioned according to

$$\begin{aligned} &P_1 A_1 + A_1^T P_1 + Q_1 - P_1 Z_1 P_1 + \epsilon_1^2 \left(A_3^T P_2^T + P_2 A_3 \right) \\ &- \varepsilon^2 \left(P_2 Z_2^T P_1 + P_1 Z_2 P_2^T \right) - \varepsilon^4 P_2 Z_4 P_2^T = 0 \end{aligned} \tag{10.68}$$

$$\begin{aligned} &P_2 A_4 + Q_2 + P_1 A_2 + \alpha \left(A_3^T P_3 - \epsilon_2^2 P_2 Z_4 P_3 \right) - \frac{1}{\alpha} P_1 Z_2 P_3 \\ &+ \epsilon_2 \left(A_1^T P_2 - P_1 Z_1 P_2 \right) - \epsilon_1 \epsilon_2^2 P_2 Z_2^T P_2 = 0 \end{aligned} \tag{10.69}$$

$$\begin{aligned} &\alpha P_3 A_4 + \alpha A_4^T P_3 + Q_3 \\ &- \varepsilon^2 P_3 Z_4 P_3 + \epsilon_1^2 \epsilon_2 \left(P_2^T A_2 + A_2^T P_2 \right) \\ &- \varepsilon^3 \left(P_3 Z_2^T P_2 + P_2^T Z_2 P_3 \right) - \varepsilon^4 P_2^T Z_1 P_2 = 0 \end{aligned} \tag{10.70}$$

Since ϵ_1 and ϵ_2 are small parameters, we can define $O(\epsilon)$ approximation of (10.68)-(10.70) as follows

$$P_1^{(0)} A_1 + A_1^T P_1^{(0)} + Q_1 - P_1^{(0)} Z_1 P_1^{(0)} = 0 \tag{10.71}$$

$$P_2^{(0)} A_4 + Q_2 + P_1^{(0)} A_2 + \alpha A_3^T P_3^{(0)} - \frac{1}{\alpha} P_1^{(0)} Z_2 P_3^{(0)} = 0 \tag{10.72}$$

$$\alpha P_3^{(0)} A_4 + \alpha A_4^T P_3^{(0)} + Q_3 = 0 \tag{10.73}$$

The unique positive semidefinite stabilizing solution $P^{(0)}$, obtained from (10.71)-(10.73), exists under Assumption 10.2.

Defining the approximation errors as before, and using the same logic, we get the following error equations

$$\begin{aligned} &E_1 D_1 + D_1^T E_1 = \varepsilon E_1 Z_1 E_1 - \epsilon_1 \alpha \left(P_2 A_3 + A_3^T P_2^T \right) \\ &+ \varepsilon \left(P_2 Z_2^T P_1 + P_1 Z_2 P_2^T \right) + \varepsilon^3 P_2 Z_4 P_2^T \end{aligned} \tag{10.74}$$

$$\begin{aligned} &E_2 A_4 + E_1 A_2 + \alpha A_3^T E_3 = \epsilon_2 E_1 Z_2 E_3 - \frac{1}{\alpha} A_1^T P_2 + \epsilon_2 P_2 Z_4 P_3 \\ &+ \epsilon_2^2 \alpha P_2 Z_2^T P_2 + \frac{1}{\alpha} P_1 Z_1 P_2 + \frac{1}{\alpha} \left(E_1 Z_2 P_3^{(0)} + P_1^{(0)} Z_2 E_3 \right) \end{aligned} \tag{10.75}$$

$$E_3 A_4 + A_4^T E_3 = -\varepsilon^2 \left(A_2^T P_2 + P_2^T A_2 \right) + \epsilon_2 P_3 Z_4 P_3$$
$$+ \epsilon_2^2 \alpha \left(P_3 Z_2^T P_2 + P_2^T Z_2 P_3 \right) + \epsilon_1 \epsilon_2^2 P_2^T Z_1 P_2 \qquad (10.76)$$

Let us propose the following reduced-order parallel synchronous algorithm for solving (10.74)-(10.76).
Algorithm 10.4:

$$E_1^{(i+1)} \mathbf{D}_1 + \mathbf{D}_1^T E_1^{(i+1)} = \varepsilon E_1^{(i)} Z_1 E_1^{(i)} - \epsilon_1 \alpha \left(P_2^{(i)} A_3 + A_3^T P_2^{(i)^T} \right)$$
$$+ \varepsilon^3 P_2^{(i)} Z_4 P_2^{(i)^T} + \varepsilon \left(P_2^{(i)} Z_2^T P_1^{(i)} + P_1^{(i)} Z_2 P_2^{(i)^T} \right) \qquad (10.77)$$

$$E_2^{(i+1)} A_4 + E_1^{(i+1)} A_2 + \alpha A_3^T E_3^{(i+1)}$$
$$= \epsilon_2 E_1^{(i+1)} Z_2 E_3^{(i+1)} - \frac{1}{\alpha} A_1^T P_2^{(i)} + \epsilon_2 P_2^{(i)} Z_4 P_3^{(i+1)} + \epsilon_2^2 \alpha P_2^{(i)} Z_2^T P_2^{(i)}$$
$$+ \frac{1}{\alpha} P_1^{(i+1)} Z_1 P_2^{(i)} + \frac{1}{\alpha} \left(E_1^{(i+1)} Z_2 P_3^{(0)} + P_1^{(0)} Z_2 E_3^{(i+1)} \right) \qquad (10.78)$$

$$E_3^{(i+1)} A_4 + A_4^T E_3^{(i+1)} = -\varepsilon^2 \left(A_2^T P_2^{(i)} + P_2^{(i)^T} A_2 \right) + \epsilon_2 P_3^{(i)} Z_4 P_3^{(i)}$$
$$+ \epsilon_2^2 \alpha \left(P_3^{(i)} Z_2^T P_2^{(i)} + P_2^{(i)^T} Z_2 P_3^{(i)} \right) + \epsilon_1 \epsilon_2^2 P_2^{(i)^T} Z_1 P_2^{(i)} \qquad (10.79)$$

with $P_j^{(i)}$ and the initial conditions given in (10.27) and (10.28).

$$\triangle$$

The following theorem indicates the features of the proposed algorithm (10.77)-(10.79).

Theorem 10.4 *Under conditions stated in Assumption 10.2, the algorithm (10.77)-(10.79) converges to the exact solution of the error term, and thus to the required solution P, with the rate of convergence of $O(\epsilon)$, that is*

$$\left\| E_j - E_j^{(i+1)} \right\| = O(\epsilon) \left\| E_j - E_j^{(i)} \right\| \qquad (10.80)$$

or equivalently

$$\left\| E_j - E_j^{(i)} \right\| = O\left(\epsilon^{(i+1)} \right), \quad i = 0, 1, 2, ...; \quad j = 1, 2, 3 \qquad (10.81)$$

$$\diamond$$

In the next section, we present several case studies of quasi singularly perturbed weakly coupled systems.

10.5 Case Studies

Case Study 3 — Case i):

The design of turbine governors of the power system considered in (Arnautovic and Skataric, 1991) is represented by the state space model of the form

$$
A = \begin{bmatrix}
-0.71 & 0 & 0 & 0 & 0 \\
0 & -2 & 0 & 0 & 0 \\
0.61 & 1.28 & -1.46 & 0.566 & 0 \\
-0.18 & -0.37 & 0.56 & -0.594 & -0.23 \\
0 & 0 & 0 & 314.16 & 0
\end{bmatrix}, \quad B = \begin{bmatrix}
0.71 \\
2 \\
0 \\
0 \\
0
\end{bmatrix}
$$

This system is partitioned with $n_1 = 3$. The penalty matrices are chosen as $Q = 0.1 \times I_5$, $R = 1$. Obtained results are presented in Table 10.3.

Case Study 4 — Case iii):

Consider the fluid catalytic cracker from Section 4.3. The matrices A and B, given in Section 4.3, are partitioned with $n_1 = 3$. The penalty matrices are chosen as $Q = 10 \times I_5$, $R = I_2$. Obtained results are presented in Table 10.4.

In both case studies we have seen very good convergence properties of the proposed parallel reduced-order algorithms.

Exercise 10.1: Derive the reduced-order parallel algorithm for the optimal control of the standard Draper/RPL satellite considered in (Keel and Bhattacharyya, 1990).

$\triangle$

10.6 Conclusion

In this chapter, we have shown that the large scale systems containing small parameters in the sense of singular perturbations and weak coupling are inherently parallel in nature, and thus, very well suited for parallel and distributed computation. Corresponding parallel synchronous algorithms are developed. The extension of these results to the asynchronous parallel algorithms is under way. It is well known that the asynchronous algorithms generate the required solution faster than the synchronous ones, but they have serious problems with convergence.

i	J_{app}
0	*
1	107.4983
2	83.9686
3	79.5014
4	79.4011
5	77.6671
6	77.0594
7	77.0955
8	76.9988
9	76.9211
10	76.9153
11	76.9157
12	76.9090
13	76.9059
14	76.9061
15	76.9059
16	76.9054
17	76.9053
18	76.9053
19	76.9052
20	$76.9052 = J_{opt}$

Table 10.3: Approximate and optimal values for criterion
$* = P^{(0)}$ is indefinite

i	J_{app}
0	2.1738
1	1.7840
2	1.7747
3	1.7463
4	1.7462
5	1.7422
6	1.7419
7	1.7415
8	1.7414
9	1.7413
10	1.7413
11	1.7412
	$1.7412 = J_{opt}$
i	$J_{app} - J_{opt}$
12	2.5376×10^{-5}
15	6.0143×10^{-6}
20	6.2229×10^{-7}
25	6.2168×10^{-8}

Table 10.4: Approximate and optimal values for criterion

Appendix 10.1

MATLAB program for parallel algorithm for solving the singularly perturbed weakly coupled linear-quadratic optimal control problem for the example presented in Case Study 10.3.

```
A=[0 1 0 0; -1 -0.01 0 0; 0 0 0 1; 0 0 -16 -0.04];
B=[0 0; 0.5878 -1; 0 0; 0.9511 2];
N1=2;
N=4;
M1=1;
M=1;
EPS1=0.1;
EPS2=0.1;
Q=eye(N);
R=eye(M);
S=B*inv(R)*B';
ALF=sqrt(EPS1/EPS2);
EPS=sqrt(EPS1*EPS2);
[K,P]=lqr(A,B,Q,R);
JOPT=trace(P);
A1=A(1:N1,1:N1);
A2=A(1:N1,N1+1:N)/EPS1;
A3=A(N1+1:N,1:N1)/ALF^2;
A4=A(N1+1:N,N1+1:N)*EPS2;
B1=B(1:N1,1:M1);
B2=B(1:N1,M1+1:M)/EPS1;
B3=B(N1+1:N,1:M1)/ALF^2;
B4=B(N1+1:N,M1+1:M)*EPS2;
Q1=Q(1:N1,1:N1);
Q2=Q(1:N1,N1+1:N)/EPS1;
Q3=Q(N1+1:N,N1+1:N);
R1=R(1:M1,1:M1);
R2=R(M1+1:M,M1+1:M);
S1=B*inv(R1)*B1';
S2=B2*inv(R2)*B2';
S3=B3*inv(R1)*B3';
S4=B4*inv(R2)*B4';
Z=B1*inv(R1)*B3'+B2*inv(R2)*B4';
[K1,P1]=lqr(A1,B1,Q1,R1);
[K3,P3]=lqr(ALF*A4,ALF*B4,Q3,R2);
P2=(-Q2-P1*A2-ALF*(A3'*P3-P1*Z*P3))*inv(A4-ALF*S4*P3);
P0=[P1 EPS^2*P2; EPS^2*P2' EPS*P3];
P0=P0
```

```
V0=lyap(A'-P0*S,Q+P0*S*P0);
J0=trace(V0);
J0=J0
D1=A1-S1*P1;
D3=A4-ALF*S4*P3;
D23=A2-ALF*Z*P3;
D21=A3-S4*P2'-Z'*P1;
E1=0*eye(N1);
E3=0*eye(N-N1);
E2=zeros(N1,N-N1);
P11=P1+EPS*E1;
P22=P2+EPS*E2;
P33=P3+EPS*E3;
for i = 1:30
QE1=EPS*E1*S1*E1+EPS1*ALF*P22*S4*P22';
QE1=QE1-EPS1*ALF*(P22*A3+A3'*P22');
QE1=QE1+EPS1^3*ALF*P22*S3*P22'
QE1=QE1+EPS1*ALF*(P11*S2*P11+P22*Z'*P11+P11*Z*P22');
E1=lyap(D1',-QE1);
P11=P1+EPS*E1;
QE3=EPS1*E3*S4*E3-EPS^2*(A2'*P22+P22'*A2);
QE3=QE3+EPS1*ALF^2*P33*S3*P33;
QE3=QE3+EPS^2*ALF*(P33*Z'*P22+P22'*Z*P33);
QE3=QE3+EPS1*EPS2^2*(P22'*S1*P22+EPS1^2*P22'*S2*P22);
E3=lyap(D3',-QE3);
P33=P3+EPS*E3;
QE2=EPS1*(E2*S4*E3+E1*Z*E3);
QE2=QE2+EPS1*EPS*(P11*S2*P22+P22*Z'*P22);
QE2=QE2+EPS1*ALF^2*P22*S3*P33-(A1-S1*P11)'*P22/ALF;
E2=(QE2-E1*D23-ALF*D21'*E3)*inv(D3);
P22=P2+EPS*E2;
PI=[P11 EPS^2*P22; EPS^2*P22' EPS*P33];
PI=PI
VI=lyap(A'-PI*S',Q+PI*S*PI);
JI=trace(VI)
DELT=JI-JOPT
i=i
pause
end
```

Chapter 11

Stochastic Output Feedback of Discrete Systems

11.1. Introduction

The problem of designing optimal controllers for linear systems with a limited number of output measurements available for control implementation has been an area of active research for many years, for example (Levine and Athans, 1970; Ermer and Vandelinde, 1973; Mendel, 1974; Kurtaran, 1975; Halyo and Broussard, 1981; Shapiro et al., 1981; Harkara et al., 1989; Gajic et al., 1990; Qureshi et al., 1992). The problem is defined as one in which the design engineer does not have a full set of state variables directly available for feedback purposes. The control engineers in such cases have two options: either to build the Kalman filter (or Luenberger observer) or to use the output feedback control. Very often it is not desirable to feedback all state variables in a complex system such as an aircraft. The design of the Kalman filter requires the dynamical system of the same order as the system under consideration. That might be costly. The output feedback control as the other alternative is much more convenient from the implementation point of view.

This chapter develops the recursive reduced-order parallel algorithm for the solution of the static output feedback control problem of the quasi weakly coupled (Skataric et al., 1990) discrete stochastic linear systems (Hogan and Gajic, 1992) and singularly perturbed discrete stochastic systems (Qureshi et al., 1992).

A discrete stochastic linear system is given by

$$x(k+1) = Ax(k) + Bu(k) + Gw(k) \qquad (11.1)$$

$$y(k) = Cx(k) + v(k) \tag{11.2}$$

where $x(k) \in \Re^n$ is the state vector, $u(k) \in \Re^m$ is the control input, $y(k) \in \Re^r$ is the measured output, $w(k) \in \Re^s$ and $v(k) \in \Re^r$ are stationary uncorrelated zero-mean Gaussian white noise stochastic processes with intensities $W > 0$ and $V > 0$, respectively. The matrices $A, B, G,$ and C are constant matrices of compatible dimensions.

With (11.1)-(11.2), consider the performance criterion

$$J = E\left\{ \sum_{k=0}^{\infty} \left[x(k)^T Q x(k) + u^T(k) R u(k) \right] \right\} \tag{11.3}$$

with positive definite R and positive semidefinite Q, which has to be minimized. In addition, the control input is constrained to

$$u(k) = Fy(k) = FCx(k) + Fv(k) \tag{11.4}$$

The optimal solution to this control problem has been obtained in terms of high-order nonlinear matrix algebraic equations, (Halyo and Broussard, 1981). The optimal feedback gain is given by

$$F = -\left(R + B^T L B\right)^{-1} B^T L A P C^T \left(C P C^T + V\right)^{-1} \tag{11.5}$$

where P and L satisfy

$$P = (A + BFC) P (A + BFC)^T + BFVF^T B^T + GWG^T \tag{11.6}$$

$$L = (A + BFC)^T L (A + BFC) + C^T F^T R F C + Q \tag{11.7}$$

The average value of the optimal performance criterion is given by

$$J = tr\left[(Q + C^T F^T R F C) P \right] + tr\left[F^T R F V \right] \tag{11.8}$$

It is shown (Halyo and Broussard, 1981) that following algorithm proposed for the numerical solution of (11.5)-(11.7) converges to a local minimum under nonrestrictive conditions.
Algorithm 11.1:

$$choose \ F \ such \ that \ A + BF^{(0)}C \ is \ stable \tag{11.9}$$

314

and solve iteratively for $i = 0, 1, 2, \ldots$

$$P^{(i+1)} = \left(A + BF^{(i)}C\right) P^{(i+1)} \left(A + BF^{(i)}C\right)^T$$
$$+ BF^{(i)}VF^{(i)^T}B^T + GWG^T \qquad (11.10)$$

$$L^{(i+1)} = \left(A + BF^{(i)}C\right)^T L^{(i+1)} \left(A + BF^{(i)}C\right) + C^T F^{(i)^T} RF^{(i)}C + Q$$
$$(11.11)$$

$$F^{(i+1)}_{new} = - \left(R + B^T L^{(i+1)}B\right)^{-1} B^T L^{(i+1)}AP^{(i+1)}C^T$$
$$\times \left(CP^{(i+1)}C^T + V\right)^{-1} \qquad (11.12)$$

$$F^{(i+1)} = F^{(i)} + \alpha_i \left(F^{(i+1)}_{new} - F^{(i)}\right) \qquad (11.13)$$

$$\triangle$$

The parameter $\alpha_i \in (0, 1]$ is chosen at each iteration to ensure that the minimum is not overshot. That is

$$J_{i+1} = tr \left[\left(Q + C^T F^{(i+1)^T}_{new} RF^{(i+1)}_{new}C\right) P^{(i+1)}\right]$$
$$+ tr \left[F^{(i+1)^T}_{new} RF^{(i+1)}_{new}V\right]$$

$$< J_i = tr \left[\left(Q + C^T F^{(i)^T}_{new} RF^{(i)}_{new}C\right) P^{(i)}\right] + tr \left[F^{(i)^T}_{new} RF^{(i)}_{new}V\right]$$
$$(11.14)$$

The block diagram of the required calculations for the full-order system, represented by formulas (11.10)-(11.13) is shown in Figure 11.1.

The next sections show that in the cases of quasi weakly coupled linear and singularly perturbed systems, equations (11.6)-(11.7) can be decomposed into six reduced-order Lyapunov equations to get the parallel algorithm (Bertsekas and Tsitsiklis, 1991) with arbitrary order of accuracy.

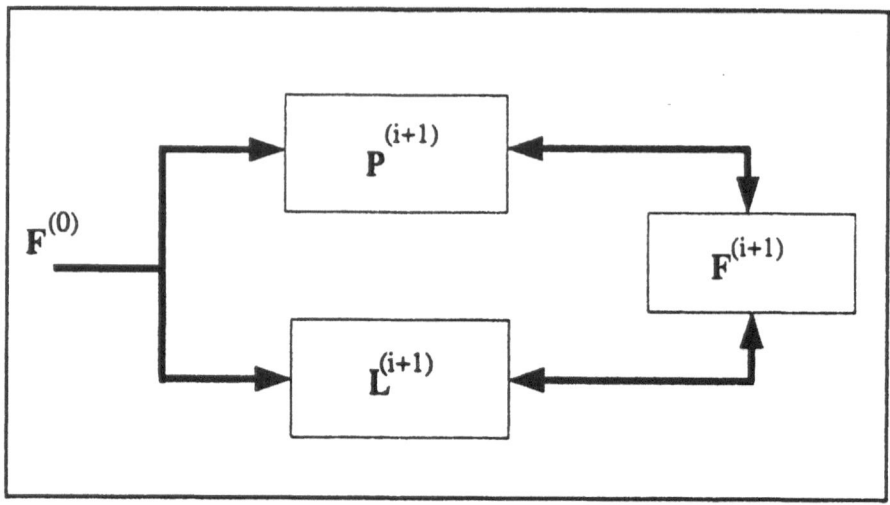

Figure 11.1: Required calculations for the full-order system

11.2 Output Feedback Control of Quasi Weakly Coupled Linear Discrete Systems

A very efficient parallel reduced-order algorithm which decomposes a high-order system into a low-order system for the case of the quasi weakly coupled discrete stochastic output feedback control problem is derived in this section. The low-order system is represented by six Lyapunov equations which may be solved in parallel to reduce computational time. The required solution can be easily obtained up to an arbitrary order of accuracy, $O\left(\epsilon^{2k}\right)$ where ϵ is a small weak coupling parameter and k represents the number of iterations. The efficiency of the proposed method is demonstrated on two real aircraft examples that possess the quasi weakly coupled structure under the assumption of a prescribed degree of stability. The aircrafts are inherently non weakly coupled systems, but since they require high degree of stability, we will demonstrate on two real aircraft examples, that the prescribed degree of stability assumption makes them quasi weakly coupled systems.

The results of this section extend the ideas of the fixed point iterations (Harkara et al., 1989; Gajic et al., 1990; Petrovic and Gajic, 1988; Gajic and Shen, 1989; Shen and Gajic, 1990a, 1990b, 990c; Su and Gajic, 1991a; Qureshi et al., 1992) to the discrete stochastic output feedback of the quasi weakly coupled (Skataric et al., 1990) linear systems, in the spirit of parallel and distributed computations (Bertsekas and Tsitsiklis, 1991).

The general weakly coupled discrete stochastic control system was studied in (Shen and Gajic, 1990c). The system is defined as

$$
\begin{aligned}
x_1(k+1) &= A_1 x_1(k) + \epsilon A_2 x_2(k) + B_1 u_1(k) + \epsilon B_2 u_2(k) \\
&\quad + G_1 w_1(k) + \epsilon G_4 w_2(k) \\
x_2(k+1) &= \epsilon A_3 x_1(k) + A_4 x_2(k) + \epsilon B_3 u_1(k) + B_4 u_2(k) \\
&\quad + \epsilon G_3 w_1(k) + G_2 w_2(k)
\end{aligned}
\tag{11.15}
$$

with corresponding measurements

$$
\begin{aligned}
y_1(k) &= C_1 x_1(k) + \epsilon C_2 x_2(k) + v_1(k) \\
y_2(k) &= \epsilon C_3 x_1(k) + C_4 x_2(k) + v_2(k)
\end{aligned}
\tag{11.16}
$$

where $x_i(k) \in \Re^{n_i}$, $i = 1, 2$, are state variables, $u_i(k) \in \Re^{m_i}$, $i = 1, 2$, are control inputs, $y_i(k) \in \Re^{r_i}$, $i = 1, 2$, are measured outputs, $w_i(k) \in \Re^{s_i}$, $i = 1, 2$, and $v_i(k) \in \Re^{r_i}$, $i = 1, 2$, are stationary uncorrelated Gaussian zero-mean white noise stochastic processes with intensities $W_i > 0$ and $V_i > 0$, respectively, and ϵ is a small coupling parameter.

The performance criterion of the weakly coupled discrete systems which has to be minimized is given by

$$
J = E\left\{ \sum_{k=0}^{\infty} \begin{bmatrix} x_1(k) \\ x_2(k) \end{bmatrix}^T Q \begin{bmatrix} x_1(k) \\ x_2(k) \end{bmatrix} + u^T(k) R u(k) \right\}
\tag{11.17}
$$

where matrices Q and R are partitioned as

$$
Q = \begin{bmatrix} Q_1 & \epsilon Q_2 \\ \epsilon Q_2^T & Q_3 \end{bmatrix} \geq 0, \quad R = \begin{bmatrix} R_1 & 0 \\ 0 & R_2 \end{bmatrix} > 0
\tag{11.18}
$$

All problem matrices are constant matrices of compatible dimensions.

The quasi weakly coupled linear stochastic discrete system differs from (11.15)-(11.16). It can be obtained from (11.1)-(11.2) with (Skataric et al., 1990)

$$
A = \begin{bmatrix} A_1 & \epsilon A_2 \\ \epsilon A_3 & A_4 \end{bmatrix}, \quad B = \begin{bmatrix} 0 \\ B_2 \end{bmatrix}, \quad G = \begin{bmatrix} 0 \\ G_2 \end{bmatrix}, \quad C = \begin{bmatrix} 0 & C_2 \end{bmatrix}
\tag{11.19}
$$

Many real physical systems such as aircrafts, flexible space structures, power systems, chemical reactors possess the quasi weakly coupled structure.

STOCHASTIC OUTPUT CONTROL

In this case the algebraic equations (11.10)-(11.11), comprising the solution to the stochastic output feedback control problem of discrete linear systems, can be decomposed subject to (11.18) and (11.19) as follows.

Partition matrices $A + BF^{(i)}C$, $BF^{(i)}VF^{(i)^T}B^T + GWG^T$, and $P^{(i)}$ as

$$A + BF^{(i)}C = \begin{bmatrix} D_1^{(i)} & \epsilon D_2^{(i)} \\ \epsilon D_3^{(i)} & D_4^{(i)} \end{bmatrix} \qquad (11.20)$$

$$BF^{(i)}VF^{(i)^T}B^T + GWG^T = \begin{bmatrix} 0 & 0 \\ 0 & S_3^{(i)} \end{bmatrix} \qquad (11.21)$$

$$P^{(i)} = \begin{bmatrix} P_1^{(i)} & \epsilon P_2^{(i)} \\ \epsilon P_2^{(i)^T} & P_3^{(i)} \end{bmatrix} \qquad (11.22)$$

where

$$D_1^{(i)} = A_1, \quad D_2^{(i)} = A_2, \quad D_3^{(i)} = A_3, \quad D_4^{(i)} = A_4 + B_2 F^{(i)} C_2$$
$$S_3^{(i)} = B_2 F^{(i)} V F^{(i)^T} B_2^T + G_2 W G_2^T$$
$$(11.23)$$

Using these partitions to expand (11.10), the following equations are obtained:

$$P_1^{(i+1)} = D_1^{(i)} P_1^{(i+1)} D_1^{(i)^T}$$
$$+\epsilon^2 \left[D_2^{(i)} P_2^{(i+1)^T} D_1^{(i)^T} + D_1^{(i)} P_2^{(i+1)} D_2^{(i)^T} + D_2^{(i)} P_3^{(i+1)} D_2^{(i)^T} \right] \qquad (11.24)$$

$$P_2^{(i+1)} = D_1^{(i)} P_2^{(i+1)} D_4^{(i)^T} + D_1^{(i)} P_1^{(i+1)} D_3^{(i)^T}$$
$$+ D_2^{(i)} P_3^{(i+1)} D_4^{(i)^T} + \epsilon^2 D_2^{(i)} P_2^{(i+1)^T} D_3^{(i)^T} \qquad (11.25)$$

$$P_3^{(i+1)} = D_4^{(i)} P_3^{(i+1)} D_4^{(i)^T} + S_3$$
$$+\epsilon^2 \left[D_3^{(i)} P_1^{(i+1)} D_3^{(i)^T} + D_4^{(i)} P_2^{(i+1)^T} D_3^{(i)^T} + D_3^{(i)} P_2^{(i+1)} D_4^{(i)^T} \right] \qquad (11.26)$$

Let us define $O\left(\epsilon^2\right)$ approximations of (11.24)-(11.26)

$$\mathbf{P}_1^{(i+1)} = D_1^{(i)} \mathbf{P}_1^{(i+1)} D_1^{(i)^T} \qquad (11.27)$$

$$\mathbf{P}_2^{(i+1)} = D_1^{(i)} \mathbf{P}_2^{(i+1)} D_4^{(i)^T}$$
$$+ D_2^{(i)} \mathbf{P}_3^{(i+1)} D_4^{(i)^T} + D_1^{(i)} \mathbf{P}_1^{(i+1)} D_3^{(i)^T} \qquad (11.28)$$

$$\mathbf{P}_3^{(i+1)} = D_4^{(i)} \mathbf{P}_3^{(i+1)} D_4^{(i)^T} + S_3 \qquad (11.29)$$

Assuming that the matrix $D_1^{(i)}$ is stable (Harkara et al., 1989), the equation (11.27) has the solution, $\mathbf{P}_1^{(i+1)} = 0$. As long as the subsystem matrix $D_4^{(i)}$ is stable, equations (11.28) and (11.29) can be solved sequentially for $\mathbf{P}_3^{(i+1)}$ and $\mathbf{P}_2^{(i+1)}$, respectively.

Defining the errors as

$$P_m^{(i+1)} = \mathbf{P}_m^{(i+1)} + \epsilon^2 E_m; \quad m = 1, 2, 3 \qquad (11.30)$$

then, subtracting (11.27)-(11.29) from (11.24)-(11.26) and doing some algebra, the error equations are obtained:

$$E_1 = D_1^{(i)} E_1 D_1^{(i)^T}$$
$$+ \left[D_2^{(i)} P_2^{(i+1)^T} D_1^{(i)^T} + D_1^{(i)} P_2^{(i+1)} D_2^{(i)^T} + D_2^{(i)} P_3^{(i+1)} D_2^{(i)^T} \right] \qquad (11.31)$$

$$E_2 = D_1^{(i)} E_2 D_4^{(i)^T} + D_1^{(i)} E_1 D_3^{(i)^T}$$
$$+ D_2^{(i)} E_3 D_4^{(i)^T} + D_2^{(i)} P_2^{(i+1)^T} D_3^{(i)^T} \qquad (11.32)$$

$$E_3 = D_4^{(i)} E_3 D_4^{(i)^T}$$
$$+ \left[D_3^{(i)} P_1^{(i+1)} D_3^{(i)^T} + D_4^{(i)} P_2^{(i+1)^T} D_3^{(i)^T} + D_3^{(i)} P_2^{(i+1)} D_4^{(i)^T} \right] \qquad (11.33)$$

The above equations can be solved efficiently by proposing the following reduced-order parallel synchronous algorithm in the spirit of those studied in (Bertsekas and Tsitsiklis, 1991).

Algorithm 11.2:

$$E_1^{(j+1)} = D_1^{(i)} E_1^{(j+1)} D_1^{(i)^T} + D_1^{(i)} \left(\mathbf{P}_2^{(i+1)} + \epsilon^2 E_2^{(j)} \right) D_2^{(i)^T}$$
$$+ D_2^{(i)} \left(\mathbf{P}_2^{(i+1)} + \epsilon^2 E_2^{(j)} \right)^T D_1^{(i)^T} + D_2^{(i)} \left(\mathbf{P}_3^{(i+1)} + \epsilon^2 E_3^{(j)} \right) D_2^{(i)^T} \qquad (11.34)$$

$$E_2^{(j+1)} = D_1^{(i)} E_2^{(j+1)} D_4^{(i)^T} + D_1^{(i)} E_1^{(j)} D_3^{(i)^T}$$
$$+ D_2^{(i)} E_3^{(j)} D_4^{(i)^T} + D_2^{(i)} \left(\mathbf{P}_2^{(i+1)} + \epsilon^2 E_2^{(j)} \right)^T D_3^{(i)^T} \qquad (11.35)$$

$$E_3^{(j+1)} = D_4^{(i)} E_3^{(j+1)} D_4^{(i)^T} + D_3^{(i)} \left(\mathbf{P}_1^{(i+1)} + \epsilon^2 E_1^{(j)} \right) D_3^{(i)^T}$$

$$+ D_4^{(i)} \left(\mathbf{P}_2^{(i+1)} + \epsilon^2 E_2^{(j)} \right)^T D_3^{(i)^T} + D_3^{(i)} \left(\mathbf{P}_2^{(i+1)} + \epsilon^2 E_2^{(j)} \right) D_4^{(i)^T}$$

$$(11.36)$$

with initial conditions chosen as $E_1^{(0)} = 0$, $E_2^{(0)} = 0$, $E_3^{(0)} = 0$.

$\triangle$

The following theorem presents the features of the proposed algorithm.

Theorem 11.1 *Algorithm 11.2 converges to the required solutions E_1, E_2, and E_3 with the rate of convergence of $O\left(\epsilon^2\right)$.*

$\diamond$

Proof: Let $e_m^{(j)} = E_m^{(j)} - E_m^{(j-1)}$, $m = 1, 2, 3$, then from (11.34)-(11.36) follows

$$D_1^{(i)} e_1^{(j+1)} D_1^{(i)^T} - e_1^{(j+1)} =$$
$$-\epsilon^2 \left[D_2^{(i)} e_2^{(j)} D_1^{(i)^T} + D_1^{(i)} e_2^{(j)} D_2^{(i)^T} + D_2^{(i)} e_3^{(j)} D_2^{(i)^T} \right] \qquad (11.37)$$

$$D_1^{(i)} e_2^{(j+1)} D_4^{(i)^T} - e_2^{(j+1)} = -\epsilon^2 D_2^{(i)} e_2^{(j)} D_3^{(i)^T}$$
$$-D_2^{(i)} e_3^{(j)} D_4^{(i)^T} - D_1^{(i)} e_1^{(j)} D_3^{(i)^T} \qquad (11.38)$$

$$D_4^{(i)} e_3^{(j+1)} D_4^{(i)^T} - e_3^{(j+1)} =$$
$$-\epsilon^2 \left[D_3^{(i)} e_1^{(j)} D_3^{(i)^T} + D_4^{(i)} e_2^{(j)^T} D_3^{(i)^T} + D_3^{(i)} e_2^{(j)} D_4^{(i)^T} \right] \qquad (11.39)$$

By stability assumption imposed on $D_1^{(i)}$ and $D_4^{(i)}$, we have from (11.37) and (11.39)

$$\left\| e_m^{(j)} \right\| = O\left(\epsilon^2\right), \qquad m = 1, 3 \qquad (11.40a)$$

and using results of (11.40a) in (11.38) we get

$$\left\| e_2^{(j)} \right\| = O\left(\epsilon^2\right) \qquad (11.40b)$$

or

$$\left\| E_m^{(j)} - E_m^{(j-1)} \right\| = O\left(\epsilon^2\right), \qquad m = 1, 2, 3; \quad j = 1, 2, \dots \qquad (11.41)$$

Continuing the same procedure, it follows by analogy that

$$\left\| E_m^{(j)} - E_m \right\| = O\left(\epsilon^{2j} \right), \quad m = 1, 2, 3; \quad j = 1, 2, \dots \qquad (11.42)$$

Thus, the proposed algorithm is convergent.

Using $E_m^{(\infty)}$, $m = 1, 2, 3$, in (11.34)-(11.36) and comparing to (11.31)-(11.33) implies that the algorithm (11.34)-(11.36) converges to the unique solution of (11.31)-(11.33). Therefore, $P^{(i+1)}$ can be solved iteratively with an arbitrary order of accuracy.

Similarly, the lower order decomposition can be obtained for equation (11.11). Partitioning matrices as follows

$$C^T F^{(i)T} R F^{(i)} C + Q = \begin{bmatrix} q_1^{(i)} & 0 \\ 0 & q_3^{(i)} \end{bmatrix}, \quad L^{(i)} = \begin{bmatrix} L_1^{(i)} & \epsilon L_2^{(i)} \\ \epsilon L_2^{(i)T} & L_3^{(i)} \end{bmatrix} \qquad (11.43)$$

where

$$q_1^{(i)} = Q_1, \quad q_3^{(i)} = C_2^T F^{(i)T} R F^{(i)} C_2 + Q_3 \qquad (11.44)$$

Using these partitions to expand (11.11), the following equations are obtained

$$L_1^{(i+1)} = D_1^{(i)T} L_1^{(i+1)} D_1^{(i)} + q_1^{(i)}$$
$$+ \epsilon^2 \left[D_3^{(i)T} L_2^{(i+1)T} D_1^{(i)} + D_1^{(i)T} L_2^{(i+1)} D_3^{(i)} + D_3^{(i)T} L_3^{(i+1)} D_3^{(i)} \right] \qquad (11.45)$$

$$L_2^{(i+1)} = D_1^{(i)T} L_2^{(i+1)} D_4^{(i)} + D_1^{(i)T} L_1^{(i+1)} D_2^{(i)}$$
$$+ D_3^{(i)T} L_3^{(i+1)} D_4^{(i)} + \epsilon^2 D_3^{(i)T} L_2^{(i+1)T} D_2^{(i)} \qquad (11.46)$$

$$L_3^{(i+1)} = D_4^{(i)T} L_3^{(i+1)} D_4^{(i)} + q_3^{(i)}$$
$$+ \epsilon^2 \left[D_2^{(i)T} L_1^{(i+1)} D_2^{(i)} + D_4^{(i)T} L_2^{(i+1)T} D_2^{(i)} + D_2^{(i)T} L_2^{(i+1)} D_4^{(i)} \right] \qquad (11.47)$$

Let us define $O\left(\epsilon^2 \right)$ approximations of (11.45)-(11.47) as

$$\mathbf{L}_1^{(i+1)} = D_1^{(i)T} \mathbf{L}_1^{(i+1)} D_1^{(i)} + q_1^{(i)} \qquad (11.48)$$

$$\mathbf{L}_2^{(i+1)} = D_1^{(i)T} \mathbf{L}_2^{(i+1)} D_4^{(i)} + D_1^{(i)T} \mathbf{L}_1^{(i+1)} D_2^{(i)} + D_3^{(i)T} \mathbf{L}_3^{(i+1)} D_4^{(i)} \qquad (11.49)$$

$$\mathbf{L}_3^{(i+1)} = D_4^{(i)T} \mathbf{L}_3^{(i+1)} D_4^{(i)} + q_3^{(i)} \qquad (11.50)$$

Defining the approximation errors as

$$L_m^{(i+1)} = \mathbf{L}_m^{(i+1)} + \epsilon^2 Z_m, \quad m = 1, 2, 3 \tag{11.51}$$

the following error equations are obtained from (11.45)-(11.47) and (11.48)-(11.51)

$$Z_1 = D_1^{(i)^T} Z_1 D_1^{(i)} + D_3^{(i)^T} L_2^{(i+1)^T} D_1^{(i)} + D_1^{(i)^T} L_2^{(i+1)} D_3^{(i)}$$
$$+ D_3^{(i)^T} L_3^{(i+1)} D_3^{(i)} \tag{11.52}$$

$$Z_2 = D_1^{(i)^T} Z_2 D_4^{(i)} + D_1^{(i)^T} Z_1 D_2^{(i)} + D_3^{(i)^T} Z_3 D_4^{(i)} + D_3^{(i)^T} L_2^{(i+1)^T} D_2^{(i)} \tag{11.53}$$

$$Z_3 = D_4^{(i)^T} Z_3 D_4^{(i)} + D_2^{(i)^T} L_1^{(i+1)} D_2^{(i)} + D_4^{(i)^T} L_2^{(i+1)^T} D_2^{(i)}$$
$$+ D_2^{(i)^T} L_2^{(i+1)} D_4^{(i)} \tag{11.54}$$

The above equations can be solved by proposing a similar kind of algorithm as (11.34)-(11.36).

Algorithm 11.3:

Initialize $Z_1^{(0)} = 0, \; Z_2^{(0)} = 0, \; Z_3^{(0)} = 0$ and calculate the following equations iteratively

$$Z_1^{(j+1)} - D_1^{(i)^T} Z_1^{(j+1)} D_1^{(i)} = D_3^{(i)^T} \left(\mathbf{L}_2^{(i+1)} + \epsilon^2 Z_2^{(j)} \right)^T D_1^{(i)}$$
$$+ D_1^{(i)^T} \left(\mathbf{L}_2^{(i+1)} + \epsilon^2 Z_2^{(j)} \right) D_3^{(i)} + D_3^{(i)^T} \left(\mathbf{L}_3^{(i+1)} + \epsilon^2 Z_3^{(j)} \right) D_3^{(i)} \tag{11.55}$$

$$Z_2^{(j+1)} - D_1^{(i)^T} Z_2^{(j+1)} D_4^{(i)} = D_1^{(i)^T} Z_1^{(j)} D_2^{(i)}$$
$$+ D_3^{(i)^T} Z_3^{(j)} D_4^{(i)} + D_3^{(i)^T} \left(\mathbf{L}_2^{(i+1)} + \epsilon^2 Z_2^{(j)} \right)^T D_2^{(i)} \tag{11.56}$$

$$Z_3^{(j+1)} - D_4^{(i)^T} Z_3^{(j+1)} D_4^{(i)} = D_2^{(i)^T} \left(\mathbf{L}_1^{(i+1)} + \epsilon^2 Z_1^{(j)} \right) D_2^{(i)}$$
$$+ D_4^{(i)^T} \left(\mathbf{L}_2^{(i+1)} + \epsilon^2 Z_2^{(j)} \right)^T D_2^{(i)} + D_2^{(i)^T} \left(\mathbf{L}_2^{(i+1)} + \epsilon^2 Z_2^{(j)} \right) D_4^{(i)} \tag{11.57}$$

$$\triangle$$

The algorithm (11.55)-(11.57) has the same properties of the algorithm (11.34)-(11.36) so we have the following theorem.

Theorem 11.2 *Algorithm 11.3 converges to the solutions of Z_m, $m = 1, 2, 3$, with the rate of convergence of $O\left(\epsilon^2\right)$, that is*

$$\left\|Z_m^{(j)} - Z_m^{(j-1)}\right\| = O\left(\epsilon^2\right)$$

$$\left\|Z_m^{(j)} - Z_m\right\| = O\left(\epsilon^{2j}\right) \tag{11.58}$$

◊

The proof of Theorem 11.2 is identical to the proof of Theorem 11.1 and thus is omitted.

To summarize, P can be found from equations (11.27)-(11.29) and equations (11.34)-(11.36), and L can be computed from equations (11.48)-(11.50) and equations (11.55)-(11.57). Since these algorithms are independent of one another the computation can be done in parallel which leads to six reduced-order Lyapunov equations.

In the remaining part of these section, we modify the quasi weakly coupled algorithm for the case when the zero elements of matrices B, G, and C, defined in (11.19), are replaced by $O\left(\epsilon\right)$ quantities. This structure results in the process of discretization of the continuous-time quasi weakly coupled systems. Namely, these matrices are given by

$$B = \begin{bmatrix} \epsilon B_1 \\ B_2 \end{bmatrix}, \quad G = \begin{bmatrix} \epsilon G_1 \\ G_2 \end{bmatrix}, \quad C = \begin{bmatrix} \epsilon C_1 & C_2 \end{bmatrix} \tag{11.59}$$

The introduced changes will produce the following modifications

$$BF^{(i)}VF^{(i)^T}B^T + GWG^T = \begin{bmatrix} \epsilon^2 S_1^{(i)} & \epsilon S_2^{(i)} \\ \epsilon S_2^{(i)^T} & S_3^{(i)} \end{bmatrix} \tag{11.60}$$

with

$$S_1^{(i)} = B_1 F^{(i)} V F^{(i)^T} B_1^T + G_1 W G_1^T$$
$$S_2^{(i)} = B_1 F^{(i)} V F^{(i)^T} B_2^T + G_1 W G_2^T \tag{11.61}$$

Also the matrices $D_m^{(i)}$, $1, 2, 3$, are changed into

$$D_1^{(i)} = A_1 + \epsilon^2 B_1 F^{(i)} C_1, \quad D_2^{(i)} = A_2 + B_1 F^{(i)} C_2$$
$$D_3^{(i)} = A_3 + B_2 F^{(i)} C_1 \tag{11.62}$$

Using these partitions to expand (11.10), the following variations are obtained for (11.24)-(11.25)

$$P_1^{(i+1)} = D_1^{(i)} P_1^{(i+1)} D_1^{(i)^T} + \epsilon^2 S_1$$
$$+ \epsilon^2 \left[D_2^{(i)} P_2^{(i+1)^T} D_1^{(i)^T} + D_1^{(i)} P_2^{(i+1)} D_2^{(i)^T} + D_2^{(i)} P_3^{(i+1)} D_2^{(i)^T} \right] \tag{11.63}$$

$$P_2^{(i+1)} = S_2 + D_1^{(i)} P_2^{(i+1)} D_4^{(i)^T} + D_1^{(i)} P_1^{(i+1)} D_3^{(i)^T}$$
$$+ D_2^{(i)} P_3^{(i+1)} D_4^{(i)^T} + \epsilon^2 D_2^{(i)} P_2^{(i+1)^T} D_3^{(i)^T} \tag{11.64}$$

The equation (11.26) is unchanged.

By perturbing equations (11.63)-(11.64) and (11.26) by an $O\left(\epsilon^2\right)$, the zeroth-order approximation can be found to be the same as in (11.27) and (11.29) except for (11.28) which becomes

$$\mathbf{P}_2^{(i+1)} - D_1^{(i)} \mathbf{P}_2^{(i+1)} D_4^{(i)^T} = S_2$$
$$+ D_2^{(i)} \mathbf{P}_3^{(i+1)} D_4^{(i)^T} + D_1^{(i)} \mathbf{P}_1^{(i+1)} D_3^{(i)^T} \tag{11.65}$$

Corresponding equations for the error terms will differ only in equation for E_1 which now has the form

$$E_1 = D_1^{(i)} E_1 D_1^{(i)^T} + D_2^{(i)} P_2^{(i+1)^T} D_1^{(i)^T}$$
$$+ D_1^{(i)} P_2^{(i+1)} D_2^{(i)^T} + D_2^{(i)} P_3^{(i+1)} D_2^{(i)^T} + S_1 \tag{11.66}$$

As a consequence of this the equation (11.34) in the algorithm (11.34)-(11.36) has to be modified into

$$E_1^{(j+1)} - D_1^{(i)} E_1^{(j+1)} D_1^{(i)^T} = D_2^{(i)} \left[\mathbf{P}_2^{(i+1)} + \epsilon^2 E_2^{(j)}\right]^T D_1^{(i)^T}$$
$$+ D_1^{(i)} \left[\mathbf{P}_2^{(i+1)} + \epsilon^2 E_2^{(j)}\right] D_2^{(i)^T} + D_2^{(i)} \left[\mathbf{P}_3^{(i+1)} + \epsilon^2 E_3^{(j)}\right] D_2^{(i)^T} + S_1 \tag{11.67}$$

Similarly the lower order decomposition can be obtained for L equations. The new structures defined in (11.59) will produce

$$C^T F^{(i)^T} R F^{(i)} C + Q = \begin{bmatrix} q_1^{(i)} & \epsilon q_2^{(i)} \\ \epsilon q_2^{(i)^T} & q_3^{(i)} \end{bmatrix} \tag{11.68}$$

with

$$q_1^{(i)} = Q_1 + \epsilon^2 C_1^T F^{(i)^T} R F^{(i)} C_1, \quad q_2^{(i)} = C_1^T F^{(i)^T} R F^{(i)} C_2 \tag{11.69}$$

which will change equations (11.46) into

$$L_2^{(i+1)} - D_1^{(i)^T} L_2^{(i+1)^T} D_4^{(i)} = D_1^{(i)^T} L_1^{(i+1)} D_2^{(i)}$$
$$+ D_3^{(i)^T} L_3^{(i+1)} D_4^{(i)} + \epsilon^2 D_3^{(i)^T} L_2^{(i+1)^T} D_2^{(i)} + q_2^{(i)} \tag{11.70}$$

so that the corresponding equation in the zeroth-order approximation equations (11.46)-(11.48) becomes

$$L_2^{(i+1)} = D_1^{(i)^T} L_2^{(i+1)^T} D_4^{(i)} + D_1^{(i)^T} L_1^{(i+1)} D_2^{(i)} + D_3^{(i)^T} L_3^{(i+1)} D_4^{(i)} + q_2^{(i)} \tag{11.71}$$

However, the error equations are found to be the same equations as (11.52)-(11.54). These equations can be solved by the same algorithm as (11.55)-(11.57).

11.3 Case Study: Flight Control Systems for Aircrafts

Case Study 1:

In order to demonstrate the efficiency of the proposed algorithm we ran a sixth-order example of a flight control system of a modern aircraft (Shapiro et al., 1981). The problem matrices are defined as follows

$$A = \begin{bmatrix} -0.746 & 0.387 & -12.9 & 6.05 & 0.952 & 0 \\ 0.024 & -0.174 & 0.4 & -0.416 & -1.76 & 0 \\ 0.006 & -0.999 & -0.058 & -0.0012 & 0.0092 & 0.0369 \\ 0 & 0 & 0 & -5 & 0 & 0 \\ 0 & 0 & 0 & 0 & -10 & 0 \\ 1 & 0 & 0 & 0 & 0 & 0 \end{bmatrix}$$

$$B^T = \begin{bmatrix} 0 & 0 & 0 & 10 & 0 & 0 \\ 0 & 0 & 0 & 0 & 20 & 0 \end{bmatrix}$$

The remaining matrices are chosen as

$$C = \begin{bmatrix} 0 & 0 & 0 & 0 & 1 & 0 \\ 0 & 0 & 0 & 0 & 0 & 1 \end{bmatrix}$$

and

$$Q = I_6, \quad R = I_2, \quad G = B, \quad V = I_2, \quad W = 1$$

The continuous-time matrices above were multiplied by a permutation matrix to obtain the form used in Section 11.2. The system was then discretized with the sampling period of 0.1. Before the system was discretized a prescribed degree of stability factor (Anderson and Moore, 1990) of $\sigma = 10$ was introduced so that the matrix $A + \sigma I$ would become diagonally dominant and thus weakly coupled. The discretized matrices are

i	$J_{red}^{(i)} = J_{opt}^{(i)}$
1	1.0988
2	1.0979
3	1.0971
4	1.0964
5	1.0958
10	1.0938
20	1.0922
30	1.0917
40	1.0915
46	1.0914
	$J_{opt} = 1.0914$

Table 11.1: The optimal and reduced criteria per iteration

$$A_D = \begin{bmatrix} 0.341 & 0.036 & -0.45 & -0.001 & 0.019 & 0.168 \\ 0.001 & 0.354 & 0.155 & 0 & -0.04 & -0.012 \\ 0 & -0.036 & 0.358 & 0.001 & 0.003 & 0.001 \\ 0.035 & 0.001 & -0.023 & 0.368 & 0.001 & 0.001 \\ 0 & 0 & 0 & 0 & 0.135 & 0 \\ 0 & 0 & 0 & 0 & 0 & 0.223 \end{bmatrix}$$

$$B_D^T = \begin{bmatrix} 0.035 & 0.07 & 0.003 & 0.001 & 0.865 & 0 \\ 0.135 & -0.009 & 0 & 0.004 & 0 & 0.518 \end{bmatrix}$$

The remaining matrices are the same in the discrete-time as they are in the continuous-time.

The obtained results are presented in Table 11.1. The number of iterations for approximating P and L are high enough so that

$$\left\| E_m^{(i+1)} \right\| - \left\| E_m^{(i)} \right\| \le 10^{-7}, \quad \left\| Z_m^{(i+1)} \right\| - \left\| Z_m^{(i)} \right\| \le 10^{-7}, \quad m = 1, 2, 3$$

For this example the typical number of iterations for such tolerance was 6.

It can be seen from Table 11.1 that the results of the global algorithm (11.10)-(11.11) and proposed reduced-order algorithm agree up to 4 decimal digits at each iteration. The additional advantages (uniqueness and choice of a good initial guess) of the reduced-order algorithms for the output feedback control problem were discussed in (Harkara et al., 1989). Parameters α and ϵ are given by $\alpha = 0.1$ and $\epsilon = 0.3$.

Case Study 2:

The second numerical example is an aircraft system considered in (Anderson and Moore, 1990). The problem matrices are shown bellow

$$A = \begin{bmatrix} -0.28 \times 10^{-8} & 0 & 0 & 0 & 0 & 0 \\ 0 & -6.76 \times 10^{-3} & 0 & 0 & 0 & 0 \\ 0 & 0 & -0.122 & 1.57 & 0 & 0 \\ 0 & 0 & -1.57 & -0.122 & 0 & 0 \\ 0 & 0 & 0 & 0 & -0071. & 21.3 \\ 0 & 0 & 0 & 0 & -21.3 & -0.71 \end{bmatrix}$$

$$B = G = \begin{bmatrix} -4.06 \times 10^{-4} & -1.65 \times 10^{-7} \\ -2.20 \times 10^{-1} & 7.61 \times 10^{-5} \\ 8.84 \times 10^{-2} & 3.05 \times 10^{-2} \\ -3.08 \times 10^{-1} & -1.05 \times 10^{-2} \\ 6.39 \times 10^{-2} & 4.46 \times 10^{-3} \\ -1.08 & -1.02 \times 10^{-2} \end{bmatrix}$$

$$C = 10^{-3} \begin{bmatrix} -0.35 & -157 & -156 & 4.61 & -337 & -0.24 \\ -0.205 & -154 & 281 & -98.5 & -1020 & -76.8 \end{bmatrix}$$

The remaining matrices are chosen as

$$Q = I_6, \quad R = I_2, \quad V = I_2, \quad W = 1$$

The continuous-time matrices above were multiplied by a permutation matrix to obtain the form used in Section 11.2. The system was then discretized with the sampling period of 0.1. Before the system was discretized a prescribed degree of stability factor of $\sigma = I$ was introduced so that the matrix $A + \sigma I$ would become diagonally dominant and thus weakly coupled. The discretized matrices are

$$A_D = \begin{bmatrix} 0.905 & 0 & 0 & 0 & 0 & 0 \\ 0 & 0.904 & 0 & 0 & 0 & 0 \\ 0 & 0 & 0.883 & 0.14 & 0 & 0 \\ 0 & 0 & -0.14 & 0.883 & 0 & 0 \\ 0 & 0 & 0 & 0 & 0.447 & 0.714 \\ 0 & 0 & 0 & 0 & -0.714 & 0.447 \end{bmatrix}$$

STOCHASTIC OUTPUT CONTROL

$$B_D = \begin{bmatrix} -0.386 \times 10^{-4} & -0.157 \times 10^{-7} \\ -0.209 \times 10^{-1} & 0.724 \times 10^{-5} \\ 0.557 \times 10^{-2} & 0.280 \times 10^{-2} \\ -0.364 \times 10^{-1} & -0.121 \times 10^{-2} \\ -0.675 \times 10^{-1} & -0.488 \times 10^{-3} \\ -0.460 \times 10^{-1} & -0.684 \times 10^{-3} \end{bmatrix}$$

The remaining matrices are the same in both the continuous and discrete-time.

The obtained results are shown in Table 11.2. Parameters α and ϵ are given by $\alpha = 0.1$ and $\epsilon = 0.3$. The number of iterations for approximating P and L are chosen according to the criterion given in previous example.

i	$J_{red}^{(i)}$	$J_{opt}^{(i)}$
1	1.1053	1.1127
2	0.81667	0.81802
3	0.62366	0.62313
4	0.48724	0.48624
5	0.38720	0.38613
10	0.14571	0.14533
20	0.04558	0.04555
30	0.03379	0.03348
40	0.03235	0.03235
50	0.03217	0.03218
71	0.03215	0.03215
J_{opt}	0.03215	0.03215

Table 11.2: The optimal and reduced criterion per iteration

All simulation results in this chapter are obtained by using the L-A-S package for the computer aided control system design (West et al., 1985).

11.4 Output Feedback of Singularly Perturbed Stochastic Discrete Systems

For continuous-time systems, the optimal solution for noise-free output feedback problem is presented at several places in the literature, for example (Levine and Athans, 1970; Mendel, 1974; Kurtaran and Sidar, 1974; Moerder and Calise, 1985a). However, for noisy output, a major difficulty is encountered in finding the optimal solution if the classical quadratic performance index is used. Due to the presence of white noise, the performance index necessarily diverges (Ermer and Vandelinde, 1973; Kurtaran and Sidar, 1974), which necessitates the use of some alternate performance measure. It was shown in (Ermer and Vandelinde, 1973) that the discrete linear stochastic output feedback control problem is well-posed. Optimal solutions for discrete stochastic output feedback control problems presented in (Ermer and Vandelinde, 1973; Kurtaran, 1975) are obtained in terms of high-order nonlinear algebraic equations.

The singularly perturbed output feedback systems did not receive much attention until 1980 (Calise and Moerder, 1985; Chemouil and Wahdam, 1980; Fossard and Magni, 1980; Moerder and Calise, 1985b, 1988; Khalil, 1981, 1987; Gajic et al., 1989), due to their inherent ill-conditioned dynamics. For noise-free output feedback, continuous-time singularly perturbed systems, a well-defined recursive algorithm is developed in (Gajic et al., 1989). The algorithm removes the inherent ill-conditioning for singularly perturbed systems by decomposing high-order nonlinear equations into low-order algebraic equations corresponding to slow and fast modes.

In this section, a recursive algorithm is developed for the discrete singularly perturbed output feedback stochastic control problem. Non-linear algebraic matrix equations are decomposed to ones corresponding to slow and fast modes, so that only low-order systems are involved in algebraic computations. Moreover, such a decomposition removes the ill-conditioning of the higher order system. The proposed algorithm gives the accuracy of $O\left(\epsilon^k\right)$, where ϵ is a small perturbation parameter and k is the number of iterations.

11.4.1 Problem Formulation

A discrete linear singularly perturbed stochastic system is given by (Gajic

STOCHASTIC OUTPUT CONTROL

et al., 1990; Gajic and Shen, 1991a)

$$x(k+1) = (I + \epsilon A_1) x_1(k) + \epsilon A_2(k) x_2(k) + \epsilon B_1 u(k) + \epsilon G_1 w(k)$$
(11.72)

$$x_2(k+1) = A_3 x_1(k) + A_4 x_2(k) + B_2 u(k) + G_2 w(k) \qquad (11.73)$$

$$y(k) = C_1 x_1(k) + C_2 x_2(k) + v(k) \qquad (11.74)$$

where $x_1(k) \in \Re^{n_1}$, $x_2(k) \in \Re^{n_2}$ are state vectors, $u \in \Re^m$ is a control input, $y \in \Re^r$ is the measured output, $w \in \Re^s$ and $v \in \Re^r$ are stationary uncorrelated Gaussian zero-mean white noise processes with intensities $W > 0$ and $V > 0$, respectively. Matrices A_i, B_j, G_j, and C_j, $i = 1, 2, 3, 4$; $j = 1, 2$, are constant matrices of compatible dimensions.

With (11.72)-(11.74), consider the performance criterion

$$J = E \left\{ \sum_{k=0}^{\infty} \begin{bmatrix} x_1(k) \\ x_2(k) \end{bmatrix}^T Q \begin{bmatrix} x_1(k) \\ x_2(k) \end{bmatrix} + u^T(k) R u(k) \right\} \qquad (11.75)$$

with positive definite matrix R and semipositive definite matrix Q, which has to be minimized. In addition, the control input $u(k)$ is constrained to

$$u(k) = Fy(k) \qquad (11.76)$$

Equations (11.72)-(11.74) can be written in the compact form as following

$$x(k+1) = Ax(k) + Bu(k) + Gw(k) \qquad (11.77)$$

$$y(k) = Cx(k) + v(k) \qquad (11.78)$$

where

$$x(k) = \begin{bmatrix} x_1(k) \\ x_2(k) \end{bmatrix}, \quad A = \begin{bmatrix} I + \epsilon A_1 & \epsilon A_2 \\ A_3 & A_4 \end{bmatrix}, \quad B = \begin{bmatrix} \epsilon B_1 \\ B_2 \end{bmatrix},$$
(11.79)

$$G = \begin{bmatrix} \epsilon G_1 \\ G_2 \end{bmatrix}, \quad C = [C_1 \quad C_2]$$

Substituting (11.78) into (11.76), we obtain

$$u(k) = FCx(k) + Fv(k) \qquad (11.80)$$

The above problem has the structure of one defined in Section 11.1. Its solution is given in terms of equations (11.5)-(11.7). Equations (11.5)-(11.7) are high-order nonlinear algebraic equations which have to be solved for P, L, and F. It is shown in (Halyo and Broussard, 1981) that the algorithm proposed for the numerical solution of (11.5)-(11.7) and given by (11.10)-(11.13) converges to a local minimum under non-restrictive assumptions.

The next section shows how we can decompose (11.10)-(11.13) in the algebraic equations corresponding to slow and fast modes, in order to get lower order well-defined algebraic equations, and to achieve any desired order of accuracy.

11.4.2 Slow-Fast Lower Order Decomposition

Equation (11.11) is the standard Lyapunov equation of the discrete singularly perturbed linear system, while (11.10) is not in the standard form. Therefore, a slight difference occurs in the lower order expressions for these two equations. First, we will decompose equation (11.10).

Partition matrices $\left(A + BF^{(i)}C\right)$, $\left(BF^{(i)}VF^{(i)^T}B^T + GWG^T\right)$, and $P^{(i)}$ as follows

$$A + BF^{(i)}C = \begin{bmatrix} I + \epsilon D_1^{(i)} & \epsilon D_2^{(i)} \\ D_3^{(i)} & D_4^{(i)} \end{bmatrix} \tag{11.81}$$

$$BF^{(i)}VF^{(i)^T}B^T + GWG^T = \begin{bmatrix} \epsilon^2 S_1^{(i)} & \epsilon S_2^{(i)} \\ \epsilon S_2^{(i)^T} & S_3^{(i)} \end{bmatrix} \tag{11.82}$$

$$P^{(i)} = \begin{bmatrix} \epsilon P_1^{(i)} & \epsilon P_2^{(i)} \\ \epsilon P_2^{(i)^T} & P_3^{(i)} \end{bmatrix}$$

where

$$D_1^{(i)} = A_1 + B_1 F^{(i)} C_1, \quad D_2^{(i)} = A_2 + B_1 F^{(i)} C_2$$
$$D_3^{(i)} = A_3 + B_2 F^{(i)} C_1, \quad D_4^{(i)} = A_4 + B_2 F^{(i)} C_2$$

$$S_1^{(i)} = B_1 F^{(i)} V F^{(i)^T} B_1^T + G_1 W G_1^T$$
$$S_2^{(i)} = B_1 F^{(i)} V F^{(i)^T} B_2^T + G_1 W G_2^T$$
$$S_3^{(i)} = B_2 F^{(i)} V F^{(i)^T} B_2^T + G_2 W G_2^T$$

(11.83)

With such partitions, expanding (11.10), we obtain

$$
\begin{aligned}
& D_1^{(i)} P_1^{(i+1)} + P_1^{(i+1)} D_1^{(i)^T} + D_2^{(i)} P_2^{(i+1)^T} \\
& + P_2^{(i+1)} D_2^{(i)^T} + D_2^{(i)} P_3^{(i+1)} D_2^{(i)^T} + S_1^{(i)} \\
& + \epsilon \left(D_1^{(i)} P_1^{(i+1)} D_1^{(i)^T} + D_2^{(i)} P_2^{(i+1)^T} D_1^{(i)^T} + D_1^{(i)} P_2^{(i+1)} D_2^{(i)^T} \right) = 0
\end{aligned}
$$

$$(11.84)$$

$$
\begin{aligned}
& P_1^{(i+1)} D_3^{(i)^T} + P_2^{(i+1)} D_4^{(i)^T} + D_2^{(i)} P_3^{(i+1)} D_4^{(i)^T} - P_2^{(i+1)} + S_2^{(i)} \\
& + \epsilon \left(D_1^{(i)} P_1^{(i+1)} D_3^{(i)^T} + D_2^{(i)} P_2^{(i+1)^T} D_3^{(i)^T} + D_1^{(i)} P_2^{(i+1)} D_4^{(i)^T} \right) = 0
\end{aligned}
$$

$$(11.85)$$

$$
\begin{aligned}
& D_4^{(i)} P_3^{(i+1)} D_4^{(i)^T} - P_3^{(i)} + S_3^{(i)} \\
& + \epsilon \left(D_3^{(i)} P_1^{(i+1)} D_3^{(i)^T} + D_3^{(i)} P_2^{(i+1)} D_4^{(i)^T} + D_4^{(i)} P_2^{(i+1)^T} D_3^{(i)^T} \right) = 0
\end{aligned}
$$

$$(11.86)$$

We can obtain an $O(\epsilon)$ approximation of (11.84)-(11.86) by setting $\epsilon = 0$, that is

$$
\begin{aligned}
& D_1^{(i)} \mathbf{P}_1^{(i+1)} + \mathbf{P}_1^{(i+1)} D_1^{(i)^T} + D_2^{(i)} \mathbf{P}_2^{(i+1)^T} + \mathbf{P}_2^{(i+1)} D_2^{(i)^T} \\
& + S_1^{(i)} + D_2^{(i)} \mathbf{P}_3^{(i+1)} D_2^{(i)^T} = 0
\end{aligned}
$$
$$(11.87)$$

$$
\mathbf{P}_1^{(i+1)} D_3^{(i)^T} + \mathbf{P}_2^{(i+1)} D_4^{(i)^T} + D_2^{(i)} \mathbf{P}_3^{(i+1)} D_4^{(i)^T} - \mathbf{P}_2^{(i+1)} + S_2^{(i)} = 0
$$
$$(11.88)$$

$$
D_4^{(i)} \mathbf{P}_3^{(i+1)} D_4^{(i)^T} - \mathbf{P}_3^{(i+1)} + S_3^{(i)} = 0
$$
$$(11.89)$$

From (11.88) we can express $\mathbf{P}_2^{(i+1)}$ in terms of $\mathbf{P}_1^{(i+1)}$ and $\mathbf{P}_3^{(i+1)}$ as

$$
\begin{aligned}
\mathbf{P}_2^{(i+1)} &= \left(\mathbf{P}_1^{(i+1)} D_3^{(i)^T} + D_2^{(i)} \mathbf{P}_3^{(i+1)} D_4^{(i)^T} + S_2^{(i)} \right) \left(I - D_4^{(i)^T} \right)^{-1} \\
&= N_1^{(i)} + \mathbf{P}_1^{(i+1)} N_2^{(i)}
\end{aligned}
$$
$$(11.90)$$

where

$$N_1^{(i)} = \left(D_2^{(i)} \mathbf{P}_3^{(i+1)} D_4^{(i)^T} + S_2^{(i)} \right) \left(I - D_4^{(i)^T} \right)^{-1}$$
$$N_2^{(i)} = D_3^{(i)^T} \left(I - D_4^{(i)^T} \right)^{-1} \tag{11.91}$$

Substituting (11.88) into (11.87) and doing some algebra, we obtain

$$\hat{A}^{(i)} \mathbf{P}_1^{(i+1)} + \mathbf{P}_1^{(i+1)} \hat{A}^{(i)^T} + \hat{Q}^{(i)} = 0 \tag{11.92}$$

where

$$\hat{A}^{(i)} = D_1^{(i)} + D_2^{(i)} N_2^{(i)^T}$$
$$\hat{Q}^{(i)} = D_2^{(i)} N_1^{(i)^T} + N_1^{(i)} D_2^{(i)^T} + D_2^{(i)} P_3^{(i+1)} D_2^{(i)^T} + S_1^{(i)} \tag{11.93}$$

Since slow and fast subsystem feedback matrices $\hat{A}^{(i)}$ and $D_4^{(i)}$ are stable, equations (11.89), (11.92), and (11.90) can be solved sequentially for $\mathbf{P}_3^{(i+1)}$, $\mathbf{P}_1^{(i+1)}$, and $\mathbf{P}_2^{(i+1)}$, respectively.

In order to make improvement over the $O(\epsilon)$ approximate solutions given above, express $P_1^{(i+1)}$, $P_2^{(i+1)}$, and $P_3^{(i+1)}$ as

$$P_1^{(i+1)} = \mathbf{P}_1^{(i+1)} + \epsilon E_1 \tag{11.94}$$

$$P_2^{(i+1)} = \mathbf{P}_2^{(i+1)} + \epsilon E_2 \tag{11.95}$$

$$P_3^{(i+1)} = \mathbf{P}_3^{(i+1)} + \epsilon E_3 \tag{11.96}$$

Subtract (11.87), (11.88), and (11.89) from corresponding equations (11.84), (11.85), and (11.86), respectively, and after some algebra, we obtain the following equations

$$D_4^{(i)} E_3 D_4^{(i)^T} - E_3 =$$
$$- \left(D_3^{(i)} P_1^{(i+1)} D_3^{(i)^T} + D_3^{(i)} P_2^{(i+1)} D_4^{(i)^T} + D_4^{(i)} P_2^{(i+1)^T} D_3^{(i)^T} \right) \tag{11.97}$$

$$E_1 \hat{A}^{(i)^T} + \hat{A}^{(i)} E_1 = -H D_2^{(i)^T} - D_2^{(i)} H^T - D_2^{(i)} E_3 D_2^{(i)^T}$$
$$- D_1^{(i)} P_1^{(i+1)} D_1^{(i)^T} - D_2^{(i)} P_2^{(i+1)^T} D_1^{(i)^T} - D_1^{(i)} P_2^{(i+1)} D_2^{(i)^T} \tag{11.98}$$

$$E_2 = E_1 D_3^{(i)^T} \left(I - D_4^{(i)^T} \right)^{-1} + H \qquad (11.99)$$

where

$$H = \left(D_2^{(i)} E_3 D_4^{(i)^T} + D_1^{(i)} P_1^{(i+1)} D_3^{(i)^T} + D_2^{(i)} P_2^{(i+1)^T} D_3^{(i)^T} \right)$$
$$\times \left(I - D_4^{(i)^T} \right)^{-1} + D_1^{(i)} P_2^{(i+1)} D_4^{(i)^T} \left(I - D_4^{(i)^T} \right)^{-1}$$

$$(11.100)$$

Equations (11.97)-(11.99) can be solved sequentially for E_3, E_1, and E_2 by proposing the following algorithm.

Algorithm 11.4:

Initialize $E_1^{(0)} = 0$, $E_2^{(0)} = 0$, $E_3^{(0)} = 0$.

For $j = 0, 1, 2, \ldots$ do the following iterations

$$D_4^{(i)} E_3^{(j+1)} D_4^{(i)^T} - E_3^{(j+1)} = -D_3^{(i)} \left(\mathbf{P}_1^{(i+1)} + \epsilon E_1^{(j)} \right) D_3^{(i)^T}$$
$$-D_3^{(i)} \left(\mathbf{P}_2^{(i+1)} + \epsilon E_2^{(j)} \right) D_4^{(i)^T} - D_4^{(i)} \left(\mathbf{P}_2^{(i+1)} + \epsilon E_2^{(j)} \right) D_3^{(i)^T}$$

$$(11.101)$$

$$H^{(j)} = \left(D_2^{(i)} E_3^{(j+1)} D_4^{(i)^T} + D_1^{(i)} \left(\mathbf{P}_1^{(i+1)} + \epsilon E_1^{(j)} \right) D_3^{(i)^T} \right)$$
$$\times \left(I - D_4^{(i)^T} \right)^{-1}$$
$$+ \left(D_2^{(i)} \left(\mathbf{P}_2^{(i+1)} + \epsilon E_2^{(j)} \right) D_3^{(i)^T} + D_1^{(i)} \left(\mathbf{P}_2^{(i+1)} + \epsilon E_2^{(j)} \right) D_4^{(i)^T} \right)$$
$$\times \left(I - D_4^{(i)^T} \right)^{-1}$$

$$(11.102)$$

$$E_1^{(j+1)} \hat{A}^{(i)^T} + \hat{A}^{(i)} E_1^{(j+1)} = -H^{(j)} D_2^{(i)^T} - D_2^{(i)} H^{(j)^T}$$
$$-D_2^{(i)} E_3^{(j+1)} D_2^{(i)^T} - D_1^{(i)} \left(\mathbf{P}_1^{(i+1)} + \epsilon E_1^{(j)} \right) D_1^{(i)^T}$$
$$-D_2^{(i)} \left(\mathbf{P}_2^{(i+1)} + \epsilon E_2^{(j)} \right)^T D_1^{(i)^T} - D_1^{(i)} \left(\mathbf{P}_2^{(i+1)} + \epsilon E_2^{(j)} \right) D_2^{(i)^T}$$

$$(11.103)$$

$$E_2^{(j+1)} = E_1^{(j+1)} D_3^{(i)^T} \left(I - D_4^{(i)^T} \right)^{-1} + H^{(j)} \qquad (11.104)$$

$$\triangle$$

The following theorem establishes the features of the proposed algorithm.

Theorem 11.3 *The algorithm described by (11.101)-(11.104) converges to the required solutions E_1, E_2, and E_3 with the rate of convergence* *of $O(\epsilon)$, that is*

$$\left\| E_m - E_m^{(j)} \right\| = O\left(\epsilon^j\right), \quad m = 1, 2, 3 \text{ and } j = 1, 2, 3, \dots \quad (11.105)$$

◇

Proof: Let $d_m^{(j)} = E_m^{(j)} - E_m^{(j-1)}$, $m = 1, 2, \dots$, then (11.101) can be written as

$$D_4^{(i)} d_3^{(j+1)} D_4^{(i)^T} - d_3^{(j+1)} =$$
$$-\epsilon \left[D_3^{(i)} d_1^{(j)} D_3^{(i)^T} + D_3^{(i)} d_2^{(j)} D_4^{(i)^T} + D_4^{(i)} d_2^{(j)^T} D_3^{(i)^T} \right] \quad (11.106)$$

from which it follows that $\| d_3^{(j+1)} \| = O(\epsilon)$.
   Similarly

$$H^{(j)} - H^{(j-1)} = \left(D_2^{(i)} d_3^{(j+1)} D_4^{(i)^T} \right) \left(I - D_4^{(i)^T} \right)^{-1}$$
$$+\epsilon \left(D_1^{(i)} d_1^{(j)} D_3^{(i)^T} + D_2^{(i)} d_2^{(j)^T} D_3^{(i)^T} + D_1^{(i)} d_2^{(j)} D_4^{(i)^T} \right) \left(I - D_4^{(i)^T} \right)^{-1} \quad (11.107)$$

   Since all terms on the right-hand side are $O(\epsilon)$, this implies that $\| H^{(j)} - H^{(j-1)} \| = O(\epsilon)$. On the same lines (11.103) gives

$$d_1^{(j+1)} \hat{A}^{(i)^T} + \hat{A}^{(i)} d_1^{(j+1)} =$$
$$- \left(H^{(j)} - H^{(j-1)} \right) D_2^{(i)^T} - D_2^{(i)} \left(H^{(j)} - H^{(j-1)} \right)^T -$$
$$D_2^{(i)} d_3^{(j+1)} D_2^{(i)^T} - \epsilon D_1^{(i)} d_1^{(j)} D_1^{(i)^T} - \epsilon D_2^{(i)} d_2^{(j)^T} D_1^{(i)^T} - \epsilon D_1^{(i)} d_2^{(j)} D_2^{(i)^T} \quad (11.108)$$

All terms on the right-hand side are $O(\epsilon)$, which implies that $\| d_1^{(j+1)} \| = O(\epsilon)$. Similarly (11.104) produces

$$d_2^{(j+1)} = d_1^{(j+1)} D_3^{(i)^T} \left(I - D_4^{(i)^T} \right)^{-1} + \left(H^{(j)} - H^{(j-1)} \right) \quad (11.109)$$

Again all terms on the right-hand side of (11.109) are $O(\epsilon)$, which implies that $\| d_2^{(j+1)} \| = O(\epsilon)$.
   Continuing the same procedure it follows that

$$\left\| E_m - E_m^{(j)} \right\| = O\left(\epsilon^j\right), \quad m = 1, 2, 3 \text{ and } j = 1, 2, 3, \dots \quad (11.110)$$

Thus, the proposed algorithm is convergent.

Using $E_m^{(\infty)}$, $m = 1, 2, 3$, in (11.101)-(11.104) and comparing it to (11.97)-(11.99), implies that the algorithm (11.101)-(11.104) converges to the unique solutions of (11.97)-(11.99). This completes the proof.

The above theorem implies that

$$\left\| P_m^{(i+1)} - \left(\mathbf{P}_m^{(i+1)} + \epsilon E_m^{(j)} \right) \right\| = O\left(\epsilon^j \right), \quad m = 1, 2, 3; \quad j = 1, 2, 3, ... \tag{11.111}$$

Therefore $P^{(i+1)}$ can be iteratively solved with an arbitrary accuracy.

Similar kind of lower order equations can be obtained for equation (11.11) also. Performing the following partitioning

$$Q + C^T F^{(i)^T} R F^{(i)} C = \begin{bmatrix} q_1^{(i)} & q_2^{(i)} \\ q_2^{(i)^T} & q_3^{(i)} \end{bmatrix} \tag{11.112}$$

and

$$L^{(i)} = \begin{bmatrix} \epsilon^{-1} L_1^{(i)} & L_2^{(i)} \\ L_2^{(i)^T} & L_3^{(i)} \end{bmatrix} \tag{11.113}$$

where

$$\begin{aligned} q_1^{(i)} &= C_1^T F^{(i)^T} R F^{(i)} C_1 + Q_1 \\ q_2^{(i)} &= C_1^T F^{(i)^T} R F^{(i)} C_2 + Q_2 \\ q_3^{(i)} &= C_2^T F^{(i)^T} R F^{(i)} C_2 + Q_3 \end{aligned} \tag{11.114}$$

the zeroth-order approximations of $L_1^{(i+1)}$, $L_2^{(i+1)}$, and $L_3^{(i+1)}$ can be found by solving the following equations

$$D_4^{(i)^T} \mathbf{L}_3^{(i+1)} D_4^{(i)} - \mathbf{L}_3^{(i+1)} = -q_3^{(i)} \tag{11.115}$$

$$\mathbf{L}_1^{(i+1)} \hat{A}^{(i)} + \hat{A}^{(i)^T} \mathbf{L}_1^{(i+1)} = -\hat{K}^{(i)} \tag{11.116}$$

$$\mathbf{L}_2^{(i+1)} = \left(\mathbf{L}_1^{(i+1)} D_2^{(i)} + D_3^{(i)^T} \mathbf{L}_3^{(i+1)} D_4^{(i)} + q_2^{(i)} \right) \left(I - D_4^{(i)} \right)^{-1} \tag{11.117}$$

where

$$\begin{aligned} \hat{K}^{(i)} &= M_1^{(i)} D_3^{(i)} + D_3^{(i)^T} M_1^{(i)^T} + D_3^{(i)^T} \mathbf{L}_3^{(i+1)} D_3^{(i)} + q_1^{(i)} \\ M_1^{(i)} &= \left(D_3^{(i)^T} \mathbf{L}_3^{(i+1)} D_4^{(i)} + q_2^{(i)} \right) \left(I - D_4^{(i)^T} \right)^{-1} \end{aligned} \tag{11.118}$$

Defining the errors as

$$L_1^{(i+1)} = \mathbf{L}_1^{(i+1)} + \epsilon\hat{E}_1 \qquad (11.119)$$

$$L_2^{(i+1)} = \mathbf{L}_2^{(i+1)} + \epsilon\hat{E}_2 \qquad (11.120)$$

$$L_3^{(i+1)} = \mathbf{L}_3^{(i+1)} + \epsilon\hat{E}_3 \qquad (11.121)$$

we obtain the following equations

$$D_4^{(i)^T}\hat{E}_3 D_4^{(i)} - \hat{E}_3 =$$
$$- \left(D_2^{(i)^T} L_1^{(i+1)} D_2^{(i)} + D_2^{(i)^T} L_2^{(i+1)} D_4^{(i)} + D_4^{(i)^T} L_2^{(i+1)^T} D_2^{(i)}\right) \qquad (11.122)$$

$$\hat{E}_1 \hat{A}^{(i)^T} + \hat{A}^{(i)} \hat{E}_1 = -\hat{H} D_3^{(i)} - D_3^{(i)^T} H - D_3^{\hat{(i)}^T} \hat{E}_3 D_3^{(i)^T}$$
$$- D_1^{(i)^T} L_1^{(i+1)} D_1^{(i)} - D_3^{(i)^T} L_2^{(i+1)} D_1^{(i)} - D_1^{(i)^T} L_2^{(i+1)} D_3^{(i)} \qquad (11.123)$$

$$\hat{E}_2 = \hat{E}_1 D_2^{(i)} \left(I - D_4^{(i)}\right)^{-1} + \hat{H} \qquad (11.124)$$

where

$$\hat{H} = \left(D_3^{(i)^T}\hat{E}_3 D_4^{(i)} + D_1^{(i)^T} L_1^{(i+1)} D_2^{(i)}\right) \left(I - D_4^{(i)}\right)^{-1}$$
$$+ \left(D_3^{(i)^T} L_2^{(i+1)^T} D_2^{(i)} + D_1^{(i)^T} L_2^{(i+1)} D_4^{(i)}\right) \left(I - D_4^{(i)}\right)^{-1} \qquad (11.125)$$

Equations (11.122)-(11.124) can be solved for $\hat{E}_3$, $\hat{E}_1$, and $\hat{E}_2$, respectively, by proposing a similar kind of algorithm as for (11.97)-(11.99), as follows.

Algorithm 11.5:

Initialize $\hat{E}_1^{(j)} = 0$, $\hat{E}_2^{(j)} = 0$, $\hat{E}_3^{(j)} = 0$.

For $j = 0, 1, 2, \ldots$ do the following iterations

$$D_4^{(i)^T} \hat{E}_3^{(j+1)} D_4^{(i)} - \hat{E}_3^{(j+1)} = -D_2^{(i)^T} \left(\mathbf{L}_1^{(i+1)} + \epsilon\hat{E}_1^{(j)}\right) D_2^{(i)}$$
$$- D_2^{(i)^T} \left(\mathbf{L}_2^{(i+1)} + \epsilon\hat{E}_2^{(j)}\right) D_4^{(i)} - D_4^{(i)^T} \left(\mathbf{L}_2^{(i+1)} + \epsilon\hat{E}_2^{(j)}\right)^T D_2^{(i)} \qquad (11.126)$$

$$\hat{H}^{(j)} = \left(D_3^{(i)^T} \hat{E}_3^{(j+1)} D_4^{(i)} + D_1^{(i)^T} \left(\mathbf{L}_1^{(i+1)} + \epsilon\hat{E}_1^{(j)}\right) D_2^{(i)}\right)$$
$$\times \left(I - D_4^{(i)}\right)^{-1}$$
$$+ \left(D_3^{(i)^T} \left(\mathbf{L}_2^{(i+1)} + \epsilon\hat{E}_2^{(j)}\right) D_2^{(i)} + D_1^{(i)^T} \left(\mathbf{L}_2^{(i+1)} + \epsilon\hat{E}_2^{(j)}\right) D_4^{(i)}\right)$$
$$\times \left(I - D_4^{(i)}\right)^{-1}$$

(11.127)

$$\hat{E}_1^{(j+1)} \hat{A}^{(i)^T} + \hat{A}^{(i)} \hat{E}_1^{(j+1)} = -\hat{H}^{(j)} D_3^{(i)} - D_3^{(i)^T} \hat{H}^{(j)^T}$$
$$- D_3^{(i)^T} \hat{E}_3^{(j+1)} D_3^{(i)} - D_1^{(i)^T} \left(\mathbf{L}_1^{(i+1)} + \epsilon\hat{E}_1^{(j)}\right) D_1^{(i)}$$
$$- D_3^{(i)^T} \left(\mathbf{L}_2^{(i+1)} + \epsilon\hat{E}_2^{(j)}\right) D_1^{(i)} - D_1^{(i)^T} \left(\mathbf{L}_2^{(i+1)} + \epsilon\hat{E}_2^{(j)}\right) D_3^{(i)}$$

(11.128)

$$\hat{E}_2^{(j+1)} = \hat{E}_1^{(j+1)} D_2^{(i)} \left(I - D_4^{(i)}\right)^{-1} + \hat{H}^{(j)}$$

(11.129)

$$\triangle$$

In summary, P can be computed by using equations (11.87)-(11.89), (11.94)-(11.96) and (11.101)-(11.104), and L can be computed by using equations (11.115)-(11.121), and (11.126)-(11.129). Furthermore, since the algorithms for P and L are independent from each other, therefore, the computation can be done in parallel. The following algorithm presents the complete solution to our problem.

Algorithm 11.6:
1. Initialize $F^{(0)}$ such that $\left(A + BF^{(0)}C\right)$ is a stable matrix.
2. For $i = 0, 1, 2, ...$ repeat steps 3-10.
3. Calculate $D_1^{(i)}, D_2^{(i)}, D_3^{(i)}, D_4^{(i)}$, and $S_1^{(i)}, S_2^{(i)}, S_3^{(i)}$, and $q_1^{(i)}, q_2^{(i)}, q_3^{(i)}$ from (11.83) and (11.109).
4. Calculate $P_3^{(i+1)}, P_1^{(i+1)}$, and $P_2^{(i+1)}$ from (11.87)-(11.89).
5. Initialize $E_1^{(0)} = 0$, $E_2^{(0)} = 0$, $E_3^{(0)} = 0$. For $j = 0, 1, 2, ...$ solve equations (11.101)-(11.104) until the desired accuracy is obtained.
6. Construct $P_1^{(i+1)}, P_2^{(i+1)}, P_3^{(i+1)}$ from equations (11.94)-(11.96).
7. Calculate $L_3^{(i+1)}, L_1^{(i+1)}, L_2^{(i+1)}$ from (11.115)-(11.117).

8. Initialize $\hat{E}_1^{(0)} = 0$, $\hat{E}_2^{(0)} = 0$, $\hat{E}_3^{(0)} = 0$. For $j = 1, 2, ...,$ solve equations (11.126)-(11.129) until the desired accuracy is achieved.

9. Construct $L_1^{(i+1)}$, $L_2^{(i+1)}$, $L_3^{(i+1)}$ from equations (11.119)-(11.121).

10. Construct P and L and calculate $F_{new}^{(i+1)}$ and $F^{(i+1)}$ from equations (11.12) and (11.16), respectively.

$\triangle$

If $F^{(0)}$ is chosen such that $A + BF^{(0)}C$ is stable, then this algorithm converges for sufficiently small α such that $0 < \alpha \leq 1$, (Halyo and Broussard, 1981). The effect of different values of α on the convergence speed and the convergence pattern, are shown in the following example.

The block diagram representation of the required calculations for the reduced-order system is shown in Figure 11.2. This block diagram shows the flow of information for the proposed asynchronous parallel algorithm.

11.5 Case Study: Discrete Model of a Steam Power System

In order to demonstrate the efficiency of the proposed algorithm which yields $O\left(\epsilon^j\right)$ approximation, we run a real world physical example (a fifth-order discrete model of a steam power system (Mahmoud, 1982)). The problem matrices are as follows

$$A = \begin{bmatrix} 0.915 & 0.0510 & 0.038 & 0.0150 & 0.0380 \\ -0.030 & 0.889 & -0.0005 & 0.046 & 0.111 \\ -0.006 & 0.468 & 0.247 & 0.014 & 0.048 \\ -0.715 & -0.022 & -0.0211 & 0.240 & -0.024 \\ -0.148 & -0.003 & -0.004 & 0.090 & 0.026 \end{bmatrix}$$

$$B^T = [\, 0.0098 \quad 0.1220 \quad 0.0360 \quad 0.5620 \quad 0.1150 \,]$$

The remaining matrices are chosen as

$$C = \begin{bmatrix} 1 & 1 & 0 & 0 & 0 \\ 0 & 0 & 1 & 1 & 1 \end{bmatrix}$$

$$Q = I_5, \; R = 1, \; G = B, \; V = I_2, \; W = 5$$

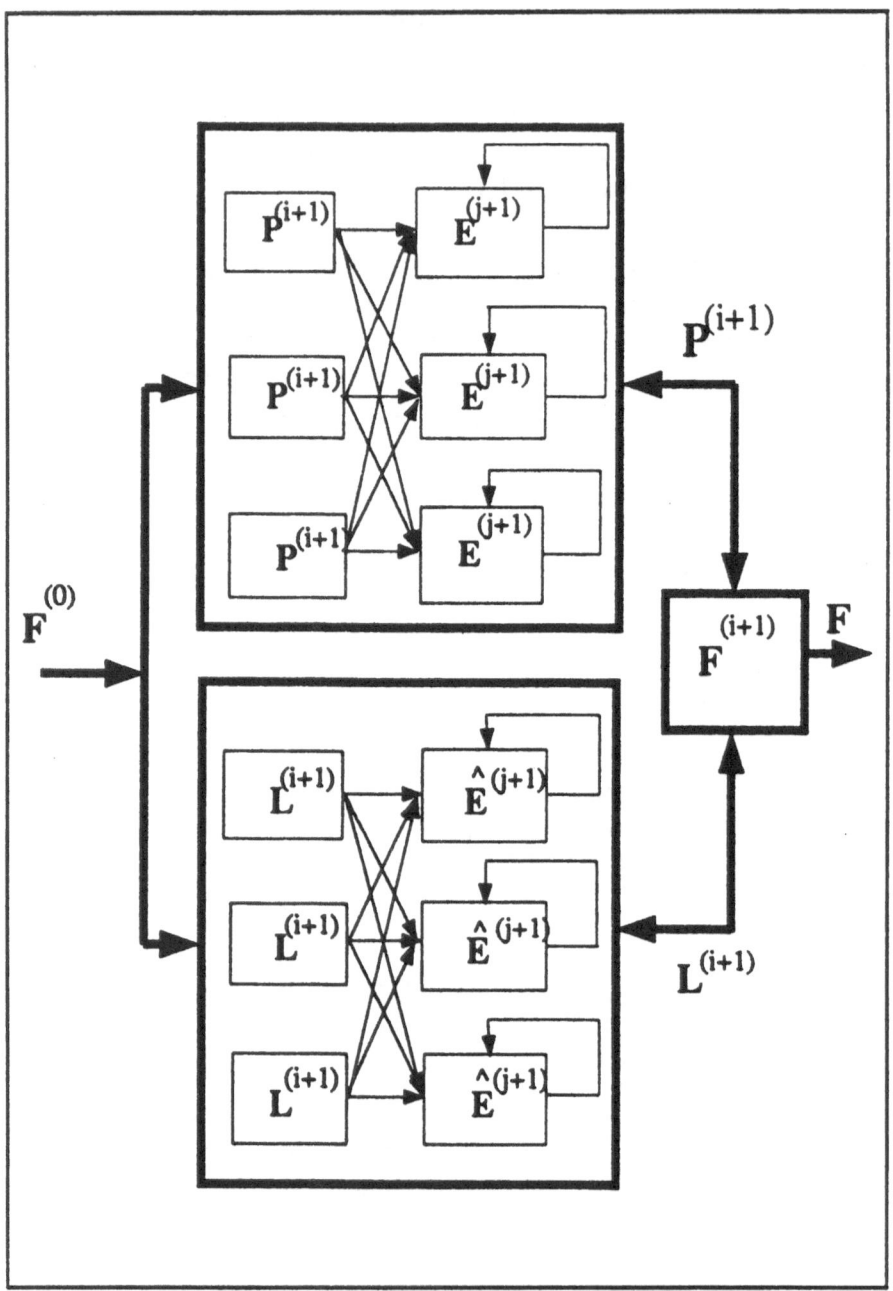

Figure 11.2: Parallelism of the required
computations for the reduced-order system

The modulus of the eigenvalues of matrix A are .9, .9, .25, .25, .03.
Thus we have two fast and three slow variables. The small parameter

ϵ is chosen to be .27 which is roughly the ratio .25/.9. The $O\left(\epsilon^j\right)$ approximation is demonstrated in Table 11.3. The number of iterations for approximating P and L are high enough so that

$$\left\|E_m^{(j+1)}\right\|_\infty - \left\|E_m^{(j)}\right\|_\infty \quad and \quad \left\|\hat{E}_m^{(j+1)}\right\|_\infty - \left\|\hat{E}_m^{(j)}\right\|_\infty ,$$

$$m = 1, 2, 3$$

are less than 10^{-5}.

Figure 11.3 shows the effect of α on the convergence speed. It is noted that the convergence is the fastest if α is chosen close to .85. Figure 11.4 shows the convergence behavior for different α.

$\epsilon = 0.27, \alpha = 0.3$ i	$J^{(i)}$	$J^{(i)} - J_{opt}^{(i)}$
1	2.86059	0.28163
2	2.68072	0.10176
3	2.61881	0.03985
4	2.59517	0.01621
5	2.58570	0.00674
6	2.58178	0.00282
7	2.58014	0.00181
8	2.57945	0.00049
9	2.57915	0.00019
10	2.57902	0.00006
11	2.57897	0.00001
12	2.57896	0.00000

Table 11.3: Performance criterion per iteration

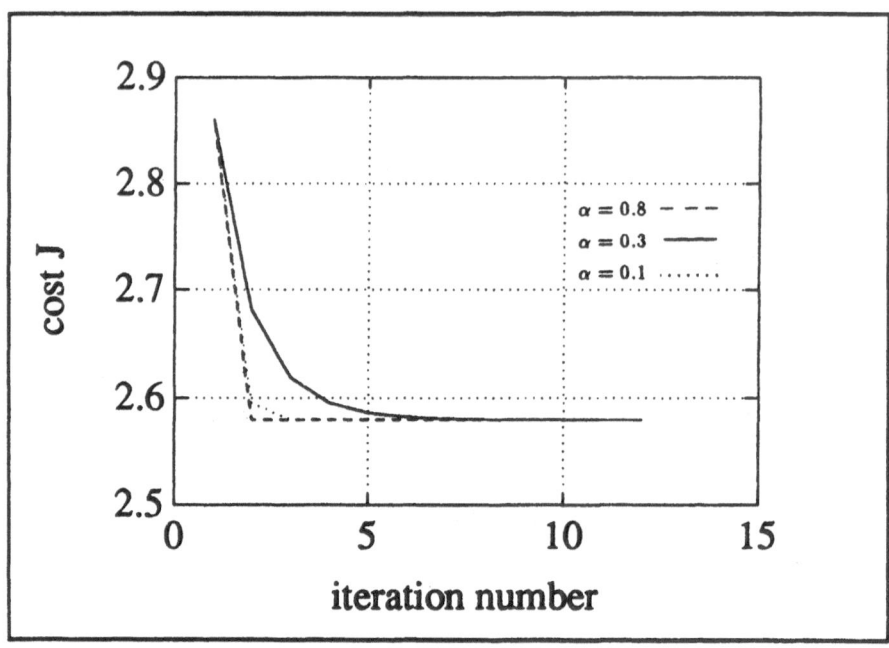

Figure 11.3: Effect of α on the convergence speed

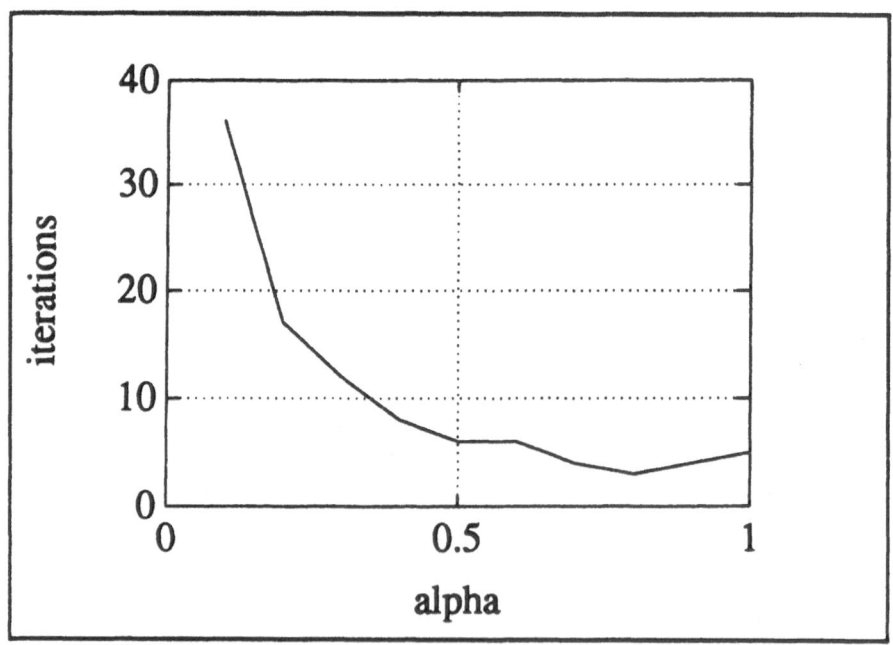

Figure 11.4: Convergence behavior for different α

Chapter 12

Applications to Differential Games

In this chapter, we present a parallel synchronous algorithm for solving the Nash differential game of weakly coupled linear systems. The singularly perturbed Nash differential games have been studied by several researchers (Gardner and Cruz, 1978; Khalil and Kokotovic, 1979c; Khalil, 1980a, 1980b). However, no algorithm has been proposed for solving the coupled algebraic Riccati equations of the singularly perturbed systems whose solutions comprise the required Nash strategies. That is why, we limit our study to the weakly coupled systems only.

12.1 Weakly Coupled Linear-Quadratic Nash Games

The linear-quadratic Nash game strategies of large scale weakly interconnected (coupled) systems were studied in (Ozguner and Perkins, 1977) by means of a power-series expansion method with respect to a small coupling parameter ϵ. This approach, originated in (Kokotovic et al., 1969), is not recursive in its application and can be inferior compared to the hierarchical type decentralized control method (especially when ϵ is not very small), as was pointed out in (Mahmoud, 1978). In this section, we develop a recursive technique which will recover the importance of ideas presented in (Kokotovic et al., 1969). Motivated by previous results for singularly perturbed systems (Gajic, 1986), we have shown that weak coupling produces algebraic problems similar to those of (Gajic, 1986) and the fixed point method used in (Gajic, 1986) is very efficient in this case also.

343

As a matter of fact, we have developed an algorithm which converges very rapidly to the exact, nonnegative definite stabilizing solution of the coupled algebraic Riccati equations and thus to the optimal linear Nash strategies, even in the case when ϵ is not small.

A controlled linear dynamic system under consideration is given by

$$\dot{x} = A(\epsilon) x + B_1(\epsilon) u_1 + B_2(\epsilon) u_2 \tag{12.1}$$

where $x \in \Re^n$ is a state vector, $u_1 \in \Re^{m_1}$ and $u_2 \in \Re^{m_2}$ are control inputs, $A(\epsilon)$, $B_i(\epsilon)$, $i = 1, 2$, are bounded matrix functions of a small parameter ϵ with compatible dimensions.

A quadratic type functional is associated with each control agent

$$J_1 = \frac{1}{2} \int_0^\infty \left[x^T Q_1(\epsilon) x + u_1^T R_1(\epsilon) u_1 + u_2^T R_{12}(\epsilon) u_2 \right] dt \tag{12.2a}$$

$$J_2 = \int_0^\infty \left[x^T Q_2(\epsilon) x + u_1^T R_{21}(\epsilon) u_1 + u_2^T R_2(\epsilon) u_2 \right] dt \tag{12.2b}$$

where the weighting matrices are symmetric satisfying $Q_i(\epsilon) \geq 0$, $R_i(\epsilon) > 0$, $R_{ij}(\epsilon) \geq 0$, $i = 1, 2$; $j = 1, 2$.

The optimal solution to the given problem with the conflict of interest and simultaneous decision making (Starr and Ho, 1969), leads to so called Nash strategies u_1^* and u_2^* satisfying

$$J_1(u_1^*, u_2^*) \leq J_1(u_1, u_2^*) \tag{12.3a}$$

$$J_2(u_1^*, u_2^*) \leq J_2(u_1^*, u_2) \tag{12.3b}$$

It was shown in (Starr and Ho, 1969) that the optimal closed-loop Nash strategies are given by

$$u_i^*(t) = R_i^{-1}(\epsilon) B_i^T(\epsilon) K_i(\epsilon) x(t), \quad i = 1, 2 \tag{12.4}$$

where K_i, $i = 1, 2$, satisfy coupled algebraic Riccati equations

$$K_1(\epsilon) A(\epsilon) + A^T(\epsilon) K_1(\epsilon) + Q_1(\epsilon) - K_1(\epsilon) S_1(\epsilon) K_1(\epsilon) -$$
$$K_1(\epsilon) S_2(\epsilon) K_2(\epsilon) - K_2(\epsilon) S_2(\epsilon) K_1(\epsilon) + K_2(\epsilon) Z_2(\epsilon) K_2(\epsilon)$$
$$= 0 = N_1(K_1, K_2)$$
$$\tag{12.5a}$$

$$K_2(\epsilon) A(\epsilon) + A^T(\epsilon) K_2(\epsilon) + Q_2(\epsilon) - K_2(\epsilon) S_2(\epsilon) K_2(\epsilon)$$
$$-K_2(\epsilon) S_1(\epsilon) K_1(\epsilon) - K_1(\epsilon) S_1(\epsilon) K_2(\epsilon) + K_1(\epsilon) Z_1(\epsilon) K_1(\epsilon)$$
$$= 0 = N_2(K_1, K_2)$$

$$\text{(12.5b)}$$

where

$$S_i(\epsilon) = B_i(\epsilon) R_i^{-1}(\epsilon) B_i^T(\epsilon), \quad i = 1, 2$$
$$Z_i(\epsilon) = B_i(\epsilon) R_i^{-1}(\epsilon) R_{ji}(\epsilon) R_i^{-1}(\epsilon) B_i^T(\epsilon)$$
$$i = 1, 2; \ j = 1, 2, \ i = j$$

The existence of the nonlinear optimal Nash strategies was established in (Basar, 1974), so that (12.4), in fact, are the best linear optimal strategies. Since a linear control law is very desirable from a practical point of view, the linear strategies (12.4) attract the attention of many researchers.

The existence of Nash strategies (12.4) and solutions of the coupled algebraic Riccati equations (12.5) has been studied in (Papavassilopoulos et al., 1979), by means of Brower's fixed point theorem and by imposing norm conditions on the given matrices. In the recent paper (Gajic and Li, 1988), under control-oriented assumptions (Kucera, 1972; Wonham, 1968), the algorithms for finding the nonnegative definite stabilizing solutions of (12.5) have been proposed (see Appendix 12.2). However, the rigorous convergence proofs have to be worked out for both of the two algorithms presented in (Gajic and Li, 1988).

It is important to point out that at the present time, there is no published general method for finding stabilizing solutions of the coupled algebraic Riccati equations (12.5). Some attempts in that direction have been made in (Bertrand, 1985; Papavassilopoulos and Olsder, 1984).

In this section, the Nash game problem is considered for a special case of weakly interconnected systems characterized by

$$A(\epsilon) = \begin{bmatrix} A_1(\epsilon) & \epsilon A_{12}(\epsilon) \\ \epsilon A_{21}(\epsilon) & A_2(\epsilon) \end{bmatrix}$$

$$B_1(\epsilon) = \begin{bmatrix} B_{11}(\epsilon) \\ \epsilon B_{21}(\epsilon) \end{bmatrix}; \quad B_2(\epsilon) = \begin{bmatrix} \epsilon B_{21}(\epsilon) \\ B_{22}(\epsilon) \end{bmatrix}$$

$$Q_1(\epsilon) = \begin{bmatrix} U_1(\epsilon) & \epsilon U_{12}(\epsilon) \\ \epsilon U_{12}^T(\epsilon) & \epsilon^2 U_2(\epsilon) \end{bmatrix}; \quad Q_2(\epsilon) = \begin{bmatrix} \epsilon^2 V_1(\epsilon) & \epsilon V_{12}(\epsilon) \\ \epsilon V_{12}^T(\epsilon) & V_2(\epsilon) \end{bmatrix}$$

This partition decomposes the state vector x into two vectors $x_1 \in \Re^{n_1}$ and $x_2 \in \Re^{n_2}$ such that $n_1 + n_2 = n$. Since the small coupling parameter ϵ can not change the basic structure of the subsystems by destroying their main properties (otherwise we can not talk about the weak coupling), it is very natural to adopt the following form for the subsystem matrices.

Assumption 12.1 (Weak coupling assumption)

$$A_i(\epsilon) = A_{i0} + \epsilon A_{0i}(\epsilon), \quad B_{ii}(\epsilon) = B_{i0} + \epsilon B_{0i}(\epsilon)$$
$$U_1(\epsilon) = U_{10} + \epsilon U_{01}(\epsilon), \quad V_2(\epsilon) = V_{20} + \epsilon V_{02}(\epsilon)$$
$$R_i(\epsilon) = R_{i0} + \epsilon R_{0i}(\epsilon), \quad i = 1, 2$$

where $A_{0i}(\epsilon)$, $B_{0i}(\epsilon)$, $R_{0i}(\epsilon)$, $i = 1, 2$; $U_{01}(\epsilon)$ and $V_{02}(\epsilon)$ are continuous functions of ϵ, whereas A_{i0}, B_{i0}, R_{i0}, $i = 1, 2$, and U_{10}, V_{20} are independent of ϵ.

$$\triangle$$

In order to simplify the algebra, we will assume, without loss of generality, that $U_{12}(\epsilon) = 0$, $V_{12}(\epsilon) = 0$, $R_{12}(\epsilon) = 0$, $R_{21}(\epsilon) = 0$, $U_2(\epsilon) = 0$, $V_1(\epsilon) = 0$, $B_{12}(\epsilon) = 0$, $B_{21}(\epsilon) = 0$. Note that we are studying a more general case that the one studied in (Ozguner and Perkins, 1977) because of the ϵ-dependence of the problem matrices. In addition, we do not need to impose the analyticity assumption with respect to ϵ, which must be done for the power-series expansion method.

The following scaling of $K_1(\epsilon)$ and $K_2(\epsilon)$ is consistent with the nature of the solution of (12.5)

$$K_1(\epsilon) = \begin{bmatrix} M_1(\epsilon) & \epsilon M_{12}(\epsilon) \\ \epsilon M_{12}^T(\epsilon) & \epsilon^2 M_2(\epsilon) \end{bmatrix}, \quad K_2(\epsilon) = \begin{bmatrix} \epsilon^2 N_1(\epsilon) & \epsilon N_{12}(\epsilon) \\ \epsilon N_{12}^T(\epsilon) & N_2(\epsilon) \end{bmatrix}$$
$$(12.6)$$

The very well-known ϵ-decoupling method (Kokotovic et al., 1969), based on the power-series expansion with respect to ϵ, will convert the given full-order problem (12.5) to a family of reduced-order problems (Ozguner and Perkins, 1977). However, the power-series expansion method is not recursive in nature and in the case when we are interested in a high order of accuracy or when ϵ is not very small, the size of the required computations can be considerable. Moreover, when the problem matrices are functions of ϵ, the power-series method demands the analyticity of all matrices. On the other hand, the expansion of quadratic terms (for example, $K_1(\epsilon) B_1(\epsilon) R_1^{-1}(\epsilon) B_1^T(\epsilon) K_1(\epsilon)$) will produce an enormous number of terms, so that the reduced-order advantage of the series expansion method becomes questionable. The presence of a small parameter ϵ will be exploited in the next section from a different point of view, leading to the recursive scheme for the solution of (12.5). Since the proposed method is of the fixed-point type, the boundness of all problem matrices over a compact set $\epsilon \in [0, \epsilon_1]$ has to be imposed. This is a much milder condition than the analyticity requirement of the power-series expansion method.

12.2 Solution of Coupled Algebraic Riccati Equations

Partitioning (12.5) compatibly with (12.6), we get the following set of equation

$$M_1(\epsilon) A_1(\epsilon) + A_1^T(\epsilon) M_1(\epsilon) + U_1(\epsilon) - M_1(\epsilon) S_{11}(\epsilon) M_1(\epsilon)$$
$$+\epsilon^2\{M_{12}(\epsilon) A_{21}(\epsilon) + A_{21}^T(\epsilon) M_{12}^T(\epsilon) - M_{12}(\epsilon) S_{22}(\epsilon) N_{12}^T(\epsilon)$$
$$-N_{12}(\epsilon) S_{22}(\epsilon) M_{12}^T(\epsilon)\} = 0$$
$$(12.7a)$$

$$M_1(\epsilon) A_{12}(\epsilon) + M_{12}(\epsilon) A_2(\epsilon) - M_1(\epsilon) S_{11}(\epsilon) M_{12}(\epsilon)$$
$$-M_{12}(\epsilon) S_{22}(\epsilon) N_2(\epsilon) - \epsilon^2\{N_{12}(\epsilon) S_{22}(\epsilon) M_2(\epsilon) - A_{21}^T(\epsilon) M_2(\epsilon)\}$$
$$+A_1^T(\epsilon) M_{12}(\epsilon) = 0$$
$$(12.7b)$$

$$M_2(\epsilon) A_2(\epsilon) + A_2^T(\epsilon) M_2(\epsilon) - M_2(\epsilon) S_{22}(\epsilon) N_2(\epsilon)$$
$$-N_2(\epsilon) S_{22}(\epsilon) M_2(\epsilon) + M_{12}^T(\epsilon) A_{21}(\epsilon) + A_{21}^T(\epsilon) M_{12}(\epsilon) \quad (12.7c)$$
$$-M_{12}^T(\epsilon) S_{11}(\epsilon) M_{12}(\epsilon) = 0$$

$$N_1(\epsilon) A_1(\epsilon) + A_1^T(\epsilon) N_1(\epsilon) - N_1(\epsilon) S_{11}(\epsilon) M_1(\epsilon)$$
$$-M_1(\epsilon) S_{11}(\epsilon) N_1(\epsilon) + N_{12}(\epsilon) A_{21}(\epsilon) + A_{21}^T(\epsilon) N_{12}^T(\epsilon) \quad (12.7d)$$
$$-N_{12}(\epsilon) S_{22}(\epsilon) N_{12}^T(\epsilon)\} = 0$$

$$\epsilon^2 N_1(\epsilon) A_{12}(\epsilon) + N_{12}(\epsilon) A_2(\epsilon) - N_{12}(\epsilon) S_{22}(\epsilon) N_2(\epsilon)$$
$$-\epsilon^2 N_1(\epsilon) S_{11}(\epsilon) M_{12}(\epsilon) - M_1(\epsilon) S_{11}(\epsilon) N_{12}(\epsilon) \quad (12.7e)$$
$$+A_{21}^T(\epsilon) N_2(\epsilon) + A_1^T(\epsilon) N_{12}(\epsilon) = 0$$

$$N_2(\epsilon) A_2(\epsilon) + A_2^T(\epsilon) N_2(\epsilon) + V_2(\epsilon) - N_2(\epsilon) S_{22}(\epsilon) N_2(\epsilon)$$
$$+\epsilon^2\{N_{12}^T(\epsilon) A_{12}(\epsilon) + A_{12}^T(\epsilon) N_{12}(\epsilon) - N_{12}^T(\epsilon) S_{11}(\epsilon) M_{12}(\epsilon)$$
$$-M_{12}^T(\epsilon) S_{11}(\epsilon) N_{12}(\epsilon)\} = 0$$
$$(12.7f)$$

where
$$S_{ii}(\epsilon) = B_{ii}(\epsilon) R_i^{-1}(\epsilon) B_{ii}^T(\epsilon), \quad i = 1, 2$$

347

12.2.1 Zeroth-Order Approximation

Let us define the $O\left(\epsilon^2\right)$ perturbation of (12.7) as

$$M_1\left(\epsilon\right) A_1\left(\epsilon\right) + A_1^T\left(\epsilon\right) M_1\left(\epsilon\right) + U_1\left(\epsilon\right) - M_1\left(\epsilon\right) S_{11}\left(\epsilon\right) M_1\left(\epsilon\right) = 0 \tag{12.8a}$$

$$M_{12}\left(\epsilon\right) D_2\left(\epsilon\right) + D_1^T\left(\epsilon\right) M_{12}\left(\epsilon\right) = -M_1\left(\epsilon\right) A_{12}\left(\epsilon\right) \tag{12.8b}$$

$$\begin{aligned} &M_2\left(\epsilon\right) D_2\left(\epsilon\right) + D_2^T\left(\epsilon\right) M_2\left(\epsilon\right) \\ = &M_{12}^T\left(\epsilon\right) S_{11}\left(\epsilon\right) M_{12}\left(\epsilon\right) - M_{12}^T\left(\epsilon\right) A_{12}\left(\epsilon\right) - A_{12}^T\left(\epsilon\right) M_{12}\left(\epsilon\right) \end{aligned} \tag{12.8c}$$

$$\begin{aligned} &N_1\left(\epsilon\right) D_1\left(\epsilon\right) + D_1^T\left(\epsilon\right) N_1\left(\epsilon\right) \\ = &N_{12}\left(\epsilon\right) S_{22}\left(\epsilon\right) N_{12}^T\left(\epsilon\right) - N_{12}\left(\epsilon\right) A_{21}\left(\epsilon\right) - A_{21}^T\left(\epsilon\right) M_{12}^T\left(\epsilon\right) \end{aligned} \tag{12.8d}$$

$$N_{12}\left(\epsilon\right) D_2\left(\epsilon\right) + D_1^T\left(\epsilon\right) N_{12}\left(\epsilon\right) = -A_{21}^T\left(\epsilon\right) N_2\left(\epsilon\right) \tag{12.8e}$$

$$N_2\left(\epsilon\right) A_2\left(\epsilon\right) + A_2^T\left(\epsilon\right) N_2\left(\epsilon\right) + V_2\left(\epsilon\right) - N_2\left(\epsilon\right) S_{22}\left(\epsilon\right) N_2\left(\epsilon\right) = 0 \tag{12.8f}$$

where

$$\begin{aligned} D_1\left(\epsilon\right) &= A_1\left(\epsilon\right) - S_{11}\left(\epsilon\right) M_1\left(\epsilon\right) \\ D_2\left(\epsilon\right) &= A_2\left(\epsilon\right) - S_{22}\left(\epsilon\right) N_2\left(\epsilon\right) \end{aligned}$$

This system of equations has decoupled form and can be solved like two lower order Riccati equations (12.8a), (12.8f) and four lower order Lyapunov equations (12.8b)-(12.8e). The nonnegative definite stabilizing solution of (12.8a) and (12.8f) exist under the well-known stabilizability-detectability assumption (Kucera, 1972; Wonham, 1968).

Assumption 12.2 The triples $\left(A_1\left(0\right), B_1\left(0\right), \sqrt{U_1\left(0\right)}\right)$ and $\left(A_2\left(0\right), B_2\left(0\right), \sqrt{V_2\left(0\right)}\right)$ are stabilizable-detectable.

$$\triangle$$

Under the same assumption, the unique solutions of (12.8b)-(12.8e) exist since $D_1\left(\epsilon\right)$ and $D_2\left(\epsilon\right)$ are stable matrices (Kucera, 1972; Wonham, 1968).

12.2.2 Solution of Higher Order of Accuracy

The zeroth-order solutions $M(\epsilon)$ and $N(\epsilon)$ are $O(\epsilon^2)$ close to the exact ones. The exact solutions can be sought in the form

$$K_1(\epsilon) = \begin{bmatrix} M_1(\epsilon) + \epsilon^2 E_1(\epsilon) & \epsilon\left[M_{12}(\epsilon) + \epsilon^2 E_{12}(\epsilon)\right] \\ \epsilon\left[M_{12}(\epsilon) + \epsilon^2 E_{12}(\epsilon)\right]^T & \epsilon^2\left[M_2(\epsilon) + \epsilon^2 E_2(\epsilon)\right] \end{bmatrix}$$
(12.9a)

$$K_2(\epsilon) = \begin{bmatrix} \epsilon^2\left[N_1(\epsilon) + \epsilon^2 G_1(\epsilon)\right] & \epsilon\left[N_{12}(\epsilon) + \epsilon^2 G_{12}(\epsilon)\right] \\ \epsilon\left[N_{12}(\epsilon) + \epsilon^2 G_{12}(\epsilon)\right]^T & N_2(\epsilon) + \epsilon^2 G_2(\epsilon) \end{bmatrix}$$
(12.9b)

Obviously, $O(\epsilon^2)$ approximations of $E(\epsilon)'s$ and $G(\epsilon)'s$ will produce $O(\epsilon^{k+2})$ approximations of required solutions, which is why we are interested in finding convenient form for these error terms and the appropriate algorithm for their solutions.

Subtracting equations (12.8) from corresponding equations (12.7) and after doing some algebra we get the following expressions for the error equations

$$E_1 D_1 + D_1^T E_1 = C_1 + \epsilon^2 F_1(E_1, E_{12}, G_{12})$$
(12.10a)

$$E_1 D_{12} + E_{12} D_2 + D_1^T E_{12} - M_{12} S_{22} G_2$$
$$= C_2 + \epsilon^2 F_2(E_1, E_{12}, G_{12}, E_2, G_2)$$
(12.10b)

$$E_{12}^T D_{12} + D_{12}^T E_{12} + E_2 D_2 + D_2^T E_2 - G_2 S_{22} M_2 - M_2 S_{22} G_2$$
$$= \epsilon^2 F_3(E_{12}, E_2, G_2)$$
(12.10c)

$$G_1 D_1 + D_1^T G_2 + G_{12} D_{21} + D_{21}^T G_{12}^T - E_1 S_{11} N_1 - N_1 S_{11} E_1$$
$$= \epsilon^2 F_4(E_1, G_{12}, G_2)$$
(12.10d)

$$G_1 D_2 + D_1^T G_{12} + D_{21}^T G_2 - E_1 S_{11} N_{12}$$
$$= C_5 + \epsilon^2 F_5(E_1, E_{12}, G_1, G_{12}, G_2)$$
(12.10e)

$$G_2 D_2 + D_2^T G_2 = C_6 + \epsilon^2 F_6(E_{12}, G_{12}, G_2)$$
(12.10f)

where

$$D_{12} = D_{12}(\epsilon) = A_{12}(\epsilon) - S_{11}(\epsilon) M_{12}(\epsilon)$$
$$D_{21} = D_{21}(\epsilon) = A_{21}(\epsilon) - S_{22}(\epsilon) N_{12}(\epsilon)$$

DIFFERENTIAL GAMES

Matrices F_i, $i = 1, 2, ...6$, and constant matrices C_j are given in Appendix 12.1. In order to simplify notation, the ϵ-dependence of the problem matrices in the equation (12.10) and in the remaining part of the chapter is omitted.

The weakly coupled and hierarchical structure of (12.10) can be exploited by proposing the following recursive scheme, which leads, after some algebra, to the six low-order completely decoupled Lyapunov equations.

Algorithm 12.1:

$$E_1^{(i+1)} D_1 + D_1^T E_1^{(i+1)}$$
$$= \epsilon^2 E_1^{(i)} S_{11} E_1^{(i)} - M_{12}^{(i)} D_{21}^{(i)} - D_{21}^{T(i)} M_{12}^{T(i)} \tag{12.11a}$$

$$E_{12}^{(i+1)} D_2 + D_1^T E_{12}^{(i+1)}$$
$$= -E_1^{(i+1)} D_{12}^{(i)} + M_{12}^{(i)} S_{22} G_2^{(i+1)} - D_{21}^{T(i)} M_2^{(i)} \tag{12.11b}$$

$$E_2^{(i+1)} D_2 + D_2^T E_2^{(i+1)} = M_2^{(i)} S_{22} G_2^{(i+1)}$$
$$+ G_2^{(i+1)} S_{22} M_2^{(i)} - E_{12}^{T(i+1)} D_{12}^{(i)} - D_{12}^{T(i)} E_{12}^{(i+1)} + \epsilon^2 E_{12}^{T(i+1)} S_{11} E_{12}^{(i+1)} \tag{12.11c}$$

$$G_1^{(i+1)} D_1 + D_1^T G_1^{(i+1)} = E_1^{(i+1)} S_{11} N_1^{(i)} + N_1^{(i)} S_{11} E_1^{(i+1)}$$
$$- G_{12}^{(i+1)} D_{21}^{(i)} - D_{21}^{T(i)} G_{12}^{T(i+1)} + \epsilon^2 G_{12}^{(i+1)} S_{22} G_{12}^{T(i+1)} \tag{12.11d}$$

$$G_{12}^{(i+1)} D_2 + D_1^T G_{12}^{(i+1)}$$
$$= -D_{21}^{T(i)} G_2^{(i+1)} + E_1^{(i+1)} S_{11} N_{12}^{(i)} - N_1^{(i)} D_{12}^{(i)} \tag{12.11e}$$

$$G_2^{(i+1)} D_2 + D_2^T G_2^{(i+1)}$$
$$= \epsilon^2 G_2^{(i)} S_{22} G_2^{(i)} - N_{12}^{T(i)} D_{12}^{(i)} - D_{12}^{T(i)} N_{12}^{(i)} \tag{12.11f}$$
$$i = 0, 1, 2, 3, ..$$

with initial conditions chosen as

$$E_1^{(0)} = E_{12}^{(0)} = E_2^{(0)} = G_1^{(0)} = G_{12}^{(0)} = G_2^{(0)} = 0$$

where

$$M_{12}^{(i)} = M_{12} + \epsilon^2 E_{12}^{(i)}, \quad N_{12}^{(i)} = N_{12} + \epsilon^2 G_{12}^{(i)}$$
$$N_1^{(i)} = N_1 + \epsilon^2 G_1^{(i)}, \quad M_2^{(i)} = M_2 + \epsilon^2 E_2^{(i)}$$
$$D_{12}^{(i)} = A_{12} - S_{11} M_{12}^{(i)}, \quad D_{21}^{(i)} = A_{21} - S_{22} N_{12}^{T(i)}$$
$$i = 1, 2, 3, ...$$

$\triangle$

These Lyapunov equations have to be solved in the given order, that is, first E_1 and G_2, then E_{12} and G_{12}, and finally E_2 and G_1.

The following theorem indicates the features of the proposed recursive scheme.

Theorem 12.1 *Under imposed weak coupling and stabilizability and detectability assumptions, given algorithm (12.11) converges to the exact solution of the error terms, and thus of $K_1(\epsilon)$ and $K_2(\epsilon)$, with the rate of convergence of $O(\epsilon^2)$, that is*

$$\| E_j(\epsilon) - E_j^{(i)}(\epsilon) \| = O(\epsilon^{2i})$$
$$\| G_j(\epsilon) - G_j^{(i)}(\epsilon) \| = O(\epsilon^{2i})$$
$$\| E_{12}(\epsilon) - E_{12}^{(i)}(\epsilon) \| = O(\epsilon^{2i}) \qquad (12.11)$$
$$\| G_{12}(\epsilon) - G_{12}^{(i)}(\epsilon) \| = O(\epsilon^{2i})$$
$$j = 1, 2; \quad i = 1, 2, 3, ...$$

and

$$\| K_j(\epsilon) - K_j^{(i)}(\epsilon) \| = O(\epsilon^{2i+2}) \qquad (12.12)$$
$$j = 1, 2; \quad i = 0, 1, 2, ..$$

$\diamond$

Proof: As a starting point, we need to show the existence of a bounded solution of (12.10) in the neighborhood of $\epsilon = 0$. By the implicit function theorem it is enough to show that the corresponding Jacobian is nonsingular at $\epsilon = 0$. The Jacobian is given by

$$J(\epsilon)_{|\epsilon=0} = \begin{bmatrix} \Gamma_1 & 0 & 0 & 0 & 0 & 0 \\ * & \Gamma_2 & 0 & 0 & 0 & * \\ 0 & * & \Gamma_3 & 0 & 0 & * \\ * & 0 & 0 & \Gamma_1 & * & 0 \\ * & 0 & 0 & 0 & \Gamma_2 & * \\ 0 & 0 & 0 & 0 & 0 & \Gamma_3 \end{bmatrix} \qquad (12.13)$$

where the asterisk denotes terms which are not important for a nonsingularity of the Jacobian. Γ's are given by the Kronecker product representation

$$\Gamma_i = I_{n_i} \otimes D_i^T(0) + D_i^T(0) \otimes I_{n_i}, \quad i = 1, 3$$
$$\Gamma_2 = I_{n_2} \otimes D_2^T(0) + D_1^T(0) \otimes I_{n_1}$$

where I_{n_i} and I_{n_2} are identity matrices. Under Assumptions 12.1 and 12.2, $D_1(0)$ and $D_2(0)$ are stable matrices for any sufficiently small

$\epsilon \in [0, \epsilon_2]$ and by the well-known properties of the Kronecker product (Lancaster and Tismenetsky, 1985), so are matrices Γ_1, Γ_2, and Γ_3. It is easy to see that the nonsingularity of the Jacobian is guaranteed by the nonsingularity of Γ_1, Γ_2, and Γ_3.

The second step in the proof of the given theorem is to give an estimate of the rate of convergence.

For $i = 0$, (12.10a) and (12.11a) imply

$$\left(E_1 - E_1^{(1)}\right) D_1 + D_1^T \left(E_1 - E_1^{(1)}\right) = \epsilon^2 F_1 \left(E_1, E_{12}, G_{12}\right)$$

which by stability of D_1 and the existence of the bounded solution of (12.10) gives

$$\| E_1 - E_1^{(1)} \| = O\left(\epsilon^2\right) \qquad (12.14a)$$

By the same arguments, from (12.10f) and (12.11f) we have

$$\| G_2 - G_2^{(1)} \| = O\left(\epsilon^2\right) \qquad (12.14f)$$

Subtracting (12.11b) from (12.10b) and using (12.14a) and (12.14f) and the expression for F_3 (from Appendix 12.1) lead to

$$\left(E_{12} - E_{12}^{(1)}\right) D_2 + D_1^T \left(E_{12} - E_{12}^{(1)}\right) = O\left(\epsilon^2\right)$$

which implies that

$$\| E_{12} - E_{12}^{(1)} \| = O\left(\epsilon^2\right) \qquad (12.14b)$$

By analogy [equations (12.10b) and (12.10e) have similar forms], (12.10e) and (12.11e) will produce

$$\| G_{12} - G_{12}^{(1)} \| = O\left(\epsilon^2\right) \qquad (12.14e)$$

Also, from (12.10c), (12.11c), (12.14a,b,c,d,e,f) and Appendix 12.1, we have

$$\left(E_2 - E_2^{(1)}\right) D_2 + D_2^T \left(E_2 - E_2^{(1)}\right) = O\left(\epsilon^2\right)$$

that is,

$$\| E_2 - E_2^{(1)} \| = O\left(\epsilon^2\right) \qquad (12.14c)$$

and, by analogy, from (12.10d) and (12.11d) we get

$$\| G_1 - G_1^{(1)} \| = O\left(\epsilon^2\right) \qquad (12.14d)$$

Using these starting observations and forms of $F'_j s$ and $C'_j s$, it can be shown that

$$\| F_j - F_j^{(i)} \| = O\left(\epsilon^{2i}\right), \quad j = 1,2; \quad i = 1,2,3,... \tag{12.15}$$

For example, for $j = 1$

$$F_1 - F_1^{(i)} = \left(E_1 - E_1^{(i)}\right) S_{11} E_1^{(i)} + E_1 S_{11}\left(E_1 - E_1^{(i)}\right)$$
$$- \left(E_{12} - E_{12}^{(i)}\right) D_{21} - D_{21}^T \left(E_{12} - E_{12}^{(i)}\right)^T$$
$$+ \left(G_{12} - G_{12}^{(i)}\right) S_{22} M_{12}^{T^{(i)}} + M_{12}^{(i)} S_{22}\left(G_{12} - G_{12}^{(i)}\right)^T$$

so that for $i = 1$, from (12.14) we have $F_1 - F_1^{(1)} = O\left(\epsilon^2\right)$, that is

$$\left(E_1 - E_1^{(2)}\right) D_1 + D_1^T\left(E_1 - E_1^{(2)}\right) = \epsilon^2\left(F_1 - F_1^{(1)}\right) = O\left(\epsilon^4\right)$$

which implies that

$$\left(E_1 - E_1^{(2)}\right) = O\left(\epsilon^4\right)$$

Continuing the same procedure, we can verify (12.15), which by the existence of the bounded solutions of $E's$ and $G's$ will imply (12.12). Note that the solution of (12.11) exist at each iteration since the corresponding Jacobian is always given by (12.13), and thus nonsingular at $\epsilon = 0$ in every iteration.

●

We would like to point out that the imposed form of solution (12.9) is an additional limiting factor for a small parameter ϵ. Since the solution of (12.10) is symmetric only (which can be easily seen from the form of corresponding equations), the small parameter ϵ has to be constrained to the set $\epsilon \in [0, \epsilon_3]$ such that $\forall \epsilon$, $K_1(\epsilon)$ and $K_2(\epsilon)$ preserve the required nonnegative definiteness. Thus, the presented method is applicable for $\epsilon \in [0, \epsilon^*]$, where $\epsilon^* = min\{\epsilon_1, \epsilon_2, \epsilon_3, ...\}$. However, the limiting condition $\epsilon^* = min\{\epsilon_1, \epsilon_2, \epsilon_3, ...\}$ is present in the entire theory of small parameters (weak coupling and singular perturbations), it is both method-dependent and problem-dependent, and not a direct consequence of the procedure studied in this chapter.

Let us compare the proposed algorithm (12.11), based on the fixed point iteration for weakly coupled systems and the power-series expansion algorithm for the same type of systems. The comparison is done for the case when the problem matrices are not functions of ϵ (which is in the favor of the power-series expansion algorithm). The equations corresponding to (12.11) are given by (Ozguner and Perkins, 1977)

$$M_1^{(i+1)} D_1 + D_1^T M_1^{(i+1)} = Z_1^{(0,1,2,...i)} \qquad (12.16a)$$

$$N_2^{(i+1)} D_2 + D_2^T N_2^{(i+1)} = Z_6^{(0,1,2,...i)} \qquad (12.16f)$$

$$M_{12}^{(i+1)} D_2 + D_1^T M_{12}^{(i+1)} = Z_2^{(0,1,2,...i)} \qquad (12.16b)$$

$$N_{12}^{(i+1)} D_2 + D_1^T G_{12}^{(i+1)} = Z_5^{(0,1,2,...i)} \qquad (112.16e)$$

$$M_2^{(i+1)} D_2 + D_2^T M_2^{(i+1)} = Z_3^{(0,1,2,...i)} \qquad (12.16c)$$

$$N_1^{(i+1)} D_1 + D_1^T N_1^{(i+1)} = Z_4^{(0,1,2,...i)} \qquad (12.16d)$$

where Z_j, $j = 1, 2, ...6$, depend on the all previously obtained terms. For example

$$Z_1^{(0,1,2,...,i)} = -(i+1)\left(M_{12}^{(i)} A_{21} + A_{21}^T M_{12}^{T(i)}\right)$$
$$+ \sum_{k=2(even)}^{i-1} \binom{i+1}{k} M_1^{(i+1-k)} S_{11} M_1^{(k)}$$
$$+ \sum_{k=1(odd)}^{i} \left\{ \binom{i+1}{k} M_{12}^{(i+1-k)} S_{22} N_{12}^{T(k)} + N_{12}^{(i+1-k)} S_{22} M_{12}^{(k)} \right\}$$
$$(12.17)$$

Both approaches produce the same type of equations (Lyapunov ones), but in order to form the right-hand side, for example of (12.11a), we have to perform only 3 matrix multiplications for every i, where for corresponding equation of the power-series expansion the number of required matrix multiplications grows very quickly as i increase (12.17). Thus, the obvious advantages of the fixed point iteration approach are:

    1) The size of required computation is considerably less, and since it does not grow per iteration, the proposed method is extremally efficient for obtaining the exact solution or the solution of very high accuracy.

    2) The fixed point method is recursive in nature (the power-series expansion method is not), and thus much easier to implement.

The approximations of the suboptimal Nash strategies (12.4) can be defined by

$$u_j^{(i)}(t) = -R_j^{-1}(\epsilon) B_j^T(\epsilon) K_j^{(i)}(\epsilon) x(t), \quad j = 1, 2; \quad i = 0, 1, 2, 3, \ldots$$
(12.18)

where

$$K_1^{(i)}(\epsilon) = \begin{bmatrix} M_1(\epsilon) + \epsilon^2 E_1^{(i)}(\epsilon) & \epsilon \left[M_{12}(\epsilon) + \epsilon^2 E_{12}^{(i)}(\epsilon) \right] \\ \epsilon \left[M_{12}(\epsilon) + \epsilon^2 E_{12}^{(i)}(\epsilon) \right]^T & \epsilon^2 \left[M_2(\epsilon) + \epsilon^2 E_2^{(i)}(\epsilon) \right] \end{bmatrix}$$
(12.19a)

$$K_2^{(i)}(\epsilon) = \begin{bmatrix} \epsilon^2 \left[N_1(\epsilon) + \epsilon^2 G_1^{(i)}(\epsilon) \right] & \epsilon \left[N_{12}(\epsilon) + \epsilon^2 G_{12}^{(i)}(\epsilon) \right] \\ \epsilon \left[N_{12}(\epsilon) + \epsilon^2 G_{12}^{(i)}(\epsilon) \right]^T & N_2(\epsilon) + \epsilon^2 G_2^{(i)}(\epsilon) \end{bmatrix}$$
(12.19b)

Then, by following the arguments of (Cruz and Chen, 1971), the cost approximations produce

$$J_j^{(i)}\left(u_1^{(i)}, u_2^{(i)}\right) = J_j(u_1^*, u_2^*) + O\left(\epsilon^{2i+2}\right), \quad j = 1, 2; \quad i = 0, 1, 2, .$$
(12.20)

The approximate cost functions for the other cases, when the control agents use the approximative strategies of the different order of accuracy (for example $u_1^{(p)}$ and $u_2^{(q)}$, $p \neq q$) can be obtained by using results of (Cruz and Chen, 1971) also. But, since the proposed method is recursive in its nature, and thus very easy to implement, and since the amount of required computations is constant per iteration (does not grow with i), the accuracy of very high order can be achieved at a very low cost, so that the proposed method can be efficient for finding the exact solution as well.

Since the proposed algorithm defines the error of approximation similarly to the power-series expansion, it can be easily seen that the approximate Nash strategies (12.18) are also well-posed in the sense of (Khalil, 1980a).

12.3 Numerical Example

In order to demonstrate the efficiency of the proposed algorithm, we have run a fourth-order example. Matrices A_1, A_{12}, A_{21}, A_2, B_{11}, and B_{22} have been chosen randomly (standard deviation = 1, mean value = 0) and the matrices $R_1 = R_2 = U_1 = V_2 = I$ are chosen such that the required stabilizability-detectability assumptions are satisfied

$$A_1 = \begin{bmatrix} -1.035 & -0.192 \\ 1.684 & -0.421 \end{bmatrix}, \quad A_{12} = \begin{bmatrix} -1.084 & 0.579 \\ 1.327 & -0.841 \end{bmatrix}$$

$$A_{21} = \begin{bmatrix} -1.370 & -0.533 \\ 1.069 & 0.835 \end{bmatrix}, \quad A_2 = \begin{bmatrix} -1.510 & -0.139 \\ 0.410 & 1.238 \end{bmatrix}$$

$$B_{11} = \begin{bmatrix} -1.019 & 0.602 \\ -0.912 & 1.329 \end{bmatrix}, \quad B_{22} = \begin{bmatrix} -1.641 & 0.330 \\ 1.068 & 0.243 \end{bmatrix}$$

$$U_1 = V_2 = R_1 = R_2 = I_2$$

Simulation results for different values of a coupling parameter ϵ are given in Table 12.1. Since we do not know the exact solution of the equation (12.5) (no method available in the literature at the present time), the error is defined as

$$e^{(i)} = max \left\{ \left\| N_1 \left(K_1^{(i)}, K_2^{(i)} \right) \right\|_\infty, \left\| N_2 \left(K_1^{(i)}, K_2^{(i)} \right) \right\|_\infty \right\}$$

In the second table, we have shown the propagation of the error per iteration when $\epsilon = 0.1$.

ϵ	i = number of required iterations such that $e^{(i)} < 10^{-10}$
0.8	16
0.6	11
0.4	8
0.2	5
0.1	4
0.05	3
0.01	2
0.001	1

Table 12.1: Dependence of number of iterations on ϵ

The results from Table 12.1 strongly support the necessity of the existence of the recursive scheme for the solution of weakly coupled linear-quadratic Nash game problem, unless ϵ is very small, the zeroth and first-order approximations are far from the optimal solution. Results from Table 12.2 verify, for this particular example, the conclusions of Theorem 12.1, that is, the rate of convergence of the proposed algorithm is $O\left(\epsilon^2\right) = O\left(10^{-2}\right)$.

(ϵ = 0.1) i	Error $e^{(i)}$
0	0.89662 x 10^{-2}
1	0.65481 x 10^{-4}
2	0.10349 x 10^{-6}
3	0.40663 x 10^{-9}
4	0.92572 x 10^{-11}

Table 12.2 Propagation of the error per iteration for a constant value of ϵ

Therefore, the solution of the Nash strategies of weakly interconnected systems can be obtained up to an arbitrary accuracy by performing iterations on the Lyapunov equations corresponding to the local subsystem problems.

Research Problem 12.1: Develop the parallel synchronous algorithm for solving the coupled algebraic Riccati equations corresponding to the singularly perturbed Nash linear-quadratic differential games. This problem can be studied for both the standard singularly perturbed systems and for the quasi singularly perturbed systems defined in Chapter 9.

$\triangle$

Appendix 12.1

$$F_1 = E_1 S_{11} E_1 + M_{12} S_{22} G_{12}^T + G_{12} S_{22} M_{12}^T$$
$$-E_{12} D_{21} - D_{21}^T E_{12}^T + \epsilon^2 \left(E_{12} S_{22} G_{12}^T + G_{12} S_{22} E_{12}^T \right)$$

$$F_2 = E_{12} S_{22} G_2 + E_1 S_{11} E_{12} + G_{12} S_{22} M_2$$
$$-D_{21}^T E_2 + \epsilon^2 G_{12} S_{22} E_2$$

$$F_3 = E_{12} S_{11} E_{12} + E_2 S_{22} G_2 + G_2 S_{22} E_2$$

$$F_4 = G_{12} S_{22} G_{12}^T + E_1 S_{11} G_1 + G_1 S_{11} E_1$$

$$F_5 = E_1 S_{11} G_{12} + G_{12} S_{22} G_2$$
$$+N_1 S_{11} G_{12} - G_1 D_{12} + \epsilon^2 G_1 S_{11} E_{12}$$

$$F_6 = G_2 S_{22} G_2 + E_{12}^T S_{11} N_{12} + N_{12}^T S_{11} E_{12}$$
$$-G_{12}^T D_{12} - D_{12}^T G_{12} + \epsilon^2 \left(E_{12}^T S_{11} E_{12} + G_{12}^T S_{11} E_{12} \right)$$

$$C_1 = -M_{12} A_{12} - A_{21}^T M_{12}^T + M_{12} S_{22} N_{12}^T + N_{12} S_{22} M_{12}^T$$

$$C_2 = -D_{21}^T M_2$$
$$C_5 = -N_1 D_{12}$$
$$C_6 = -N_{12}^T A_{12} - A_{12}^T N_{12} + M_{12}^T S_{11} N_{12} + N_{12}^T S_{11} M_{12}$$

Appendix 12.2

Algorithm for Solving Coupled Algebraic
Riccati Equations of Nash Differential Games

The considered coupled algebraic Riccati equations have the forms

$$K_1 A + A^T K_1 + Q_1 - K_1 S_1 K_1 - K_2 S_2 K_1 - K_1 S_2 K_2 + K_2 Z_2 K_2 = \mathcal{N}_1 (K_1; K_2) = 0$$

(a.1)

$$K_2 A + A^T K_2 + Q_2 - K_2 S_2 K_2 - K_2 S_1 K_1 - K_1 S_1 K_2 + K_1 Z_1 K_1 = \mathcal{N}_2 (K_1; K_2) = 0$$

(a.2)

with

$$S_i = B_i R_{ii}^{-1} B_i^T, \; i = 1, 2 \; ; \quad Z_i = B_i R_{ii}^{-1} R_{ji} R_{ii}^{-1} B_i^T, \; i, j = 1, 2; \quad i = j$$

The proposed algorithm is based on the simulation results presented in (Gajic and Li, 1988). It seems, that the numerical method proposed is valid under the following assumption.

Assumption 12.3 Either the triple $(A, B_1, \sqrt{Q_1})$ or $(A, B_2; \sqrt{Q_2})$ is stabilizable-detectable.

$\triangle$

These conditions are quite natural since at least one control agent has to be able to control and observe unstable modes. Because the game is a noncooperative one, the assumption that their joint effect will take care of unstable modes seems to be very idealistic.

Let us suppose that $(A, B_1, \sqrt{Q_1})$ is stabilizable-detectable. Then a unique positive definite solution of an auxiliary algebraic Riccati equation

$$K_1^{(0)} A + A^T K_1^{(0)} + Q_1 - K_1^{(0)} S_1 K_1^{(0)} = 0 \qquad (a.3)$$

exists such that $\left(A - S_1 K_1^{(0)}\right)$ is a stable matrix. By plugging $K_1 = K_1^{(0)}$ in (a.2) we get the second auxiliary Riccati equation as

$$K_2^{(0)} \left(A - S_1 K_1^{(0)}\right) + \left(A - S_1 K_1^{(0)}\right)^T K_2^{(0)} + \left(Q_2 + K_1^{(0)} Z_1 K_1^{(0)}\right) - K_2^{(0)} S_2 K_2^{(0)} = 0$$

(a.4)

Since $\left(A - S_1 K_1^{(0)}\right)$ is a stable matrix and $Q_2 + K_1^{(0)} Z_1 K_1^{(0)}$ is a positive semidefinite matrix, the corresponding closed-loop matrix $\left(A - S_1 K_1^{(0)} - S_2 K_2^{(0)}\right)$ is stable. In fact, the triple $\left(A - S_1 K_1^{(0)}, B_2, \sqrt{Q_2 + K_1^{(0)} S_1 K_1^{(0)}}\right)$ is stabilizable-detectable and $K_2^{(0)}$ is uniquely determined.

Let us now propose the iterative scheme for solving (a.1)-(a.2). By decoupling these equations by using appropriately one step delay, we get the Lyapunov type iterative scheme similarly to (Kleinman, 1968), with $K_1^{(0)}$ and $K_2^{(0)}$ playing the role of the initial points (Gajic and Li, 1988). Algorithm 12.2:

$$
\left(A - S_1 K_1^{(i)} - S_2 K_2^{(i)}\right)^T K_1^{(i+1)} + K_1^{(i+1)} \left(A - S_1 K_1^{(i)} - S_2 K_2^{(i)}\right) =
$$
$$
= \overline{Q_1^{(i)}} = - \left(Q_1 + K_1^{(i)} S_1 K_1^{(i)} + K_2^{(i)} Z_2 K_2^{(i)}\right), \quad i = 0, 1, 2, \dots
$$
$$
\text{(a.5)}
$$

$$
\left(A - S_1 K_1^{(i)} - S_2 K_2^{(i)}\right)^T K_2^{(i+1)} + K_2^{(i+1)} \left(A - S_1 K_1^{(i)} - S_2 K_2^{(i)}\right) =
$$
$$
= \overline{Q_2^{(i)}} = - \left(Q_2 + K_1^{(i)} Z_1 K_1^{(i)} + K_2^{(i)} S_2 K_2^{(i)}\right), \quad i = 0, 1, 2, \dots
$$
$$
\text{(a.6)}
$$
$$
\triangle
$$

Example 12.1

In order to demonstrate the efficiency of the proposed algorithm we have run a tenth-order example, which is in fact a system of 110 nonlinear algebraic equations. Matrices A, B_1, and B_2 have been chosen randomly whereas the choice of matrices Q_1, Q_2, R_{11}, R_{12}, and R_{22} assures that Assumption 12.3 is satisfied. These matrices are given by

$$
A = \begin{bmatrix}
-1.944 & 0.572 & 1.446 & -0.576 & 0.736 & -0.601 & -0.722 & -0.088 & 0.977 & 0.380 \\
1.440 & 0.393 & 1.023 & -0.711 & 1.282 & -0.679 & 0.010 & 0.588 & 1.281 & -1.414 \\
-0.881 & 1.058 & -1.492 & 1.113 & -1.728 & 0.498 & 0.313 & 1.509 & -1.536 & -0.264 \\
-1.170 & -1.055 & -0.058 & -0.723 & -0.939 & 1.453 & -1.087 & -0.486 & 1.066 & 0.235 \\
0.736 & -0.569 & 1.449 & -1.383 & 0.116 & -0.052 & 1.387 & 0.659 & -1.658 & -1.437 \\
0.014 & 0.658 & 0.586 & -0.850 & -0.074 & -1.335 & -0.261 & -1.021 & -0.449 & 1.444 \\
-0.734 & 0.621 & 0.422 & -0.369 & -0.395 & -0.453 & 1.228 & 0.213 & -1.380 & 1.307 \\
0.820 & -1.746 & 0.178 & -0.860 & -1.235 & -0.902 & 0.390 & -0.656 & -1.658 & 1.329 \\
0.831 & 0.569 & 1.408 & 1.500 & 1.396 & -0.605 & 0.387 & -0.729 & 1.717 & 1.309 \\
0.051 & -0.224 & 1.394 & 0.104 & -1.742 & -0.386 & -0.047 & -0.505 & -1.135 & 1.392
\end{bmatrix}
$$

$$
R_{11} = \begin{bmatrix} 1 & 0 \\ 0 & 2 \end{bmatrix}, \quad R_{12} = \begin{bmatrix} 3 & 0 \\ 0 & 4 \end{bmatrix}, \quad R_{21} = \begin{bmatrix} 5 & 0 \\ 0 & 6 \end{bmatrix}, \quad R_{22} = \begin{bmatrix} 7 & 0 \\ 0 & 8 \end{bmatrix}
$$
$$
Q_1 = I_{10}, \qquad Q_2 = I_{10}
$$

$$B_1 = \begin{bmatrix} -2.036 & 0.637 \\ 1.560 & 0.447 \\ -0.907 & 1.154 \\ -1.214 & -1.091 \\ 0.813 & -0.575 \\ 0.044 & 0.729 \\ -0.750 & 0.690 \\ 0.901 & -1.826 \\ 0.913 & 0.635 \\ 0.084 & -0.209 \end{bmatrix}, \quad B_2 = \begin{bmatrix} -1.648 & -0.759 \\ 0.171 & 1.256 \\ -0.380 & -1.076 \\ -1.465 & -0.101 \\ 1.854 & 0.745 \\ 0.015 & 1.717 \\ 0.458 & -0.091 \\ 0.255 & -1.304 \\ 0.274 & -0.763 \\ -0.502 & 1.345 \end{bmatrix}$$

The obtained results are really remarkable since only after 8 iterations we got very good convergence. These results are presented in Table 12.3. The errors are defined as the absolute values of the largest elements in matrices $\mathcal{N}_1\left(K_1^{(i)}, K_2^{(i)}\right)$ and $\mathcal{N}_2\left(K_1^{(i)}, K_2^{(i)}\right)$ where i stands for the number of iterations.

Iteration	*error 1*	*error 2*
1	1.5283 x 10^{+2}	1.4193 x 10^{+2}
2	1.6726 x 10^{+1}	4.0585 x 10^{+1}
3	3.1057 x 10^{+0}	1.2188 x 10^{+1}
4	2.3207 x 10^{-1}	2.4337 x 10^{+0}
5	1.2386 x 10^{-1}	7.6489 x 10^{-2}
6	4.1600 x 10^{-3}	6.9948 x 10^{-5}
7	7.0661 x 10^{-4}	3.0096 x 10^{-7}
8	2.4374 x 10^{-5}	9.2183 x 10^{-8}

Table 12.3: Simulation results for a
system of 110 nonlinear scalar equations

△

The second algorithm for solving the coupled algebraic Riccati equations presented in (Gajic and Li, 1988) is based on the discrete homotopy. It is numerically less efficient than Algorithm 12.2, but it seems that it would be easier to prove its convergence. The study in that direction is underway.

Research Problem 12.2: Prove that the algorithm (a.3)-(a.6) converges under Assumption 12.3 to the positive semidefinite stabilizing solutions of the coupled algebraic Riccati equations (a.1)-(a.2).

$\triangle$

Chapter 13

Recursive Approach to High Gain and Cheap Control Problems

In the first part of this chapter we present a parallel synchronous reduced-order algorithms for solving the algebraic Riccati equation corresponding to both the high gain feedback and cheap control optimal problems. In the subsequent sections, we study the open-loop optimal control problem and the problem of the complete decomposition of the algebraic "cheap/high gain" Riccati equation into the reduced-order pure-slow and pure-fast Riccati equations.

13.1 Linear-Quadratic High Gain and Cheap Control Problems

A primary goal in studying optimal control of high gain and cheap control problems, as with any optimal control problem, is to determine the control which minimizes the value of some performance index. In general, high gain systems are those in which the norm of the feedback control matrix is of high magnitude, usually one or more orders of magnitude greater than that of the norm of the system matrix. A cheap control problem is characterized by a small penalty imposed on the control term in the performance index usually one or more orders of magnitude smaller that the state weighting term. The difference in magnitude scales can be quantitatively described by a parameter ϵ, where ϵ is some constant less than one.

High gain and cheap control problems have been studied extensively by a number of researchers (Jameson and O'Malley, 1975; O'Malley and Jameson, 1975, 1977; O'Malley, 1976; Young et al., 1977; Francis and Glover, 1978; Francis, 1979; Sannuti, 1983; Sannuti and Wason, 1983; O'Reilly, 1983; Priel and Shaked, 1983; Saberi and Sannuti, 1986, 1987; Kokotovic et al., 1986; Petersen, 1986; Murata et al., 1990). The modern approach to the analysis of high gain and cheap control problems involves the use of singular perturbation method. The singular perturbation technique offers an intuitive understanding into the behavior of high gain and cheap control problems.

In general, the application of singular perturbation method involves a suitable representation and partitioning of the problem matrices, and explicitly introduces a small positive parameter ϵ. The role of the parameter ϵ varies with the type of system under investigation; however, once introduced, its exact nature is unimportant. The solution is found for $\epsilon = 0$, and a Taylor series expansion is then taken about this zero-order solution to find a higher order solution to any prescribed degree of accuracy. Several problems arise as a result of the application of this technique. The problem matrices must be analytical functions of ϵ, and for certain systems this may not be so. A solution of higher order requires an enormous number of terms. Furthermore, when the value of ϵ is not small enough, the obtained solution may fail to yield an accurate solution.

In this section, a recursive fixed point method is presented to find the solution of the linear cheap control and high gain problems. The method is advantageous to the traditional power-series expansion method which is not recursive in nature, since when a high order of accuracy is desired, the size of computations can be considerably high with the traditional method. The results of this study give the numerical decomposition so that only low-order systems are involved in algebraic computations. The solution by the proposed algorithm converges to the exact solution with the rate of $O(\epsilon)$, where ϵ is a small positive parameter. The proposed parallel reduced-order algorithm is applicable under some mild conditions which are fully stated.

The structures of the problem matrices and the methodology used do not allow either impulsive behavior or singular controls in the problems under considerations. These two undesired limiting phenomena appear very often in the cases when the classical singular perturbation approach is used to study the cheap control and high gain feedback problems. As a matter of fact, we study these two problems for ϵ small and positive — which is physical reality, but not for $\epsilon \rightarrow 0$ — which is mathematical fiction that produces impulsive behavior and singular controls.

The efficiency of this algorithm is demonstrated on a real example, a flexible space structure. The simulation results support the conclusion of the theorem stated and proved in this section.

We study the optimal control problem of high gain and cheap control problems at steady state. These results can be extended to the finite continuous-time optimization problems by using results on the slow-fast time scale decomposition of the differential Riccati equation from (Grodt and Gajic, 1988). An extension of the presented results to the discrete-time cheap control problems (Priel and Shaked, 1983; Sen and Datta, 1992) might be an interesting area for future research. All necessary results for studying cheap control and high gain feedback control problems in the discrete-time domain are obtained in the proceeding chapters of this book.

13.1.1 High Gain Feedback Control

Consider a system given by

$$\dot{x} = Ax + \hat{B}u \qquad (13.1)$$

partitioned as

$$\begin{bmatrix} \dot{x}_1 \\ \dot{x}_2 \end{bmatrix} = \begin{bmatrix} A_{11} & A_{12} \\ A_{21} & A_{22} \end{bmatrix} \begin{bmatrix} x_1 \\ x_2 \end{bmatrix} + \begin{bmatrix} 0 \\ \frac{B_2}{\epsilon} \end{bmatrix} u \qquad (13.2)$$

where $x_1 \in \Re^n$, $x_2 \in \Re^m$ are state variables, $u \in \Re^m$ is a control input, and ϵ is a small positive parameter. No loss of generality is incurred, since the system model can always be transformed to (13.2) provided $B_2 \in \Re^{m \times m}$ is of full rank m (Jameson and O'Malley, 1975). Thus, the problem is studied under the following standard assumption (Kokotovic et al., 1986).

Assumption 13.1 $det\ B_2 \neq 0$.

$\triangle$

The scalar cost functional associated with (13.2), defined by

$$J(\epsilon) = \frac{1}{2} \int_0^\infty [x^T Q x + u^T R u]\ dt \qquad (13.3)$$

is minimized by the well-known optimal control law

$$u(t) = -R^{-1}\widehat{B}^T K x(t) = -\frac{1}{\epsilon} R^{-1} B^T K x(t) \qquad (13.4)$$

where K is a symmetric and positive semidefinite solution of the quadratic matrix algebraic Riccati equation

$$KA + A^T K + Q = \frac{1}{\epsilon^2} KBR^{-1} B^T K, \quad B = \epsilon \widehat{B} \qquad (13.5)$$

where K and Q are partitioned as

$$K = \begin{bmatrix} K_{11} & \epsilon K_{12} \\ \epsilon K_{12}^T & \epsilon K_{22} \end{bmatrix}, \quad Q = \begin{bmatrix} Q_{11} & Q_{12} \\ Q_{12}^T & Q_{22} \end{bmatrix} \qquad (13.6)$$

Due to the presence of $O\left(\frac{1}{\epsilon}\right)$ term in (13.2), which multiplies the control input $u(t)$, this problem is known in the literature as the high gain feedback problem.

13.1.2 Cheap Control Problem

Consider a system model given by

$$\begin{bmatrix} \dot{x}_1 \\ \dot{x}_2 \end{bmatrix} = \begin{bmatrix} A_{11} & A_{12} \\ A_{21} & A_{22} \end{bmatrix} \begin{bmatrix} x_1 \\ x_2 \end{bmatrix} + \begin{bmatrix} 0 \\ B_2 \end{bmatrix} u \qquad (13.7)$$

where $x_1 \in \Re^n$, $x_2 \in \Re^m$, $u \in \Re^m$, and B_2 is a nonsingular $m \times m$ constant matrix.

The scalar cost functional associated with (13.7), which represents the optimal cheap control problem, is given by

$$J(\epsilon) = \frac{1}{2} \int\limits_0^\infty \left[x^T Q x + \epsilon^2 u^T R u \right] dt \qquad (13.8)$$

This functional has to be minimized by selecting the m-dimensional control vector u. Q and R are symmetric positive semidefinite and positive definite matrices, respectively, and ϵ is a small positive parameter.

The feedback control law for the optimal cheap control problem defined by (13.7)-(13.8) is given by

$$u(t) = -\frac{1}{\epsilon^2} R^{-1} B^T K x(t) \qquad (13.9)$$

where K is a symmetric and positive semidefinite solution of the quadratic matrix algebraic Riccati equation (13.5) and

$$A = \begin{bmatrix} A_{11} & A_{12} \\ A_{21} & A_{22} \end{bmatrix}, x = \begin{bmatrix} x_1 \\ x_2 \end{bmatrix}$$

$$S = BR^{-1}B^T = \begin{bmatrix} S_{11} & S_{12} \\ S_{12}^T & S_{22} \end{bmatrix} = \begin{bmatrix} 0 & 0 \\ 0 & B_2 R^{-1} B_2^T \end{bmatrix} \qquad (13.10)$$

Although the initial problem statements may differ, the forms of the Riccati equations are identical for both the high gain and cheap control problems, assuming that the problem matrices are defined consistently. The recursive parallel algorithm for finding the solution of this Riccati equation in terms of the reduced-order problems is presented in the next section.

13.1.3 Parallel Algorithm for Solving Algebraic Riccati Equation

In this section, we first obtain the zero-order solution (which is an $O(\epsilon)$ close to the exact one) and then derive a parallel algorithm that produces an arbitrary order of accuracy, that is the approximate solution $O(\epsilon^k)$ close to the exact one, where k stands for the number of iterations of the proposed algorithm.

Zero-order solution:
The partitioned form of equation (13.5) subject to (13.6) and (13.10) is

$$K_{11}A_{11} + A_{11}^T K_{11} + \epsilon K_{12}A_{21} + \epsilon A_{21}^T K_{12}^T + Q_{11} = K_{12}S_{22}K_{12}^T$$
$$K_{11}A_{12} + \epsilon K_{12}A_{22} + \epsilon A_{11}^T K_{12} + \epsilon A_{21}^T K_{22} + Q_{12} = K_{12}S_{22}K_{22}$$
$$\epsilon K_{12}^T A_{12} + \epsilon A_{12}^T K_{12} + \epsilon K_{22}A_{22} + \epsilon A_{22}^T K_{22} + Q_{22} = K_{22}S_{22}K_{22}$$
$$\qquad (13.11)$$

The zero-order solution $K_j^{(0)}$, for $j = 1, 2, 3$, can be found by setting $\epsilon = 0$ in (13.11), giving

$$K_{11}^{(0)}A_{11} + A_{11}^T K_{11}^{(0)} + Q_{11} = K_{12}^{(0)}S_{22}K_{12}^{T(0)} \qquad (13.12a)$$

$$K_{11}^{(0)}A_{12} + Q_{12} = K_{12}^{(0)}S_{22}K_{22}^{(0)} \qquad (13.12b)$$

$$K_{22}^{(0)}S_{22}K_{22}^{(0)} = Q_{22} \qquad (13.12c)$$

In order to be able to solve (13.12) in the spirit of theory of singular perturbations, we have to impose the following assumptions (the standard assumptions in the singular perturbation approach to the cheap control and high gain feedback optimal linear-quadratic problems).

Assumption 13.2 Q_{22} is a positive definite matrix.

$$\triangle$$

With Q_{22} being positive definite, a two time-scale decomposition can be made. This is necessary to assure the existence of the positive define solution for $K_{22}^{(0)}$ (Kokotovic et al., 1986).

Assumption 13.3 $K_{22}^{(0)}$ is a positive definite matrix.

$$\triangle$$

The positive definite solution for $K_{22}^{(0)}$ can be obtained from (13.12c) as

$$S_{22}^{1/2} K_{22}^{(0)} S_{22}^{1/2} S_{22}^{1/2} K_{22}^{(0)} S_{22}^{1/2} = \left(S_{22}^{1/2} K_{22}^{(0)} S_{22}^{1/2} \right)^2 = S_{22}^{1/2} Q_{22} S_{22}^{1/2} \tag{13.13}$$

so that

$$K_{22}^{(0)} = S_{22}^{-1/2} \left(S_{22}^{1/2} Q_{22} S_{22}^{1/2} \right)^{1/2} S_{22}^{-1/2} > 0 \tag{13.14}$$

Rearranging (13.12b) produces

$$K_{12}^{(0)} = \left(Q_{12} + K_{11}^{(0)} A_{12} \right) \left(S_{22} K_{22}^{(0)} \right)^{-1} \tag{13.15}$$

The remaining quadratic equation (13.12a) becomes

$$K_{11}^{(0)} A_{11} + A_{11}^T K_{11}^{(0)} + Q_{11}$$
$$= \left(Q_{12} + K_{11}^{(0)} A_{12} \right) K_{22}^{(0)-1} S_{22}^{-1} K_{22}^{(0)-1} \left(Q_{12} + K_{11}^{(0)} A_{12} \right)^T \tag{13.16}$$

Eliminating $K_{22}^{(0)}$ from the last equation by using (13.15), we get the n-th order algebraic Riccati equation

$$K_{11}^{(0)} \left(A_{11} - A_{12} Q_{22}^{-1} Q_{12}^T \right) + \left(A_{11} - A_{12} Q_{22}^{-1} Q_{12}^T \right)^T K_{11}^{(0)}$$
$$- K_{11}^{(0)} A_{12} Q_{22}^{-1} A_{12}^T K_{11}^{(0)} + \left(Q_{11} - Q_{12} Q_{22}^{-1} Q_{12}^T \right) = 0 \tag{13.17}$$

The following assumption guarantees the existence of the positive semidefinite stabilizing solutions for the algebraic Riccati equation (13.17).

Assumption 13.4 The triple (A_0, B_0, Q_0) is stabilizable-detectable, where

$$A_0 = A_{11} - A_{12} Q_{22}^{-1} Q_{12}^T, \quad B_0 B_0^T = A_{12} Q_{22}^{-1} A_{21}^T$$
$$Q_0 Q_0^T = Q_{11} - Q_{12} Q_{22}^{-1} Q_{12}^T$$

$$\triangle$$

The obtained zero-order solution $K^{(0)}$, defined by

$$K^{(0)} = \begin{bmatrix} K_{11}^{(0)} & \epsilon K_{12}^{(0)} \\ \epsilon K_{12}^T & \epsilon K_{22}^{(0)} \end{bmatrix} \tag{13.18}$$

in an $O(\epsilon)$ close to the exact one, K.

Solution of Higher Order of Accuracy:
Define the difference between the exact solution K and the zero-order solution $K^{(0)}$

$$\epsilon E = K - K^{(0)} \tag{13.19}$$

where E stands for an error term partitioned consistently as

$$E = \begin{bmatrix} E_{11} & \epsilon E_{12} \\ \epsilon E_{12}^T & \epsilon E_{22} \end{bmatrix} \tag{13.20}$$

Substituting (13.18)-(13.20) into (13.11) yields

$$\left(K_{11}^{(0)} + \epsilon E_{11}\right) A_{11} + A_{11}^T \left(K_{11}^{(0)} + \epsilon E_{11}\right) + Q_{11}$$
$$+ \epsilon \left(K_{12}^{(0)} + \epsilon E_{12}\right) A_{21} + \epsilon A_{21}^T \left(K_{12}^{(0)} + \epsilon E_{12}\right)^T \tag{13.21}$$
$$= \left(K_{12}^{(0)} + \epsilon E_{12}\right) S_{22} \left(K_{12}^{(0)} + \epsilon E_{12}\right)^T$$

$$\left(K_{11}^{(0)} + \epsilon E_{11}\right) A_{12} + \epsilon \left(K_{12}^{(0)} + \epsilon E_{12}\right) A_{22} + Q_{12}+$$
$$\epsilon A_{21}^T \left(K_{12}^{(0)} + \epsilon E_{12}\right) + \epsilon A_{11}^T \left(K_{12}^{(0)} + \epsilon E_{12}\right) \tag{13.22}$$
$$= \left(K_{12}^{(0)} + \epsilon E_{12}\right) S_{22} \left(K_{22}^{(0)} + \epsilon E_{22}\right)$$

$$\epsilon \left(K_{12}^{(0)} + \epsilon E_{12}\right)^T A_{12} + \epsilon A_{12}^T \left(K_{12}^{(0)} + \epsilon E_{12}\right) + Q_{22}$$
$$+ \epsilon \left(K_{22}^{(0)} + \epsilon E_{22}\right) A_{22} + \epsilon A_{22}^T \left(K_{22}^{(0)} + \epsilon E_{22}\right) \tag{13.23}$$
$$= \left(K_{22}^{(0)} + \epsilon E_{22}\right) S_{22} \left(K_{22}^{(0)} + \epsilon E_{22}\right)$$

Subtracting (13.12) from (13.21)-(13.23) produces

$$E_{11} A_{11} + A_{11}^T E_{11} + \left(K_{12}^{(0)} + \epsilon E_{12}\right) A_{21} + A_{21}^T \left(K_{12}^{(0)} + \epsilon E_{12}\right)^T$$
$$= E_{12} S_{22} \left(K_{12}^{(0)} + \epsilon E_{12}\right)^T + \left(K_{12}^{(0)} + \epsilon E_{12}\right) S_{22} E_{12}^T - \epsilon E_{12} S_{22} E_{12}^T \tag{13.24}$$

$$E_{11}A_{12} + \left(K_{12}^{(0)} + \epsilon E_{12}\right)A_{22}$$
$$+ A_{11}^T\left(K_{12}^{(0)} + \epsilon E_{12}\right) + A_{21}^T\left(K_{22}^{(0)} + \epsilon E_{22}\right)$$
$$= E_{12}S_{22}\left(K_{22}^{(0)} + \epsilon E_{22}\right) + \left(K_{12}^{(0)} + \epsilon E_{12}\right)S_{22}E_{22} - \epsilon E_{12}S_{22}E_{22}$$

$$(13.25)$$

$$\left(K_{12}^{(0)} + \epsilon E_{12}\right)^T A_{12} + A_{12}^T\left(K_{12}^{(0)} + \epsilon E_{12}\right)$$
$$+ \left(K_{22}^{(0)} + \epsilon E_{22}\right)A_{22} + A_{22}^T\left(K_{22}^{(0)} + \epsilon E_{22}\right)$$
$$= E_{22}S_{22}\left(K_{22}^{(0)} + \epsilon E_{22}\right) + \left(K_{22}^{(0)} + \epsilon E_{22}\right)S_{22}E_{22} - \epsilon E_{22}S_{22}E_{22}$$

$$(13.26)$$

Equation (13.26) can be expressed with respect to E_{22} as

$$E_{22}S_{22}K_{22}^{(0)} + K_{22}^{(0)}S_{22}E_{22} = -\epsilon E_{22}S_{22}E_{22}$$
$$+ \left(K_{12}^{(0)} + \epsilon E_{12}\right)^T A_{12} + A_{12}^T\left(K_{12}^{(0)} + \epsilon E_{12}\right) \qquad (13.27)$$
$$+ \left(K_{22}^{(0)} + \epsilon E_{22}\right)A_{22} + A_{22}^T\left(K_{22}^{(0)} + \epsilon E_{22}\right)$$

Since this equation is an $O\left(\epsilon\right)$ perturbation of the algebraic Lyapunov equation and since the product $S_{22}K_{22}^{(0)}$ is positive definite (by Assumptions 13.1 and 13.3), it follows that the unique solution E_{22} exists for sufficiently small values of ϵ.

Equation (13.25) can be directly solved for E_{12} as

$$E_{12} = E_{11}A_{12}\left(S_{22}K_{22}^{(0)}\right)^{-1}$$
$$+ \left[A_{21}^T\left(K_{22}^{(0)} + \epsilon E_{22}\right) - K_{12}^{(0)}S_{22}E_{22} - \epsilon E_{12}S_{22}E_{22}\right]\left(S_{22}K_{22}^{(0)}\right)^{-1}$$
$$- \left[\left(K_{12}^{(0)} + \epsilon E_{12}\right)A_{22} - A_{11}^T\left(K_{12}^{(0)} + \epsilon E_{12}\right)\right]\left(S_{22}K_{22}^{(0)}\right)^{-1}$$

$$(13.28)$$

Equation (13.24) exhibits terms in E_{12} which are not multiplied by ϵ. Substituting (13.28) for these terms in (13.24) and doing some algebra we get

$$E_{11}\left(A_{11} - A_{12}K_{22}^{(0)^{-1}}K_{12}^{(0)^T}\right) + \left(A_{11} - A_{12}K_{22}^{(0)^{-1}}K_{12}^{(0)^T}\right)^T E_{11}$$

$$= \epsilon E_{12}S_{22}E_{12}^T + H\left(K_{12}^{(0)}, K_{22}^{(0)}, E_{12}, E_{22}, \epsilon\right)$$

$$+ H^T\left(K_{12}^{(0)}, K_{22}^{(0)}, E_{12}, E_{22}, \epsilon\right)$$

$$(13.29)$$

where

$$H\left(K_{12}^{(0)}, K_{22}^{(0)}, E_{12}, E_{22}, \epsilon\right) = [\left(K_{12}^{(0)} + \epsilon E_{12}\right)A_{22}$$

$$+ A_{11}^T\left(K_{12}^{(0)} + \epsilon E_{12}\right) + A_{21}^T\left(K_{22}^{(0)} + \epsilon E_{22}\right) - \epsilon E_{12}S_{22}E_{22}$$

$$- K_{12}^{(0)}S_{22}E_{22}]K_{22}^{(0)^{-1}}K_{12}^{(0)^T} - \left(K_{12}^{(0)} + \epsilon E_{12}\right)A_{12}$$

In order to get the unique solution for E_{11} from the algebraic Lyapunov equation (13.29), it is required that the matrix $A_{11} - A_{12}K_{22}^{(0)^{-1}}K_{12}^{(0)^T}$ is stable. The stability of this matrix follows from Assumptions 13.1-13.4. This can be observed as follows. In the first-order approximation the stabilizing feedback control is given by

$$u^{(0)} = -\frac{1}{\epsilon^2}R^{-1}[0 \quad B_2^T]\begin{bmatrix} K_{11}^{(0)} & \epsilon K_{12}^{(0)} \\ \epsilon K_{12}^{(0)^T} & \epsilon K_{22}^{(0)} \end{bmatrix}\begin{bmatrix} x_1^{(0)} \\ x_2^{(0)} \end{bmatrix}$$

$$= -\frac{1}{\epsilon}R^{-1}K_{12}^{(0)^T}x_1^{(0)} - \frac{1}{\epsilon}R^{-1}K_{22}^{(0)}x_2^{(0)} \qquad (13.30)$$

so that the approximate trajectories are given by

$$\dot{x}_1^{(0)} = A_{11}x_1^{(0)} + A_{12}x_2^{(0)}$$

$$\dot{x}_2^{(0)} = A_{21}x_1^{(0)} + A_{22}x_2^{(0)} - \frac{1}{\epsilon}S_{22}K_{12}^{(0)^T}x_1^{(0)} - \frac{1}{\epsilon}S_{22}K_{22}^{(0)}x_2^{(0)} \qquad (13.31)$$

This system has the singularly perturbed form. The fast variable $x_2^{(0)}$, represented by

$$\epsilon\dot{x}_2^{(0)} = \epsilon A_{21}x_1^{(0)} + \epsilon A_{22}x_2^{(0)} - S_{22}K_{12}^{(0)^T}x_1^{(0)} - S_{22}K_{22}^{(0)}x_2^{(0)} \qquad (13.32)$$

has the quasi-steady state value $x_2^{(0)}$ obtained from

$$0 = -S_{22}K_{12}^{(0)^T}x_1^{(0)} - S_{22}K_{22}^{(0)}x_2^{(0)} \qquad (13.33)$$

Solving this equation with respect to $x_2^{(0)}$ produces

$$x_2^{(0)} = -\left(S_{22}K_{22}^{(0)}\right)^{-1} S_{22}K_{12}^{(0)^T} x_1^{(0)} = -K_{22}^{(0)^{-1}} K_{12}^{(0)} x_1^{(0)} \qquad (13.34)$$

Substituting (13.34) into the differential equation describing the motion of the slow variable, we get

$$\dot{x}_1^{(0)} = \left(A_{11} - A_{12}K_{22}^{(0)^{-1}} K_{12}^{(0)^T}\right) x_1^{(0)} = A_s x_1^{(0)} \qquad (13.35)$$

Since in the previous analysis we have been neglecting only $O\left(\epsilon\right)$ quantities and by the fact that the optimal control stabilizes the system, it follows that the matrix A_s is stable.

The structures of equations (13.27)-(13.29) suggest the following parallel, synchronous, reduced-order algorithm (in the spirit of those developed in Gajic et al., 1990) for finding the solution of the error terms.

Algorithm 13.1:

$$E_{11}^{(i+1)} A_s + A_s^T E_{11}^{(i+1)} = -\epsilon E_{12}^{(i)} S_{22} E_{12}^{(i)^T}$$
$$+ H\left(K_{12}^{(0)}, K_{22}^{(0)}, E_{12}^{(i)}, E_{22}^{(i+1)}, \epsilon\right) + H^T\left(K_{12}^{(0)}, K_{22}^{(0)}, E_{12}^{(i)}, E_{22}^{(i+1)}, \epsilon\right) \qquad (13.36)$$

$$E_{22}^{(i+1)} S_{22} K_{22}^{(0)} + K_{22}^{(0)} S_{22} E_{22}^{(i+1)}$$
$$= K_{12}^{(i)^T} A_{12} + A_{12}^T K_{12}^{(i)} + K_{22}^{(i)} A_{22} + A_{22}^T K_{22}^{(i)} - \epsilon E_{22}^{(i)} S_{22} E_{22}^{(i)} \qquad (13.37)$$

$$E_{12}^{(i+1)} = \left[E_{11}^{(i+1)} A_{12} + K_{12}^{(i)} A_{22} + A_{11}^T K_{12}^{(i)} + A_{21}^T K_{22}^{(i)}\right]$$
$$\times \left(S_{22}K_{22}^{(0)}\right)^{-1} - K_{12}^{(0)} S_{22} E_{22}^{(i+1)} \left(S_{22}K_{22}^{(0)}\right)^{-1} \qquad (13.38)$$
$$-\epsilon E_{12}^{(i)} S_{22} E_{22}^{(i+1)} \left(S_{22}K_{22}^{(0)}\right)^{-1}$$

with initial conditions chosen as $E_{11}^{(0)} = 0; E_{12}^{(0)} = 0; E_{22}^{(0)} = 0.$

$$\triangle$$

The i-th order solution $K^{(i)}$ of the Riccati equation (13.5) is defined as

$$K^{(i)} = K^{(0)} + \epsilon E^{(i)} \qquad (13.39)$$

with

$$E^{(i)} = \begin{bmatrix} E_{11}^{(i)} & \epsilon E_{12}^{(i)} \\ \epsilon E_{12}^{T(i)} & \epsilon E_{22}^{(i)} \end{bmatrix} \qquad (13.40)$$

consistently to the previous partitioning of K and $K^{(0)}$.

The following theorem indicates features of the proposed algorithm (13.36)-(13.40).

Theorem 13.1 *Under Assumptions 13.1–13.4, the algorithm (13.36)-(13.40) converges to the required solution E with the rate of convergence of $O(\epsilon)$, that is,*

$$\| E^{(i)} - E^{(i+1)} \| = O(\epsilon) \tag{13.41}$$

or

$$\| E - E^{(i)} \| = O(\epsilon^i), \quad i = 1, 2, \dots. \tag{13.42}$$

◊

Proof: In the first step of this proof, we have to establish the existence of the bounded unique solution of the error equations (13.24)-(13.26). By the implicit function theorem, (Khalil, 1992), the corresponding Jacobian has to be nonsingular at $\epsilon = 0$. It can be shown that the Jacobian of (13.24)-(13.26) is given by

$$J = \begin{bmatrix} \Gamma_{11} & 0 & 0 \\ \Gamma_{21} & \Gamma_{22} & \Gamma_{23} \\ 0 & 0 & \Gamma_{33} \end{bmatrix} + O(\epsilon) \tag{13.43}$$

with

$$\begin{aligned} \Gamma_{11} &= A_s^T \otimes I_{n_1} + I_{n_1} \otimes A_s^T \\ \Gamma_{22} &= \left(S_{22} K_{22}^{(0)} \right)^T \otimes I_{n_2} + I_{n_2} \otimes \left(S_{22} K_{22}^{(0)} \right)^T \\ \Gamma_{33} &= \Gamma_{22} \end{aligned} \tag{13.44}$$

where $\otimes$ stands for the Kronecker product. For the Jacobian to be nonsingular Γ_{ii}, $i = 1, 2, 3$, have to be nonsingular. It follows from the properties of the Kronecker product (Lancaster and Tismenetsky, 1985) and Assumptions 13.1, 13.3, and 13.4, that matrices Γ_{11} and Γ_{22} are respectively negative and positive definite, and thus are nonsingular.

In the second step, we have to prove convergence and give an estimate of the rate of convergence. Equations (13.24)-(13.26) have the following forms

$$E_{11} A_{11} + A_{11}^T E_{11} = E_{12} S_{22} K_{12}^{(0)^T} + K_{12}^{(0)} S_{22} E_{12}^T + const + O(\epsilon) \tag{13.45}$$

$$E_{11} A_{12} - E_{12} S_{22} K_{22}^{(0)^T} - K_{12}^{(0)} S_{22} E_{22} = const + O(\epsilon) \tag{13.46}$$

CHEAP CONTROL AND HIGH GAIN

$$E_{22}S_{22}K_{22}^{(0)^T} + K_{22}^{(0)}S_{22}E_{22} = const + O\left(\epsilon\right) \qquad (13.47)$$

or, after some algebra, (eliminate E_{12} from (13.45) by using (13.46); then, eliminate E_{22} from modified equation (13.45) by using (13.47)), we get

$$E_{11}A_s + A_s^T E_{11} = const + O\left(\epsilon\right)$$
$$E_{22}A_f + A_f^T E_{22} = const + O\left(\epsilon\right) \qquad (13.48)$$
$$E_{12} = \left(E_{11}A_{12} - K_{12}^{(0)}S_{22}E_{22} + const + O\left(\epsilon\right)\right)A_f^{-1}$$

where $A_f = S_{22}K_{22}^{(0)}$. The algorithm (13.36)-(13.38) basically has the form

$$E_{11}^{(i+1)}A_s + A_s^T E_{11}^{(i+1)} = const + O\left(\epsilon\right)$$
$$E_{22}^{(i+1)}A_f + A_f^T E_{22}^{(i+1)} = const + O\left(\epsilon\right)$$
$$E_{12}^{(i+1)} = \left(E_{11}^{(i+1)}A_{12} - K_{12}^{(0)}S_{22}E_{22}^{(i+1)} + const + O\left(\epsilon\right)\right)A_f^{-1}$$
$$i = 0, 1, 2, \ldots.$$

$$(13.49)$$

By nonsingularity of A_s and A_f, it follows that

$$\left\|E_{jj}^{(i+1)} - E_{jj}^{(i)}\right\| = O\left(\epsilon\right), \quad j = 1, 2 \qquad (13.50)$$

which together with (13.49) implies

$$\left\|E_{12}^{(i+1)} - E_{12}^{(i)}\right\| = O\left(\epsilon\right) \qquad (13.51)$$

The order of $O\left(\epsilon^i\right)$ accuracy is proved by induction.

Clearly $O\left(\epsilon^i\right)$ approximation of E will produce $O\left(\epsilon^{i+1}\right)$ approximation of the sought solution K, that is,

$$K - K^{(i)} = O\left(\epsilon^{i+1}\right) \qquad (13.52)$$

where i is the number of iteration.

The near-optimal control law is now given by

$$u^{(i)}(t) = -\frac{1}{\epsilon}R^{-1}[0 \quad B_2^T]K^{(i)}x^{(i)}(t) = -F^{(i)}x^{(i)}(t) \qquad (13.53)$$

where the approximate state trajectories satisfy

$$\dot{x}^{(i)}(t) = Ax^{(i)}(t) + Bu^{(i)}(t) \qquad (13.54)$$

with $K^{(i)}$ obtained from (13.39).

The approximate performance criterion is obtained from

$$J^{(i)} = trV^{(i)} \qquad (13.55)$$

where $V^{(i)}$ satisfies

$$\left(A - BF^{(i)}\right)^T V^{(i)} + V^{(i)}\left(A - BF^{(i)}\right) + Q + F^{(i)^T}RF^{(i)} = 0 \qquad (13.56)$$

This algebraic Lyapunov equation is ill-defined due to presence of $O\left(\frac{1}{\epsilon}\right)$ term in $F^{(i)}$. However, introducing a proper scaling for $V^{(i)}$ as

$$V^{(i)} = \begin{bmatrix} V_{11}^{(i)} & \epsilon V_{12}^{(i)} \\ \epsilon V_{12}^{(i)^T} & \epsilon V_{22}^{(i)} \end{bmatrix} \qquad (13.57)$$

and partitioning the algebraic Lyapunov equation (13.56) according to (13.6), (13.10), (13.53), and (13.57) produces a well-defined system of reduced-order Lyapunov or Lyapunov-like algebraic equations, where $K_{11}^{(i)}, K_{12}^{(i)}$, and $K_{22}^{(i)}$ play the role of system coefficients. Perturbing these coefficients by an $O\left(\epsilon^i\right)$ produces the same perturbation in the required solutions for $V_{11}^{(i)}, V_{12}^{(i)}$, and $V_{22}^{(i)}$. Thus, we can conclude that

$$J_{opt} - J^{(i)} = O\left(\epsilon^i\right) \qquad (13.58)$$

It is left as an exercise to the reader to derive the reduced-order parallel algorithm for solving the algebraic Lyapunov equation (13.56).

Exercise 13.1: Find the zeroth-order approximation of the algebraic Lyapunov equation (13.56) in terms of completely decomposed reduced-order Lyapunov or Lyapunov-like algebraic equations. Then, develop a parallel reduced-order algorithm for solving (13.56) with an arbitrary order of accuracy.

$\triangle$

Example 13.1

The algorithm developed in the proceeding section is applied to the following system, where matrices are chosen arbitrarily so as to satisfy the required assumptions.

$$A = \begin{bmatrix} -2 & 0 & 0 & 1 \\ 1 & -2 & 1 & 0 \\ 0 & 0 & -1 & 2 \\ 1 & 0 & 0 & -3 \end{bmatrix}, \quad B = \begin{bmatrix} 0 & 0 \\ 0 & 0 \\ 2 & 1 \\ 1 & 2 \end{bmatrix}$$

$$Q = \begin{bmatrix} 1 & 0 & 0 & 0 \\ 0 & 1 & 0 & 0 \\ 0 & 0 & 6 & 5 \\ 0 & 0 & 5 & 6 \end{bmatrix}, \quad R = I_2$$

For different values of ϵ, the exact rate of convergence varies. Results of computations are shown in Table 13.1, where the number of iterations needed to achieve convergence accurate to five decimal places is summarized for $\epsilon = 0.01, \ 0.1, \ 0.2, \ 0.5$.

Table 13.2 contains results of computations for $\epsilon = 0.1$. The zero-order solution, $K^{(0)}$ is given in the first column. Also shown is the computed solution after $n = 9$ iterations, $K^{(9)}$, which is equal to the exact solution K with an accuracy of five decimal places. The difference between the exact solution K and the n-th order solution $K^{(n)}$ is also given in Table 13.2 for various n, simplified to 2 decimal places, in order to illustrate convergence of the algorithm.

Several points are supported by this example. The zero-order solution $K^{(0)}$ is within $O(\epsilon)$ of the exact solution K. The k-th order solution $K^{(k)}$ converges to the exact solution with a rate approximated by $O(\epsilon^k)$.

ϵ	number of required iterations
0.01	3
0.1	9
0.2	>10
0.5	divergent

Table 13.1: Convergence as a function of ϵ

Research Problem 13.1: Study the continuous-time optimal cheap control (or high gain feedback problem) on a finite time interval. Obtain the required solution in terms of the reduced-order differential Riccati equations.

△

i,j	$K_{ij}^{(0)}$	$K_{ij}^{(9)} = K_{ij}$	$K_{ij}^{(2)} - K_{ij}$	$K_{ij}^{(7)} - K_{ij}$
1,1	0.27551×10^0	0.27982×10^0	-0.96×10^{-3}	0.14×10^{-5}
1,2	0.64761×10^{-1}	0.62674×10^0	0.34×10^{-3}	-0.56×10^{-6}
1,3	-0.88276×10^{-2}	-0.91763×10^{-2}	-0.46×10^{-3}	0.56×10^{-6}
1,4	0.12248×10^{-1}	0.13235×10^{-1}	0.52×10^{-3}	-0.65×10^{-6}
2,2	0.24486×10^0	0.24624×10^0	-0.55×10^{-5}	0.15×10^{-6}
2,3	0.10561×10^{-1}	0.87739×10^{-2}	0.46×10^{-4}	-0.97×10^{-7}
2,4	-0.74489×10^{-2}	-0.54706×10^{-2}	-0.38×10^{-4}	0.11×10^{-6}
3,3	0.10528×10^0	0.94130×10^{-1}	-0.39×10^{-4}	0.12×10^{-6}
3,4	0.52771×10^{-2}	0.16722×10^{-1}	0.12×10^{-3}	-0.16×10^{-6}
4,4	0.10528×10^0	0.91449×10^{-1}	-0.21×10^{-3}	0.20×10^{-6}

Table 13.2: Computed solution of the Riccati equation

Research Problem 13.2: Extend the methodology presented in this chapter to the discrete-time domain and study corresponding cheap control and high gain feedback problems. Consider both the finite time optimal control problem and its steady state behavior.

△

13.2 Case Study: Large Space Structure

The algorithm developed in the proceeding section is applied to the numerical example given in (Moerder and Calise, 1985b), where the problem of controlling a large space structure is addressed. This control system has the structure of the high gain feedback problems.

The original problem matrices given in Appendix 13.1 are first transformed into the form given by (13.7), yielding

$$
A_{11} = 0, \qquad A_{12} = \begin{bmatrix} 1 & 0 & 0 & 0 \\ 0 & 1 & 0 & 0 \\ 0 & 0 & 1 & 0 \\ 0 & 0 & 0 & 1 \end{bmatrix}
$$

$$
A_{21} = \begin{bmatrix} -0.176 & 0 & 0 & 0 \\ 0 & -0.176 & 0 & 0 \\ 0 & 0 & -4.41 & 0 \\ 0 & 0 & 0 & -4.41 \end{bmatrix}, \qquad A_{22} = 0
$$

$$
B_2 = \begin{bmatrix} -9.20 & -1.40 & 0.92 & -1.40 \\ 0.65 & 1.60 & 0.65 & -1.60 \\ 1.40 & -1.00 & 1.40 & 1.00 \\ 2.05 & -0.80 & -2.00 & -0.80 \end{bmatrix}
$$

$$
Q_{22} = \begin{bmatrix} 21.06 & 0 & 0 & -6.12 \\ 0 & 23.86 & -5.90 & 0 \\ 0 & -5.90 & 38.74 & 0 \\ -6.12 & 0 & 0 & 38.74 \end{bmatrix}, \qquad Q_{12} = 0_{4 \times 4}; \quad R = I_4
$$

In addition, to satisfy Assumption 13.4, Q_{11} is explicitly chosen as an identity matrix. Table 13.3 contains the results of the computation for $\epsilon = 0.2$. Some components of the zero-order solution $K^{(0)}$ are given in the first column of the table. Also shown is the computed solution after $n = 6$ iterations, $K^{(6)}$, which is equal to the exact solution K with an accuracy of five decimal places.

As in the previous example, the zero-order solution $K^{(0)}$ is within $O(\epsilon)$ of the exact solution K. Also, the k-th order solution $K^{(k)}$ converges to the exact solution with a rate approximated by $O(\epsilon^k)$.

i,j	$K_{i,j}^{(0)} - K_{i,j}^{(1)}$	$K_{i,j}^{(0)} - K_{i,j}^{(3)}$	$K_{i,j}^{(0)} - K_{i,j}^{(5)}$	$K_{i,j}^{(6)}$
1,1	0.11×10^{-2}	0.32×10^{-5}	-0.43×10^{-8}	$0.45740 \times 10^{+1}$
1,2	-0.10×10^{-4}	-0.23×10^{-6}	0.25×10^{-7}	-0.24621×10^{-4}
1,3	0.12×10^{-3}	0.17×10^{-5}	0.30×10^{-7}	0.10077×10^{-3}
1,4	-0.12×10^{-1}	-0.28×10^{-4}	-0.52×10^{-7}	-0.57794×10^{0}
1,5	-0.39×10^{-5}	-0.17×10^{-7}	-0.34×10^{-10}	0.84580×10^{-1}
1,6	0.54×10^{-6}	0.25×10^{-8}	0.23×10^{-11}	-0.54389×10^{-4}
1,7	0.25×10^{-5}	0.28×10^{-7}	0.16×10^{-9}	-0.93145×10^{-4}
1,8	-0.12×10^{-3}	-0.29×10^{-6}	-0.55×10^{-9}	0.74500×10^{-2}

Table 13.3 Solution of the algebraic Riccati equation

13.3 Decomposition of the Open-Loop Cheap Control Problem

In this section, the singular perturbation approach is used to obtain an alternate and more efficient method of solving the two-point boundary value problem for the optimal open-loop cheap control. The original two-point boundary value problem is transformed into completely decoupled initial value problems. The solution obtained in this manner clearly exhibits both the singular arc and the fast transients, separately.

Consider the cheap control problem defined by

$$\begin{bmatrix} \dot{x}_1(t) \\ \dot{x}_2(t) \end{bmatrix} = \begin{bmatrix} A_{11} & A_{12} \\ A_{21} & A_{22} \end{bmatrix} \begin{bmatrix} x_1(t) \\ x_2(t) \end{bmatrix} + \begin{bmatrix} 0 \\ B_2 \end{bmatrix} u(t), \quad x_0 = x(t_0)$$
(13.59)

where $x_1(t) \in \Re^n$, $x_2(t) \in \Re^m$, $u(t) \in \Re^m$, and B_{22} is a nonsingular $m \times m$ matrix. The performance criterion of the cheap control problem is defined as

$$J = \frac{1}{2}x^T(T)Fx(T) + \frac{1}{2}\int_{t_0}^{T}[x^T(t)Qx(t) + \epsilon^2 u^T(t)Ru(t)]\,dt$$

(13.60)

with positive definite R and positive semidefinite Q and F. It is obvious that for $\epsilon = 0$, the optimal control problem (13.59)-(13.60) is singular.

The open-loop optimal cheap control problem has the solution given by

$$u(t) = -\frac{1}{\epsilon^2}R^{-1}B^T p(t)$$

(13.61)

where $p(t) \in \Re^{n+m}$ is a costate variable satisfying (Jameson and O'Malley, 1975)

$$\begin{bmatrix} \dot{x}(t) \\ \dot{p}(t) \end{bmatrix} = \begin{bmatrix} A & -S \\ -Q & -A^T \end{bmatrix} \begin{bmatrix} x(t) \\ p(t) \end{bmatrix}$$

(13.62)

The boundary conditions are expressed in the standard form as

$$M\begin{bmatrix} x(t_0) \\ p(t_0) \end{bmatrix} + N\begin{bmatrix} x(T) \\ p(T) \end{bmatrix} = c$$

(13.63)

where

$$M = \begin{bmatrix} I & 0 \\ 0 & 0 \end{bmatrix}, \quad N = \begin{bmatrix} 0 & 0 \\ -F & I \end{bmatrix}, \quad c = \begin{bmatrix} x(t_0) \\ 0 \end{bmatrix}$$

(13.64)

for the free endpoint problem; and

$$M = \begin{bmatrix} I & 0 \\ 0 & 0 \end{bmatrix}, \quad N = \begin{bmatrix} 0 & 0 \\ I & 0 \end{bmatrix}, \quad c = \begin{bmatrix} x(t_0) \\ x(T) \end{bmatrix}$$

(13.65)

for the fixed endpoint problem.

Matrices A, Q, B, S, and F, and vectors x and p, respectively, have the forms

$$A = \begin{bmatrix} A_{11} & A_{12} \\ A_{21} & A_{22} \end{bmatrix}, \quad Q = \begin{bmatrix} Q_{11} & Q_{12} \\ Q_{12}^T & Q_{22} \end{bmatrix}, \quad B = \begin{bmatrix} 0 \\ B_{22} \end{bmatrix}$$

$$S = \frac{1}{\epsilon^2}BR^{-1}B^T = \begin{bmatrix} 0 & 0 \\ 0 & \frac{1}{\epsilon^2}B_{22}R^{-1}B_{22}^T \end{bmatrix}; \quad F = \begin{bmatrix} F_1 & \epsilon F_2 \\ \epsilon F_2^T & \epsilon F_3 \end{bmatrix}$$

$$x(t) = \begin{bmatrix} x_1(t) \\ x_2(t) \end{bmatrix}, \quad p(t) = \begin{bmatrix} p_1(t) \\ \epsilon p_2(t) \end{bmatrix}$$

The main purpose of this section is to obtain the solution of the open-loop cheap control problem by singular perturbation approach, so that the solution can clearly exhibit both the singular arc and the fast transients away from it.

If we partition $p(t) = \left[p_1^T(t) \; \epsilon p_2^T(t) \right]^T$ with $p_1(t) \in \Re^n$ and $p_2(t) \in \Re^m$, and interchange the second and third rows in (13.62) we will get

$$
\begin{bmatrix} \dot{x}_1(t) \\ \dot{p}_1(t) \\ \epsilon \dot{x}_2(t) \\ \epsilon \dot{p}_2(t) \end{bmatrix} = \begin{bmatrix} T_1 & T_2 \\ T_3 & T_4 \end{bmatrix} \begin{bmatrix} x_1(t) \\ p_1(t) \\ x_2(t) \\ p_2(t) \end{bmatrix}
\tag{13.66}
$$

where

$$
T_1 = \begin{bmatrix} A_{11} & 0 \\ -Q_{11} & -A_{11}^T \end{bmatrix}, \quad T_2 = \begin{bmatrix} A_{12} & 0 \\ -Q_{12} & -\epsilon A_{21}^T \end{bmatrix}
$$

$$
\tag{13.67}
$$

$$
T_3 = \begin{bmatrix} \epsilon A_3 & 0 \\ -Q_{12}^T & -A_{12}^T \end{bmatrix}, \quad T_4 = \begin{bmatrix} \epsilon A_{22} & -B_{22}R^{-1}B_{22}^T \\ -Q_{22} & -\epsilon A_{22}^T \end{bmatrix}
$$

Note that equation (13.66) has the singular perturbation form and the matrix T_4 is the Hamiltonian matrix of the 'fast' subsystem.

In the sequel, we use the following transformation (Chang, 1972) defined by

$$
\mathbf{T}_1 = \begin{bmatrix} I_{2n} - \epsilon HL & -\epsilon H \\ L & I_{2m} \end{bmatrix}, \quad \mathbf{T}_1^{-1} = \begin{bmatrix} I_{2n} & \epsilon H \\ -L & I_{2m} - \epsilon LH \end{bmatrix} \tag{13.68}
$$

where L and H satisfy

$$
T_4 L - T_3 - \epsilon L (T_1 - T_2 L) = 0 \tag{13.69}
$$

$$
-H(T_4 + \epsilon LT_2) + T_2 + \epsilon(T_1 - T_2 L)H = 0 \tag{13.70}
$$

Equations (13.69) and (13.70) have unique solutions under condition that T_4 is nonsingular at $\epsilon = 0$. These equations can be solved by using the recursive algorithm developed in Chapter 3. The transformation (13.68) is then applied to (13.66) to produce two completely decoupled subsystems

$$\begin{bmatrix} \dot{\eta}_1 \\ \dot{\xi}_1 \end{bmatrix} = (T_1 - T_2 L) \begin{bmatrix} \eta_1 \\ \xi_1 \end{bmatrix} \tag{13.71}$$

$$\epsilon \begin{bmatrix} \dot{\eta}_2 \\ \dot{\xi}_2 \end{bmatrix} = (T_4 + \epsilon L T_2) \begin{bmatrix} \eta_2 \\ \xi_2 \end{bmatrix} \tag{13.72}$$

where

$$\begin{bmatrix} \eta_1 \\ \xi_1 \\ \eta_2 \\ \xi_2 \end{bmatrix} = \mathbf{T}_1 \begin{bmatrix} x_1 \\ p_1 \\ x_2 \\ p_2 \end{bmatrix} \tag{13.73}$$

The boundary conditions are changed by interchanging p_1 and x_2. The modified matrices in (13.63) are

$$M_1 \begin{bmatrix} x_1(t_0) \\ p_1(t_0) \\ x_2(t_0) \\ p_2(t_0) \end{bmatrix} + N_1 \begin{bmatrix} x_1(T) \\ p_1(T) \\ x_2(T) \\ p_2(T) \end{bmatrix} = c_1 \tag{13.74}$$

where

$$M_1 = \begin{bmatrix} I_n & 0 & 0 & 0 \\ 0 & 0 & 0 & 0 \\ 0 & 0 & I_m & 0 \\ 0 & 0 & 0 & 0 \end{bmatrix}, \quad N_1 = \begin{bmatrix} 0 & 0 & 0 & 0 \\ -F_{11} & I_n & -\epsilon F_{12} & 0 \\ 0 & 0 & 0 & 0 \\ -F_{12}^T & 0 & -F_{22} & I_m \end{bmatrix}$$

$$c_1 = \begin{bmatrix} x_1(t_0) \\ 0 \\ x_2(t_0) \\ 0 \end{bmatrix} \tag{13.75}$$

for the free ending problem; and

$$M_1 = \begin{bmatrix} I_n & 0 & 0 & 0 \\ 0 & 0 & 0 & 0 \\ 0 & 0 & I_m & 0 \\ 0 & 0 & 0 & 0 \end{bmatrix}, \quad N_1 = \begin{bmatrix} 0 & 0 & 0 & 0 \\ I_n & 0 & 0 & 0 \\ 0 & 0 & 0 & 0 \\ 0 & 0 & I_m & 0 \end{bmatrix}$$

$$c_1 = \begin{bmatrix} x_1(t_0) \\ x_1(T) \\ x_2(t_0) \\ x_2(T) \end{bmatrix} \tag{13.76}$$

for the fixed ending problem.

The nonsingular transformation (13.68) applied to (13.74) produces

$$M_2 \begin{bmatrix} \eta_1(t_0) \\ \xi_1(t_0) \\ \eta_2(t_0) \\ \xi_2(t_0) \end{bmatrix} + N_2 \begin{bmatrix} \eta_1(T) \\ \xi_1(T) \\ \eta_2(T) \\ \xi_2(T) \end{bmatrix} = c_1 \qquad (13.77)$$

where

$$M_2 = M_1 \mathbf{T}_1^{-1}, \quad N_2 = N_1 \mathbf{T}_1^{-1} \qquad (13.78)$$

Since solutions of (13.71) and (13.72) are given by

$$\begin{bmatrix} \eta_1(t) \\ \xi_1(t) \end{bmatrix} = e^{(T_1 - T_2 L)(t - t_0)} \begin{bmatrix} \eta_1(t_0) \\ \xi_1(t_0) \end{bmatrix} \qquad (13.79)$$

$$\begin{bmatrix} \eta_2(t) \\ \xi_2(t) \end{bmatrix} = e^{\frac{1}{\epsilon}(T_4 + \epsilon L T_2)(t - t_0)} \begin{bmatrix} \eta_2(t_0) \\ \xi_2(t_0) \end{bmatrix} \qquad (13.80)$$

we can eliminate $\eta_1(T)$, $\xi_1(T)$, $\eta_2(T)$, and $\xi_2(T)$ from (13.77), which yields

$$\left\{ M_2 + N_2 \begin{bmatrix} e^{(T_1 - T_2 L)(T - t_0)} & 0 \\ 0 & e^{\frac{1}{\epsilon}(T_4 + \epsilon L T_2)(T - t_0)} \end{bmatrix} \right\} \begin{bmatrix} \eta_1(t_0) \\ \xi_1(t_0) \\ \eta_2(t_0) \\ \xi_2(t_0) \end{bmatrix} = c_1 \qquad (13.81)$$

Equation (13.81) can be represented in the form

$$\beta(\epsilon) \begin{bmatrix} \eta_1(t_0) \\ \xi_1(t_0) \\ \eta_2(t_0) \\ \xi_2(t_0) \end{bmatrix} = c_1 \qquad (13.82)$$

It has been shown in Lemma 6.1 that the matrix $\beta(\epsilon)$ is invertible, hence equation (13.82) can be solved to obtain $\eta_1(t_0)$, $\xi_1(t_0)$, $\eta_2(t_0)$, and $\xi_2(t_0)$.

Equation (13.79) gives the solution of singular arc $\eta_1(t)$, and equation (13.80) gives the solution of fast transient $\eta_2(t)$ of the cheap control problem.

After getting the solutions of (13.79) and (13.80), using (13.73), we obtain values for $x_1(t)$, $x_2(t)$, $p_1(t)$, and $p_2(t)$. The costate variable $p(t)$ and the optimal control law $u(t)$ are therefore found.

13.4 Numerical Example

In order to illustrate the proposed method, we consider the following system which is in the form of (13.59)

$$\dot{x} = \begin{bmatrix} -2 & 0 & 0 & 1 \\ 1 & -2 & 1 & 0 \\ 0 & 0 & -1 & 2 \\ 1 & 0 & 0 & -3 \end{bmatrix} x(t) + \begin{bmatrix} 0 & 0 \\ 0 & 0 \\ 1 & 0 \\ 0 & 1 \end{bmatrix} u(t)$$

with the initial condition

$$x^T(t_0) = [2 \quad 2 \quad 2 \quad 2]$$

and $\epsilon = 0.1$.

For the free endpoint problem we take the weighting matrices as

$$F = \begin{bmatrix} 1 & 0 & 0 & 0 \\ 0 & 1 & 0 & 0 \\ 0 & 0 & 0.1 & 0 \\ 0 & 0 & 0 & 0.1 \end{bmatrix}, \quad Q = \begin{bmatrix} 1 & 0 & 0 & 0 \\ 0 & 1 & 0 & 0 \\ 0 & 0 & 1 & 0 \\ 0 & 0 & 0 & 1 \end{bmatrix}, \quad R = \begin{bmatrix} 1 & 0 \\ 0 & 1 \end{bmatrix}$$

For the fixed endpoint problem we use the following matrices

$$Q = I_4, \quad R = I_2, \quad x^T(T) = [1 \quad 1 \quad 1 \quad 1]$$

Simulation results are obtained by using the software package MAT-LAB (Hill, 1988). The time interval is specified by $t_0 = 0$ and $T = 1$.

Figures 13.1–13.4 (free endpoint) and 13.5–13.8 (fixed endpoint) give the comparison of the optimal state variables of the original system (13.59) with the optimal state variables of the transformed system (13.71)-(13.72). From these figures, it is clear that the singular arcs and fast transients of the cheap control problem are completely separated. The following notation has been used in Figures 13.1–13.8: $x_1(t) = [x_{11}(t) \ x_{12}(t)]^T$, $x_2(t) = [x_{21}(t) \ x_{22}(t)]^T$, $\eta_1(t) = [\eta_{11}(t) \ \eta_{12}(t)]^T$, and $\eta_2(t) = [\eta_{21}(t) \ \eta_{22}(t)]^T$, where the variables $x(t)$ are represented by the solid lines, and the variables $\eta(t)$ are represented by the dashed lines.

Research Problem 13.3: Study the discrete-time version of the optimal cheap control open-loop problem. Consider both the finite time and the steady state solutions.

△

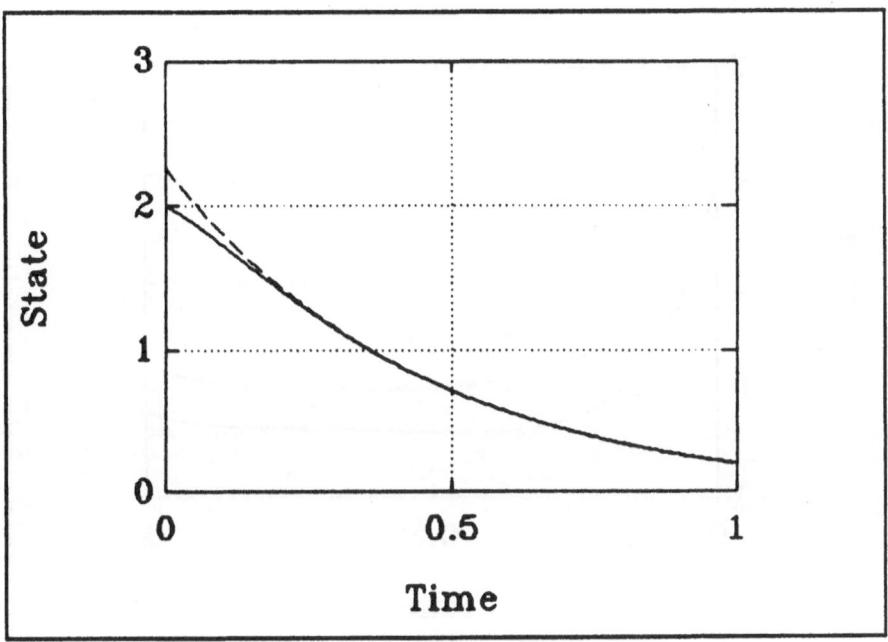

Figure 13.1: Free endpoint problem $x_{11}(t)$, $\eta_{11}(t)$

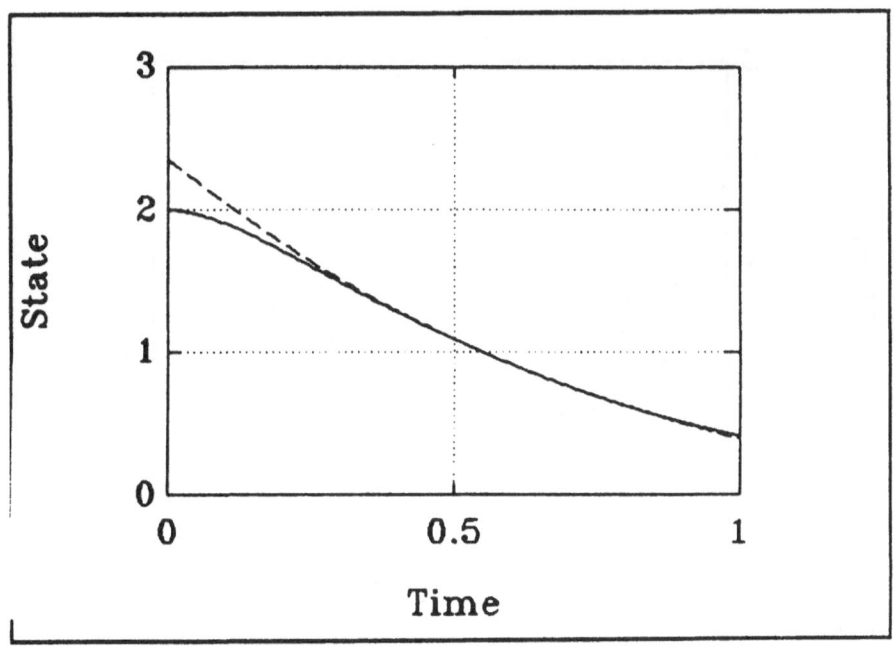

Figure 13.2: Free endpoint problem $x_{12}(t)$, $\eta_{12}(t)$

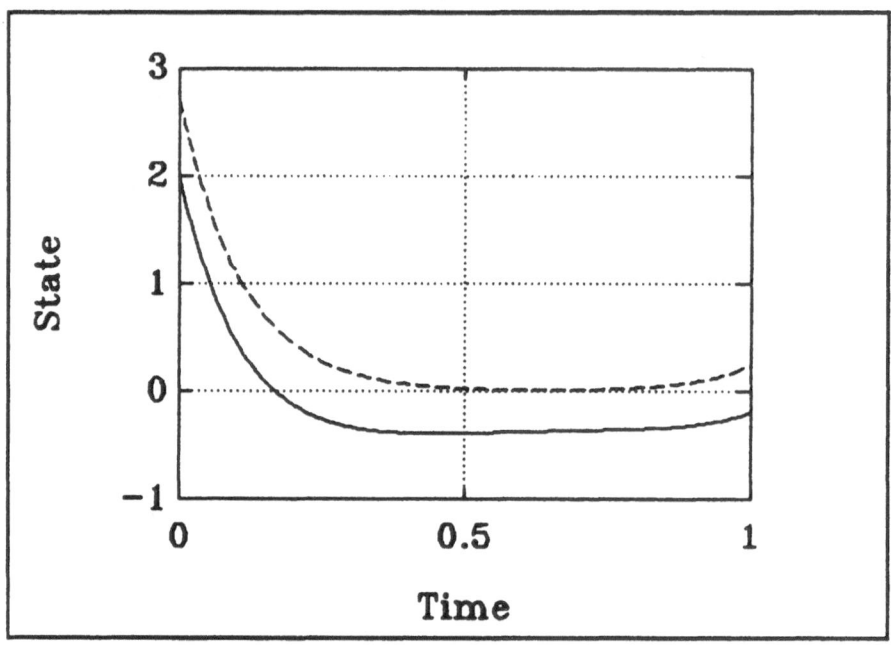

Figure 13.3: Free endpoint problem $x_{21}(t)$, $\eta_{21}(t)$

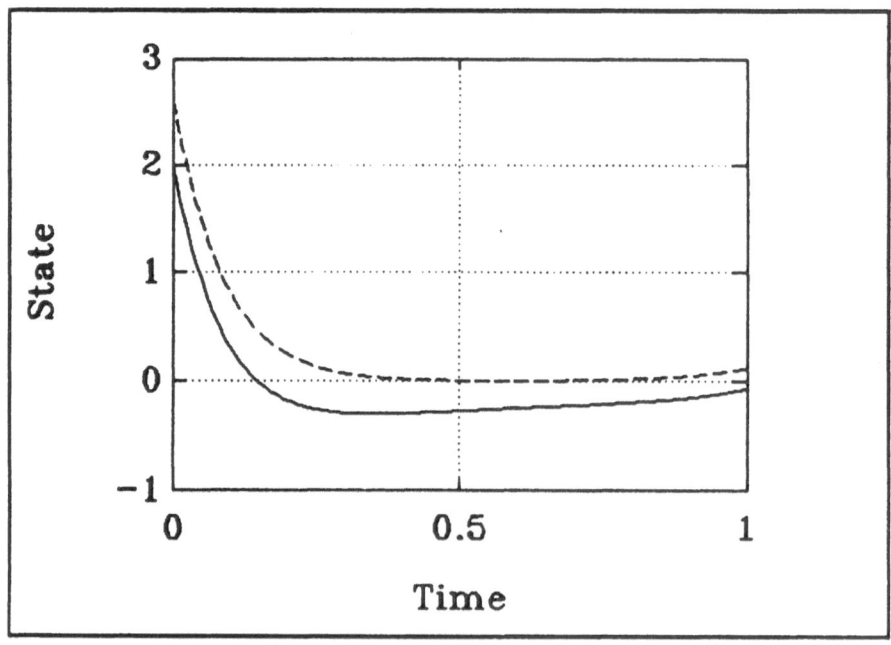

Figure 13.4: Free endpoint problem $x_{22}(t)$, $\eta_{22}(t)$

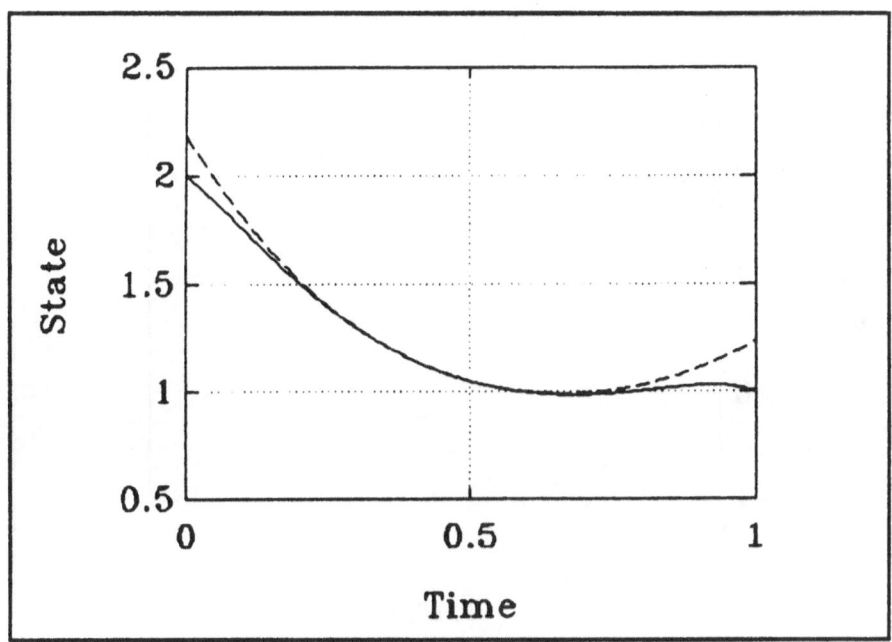

Figure 13.5: Fixed endpoint problem $x_{11}(t)$, $\eta_{11}(t)$

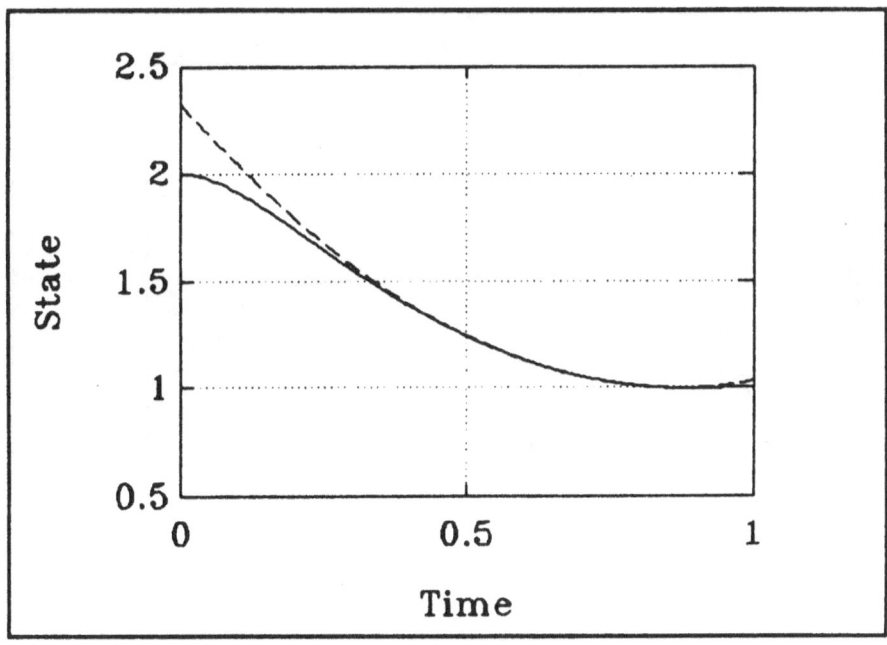

Figure 13.6: Fixed endpoint problem $x_{12}(t)$, $\eta_{12}(t)$

387

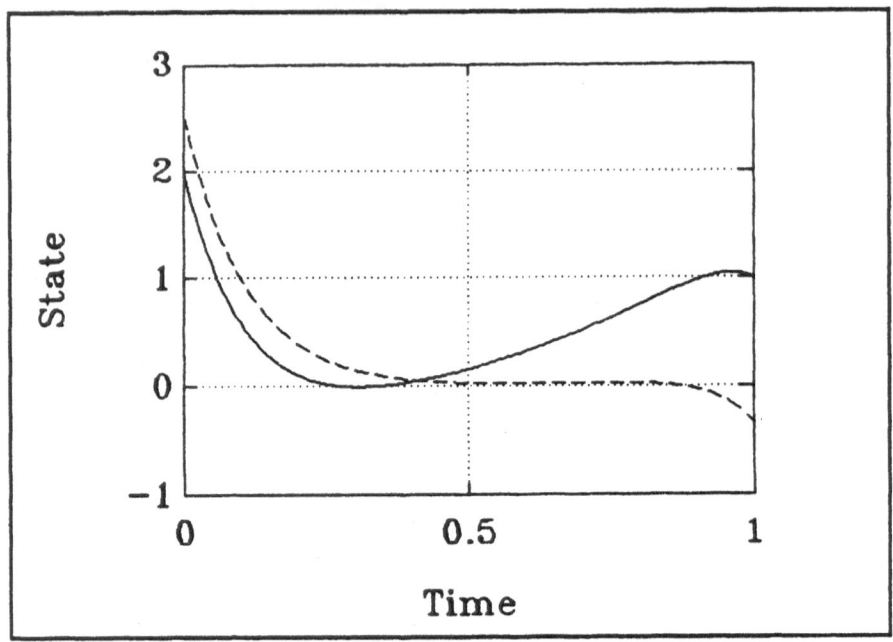

Figure 13.7: Fixed endpoint problem $x_{21}(t)$, $\eta_{21}(t)$

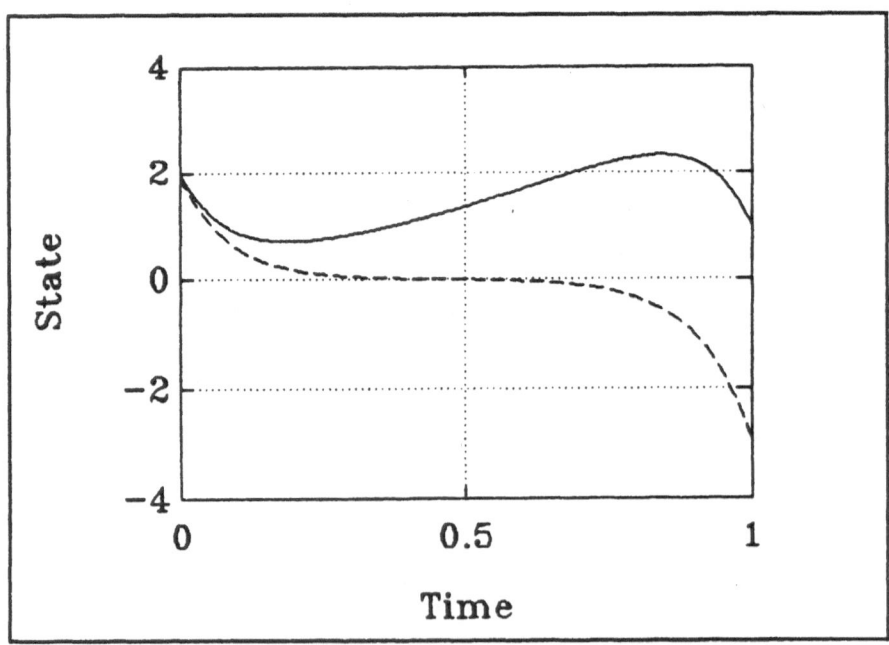

Figure 13.8: Fixed endpoint problem $x_{22}(t)$, $\eta_{22}(t)$

13.5 Exact Decomposition of the Algebraic Riccati Equation for Cheap Control Problem

In this section, we study the linear-quadratic regulator problem of cheap control problem by using the approach given in Chapter 7. The ill-defined algebraic Riccati equation of cheap control problem is completely and exactly decomposed into two reduced-order nonsymmetric well-defined algebraic Riccati equations, and the optimal solution of the Riccati equation is obtained in terms of the reduced-order problems.

In equation (13.5), we have obtained the form of the algebraic Riccati equation for the cheap control and high gain feedback problems

$$PA + A^T P + Q = \frac{1}{\epsilon^2} PBR^{-1}B^T P$$

The state and costate variables are related by $p = Px$. Partitioning p such that $p = [p_1^T \quad \epsilon p_2^T]^T$ with $p_1 \in \Re^n$, $p_2 \in \Re^m$ and interchanging second and third rows in (13.62), we can get (13.66), that is

$$\begin{bmatrix} \dot{x}_1 \\ \dot{p}_1 \\ \epsilon \dot{x}_2 \\ \epsilon \dot{p}_2 \end{bmatrix} = \begin{bmatrix} T_1 & T_2 \\ T_3 & T_4 \end{bmatrix} \begin{bmatrix} x_1 \\ p_1 \\ x_2 \\ p_2 \end{bmatrix}$$

where $T_i's$ are defined in (13.67).

The transformation (13.68) applied to (13.66) produces two completely decoupled subsystems (13.71)-(13.72)

$$\begin{bmatrix} \dot{\eta}_1 \\ \dot{\xi}_1 \end{bmatrix} = (T_1 - T_2 L) \begin{bmatrix} \eta_1 \\ \xi_1 \end{bmatrix}$$

$$\epsilon \begin{bmatrix} \dot{\eta}_2 \\ \dot{\xi}_2 \end{bmatrix} = (T_4 + \epsilon L T_2) \begin{bmatrix} \eta_2 \\ \xi_2 \end{bmatrix}$$

where the corresponding transformation is defined by (13.73), with

$$\begin{bmatrix} \eta_1 \\ \xi_1 \\ \eta_2 \\ \xi_2 \end{bmatrix} = T_1 \begin{bmatrix} x_1 \\ p_1 \\ x_2 \\ p_2 \end{bmatrix}$$

In order to find the optimal solution of the cheap control problem in terms of the reduced-order subsystems, we have to find the relations between full-order Riccati equation (13.5) and the decomposed reduced-order Riccati equations corresponding to subsystems (13.71) and (13.72).

The rearrangement and modification of variables in (13.66) is done by using the permutation matrix E_1 of the form

$$\begin{bmatrix} x_1 \\ p_1 \\ x_2 \\ p_2 \end{bmatrix} = \begin{bmatrix} I_n & 0 & 0 & 0 \\ 0 & 0 & I_n & 0 \\ 0 & I_m & 0 & 0 \\ 0 & 0 & 0 & \frac{I_m}{\epsilon} \end{bmatrix} \begin{bmatrix} x_1 \\ x_2 \\ p_1 \\ \epsilon p_2 \end{bmatrix} = E_1 \begin{bmatrix} x \\ p \end{bmatrix} \qquad (13.83)$$

Combining equations (13.73) and (13.83), we obtain the relationship between the original coordinates and the new ones

$$\begin{bmatrix} \eta_1 \\ \eta_2 \\ \xi_1 \\ \xi_2 \end{bmatrix} = E_2^T T_1 E_1 \begin{bmatrix} x \\ p \end{bmatrix} = \Pi \begin{bmatrix} x \\ p \end{bmatrix} = \begin{bmatrix} \Pi_1 & \Pi_2 \\ \Pi_3 & \Pi_4 \end{bmatrix} \begin{bmatrix} x \\ p \end{bmatrix} \qquad (13.84)$$

where E_2 is a permutation matrix in the form

$$E_2 = \begin{bmatrix} I_n & 0 & 0 & 0 \\ 0 & 0 & I_n & 0 \\ 0 & I_m & 0 & 0 \\ 0 & 0 & 0 & I_m \end{bmatrix} \qquad (13.85)$$

Since $p = Px$, where P satisfies the algebraic Riccati equation (13.5), it follows that

$$\begin{bmatrix} \eta_1 \\ \eta_2 \end{bmatrix} = (\Pi_1 + \Pi_2 P)\, x, \qquad \begin{bmatrix} \xi_1 \\ \xi_2 \end{bmatrix} = (\Pi_3 + \Pi_4 P)\, x \qquad (13.86)$$

In the original coordinates, the required optimal solution has a closed-loop nature. We have the same attribute for the new systems (13.71) and (13.72); that is

$$\begin{bmatrix} \xi_1 \\ \xi_2 \end{bmatrix} = \begin{bmatrix} P_1 & 0 \\ 0 & P_2 \end{bmatrix} \begin{bmatrix} \eta_1 \\ \eta_2 \end{bmatrix} \qquad (13.87)$$

Then, (13.71) and (13.72) yield

$$\begin{bmatrix} P_1 & 0 \\ 0 & P_2 \end{bmatrix} = (\Pi_3 + \Pi_4 P)(\Pi_1 + \Pi_2 P)^{-1} \qquad (13.88)$$

Following the same logic, we can find P reversely by introducing

$$E_1^{-1}T_1^{-1}E_2 = \Omega = \begin{bmatrix} \Omega_1 & \Omega_2 \\ \Omega_3 & \Omega_4 \end{bmatrix} \tag{13.89}$$

with

$$E_1^{-1} = \begin{bmatrix} I_n & 0 & 0 & 0 \\ 0 & 0 & I_n & 0 \\ 0 & I_m & 0 & 0 \\ 0 & 0 & 0 & \epsilon I_m \end{bmatrix} \tag{13.90}$$

which yields to

$$P = \left(\Omega_3 + \Omega_4 \begin{bmatrix} P_1 & 0 \\ 0 & P_2 \end{bmatrix}\right)\left(\Omega_1 + \Omega_2 \begin{bmatrix} P_1 & 0 \\ 0 & P_2 \end{bmatrix}\right)^{-1} \tag{13.91}$$

It is shown in Appendix 7.1 that the required matrices in (13.88) and (13.91) are invertible for sufficiently small values of ϵ.

Partitioning (13.71) and (13.72) as

$$\begin{bmatrix} \dot{\eta}_1 \\ \dot{\xi}_1 \end{bmatrix} = \begin{bmatrix} a_1 & a_2 \\ a_3 & a_4 \end{bmatrix}\begin{bmatrix} \eta_1 \\ \xi_1 \end{bmatrix} = (T_1 - T_2 L)\begin{bmatrix} \eta_1 \\ \xi_1 \end{bmatrix} \tag{13.92}$$

$$\epsilon\begin{bmatrix} \dot{\eta}_2 \\ \dot{\xi}_2 \end{bmatrix} = \begin{bmatrix} b_1 & b_2 \\ b_3 & b_4 \end{bmatrix}\begin{bmatrix} \eta_2 \\ \xi_2 \end{bmatrix} = (T_4 + \epsilon L T_2)\begin{bmatrix} \eta_2 \\ \xi_2 \end{bmatrix} \tag{13.93}$$

and using (13.87) yield to two reduced-order nonsymmetric algebraic Riccati equations

$$0 = P_1 a_1 - a_4 P_1 - a_3 + P_1 a_2 P_1 \tag{13.94}$$

$$0 = P_2 b_1 - b_4 P_2 - b_3 + P_2 b_2 P_2 \tag{13.95}$$

where

$$\begin{bmatrix} a_1 & a_2 \\ a_3 & a_4 \end{bmatrix} = \begin{bmatrix} A_{11} - A_{12}L_1 & -A_{12}L_2 \\ -Q_{11} + Q_{12}L_1 + \epsilon A_{21}^T L_3 & -A_{11}^T + Q_{12}L_2 + \epsilon A_{21}^T L_4 \end{bmatrix}$$

$$\begin{bmatrix} b_1 & b_2 \\ b_3 & b_4 \end{bmatrix} =$$
$$\begin{bmatrix} \epsilon A_{22} + \epsilon(L_1 A_{12} - L_2 Q_{12}) & -B_{22}R^{-1}B_{22}^T - \epsilon^2 L_2 A_{21}^T \\ -Q_{22} + \epsilon(L_3 A_{12} - L_4 Q_{12}) & -\epsilon A_{22}^T - \epsilon^2 L_4 A_{21}^T \end{bmatrix} \tag{13.96}$$

with

$$L = \begin{bmatrix} L_1 & L_2 \\ L_3 & L_4 \end{bmatrix}, \quad H = \begin{bmatrix} H_1 & H_2 \\ H_3 & H_4 \end{bmatrix} \qquad (13.97)$$

The reduced-order algebraic Riccati equation (13.94) is nonsymmetric and it is given by

$$\begin{aligned} P_1 \left(A_{11} - A_{12}L_1 \right) + \left(A_{11}^T - Q_{12}L_2 - \epsilon A_{21}^T L_4 \right)^T P_1 \\ + \left(Q_{11} - Q_{12}L_1 - \epsilon A_{21}^T L_3 \right) - P_1 A_{12} L_2 P_1 = 0 \end{aligned} \qquad (13.98)$$

The reduced-order algebraic Riccati equation (13.95) is also non-symmetric

$$\begin{aligned} \epsilon P_2 \left(A_{22} + L_1 A_{12} - L_2 Q_{12} \right) + \epsilon \left(A_{22}^T + \epsilon L_4 A_{21}^T \right) P_2 \\ + \left[Q_{22} - \epsilon \left(L_3 A_{12} - L_4 Q_{12} \right) \right] - P_2 \left(B_{22} R^{-1} B_{22}^T + \epsilon^2 L_2 A_{21}^T \right) P_2 = 0 \\ (13.99) \end{aligned}$$

but its $O\left(\epsilon\right)$ approximation is a symmetric one, that is

$$P_2 B_{22} R^{-1} B_{22}^T P_2 - Q_{22} = 0 \qquad (13.100)$$

Under Assumptions 13.1 and 13.2, the unique positive definite solution of equation (13.100) is given by

$$P_2 = S_{22}^{-\frac{1}{2}} \left(S_{22}^{\frac{1}{2}} Q_{22} S_{22}^{\frac{1}{2}} \right)^{\frac{1}{2}} S_{22}^{-\frac{1}{2}} \qquad (13.101)$$

where

$$S_{22} = B_{22} R^{-1} B_{22}^T$$

It is left as an exercise to the reader to show that an $O\left(\epsilon\right)$ of (13.98) can be obtained by solving the following Riccati equation

$$\begin{aligned} P_1 \left(A_{11} - A_{12} Q_{22}^{-1} Q_{12}^T \right) + \left(A_{11} - A_{12} Q_{22}^{-1} Q_{12}^T \right)^T P_1 \\ - P_1 A_{12} Q_{22}^{-1} A_{12}^T P_1 + \left(Q_{11} - Q_{12} Q_{22}^{-1} Q_{12}^T \right) = 0 \end{aligned} \qquad (13.102)$$

Exercise 13.1: Show that an $O\left(\epsilon\right)$ approximation of the nonsymmetric Riccati equation (19.98) is given by (13.102). Hint: (Kokotovic et al., 1986 — section on the cheap control problem).

$\triangle$

The unique positive semidefinite stabilizing solution of the algebraic Riccati equation (13.102) exists under the following assumption (Kokotovic et al., 1986).

Assumption 13.5 The pair (A, B) is stabilizable and the pair (C, A_{11}) is detectable, where

$$C^T C = Q_{11} - Q_{12} Q_{22}^{-1} Q_{12}^T$$

$\triangle$

Using the facts that the unique equations (13.101) and (13.102) exist, and that these equations are obtained by perturbing the original equations (13.98)-(13.99) by an $O(\epsilon)$, the existence of the unique solutions of (13.98) and (13.99) is guaranteed by the following lemma.

Lemma 13.1 *Under Assumptions 13.1, 13.2, and 13.5* $\exists \epsilon_0 > 0$ *such that* $\forall \epsilon \leq \epsilon_0$ *unique solutions of (13.98) and (13.99) exist.*

$\diamond$

Proof: It follows by the direct application of the implicit function theorem (Ortega and Rheinboldt, 1970) and by the facts that the corresponding Jacobians of (13.98)-(13.99) are nonsingular at $\epsilon = 0$.

●

Solutions of equations (13.100) and (13.102) represent very good choices of the initial conditions for the Newton method to be used for solving the original equations (13.98) and (13.99).

It can be shown, like in Chapter 7, that the Newton algorithm in this case is given by

$$P_1^{(i+1)} \left(a_1 + a_2 P_2^{(i)} \right) - \left(a_4 - P_1^{(i)} a_2 \right) P_2^{(i+1)} = a_3 + P_1^{(i)} a_2 P_1^{(i)}$$

$$P_1^{(0)} = P_1, \quad i = 0, 1, 2, ...$$

(13.103)

$$P_2^{(i+1)} = S_{22}^{-\frac{1}{2}} \left(S_{22}^{\frac{1}{2}} M^{(i)} S_{22}^{\frac{1}{2}} \right)^{\frac{1}{2}} S_{22}^{-\frac{1}{2}}$$

(13.104)

$$P_2^{(0)} = P_2$$

where

$$M^{(i)} = P_2^{(i)} b_1 - b_4 P_2^{(i)} - b_3^{(i)} - \epsilon^2 P_2^{(i)} L_2 A_{21}^T P_2^{(i)}$$

Note that equation (13.94) is a nonsymmetric algebraic equation, so that we need to solve n^2 equations in (13.103) in order to get the solution P_1. Iterating (13.104) is producing the solution for P_2. Then, the global solution P is obtained from (13.91).

Using solutions of both Riccati equations (13.98) and (13.99), and formulas (13.87), (13.92), and (13.93), we can get completely decoupled slow and fast subsystems in the new coordinates as

$$\dot{\eta}_1 = (a_1 + a_2 P_1) \eta_1$$

(13.105)

$$\epsilon \dot{\eta}_2 = (b_1 + b_2 P_2)\, \eta_2 \qquad\qquad (13.106)$$

The interpretation of the result presented by (13.105) and (13.106) is that the optimal processing (control and/or filtering) might be completely performed at the subsystem levels. In addition, considerable reduction in computational requirements is achieved, since we only need to solve the reduced-order equations independently.

13.6 Numerical Example

In order to illustrate the proposed method, we consider a system in the form of (13.59)

$$\dot{x} = \begin{bmatrix} -2 & 0 & 0 & 1 \\ 1 & -2 & 1 & 0 \\ 0 & 0 & -1 & 2 \\ 1 & 0 & 0 & -3 \end{bmatrix} x + \begin{bmatrix} 0 & 0 \\ 0 & 0 \\ 2 & 1 \\ 1 & 2 \end{bmatrix} u$$

The weighting matrices are

$$Q = \begin{bmatrix} 1 & 0 & 0 & 0 \\ 0 & 1 & 0 & 0 \\ 0 & 0 & 6 & 5 \\ 0 & 0 & 5 & 6 \end{bmatrix}, \quad R = I_2$$

and $\epsilon = 0.1$.

The solution of (13.83) is obtained by using the MATLAB function lqr (Hill, 1988)

$$P_{exact} = \begin{bmatrix} 0.2798 & 0.0627 & -0.0092 & 0.0132 \\ 0.0627 & 0.2462 & 0.0088 & -0.0055 \\ -0.0092 & 0.0088 & 0.0941 & 0.0167 \\ 0.0132 & -0.0055 & 0.0167 & 0.0914 \end{bmatrix}$$

Solutions of lower order Riccati equations P_1 and P_2, and the solution for P obtained from equation (13.97) are given in Table 13.4. $P^{(7)}$ is identical to P_{exact}.

i	$P_1^{(i)}$	$P_2^{(i)}$	$P^{(i)}$
0	-0.0899 0.1041 0.1208 -0.0760	1.0528 0.0528 0.0528 1.0528	-0.0881 0.1028 0.0088 -0.0076 0.1183 -0.0741 -0.0114 0.0117 0.0098 -0.0105 0.1041 0.0065 -0.0084 0.0107 0.0065 0.1042
1	0.2938 0.0518 0.0488 0.2630	0.8759 0.2381 0.2375 0.8372	-0.0860 0.1015 0.0168 -0.0155 0.1162 -0.0728 -0.0195 0.0194 0.0175 -0.0169 0.1027 0.0078 -0.0160 0.0167 0.0078 0.1028
4	0.2780 0.0646 0.0638 0.2449	0.9337 0.1742 0.1740 0.9076	0.2798 0.0627 -0.0090 0.0131 0.0627 0.2462 0.0086 -0.0053 -0.0090 0.0086 0.0932 0.0178 0.0131 -0.0053 0.0178 0.0903
7	0.2780 0.0646 0.0638 0.2449	0.9298 0.1786 0.1783 0.9028	0.2798 0.0627 -0.0092 0.0132 0.0627 0.2462 0.0088 -0.0055 -0.0092 0.0088 0.0941 0.0167 0.0132 -0.0055 0.0167 0.0914
	$P_1^{opt} = P_1^{(7)}$	$P_2^{opt} = P_2^{(7)}$	$P^{opt} = P^{(7)}$

Table 13.4: Decomposition of the cheap control Riccati equation

Appendix 13.1

In (Moerder and Calise, 1985b) the control problem of damping the vibratory modes of a large space platform is addressed. Data used are taken from (Sesak et al., 1979), where the various model characteristics are defined.

The platform in question is controlled by a pair of actuator-sensors which control mechanical pitch and roll. The structure is modeled by its normal (four-mode) modal coordinates, so that the system matrix is block diagonal, with diagonal blocks given by

$$A_{jj} = \begin{bmatrix} 0 & 1 \\ -\omega_j^2 & 0 \end{bmatrix}$$

where ω_j represents the j-th modal frequency in rad/sec. With data for various modal frequences given in (Sesak et al., 1979), the system matrix is given by

$$A_1 = \begin{bmatrix} 0 & 1 & 0 & 0 \\ -(0.42)^2 & 0 & 0 & 0 \\ 0 & 0 & 0 & 1 \\ 0 & 0 & -(0.42)^2 & 0 \end{bmatrix}, \quad A_2 = 0$$

$$A_4 = \begin{bmatrix} 0 & 1 & 0 & 0 \\ -(2.1)^2 & 0 & 0 & 0 \\ 0 & 0 & 0 & 1 \\ 0 & 0 & -(2.2)^2 & 0 \end{bmatrix}, \quad A_3 = 0$$

The input and output matrices are given by

$$B_{km} = \frac{\sigma_{jm}}{\mu_j}, \quad C_{mk} = \sigma_{jm}$$

where μ and σ represent the modal mass and slope, respectively. The input and output matrices are given by

$$B_1 = \begin{bmatrix} 0 & 0 & 0 & 0 \\ -0.92 & -1.4 & 0.92 & -1.4 \\ 0 & 0 & 0 & 0 \\ 0.65 & 1.6 & 0.65 & -1.6 \end{bmatrix}$$

$$B_2 = \begin{bmatrix} 0 & 0 & 0 & 0 \\ 1.4 & -1 & 1.4 & 1 \\ 0 & 0 & 0 & 0 \\ 2.05 & -0.80 & -2 & -0.8 \end{bmatrix}$$

and

$$C_1 = \begin{bmatrix} 0 & -1.8 & 0 & 1.3 \\ 0 & -2.7 & 0 & 3.2 \\ 0 & 1.8 & 0 & 1.3 \\ 0 & -2.7 & 0 & -3.2 \end{bmatrix}$$

$$C_2 = \begin{bmatrix} 0 & 2.9 & 0 & 4.1 \\ 0 & -2.1 & 0 & -1.6 \\ 0 & 2.9 & 0 & -4.1 \\ 0 & 2.1 & 0 & -1.6 \end{bmatrix}$$

The permutation matrix used to transform the system in the desired form is

$$P = \begin{bmatrix} 1 & 0 & 0 & 0 & 0 & 0 & 0 & 0 \\ 0 & 0 & 1 & 0 & 0 & 0 & 0 & 0 \\ 0 & 0 & 0 & 0 & 1 & 0 & 0 & 0 \\ 0 & 0 & 0 & 0 & 0 & 0 & 1 & 0 \\ 0 & 1 & 0 & 0 & 0 & 0 & 0 & 0 \\ 0 & 0 & 0 & 1 & 0 & 0 & 0 & 0 \\ 0 & 0 & 0 & 0 & 0 & 1 & 0 & 0 \\ 0 & 0 & 0 & 0 & 0 & 0 & 0 & 1 \end{bmatrix}$$

The matrices used in Section 13.2 are then derived according to

$$A = PAP^{-1}, \quad B = PB$$

Chapter 14

Optimal Control of Singularly Perturbed and Weakly Coupled Bilinear Systems

14.1 Introduction

The theory of singular perturbations has been a highly recognized and rapidly developing area of control systems research in the last twenty five years (Kokotovic et al., 1986; Kokotovic and Khalil, 1986; Gajic et al., 1990). Almost all important control aspects for linear systems have been studied so far and valuable and practically implementable results have been obtained. The extension of these results to the nonlinear systems happened to be a very difficult task. Only under very restrictive conditions and for very limited classes of nonlinear systems some results were obtained (Saberi and Khalil, 1984, 1985; O'Malley, 1974a, 1974b; Chow and Kokotovic, 1978a, 1978b, 1981; Suzuki, 1981).

In between of linear and nonlinear systems lies a very large class of so-called bilinear systems (Mohler, 1991). The importance of bilinear systems has been recognized at least since the work of Wiener (Wiener, 1948), who believes that they are in the essence of understanding the behavior of neural and biological computing networks. They represent an enormous number of the real world phenomena (Mohler, 1970, 1973, 1991; Mohler and Kolodziej, 1980). This class of "nearly linear" systems has not been studied so far in the context of singular perturbations, except for a few minor attempts (Guillen and Armada, 1980; Tzafestas and Anagnostou, 1984a; Asamoah and Jamshidi, 1987).

Many real physical systems possess the structure of the singularly perturbed bilinear control systems such as: neutron level control problem in a fission reactor (Mohler, 1973), dc-motor (Bruni et al., 1974), induction motor drives (Figalli et al., 1984), regulation of carbon-dioxide in the respiratory system (Mohler, 1970), mechanical brake system (Mohler, 1970), and distillation columns (Espana and Landau, 1978).

The purpose of this chapter is to study the optimal control problem of singularly perturbed and weakly coupled bilinear systems with a quadratic performance criterion. We study both the open-loop and closed-loop optimal control problems.

In the first part of this chapter, we study the optimal open-loop control problem of singularly perturbed bilinear systems with a quadratic performance criterion. The obtained results utilize the recursive scheme for the optimal control of a general bilinear system with a quadratic performance criterion (Hofer and Tibken, 1988) and the time varying version of the reduced-order method with an arbitrary degree of accuracy for solving the linear-quadratic optimal open-loop singularly perturbed control problem (Su et al., 1992a). This problem is solved as a sequence of linear two-point boundary value singularly perturbed problems. At each iteration step the ill-conditioned linear time varying two-point boundary value problem is transformed in the pure-slow and pure-fast completely decoupled initial value problems. By doing this, the stiffness of the singularly perturbed two-point boundary value problem is converted in the problem of an ill-defined linear system of algebraic equations. However, the latter problem is much easier to handle. The size of required computations is reduced since the introduced transformation allows parallel processing of information.

In Section 14.3, we utilize the idea of the composite control law for singularly perturbed systems (Saberi and Khalil, 1985; Suzuki, 1981; Chow and Kokotovic, 1976), and the recursive scheme for the optimal control of a general bilinear system with a quadratic performance criterion (Cebuhar and Constanza, 1984). The obtained composite control law for singularly perturbed bilinear systems is represented by a linear combination of the slow and fast variables. The matrix coefficients for this linear combination are obtained from the recursive scheme applied to the two reduced-order independent time varying linear-quadratic control problems. The composite control law is $O(\epsilon)$ close to the optimal one, which implies the $O(\epsilon)$ closeness of the near-optimal trajectories to the optimal ones, and the $O(\epsilon)$ approximation for the performance criterion. A real world numerical example, an induction motor drives, is used to demonstrate the efficiency of the obtained composite control. In addition, an algorithm for achieving higher order approximations is proposed in the spirit of the recursive methods for singularly perturbed control systems presented in (Gajic et al., 1990).

In the last part of this chapter, the weakly coupled bilinear control systems are considered. This class of systems has been studied so far only in the paper (Tzafestas and Anagnostou, 1984b), where the stabilization problem has been considered. Corresponding results given in terms of the reduced-order problems (similarly to the singularly perturbed bilinear systems) are obtained for both the optimal open-loop and closed-loop control of weakly coupled bilinear systems.

14.2 Open-Loop Optimal Control of Singularly Perturbed Bilinear Systems

Consider the optimal control problem of a bilinear system represented by

$$\dot{x} = Ax + Bu + \{xN\}u, \qquad x(0) = x_0 \qquad (14.1)$$

with a performance criterion

$$J = \frac{1}{2} \int_{t_0}^{t_f} \left(x^T Q x + u^T R u \right) dt \qquad (14.2)$$

where $x \in \Re^n$ are state variables, $u \in \Re^m$ is a control input, $A, B, N, Q,$ and R are constant matrices of appropriate dimensions, with $R = R^T > 0$, and $Q = Q^T \geq 0$. The notation used for the bilinear term in (14.1) means

$$\{xN\} = \sum_{j=1}^{n} x_j N_j, \qquad N_j \in \Re^{n \times m} \qquad (14.3)$$

From the Hamiltonian of (14.1)-(14.2) given by

$$H(x, u, p) = \frac{1}{2} \left(x^T Q x + u^T R u \right) + p^T \left(Ax + Bu + \{xN\}u \right) \quad (14.4)$$

we get the expression for the open-loop optimal control as

$$u^* = -R^{-1} (B + \{xN\})^T p(t) \qquad (14.5)$$

where $p(t) \in \Re^n$ stands for the costate variables. The costate variable can be obtained from the following system of equations (Hofer and

Tibken, 1988)

$$\dot{x}_i = [Ax]_i - \left[(B + \{xN\})\, R^{-1}\, (B + \{xN\})^T \rho\right]_i, \qquad x_i\,(t_0) = x_i^0,$$

$$\dot{\rho}_i = -[Qx]_i - [A^T\rho]_i$$

$$+ \frac{1}{2}\rho^T \left[N_i R^{-1} (B + \{xN\})^T + (B + \{xN\})\, R^{-1} N_i^T \right] \rho,$$

$$\rho_i\,(t_f) = [Fx\,(t_f)]_i$$

$$\tag{14.6}$$

where $[...]_i$, $i = 1, \ldots , n$, is the i-th component of the corresponding vector. This two-point boundary value problem of the coupled nonlinear differential equations is not easy solvable. It is shown in (Hofer and Tibken, 1988) that the system (14.6) can be rewritten in the compact form of the state-costate equations resembling to those of a linear-quadratic optimal control problem

$$\dot{x} = \widetilde{A}x - \widetilde{B}R^{-1}\widetilde{B}^T\rho, \qquad x\,(t_0) = x^0,$$

$$\dot{\rho} = -\widetilde{Q}x - \widetilde{A}^T\rho, \qquad\qquad \rho(t_f) = Fx\,(t_f) \tag{14.7}$$

where $\widetilde{A}$, $\widetilde{Q}$, $\widetilde{B}R^{-1}\widetilde{B}^T$ are time varying matrices. Note that these matrices are functions of $x\,(t)$ and $\rho\,(t)$ so that the right-hand side of (14.7) is nonlinear. The following linear two-point boundary value scheme has been proposed for solving (14.7), (Hofer and Tibken, 1988)

$$\dot{x}^{(k+1)} = \widetilde{A}^{(k)}x^{(k+1)} - \widetilde{B}^{(k)}R^{-1}\widetilde{B}^{(k)^T}\rho^{(k+1)}$$

$$\dot{\rho}^{(k+1)} = -\widetilde{Q}^{(k)}x^{(k+1)} - \widetilde{A}^{(k)^T}\rho^{(k+1)} \tag{14.8}$$

with boundary conditions expressed in the standard form as

$$U \begin{bmatrix} x^{(k+1)}\,(t_0) \\ \rho^{(k+1)}\,(t_0) \end{bmatrix} + V \begin{bmatrix} x^{(k+1)}\,(t_f) \\ \rho^{(k+1)}\,(t_f) \end{bmatrix} = c \tag{14.9}$$

where

$$U = \begin{bmatrix} I & 0 \\ 0 & 0 \end{bmatrix}, \qquad V = \begin{bmatrix} 0 & 0 \\ -F & I \end{bmatrix}, \qquad c = \begin{bmatrix} x^0 \\ 0 \end{bmatrix} \tag{14.10}$$

The time varying matrices are given by

$$\widetilde{A}_{ij}^{(k)} = A_{ij} - \frac{1}{2}\left[(N_j R^{-1} B^T + B R^{-1} N_j^T)\, \rho^{(k)}\,(t) \right]_i, \qquad i,j = 1, ..., n$$

$$\tilde{Q}_{ij}^{(k)} = Q_{ij} - \frac{1}{2}\rho^{(k)}(t)^T \left(N_i R^{-1} N_j^T + N_j R^{-1} N_i^T\right) \rho^{(k)}(t); \quad (14.11)$$

$$i, j = 1, ..., n$$

$$\tilde{B}^{(k)} R^{-1} \tilde{B}^{(k)^T} = \left(B + \left\{x^{(k)} N\right\}\right) R^{-1} \left(B + \left\{x^{(k)} N\right\}\right)^T$$

$$-\frac{1}{2}\left(\left\{x^{(k)} N\right\} R^{-1} B^T + B R^{-1} \left\{x^{(k)} N\right\}^T\right)$$

The convergence of the above algorithm to the solution of (14.7) was proved in (Hofer and Tibken, 1988).

In this section, we exploit the iterative scheme (14.8), comprising a sequence of linear two-point boundary value problems, in order to derive the solution for the optimal open-loop control of singularly perturbed bilinear systems. The solution is obtained in the spirit of the general theory of singular perturbations, namely the problem is decomposed and studied in slow and fast time scales. The open-loop optimal control of singularly perturbed linear systems was studied in (Su et al., 1992a) and (Wilde and Kokotovic, 1973). The approach taken in (Wilde and Kokotovic, 1973) is efficient for an $O(\epsilon)$ of accuracy. In (Su et al., 1992a) a recursive approach is obtained such that an arbitrary order of accuracy, $O(\epsilon^k)$, $k = 1, 2, 3, ...$, can be obtained. The importance of the results reported in (Su et al., 1992a) is in the fact that the stiffness of the singularly perturbed two-point boundary value problem is converted into the problem of an ill-defined system of linear algebraic equations. The latter problem is much easier to handle. The study in (Su et al., 1992a) was limited to the time invariant systems. In this section, we show that following the ideas of (Su et al., 1992a) we will be able to handle in the same manner the time varying singularly perturbed two-point boundary value problem, such that an arbitrary order of accuracy can be obtained, and that the stiffness of the original problem is replaced by an ill-defined system of linear algebraic equations of order $2n$.

The singularly perturbed bilinear control system under consideration is represented by

$$\begin{bmatrix} \dot{y} \\ \epsilon\dot{z} \end{bmatrix} = \begin{bmatrix} A_1 & A_2 \\ A_3 & A_4 \end{bmatrix} \begin{bmatrix} y \\ z \end{bmatrix} + \begin{bmatrix} B_1 \\ B_2 \end{bmatrix} u + \left\{ \begin{bmatrix} y \\ z \end{bmatrix} \begin{bmatrix} N_s \\ N_f \end{bmatrix} \right\} u \quad (14.12)$$

with initial conditions

$$\begin{bmatrix} y(t_0) \\ z(t_0) \end{bmatrix} = \begin{bmatrix} y^0 \\ z^0 \end{bmatrix}$$

where $y \in \Re^{n_1}$, $z \in \Re^{n_2}$ are, respectively, slow and fast state variables, ϵ is a small positive parameter, and

$$\left\{ \begin{bmatrix} y \\ z \end{bmatrix} \begin{bmatrix} N_s \\ N_f \end{bmatrix} \right\} = \sum_{j=1}^{n_1} y_j \begin{bmatrix} N_{sj} \\ N_{fj} \end{bmatrix} + \sum_{j=n_1+1}^{n_1+n_2} z_j \begin{bmatrix} N_{sj} \\ N_{fj} \end{bmatrix} \qquad (14.13)$$

A quadratic cost functional associated with (14.10) has the form

$$J = \frac{1}{2} \int_{t_0}^{t_f} \left(\begin{bmatrix} x \\ z \end{bmatrix}^T Q \begin{bmatrix} x \\ z \end{bmatrix} + u^T R u \right) dt \qquad (14.14)$$

The following notation is used in order to relate (14.1)-(14.3) and (14.10)-(14.12)

$$A = \begin{bmatrix} A_1 & A_2 \\ \frac{A_3}{\epsilon} & \frac{A_4}{\epsilon} \end{bmatrix}, \qquad Q = \begin{bmatrix} Q_1 & Q_2 \\ Q_2^T & Q_3 \end{bmatrix}$$

$$B = \begin{bmatrix} B_1 \\ \frac{B_2}{\epsilon} \end{bmatrix}, \quad N = \begin{bmatrix} N_s \\ \frac{N_f}{\epsilon} \end{bmatrix}, \quad x(t) = \begin{bmatrix} y(t) \\ z(t) \end{bmatrix} \qquad (14.15)$$

In the following, we will utilize the recursive scheme (14.8)-(14.11) in order to find the optimal open-loop control law for the singularly perturbed bilinear-quadratic optimal control problem represented by (14.12)-(14.15) in terms of the reduced-order slow and fast subsystems (Kokotovic et al., 1986).

It can be shown that the system of equations (14.8) preserves the singularly perturbed structure. Namely, the use of (14.12)-(14.15) in (14.4)-(14.5) and (14.8)-(14.11) produces

$$\begin{bmatrix} \dot{y} \\ \epsilon \dot{z} \end{bmatrix}^{(k+1)} = \tilde{A}^{(k)} \begin{bmatrix} y \\ z \end{bmatrix}^{(k+1)} - \tilde{B}^{(k)} R^{-1} \tilde{B}^{(k)T} \begin{bmatrix} p_s \\ p_f \end{bmatrix}^{(k+1)}$$

$$\begin{bmatrix} \dot{p}_s \\ \dot{p}_f \end{bmatrix}^{(k+1)} = -\tilde{Q}^{(k)} \begin{bmatrix} y \\ z \end{bmatrix}^{(k+1)} - \tilde{A}^{(k)T} \begin{bmatrix} p_s \\ p_f \end{bmatrix}^{(k+1)} \qquad (14.16)$$

where $p_s \in \Re^{n_1}$ and $p_f \in \Re^{n_2}$ are costate vectors corresponding, respectively, to the slow variables $y^{(k+1)}$ and the fast variables $z^{(k+1)}$. The time varying matrices in (14.16) are given by

$$\widetilde{A}^{(k)} = \begin{bmatrix} \widetilde{A}_1^{(k)} & \widetilde{A}_2^{(k)} \\ \dfrac{\widetilde{A}_3^{(k)}}{\epsilon} & \dfrac{\widetilde{A}_4^{(k)}}{\epsilon} \end{bmatrix}, \qquad \widetilde{Q}^{(k)} = \begin{bmatrix} \widetilde{Q}_1^{(k)} & \widetilde{Q}_2^{(k)} \\ \widetilde{Q}_2^{(k)T} & \widetilde{Q}_3^{(k)} \end{bmatrix}$$

$$\widetilde{S}^{(k)} = \widetilde{B}^{(k)} R^{-1} \widetilde{B}^{(k)T} = \begin{bmatrix} \widetilde{S}_1^{(k)} & \dfrac{\widetilde{S}_2^{(k)}}{\epsilon} \\ \dfrac{\widetilde{S}_2^{(k)T}}{\epsilon} & \dfrac{\widetilde{S}_3^{(k)}}{\epsilon^2} \end{bmatrix} \tag{14.17}$$

After some algebra the state-costate equation (14.16) can be written in the form:

$$\begin{bmatrix} \dot{w}^{(k+1)} \\ \dot{\lambda}^{(k+1)} \end{bmatrix} = \begin{bmatrix} \widetilde{T}_1^{(k)} & \widetilde{T}_2^{(k)} \\ \dfrac{\widetilde{T}_3^{(k)}}{\epsilon} & \dfrac{\widetilde{T}_4^{(k)}}{\epsilon} \end{bmatrix} \begin{bmatrix} w^{(k+1)} \\ \lambda^{(k+1)} \end{bmatrix} \tag{14.18}$$

where the new notation is

$$w^{(k+1)} = \begin{bmatrix} y^{(k+1)} \\ \rho_s^{(k+1)} \end{bmatrix}, \quad \lambda^{(k+1)} = \begin{bmatrix} z^{(k+1)} \\ \rho_f^{(k+1)} \end{bmatrix}, \quad \rho^{(k+1)} = \begin{bmatrix} \rho_s^{(k+1)} \\ \rho_f^{(k+1)} \end{bmatrix} \tag{14.19}$$

The time varying matrices $\widetilde{T}_i^{(k)}$ introduced in (14.18) are given by

$$\widetilde{T}_1^{(k)} = \begin{bmatrix} \widetilde{A}_1^{(k)} & -\widetilde{S}_1^{(k)} \\ -\widetilde{Q}_1^{(k)} & -\widetilde{A}_1^{(k)T} \end{bmatrix}, \quad \widetilde{T}_2^{(k)} = \begin{bmatrix} \widetilde{A}_2^{(k)} & -\widetilde{S}_2^{(k)} \\ -\widetilde{Q}_2^{(k)} & -\widetilde{A}_3^{(k)T} \end{bmatrix}$$

$$\widetilde{T}_3^{(k)} = \begin{bmatrix} \widetilde{A}_3^{(k)} & -\widetilde{S}_2^{(k)T} \\ -\widetilde{Q}_2^{(k)T} & -\widetilde{A}_2^{(k)T} \end{bmatrix}, \quad \widetilde{T}_4^{(k)} = \begin{bmatrix} \widetilde{A}_4^{(k)} & -\widetilde{S}_2^{(k)} \\ -\widetilde{Q}_3^{(k)} & -\widetilde{A}_4^{(k)T} \end{bmatrix} \tag{14.20}$$

The expression for the boundary conditions is changed due to an interchange of rows corresponding to $\rho_s^{(k+1)}$ and $z^{(k+1)}$, which modifies matrices defined in (14.9) as follows

$$U_1 \begin{bmatrix} w^{(k+1)}(t_0) \\ \lambda^{(k+1)}(t_0) \end{bmatrix} + V_1 \begin{bmatrix} w^{(k+1)}(t_f) \\ \lambda^{(k+1)}(t_f) \end{bmatrix} = c_1 \tag{14.21}$$

where

$$U_1 = \begin{bmatrix} I_{n_1} & 0 & 0 & 0 \\ 0 & 0 & 0 & 0 \\ 0 & 0 & I_{n_2} & 0 \\ 0 & 0 & 0 & 0 \end{bmatrix}, \quad c_1 = \begin{bmatrix} y^0 \\ 0 \\ z^0 \\ 0 \end{bmatrix}$$

$$V_1 = \begin{bmatrix} 0 & 0 & 0 & 0 \\ -F_1 & I_{n_1} & -\epsilon F_2 & 0 \\ 0 & 0 & 0 & 0 \\ -F_2^T & 0 & -F_3 & I_{n_2} \end{bmatrix} \tag{14.22}$$

In order to obtain the slow and fast decoupled subsystems from (14.18), we apply the Chang transformation (Chang, 1972). In this section, we use a new version of the Chang transformation given by (Qureshi and Gajic, 1992)

$$\mathbf{T}_3^{(k)}(t, \epsilon) = \begin{bmatrix} I_1 & -\epsilon P^{(k)} \\ -M^{(k)} & I_2 \end{bmatrix}$$

$$\mathbf{T}_3^{(k)^{-1}}(t, \epsilon) = \begin{bmatrix} I_1 - \epsilon P^{(k)} W^{(k)} M^{(k)} & \epsilon P^{(k)} W^{(k)} \\ W^{(k)} M^{(k)} & W^{(k)} \end{bmatrix} \tag{14.23}$$

with

$$W^{(k)} = \left(I_2 - \epsilon Q^{(k)} P^{(k)} \right)^{-1} \tag{14.24}$$

where I_1 and I_2 are identity matrices of order $2n_1$ and $2n_2$, respectively. The matrices $P^{(k)}$ and $M^{(k)}$ are the solutions of the following completely decoupled stiff matrix differential equations

$$\begin{aligned} \epsilon \dot{P}^{(k)} &= -P^{(k)} \tilde{T}_4^{(k)} + \tilde{T}_2^{(k)} + \epsilon \left(\tilde{T}_1^{(k)} P^{(k)} - P^{(k)} \tilde{T}_3^{(k)} P^{(k)} \right) \\ \epsilon \dot{M}^{(k)} &= \tilde{T}_4^{(k)} M^{(k)} + \tilde{T}_3^{(k)} + \epsilon \left(M^{(k)} \tilde{T}_1^{(k)} + M^{(k)} \tilde{T}_2^{(k)} M^{(k)} \right) \end{aligned} \tag{14.25}$$

The initial conditions for differential equations (14.25) are arbitrary (Chang, 1972). The existence of the solutions of (14.25) for sufficiently small values of ϵ is established in (Chang 1972; Qureshi and Gajic, 1992).

The transformation (14.23) applied to the system (14.18) produces two completely decoupled subsystems

$$\dot{\eta}^{(k)} = \left(\tilde{T}_1^{(k)} - P^{(k)} \tilde{T}_3^{(k)} \right) \eta^{(k)} \tag{14.26}$$

$$\dot{\xi}^{(k)} = \frac{1}{\epsilon} \left(\tilde{T}_4^{(k)} - \epsilon M^{(k)} \tilde{T}_2^{(k)} \right) \xi^{(k)} \tag{14.27}$$

with

$$\begin{bmatrix} \eta^{(k)} \\ \xi^{(k)} \end{bmatrix} = \mathbf{T}_3^{(k)}(t, \epsilon) \begin{bmatrix} w^{(k)} \\ \lambda^{(k)} \end{bmatrix} \tag{14.28}$$

Consequently, the change of variables transforms the boundary conditions

$$U_2 \begin{bmatrix} \eta^{(k)}(t_0) \\ \xi^{(k)}(t_0) \end{bmatrix} + V_2 \begin{bmatrix} \eta^{(k)}(t_f) \\ \xi^{(k)}(t_f) \end{bmatrix} = c_1 \qquad (14.29)$$

where

$$U_2 = U_1 T_3^{(k)^{-1}}(t_0, \epsilon), \qquad V_2 = V_1 T_3^{(k)^{-1}}(t_f, \epsilon) \qquad (14.30)$$

The solutions of equations (14.26) and (14.27) are

$$\eta^{(k)}(t) = \Phi^{(k)}(t, t_0, \epsilon) \eta^{(k)}(t_0)$$
$$\xi^{(k)}(t) = \Psi^{(k)}(t, t_0, \epsilon) \xi^{(k)}(t_0) \qquad (14.31)$$

where $\Phi(t, t_0, \epsilon)$ and $\Psi(t, t_0, \epsilon)$ are the transition matrices of (14.26) and (14.27), respectively.

The initial conditions $\eta^{(k)}(t_0)$ and $\xi^{(k)}(t_0)$ have to be determined. Substitution of (14.31) into (14.29) yields

$$\Delta(\epsilon) \begin{bmatrix} \eta^{(k)}(t_0) \\ \xi^{(k)}(t_0) \end{bmatrix} = c_1 \qquad (14.32)$$

where

$$\Delta(\epsilon) = U_2(\epsilon) + V_2(\epsilon) \begin{bmatrix} \Phi(t_f, t_0, \epsilon) & 0 \\ 0 & \Psi(t_f, t_0, \epsilon) \end{bmatrix} \qquad (14.33)$$

If $\Delta^{-1}(\epsilon)$ exists then the solution of (14.32) will be

$$\begin{bmatrix} \eta^{(k)}(t_0) \\ \xi^{(k)}(t_0) \end{bmatrix} = \Delta^{-1}(\epsilon) c_1 \qquad (14.34)$$

Note that as $\epsilon \to 0$

$$\left\{ T_3^{(k)}(t, 0) \right\}^{-1} = \begin{bmatrix} I_1 & 0 \\ M^{(k)}(t_0) & I_2 \end{bmatrix} \qquad (14.35)$$

and therefore

$$U_2 = U_1 \begin{bmatrix} I_1 & 0 \\ M^{(k)}(t_0) & I_2 \end{bmatrix}, \qquad V_2 = V_1 \begin{bmatrix} I_1 & 0 \\ M^{(k)}(t_f) & I_2 \end{bmatrix} \qquad (14.36)$$

After partitioning the transition matrices $\Phi(t, t_0, 0)$ and $\Psi(t, t_0, 0)$ as

$$\Phi(t, t_0, 0) = \begin{bmatrix} \Phi_{11}(t, t_0, 0) & \Phi_{12}(t, t_0, 0) \\ \Phi_{21}(t, t_0, 0) & \Phi_{22}(t, t_0, 0) \end{bmatrix}$$

$$\Psi(t, t_0, 0) = \begin{bmatrix} \Psi_{11}(t, t_0, 0) & \Psi_{12}(t, t_0, 0) \\ \Psi_{21}(t, t_0, 0) & \Psi_{22}(t, t_0, 0) \end{bmatrix}$$

and after some algebra the matrix $\Delta(\epsilon)$ is obtained in the form

$$\Delta(\epsilon) = \begin{bmatrix} I_{n_1} & 0 & 0 & 0 \\ * & \Delta_{22} & 0 & 0 \\ * & * & I_{n_2} & 0 \\ * & * & * & \Delta_{44} \end{bmatrix} + O(\epsilon) \qquad (14.37)$$

where

$$\Delta_{22} = \Phi_{22}(t_f, t_0, 0) - F_1 \Phi_{12}(t_f, t_0, 0)$$

$$\Delta_{44} = \Psi_{22}(t_f, t_0, 0) - F_3 \Psi_{12}(t_f, t_0, 0)$$

The asterisks denote the terms which are not important for the nonsingularity of $\Delta(\epsilon)$.

Since the matrices Δ_{22} and Δ_{44} are nonsingular (Kirk, 1970, page 211), so does $\Delta(\epsilon)$ for sufficiently small values of ϵ, with $0 < \epsilon \le \epsilon_1$ and ϵ sufficiently small.

Note that due to presense of the $\frac{1}{\epsilon}$ term in $\Psi(t_f, t_0, 0)$, the system of linear algebraic equations (14.32) is ill-conditioned. However, this problem is much easier then the original two-point stiff boundary value problem. In summary, we have established the following theorem.

Theorem 14.1 *Let the problem matrices be continuous functions of t on the time interval $t_0 \le t \le t_f$, then for all sufficiently small ϵ the boundary value problem (14.26)–(14.27) and (14.29) has the solution given by*

$$\begin{bmatrix} \eta(t, \epsilon) \\ \xi(t, \epsilon) \end{bmatrix} = \begin{bmatrix} \Phi(t, t_0, \epsilon) & 0 \\ 0 & \Psi(t, t_0, \epsilon) \end{bmatrix} \Delta^{-1}(\epsilon) c_1$$

◇

Consequently, the solution of the boundary problem (14.18)-(14.21) is

$$\begin{bmatrix} w^{(k+1)}(t, \epsilon) \\ \lambda^{(k+1)}(t, \epsilon) \end{bmatrix} = \left\{ T_3^{(k)}(t, \epsilon) \right\}^{-1} \begin{bmatrix} \eta^{(k+1)}(t, \epsilon) \\ \xi^{(k+1)}(t, \epsilon) \end{bmatrix} \qquad (14.38)$$

so that the required variables $y^{(k+1)}$ and $z^{(k+1)}$ are obtained by partitioning the vectors $w^{(k+1)}$ and $\lambda^{(k+1)}$ according to (14.19).

The main problem that we are faced with in the presented method is the problem of finding the transition matrices $\Phi(t, t_0, \epsilon)$ and $\Psi(t, t_0, \epsilon)$ of the corresponding time varying systems. One way to overcome this problem is to study the optimal open-loop control of singularly perturbed system in the discrete-time domain. This indicates an important research problem, which can be formulated as follows.

Research Problem 13.1: Study the open-loop optimal control problem of bilinear systems in the discrete-time domain by using the discrete-time version of the results presented in Section 14.2.

$$\triangle$$

14.3 Closed-Loop Optimal Control of Singularly Perturbed Bilinear Systems

Consider the optimal control problem of a bilinear system represented by (14.1)-(14.3). The closed-loop solution of the optimization problem (14.1)-(14.2) at steady state ($t_f = \infty$) yields to the optimal control in the form

$$u^* = -R^{-1}(B + \{xN\})^T P(x) x \qquad (14.39)$$

where $P(x)$ is the solution of the following equation (Cebuhar and Constanza, 1984)

$$
\begin{aligned}
& Q + P(x)A + A^T P(x) \\
& -P(x)(B + \{xN\})R^{-1}(B + \{xN\})^T P(x) = 0
\end{aligned}
\qquad (14.40)
$$

This nonlinear system of algebraic matrix equations is very hard to solve. However, it has been shown in (Cebuhar and Constanza, 1984) that the sequence of linear systems

$$
\begin{aligned}
\dot{x}_0 &= Ax_0 + Bu, \qquad x_0(t_0) = x^0 \\
\dot{x}_i &= Ax_i + B_i(t)u, \qquad B_i(t) \triangleq B + \{x_{i-1}(t)N\}, \qquad x_i(t_0) = x^0
\end{aligned}
\qquad (14.41)
$$

and the sequence of the time varying algebraic Riccati equations

$$Q + P_i(t)A + A^T P_i(t) - P_i(t)B_i(t)R^{-1}B_i^T(t)P_i(t) = 0 \qquad (14.42)$$

produce the sequence of the feedback controls

$$u_i^* = -R^{-1}B_i^T(t)P_i(t)x_i \qquad (14.43)$$

such that

$$u_i^*(t) \to u^*(t), \qquad x_i^*(t) \to x^*(t) \qquad (14.44)$$

The convergence stated in (14.44) is uniform in t, and is guaranteed under the following assumption.

Assumption 14.1 The pair (A, B) is controllable and x stays in the controllability domain $X_c = \{x \in R^n | (A, B + \{xN\}) \; controllable\}$.

$$\triangle$$

It is important to point out that (14.42) and (14.43) establish in some sense the optimal linear feedback law. Namely, using the feedback coefficient from (14.42) in the linear feedback loop around the bilinear system (14.1) produces the approximate linear feedback law. This is a very strong result since it is known that is almost impossible to get, in general, the optimal feedback control of nonlinear (and thus bilinear) systems due to fact that the partial differential Hamilton-Bellman-Jacobi equation has no analytical solutions.

In this chapter, we will relax the controllability assumption into the stabilizability assumption (Kucera, 1972). Also, since the matrix Q in (14.42) does not change per iteration it is convenient to assume that the pair $(A, \sqrt{Q})$ is detectable. This will lead the the existence of the unique stabilizing solution $P_i(t)$, in order words, the matrix $A - B_i(t) R^{-1} B_i(t) P_i(t)$ will be stable for every frozen $t \in [0, \infty)$. Due to stability of the closed loop system matrix, at steady state we have $0 = \left(A - B_i(t) R^{-1} B_i(t) P_i(t)\right) x_e(t)$, that is, the unique equilibrium point of the bilinear system is the origin, so that $B_i(t) \rightarrow B = const$, and the equation (14.42) tends to the time-invariant algebraic Riccati equation. The required optimal feedback control (14.43) in that case tends to zero, so there is no need to solve the equation (14.42) over an infinite period of time. Thus, we will use the following assumption.

Assumption 14.2 The pair (A, B) is stabilizable, x stays in the stabilizability domain $X_s = \{x \in R^n | (A, B + \{xN\}) \; stabilizable\}$, and the pair $(A, \sqrt{Q})$ is detectable.

$$\triangle$$

The main goal of this section is to exploit the iterative procedure (14.41)-(14.43) for the singularly perturbed bilinear structure (14.12)-(14.15) in order to get an expression for the near-optimal control in terms of the reduced-order slow and fast subsystems (Kokotovic et al., 1986). There are two important reasons for this study: 1) to avoid an ill-defined numerical problem associated with the equation (14.42) subject to (14.15); and, 2) to reduce the size of required computations and generate the near-optimal solution in parallel — in slow and fast time scales, and speed up the optimization process.

14.3.1 Composite Control of Bilinear Singularly Perturbed Systems

Following the result of (Chow and Kokotovic, 1976), the composite control of the sequence of the linear-quadratic optimal control problems (14.41)-(14.44), subject to the singularly perturbed structure (14.12)-(14.15), can be obtained from the slow and fast time scales linear-quadratic optimal control problems. Note that on the contrary to (Chow and Kokotovic, 1976), we are faced with the time varying problem. The slow time scale problem of order n_1 for the block diagonal structure of the penalty matrix Q (it has been assumed without loss of generality that $Q_2 = 0$), is given by

$$\dot{y}_s = A_0 y_s + B_s(t) u_s, \quad y_s(t_0) = y^0$$
$$z_s(t) = -A_4^{-1}(A_3 y_s + B_{2i}(t) u_s)$$
$$J_s = \frac{1}{2} \int_{t_0}^{\infty} (y_s^T Q_0 y_s + 2u_s^T D_s y_s + u_s^T R_s u_s) \, dt \qquad (14.45)$$

where

$$B_i(t) = B + \{x_{i-1}(t) N\} = \begin{bmatrix} B_{1i}(t) \\ \frac{B_{2i}(t)}{\epsilon} \end{bmatrix}$$

$$A_0 \triangleq A_1 - A_2 A_4^{-1} A_3, \quad B_s(t) \triangleq B_{1i}(t) - A_2 A_4^{-1} B_{2i}(t)$$
$$Q_0 = Q_1 + A_3^T A_4^{-T} Q_3 A_4^{-1} A_3 \qquad (14.46)$$
$$D_s(t) = B_{2i}^T(t) A_4^{-T} Q_3 A_4^{-1} A_3$$
$$R_s(t) = R + B_{2i}^T(t) A_4^{-T} Q_3 A_4^{-1} B_{2i}(t)$$

The optimal slow control strategy is

$$u_s(t) = -R_s^{-1}(t) (D_s(t) + B_s^T(t) P_s(t)) y_s(t) = G_0(t) y_s(t) \qquad (14.47)$$

where $P_s(t)$ satisfies the algebraic Riccati equation

$$P_s(t) A_s(t) + A_s^T(t) P_s(t) + Q_s(t)$$
$$-P_s(t) B_s(t) R_s^{-1}(t) B_s^T(t) P_s(t) = 0 \qquad (14.48)$$

with

$$A_s(t) = (A_0(t) - B_s(t) R_s^{-1}(t) D_s(t))$$
$$Q_s(t) = Q_0 - D_s^T(t) R_s^{-1}(t) D_s(t) \qquad (14.49)$$

The fast time scale optimization problem of order n_2 is given by

$$\epsilon \dot{z}_f = A_4 z_f + B_{2i}(t) u_f, \quad z_f(t_0) = z^0 - z_s(t_0)$$

$$J_f = \frac{1}{2} \int_{t_0}^{\infty} \left(z_f^T Q_2 z_f + u_f^T R u_f \right) dt \qquad (14.50)$$

where $z_f = z - z_s$ and $u_f = u - u_s$ denote fast parts of the corresponding variables. The optimal control for the fast subsystem is

$$u_f(t) = -R^{-1} B_{2i}^T(t) P_f(t) z_f(t) = G_2(t) z_f(t) \qquad (14.51)$$

where $P_f(t)$ is the solution of the "fast" algebraic Riccati equation

$$P_f(t) A_4 + A_4^T P_f(t) - P_f(t) B_{2i}(t) R^{-1} B_{2i}^T(t) P_f(t) + Q_3 = 0 \qquad (14.52)$$

A realizable composite control requires that the system states x_s and z_f be expressed in terms of the actual system states x and z. This can be achieved by replacing x_s by x and z_f by $z - z_s$ so that

$$u_c(t) = G_2(t) \left[z(t) + A_4^{-1} \left(A_3 x(t) + B_{2i}(t) G_0(t) x(t) \right) \right]$$
$$+ G_0(t) x = (t) G_1(t) x(t) + G_2(t) z(t) \qquad (14.53)$$

where

$$G_1 = \left(I_r + G_2 A_4^{-1} B_{2i} \right) G_0 + G_2 A_4^{-1} A_3 \qquad (14.54)$$

The near optimality of the composite control law is stated in the following lemma.

Lemma 14.1 *Under the stability assumptions the composite control law (14.53) is suboptimal in the sense*

$$u_{opt}(t) = u_c(t) + O(\epsilon), \quad t \geq t_0$$
$$y(t) = y_s(t) + O(\epsilon), \quad t \geq t_0 \qquad (14.55)$$
$$z(t) = z_s(t) + z_f(t) + O(\epsilon), \quad t \geq t_0$$

◇

The proof of this lemma follows from (Chow and Kokotovic, 1976; Kokotovic et al., 1986).

Thus, instead of solving at each iteration the global full-order numerically ill-defined algebraic Riccati equation (14.42), in the presented slow-fast decomposition technique we are faced with the problem of solving two reduced-order well-defined algebraic Riccati equations (14.48) and (14.52).

The unique solutions of (14.48) and (14.52) exist under the following assumptions.

Assumption 14.3 The pairs $(A_s(t), B_s(t))$ and (A_4, B_{2i}) are stabilizable, the pairs $\left(A_s(t), \sqrt{Q_s(t)}\right)$ and $(A_4, \sqrt{Q_3})$ are detectable, and $x_i(t)$ stay in the stabilizability domains of the slow and fast subsystems for every $t \geq t_0$.

$$\triangle$$

An $O(\epsilon)$ perturbation in each iteration of the presented slow-fast iterative scheme given by (14.54) will propagate into the next iteration, but due to the continuous dependence of the solution of the sequence of linear differential equations (14.41) with respect to perturbations in system coefficients, the presented method produces

$$u_{ci}^*(t) \rightarrow u^*(t) + O(\epsilon)$$
$$\begin{bmatrix} y_{si}^*(t) \\ z_{si}^*(t) + z_{fi}^*(t) \end{bmatrix} \rightarrow x^*(t) + O(\epsilon) \tag{14.56}$$

where i stands for the iteration number.

14.4 Case Study: Induction Motor Drives

In order to demonstrate the proposed method we have solved a fourth-order example representing the model of induction motor drives (Figalli et al., 1984). A frequency controlled two phase induction motor can be represented in the bilinear singularly perturbed form (14.12). The state and control variable are

$$x = \begin{bmatrix} y_1 \\ y_2 \\ z_1 \\ z_2 \end{bmatrix} = \begin{bmatrix} \phi_{ds} \\ \phi_{qs} \\ i_{ds} \\ i_{qs} \end{bmatrix}, \quad u = \begin{bmatrix} u_1 \\ u_2 \\ u_3 \end{bmatrix} = \begin{bmatrix} v_{ds} \\ v_{qs} \\ \omega_s \end{bmatrix}$$

where

ϕ_{ds} and ϕ_{qs} — projections of the stator flux
i_{ds} and i_{qs} — projections of stator current
v_{ds} and v_{qs} — projections of the supply voltage
ω_s — slip angular frequency.

The problem matrices have the following values

$$A = \begin{bmatrix} 0 & 321.57 & -.312 & 0 \\ -312.57 & 0 & 0 & -.312 \\ 98.87 & 27059 & -44.93 & 2.57 \\ -27059 & 98.87 & -2.57 & -44.93 \end{bmatrix}$$

$$B = \begin{bmatrix} 1 & 0 & 0 \\ 0 & 1 & -7.3 \\ 87.3 & 0 & 87.8 \\ 0 & 87.3 & -53 \end{bmatrix}, \quad x(t_0) = \begin{bmatrix} -.07 \\ .04 \\ 15 \\ 47 \end{bmatrix}$$

$$N_1 = \begin{bmatrix} 0 & 0 & 0 \\ 0 & 0 & -1 \\ 0 & 0 & 0 \\ 0 & 0 & 0 \end{bmatrix}, \quad N_2 = \begin{bmatrix} 0 & 0 & 1 \\ 0 & 0 & 0 \\ 0 & 0 & 0 \\ 0 & 0 & 0 \end{bmatrix}, \quad N_3 = \begin{bmatrix} 0 & 0 & 0 \\ 0 & 0 & 0 \\ 0 & 0 & 0 \\ 0 & 1 & 0 \end{bmatrix}$$

$$N_4 = \begin{bmatrix} 0 & 0 & 0 \\ 0 & 0 & 0 \\ 0 & 0 & 1 \\ 0 & 0 & 0 \end{bmatrix}, \quad Q = I_4, \quad R = \begin{bmatrix} .1 & 0 & 0 \\ 0 & .1 & 0 \\ 0 & 0 & 50 \end{bmatrix}$$

The simulation results are presented in Figures 14.1–14.4. In these figures the solid lines represent the optimal control and the dashed lines represent the composite control. It can be seen that the approximate trajectories are $O(\epsilon)$ close to the optimal ones.

14.5 Near-Optimal Control of Singularly Perturbed Bilinear Systems

In the previous section, we have obtained results for the composite control law, which produces the accuracy of an $O(\epsilon)$. In some applications of singularly and regularly perturbed systems an $O(\epsilon)$ accuracy may not be sufficient, (see, for example, (Shen and Gajic, 1990a; Gajic et al., 1989). The iterative refinement of (Gajic et al., 1990), to be performed at each discrete-time instant along time axis, can be used to increase the accuracy in (14.55)-(14.56) up to $O(\epsilon^k)$. The corresponding algorithm applied to the problem under consideration is given bellow.

Define the approximations of the required solution of (14.42) as

$$P_i^{(k)}(\epsilon) = \mathbf{P}_i(\epsilon) + \epsilon E_i^{(k)}(\epsilon) \tag{14.57}$$

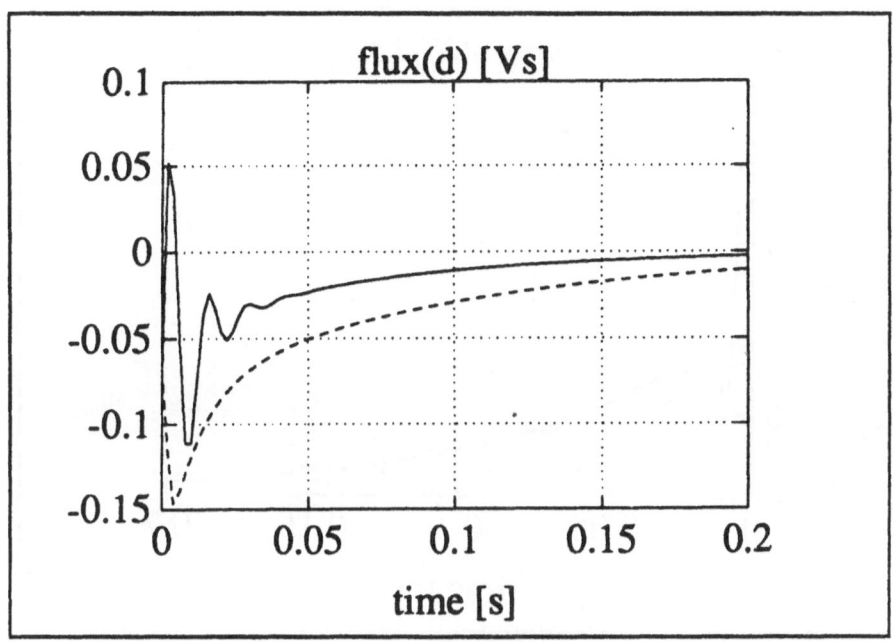

Figure 14.1: Optimal and approximate trajectories for flux ϕ_{ds}

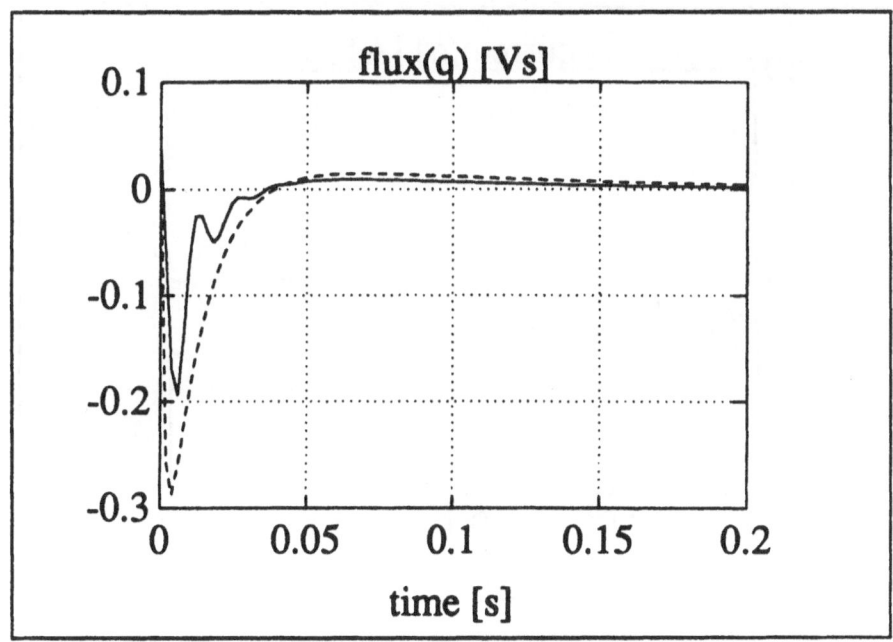

Figure 14.2: Optimal and approximate trajectories for flux ϕ_{qs}

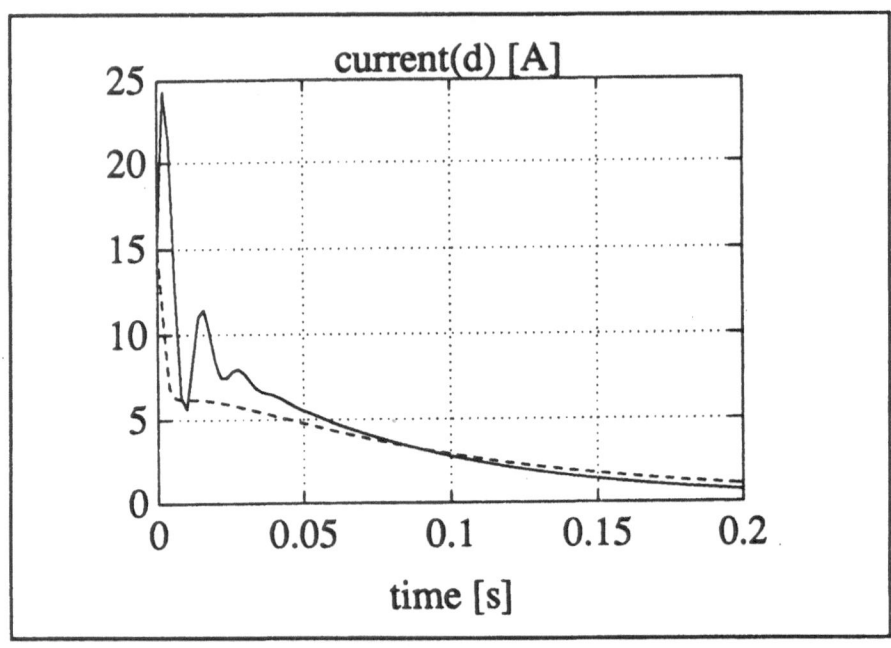

Figure 14.3: Optimal and approximate trajectories for current i_{ds}

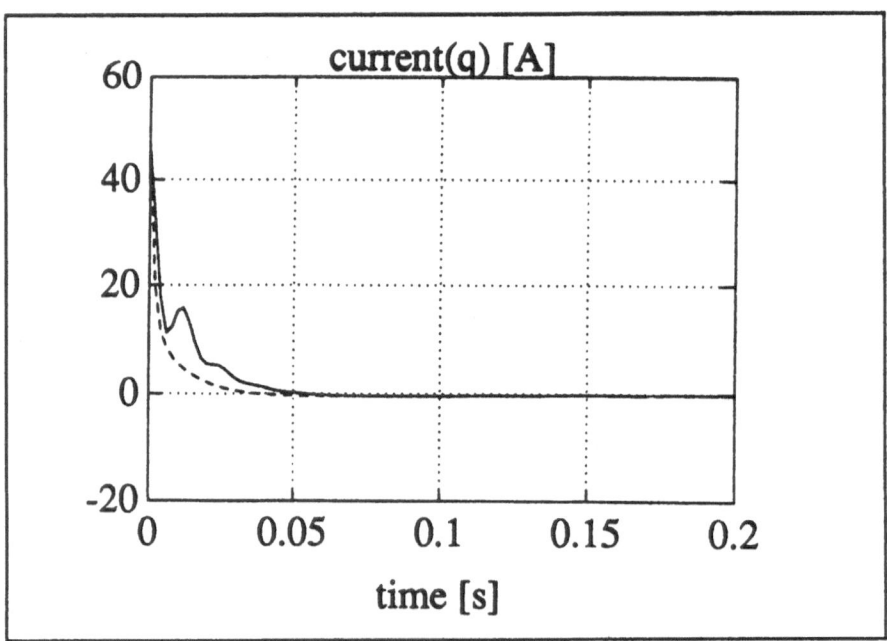

Figure 14.4: Optimal and approximate trajectories for current i_{qs}

where $E_i^{(k)}$ stand for the approximation errors, k for the order of approximation, and i is the iteration number with respect to (14.42). The zeroth-order solution, $\mathbf{P}_i(t)$, is partitioned according to

$$\mathbf{P}_i(t) = \begin{bmatrix} \mathbf{P}_{1i}(t) & \epsilon \mathbf{P}_{2i}(t) \\ \epsilon \mathbf{P}_{2i}^T(t) & \epsilon \mathbf{P}_{3i}(t) \end{bmatrix} \tag{14.58}$$

The elements $\mathbf{P}_{ji}$, $j = 1, 2, 3$, are obtained from (14.42) by setting $\epsilon = 0$, that is

$$\mathbf{P}_{1i}(t) = P_{si}(t)$$
$$\mathbf{P}_{2i}(t) = -\left(\mathbf{P}_{1i}A_2 + A_3^T\mathbf{P}_{3i} - \mathbf{P}_{1i}S_i\mathbf{P}_{3i}\right)(A_4 - S_{2i}\mathbf{P}_{3i})^{-1}$$
$$\mathbf{P}_{3i}(t) = P_{fi}(t)$$
$$\tag{14.59}$$

Note that $P_{si}(t)$ and $P_{fi}(t)$ are obtained from (14.48) and (14.52), where i stands for the given iteration of (14.42), and newly defined matrices are

$$S_i = B_{1i}R^{-1}B_{2i}^T, \quad S_{ji} = B_{ji}R^{-1}B_{ji}^T, \quad j = 1, 2 \tag{14.60}$$

The approximation errors partitioned as

$$E_i^{(k)} = \begin{bmatrix} E_{1i}^{(k)} & \epsilon E_{2i}^{(k)} \\ \epsilon E_{2i}^{(k)T} & \epsilon E_{3i}^{(k)} \end{bmatrix} \tag{14.61}$$

can be obtained from the following algorithm (Gajic et al., 1990)

$$E_{3i}^{(k+1)}D_{3i} + D_{3i}^T E_{3i}^{(k+1)} = H_{3i}^{(k)}$$
$$E_{2i}^{(k+1)}D_{3i} + E_{1i}^{(k+1)}D_{21i} + D_{22i}^T E_{3i}^{(k+1)} = -H_{1i}^{(k)}$$
$$E_{1i}^{(k+1)}D_{1i} + D_{1i}^T E_{1i}^{(k+1)} = D^T H_{1i}^{(k)T} + H_{1i}^{(k)}D + D_i^T H_{3i}^{(k)}D_i + \epsilon H_{2i}^{(k)}$$
$$E_{ji}^{(0)} = 0, \quad j = 1, 2, 3$$
$$\tag{14.62}$$

where

$$D_{1i} = D_{11i} - D_{21i}D_{3i}^{-1}D_{22i}, \quad D_{11i} = A_1 - S_{1i}\mathbf{P}_{1i} - S_i\mathbf{P}_{2i}^T$$
$$D_{21i} = A_2 - S_i\mathbf{P}_{3i}, \quad D_{22i} = A_3 - S_i^T\mathbf{P}_{1i} - S_{2i}\mathbf{P}_{2i}^T$$
$$D_{3i} = A_4 - S_{2i}\mathbf{P}_{3i}, \quad D_i = D_{3i}^{-1}D_{22i}$$
$$\tag{14.63}$$

and

$$H_{1i}^{(k)} = A_1^T P_{2i}^{(k)} - P_{1i}^{(k)} S_{1i} P_{2i}^{(k)} - P_{2i}^{(k)} S_i^T P_{2i}^{(k)}$$
$$-\epsilon \left(E_{1i}^{(k)} S_i E_{3i}^{(k)} + E_{2i}^{(k)} S_{2i} E_{3i}^{(k)} \right) \qquad (14.64)$$

$$H_{2i}^{(k)} = E_{1i}^{(k)} S_{1i} E_{1i}^{(k)} + E_{1i}^{(k)} S_i E_{2i}^{(k)^T} + E_{2i}^{(k)} S_i^T E_{1i}^{(k)} + E_{2i}^{(k)} S_{2i} E_{2i}^{(k)^T}$$
$$H_{3i}^{(k)} = -P_{2i}^{(k)^T} - A_2^T P_{2i}^{(k)} + \epsilon P_{2i}^{(k)^T} S_{1i} P_{2i}^{(k)} + \epsilon E_{3i}^{(k)} S_{2i} E_{3i}^{(k)}$$
$$+ P_{2i}^{(k)^T} S_i P_{3i}^{(k)} + P_{3i}^{(k)} S_i^T P_{2i}^{(k)}$$
$$(14.65)$$

Note that matrices D_{1i} and D_{3i} are stable (Gajic et al., 1990). This algorithm converges with the rate of convergence of $O(\epsilon)$, that is (Gajic et al., 1990)

$$\left\| E_{ji}^{(k+1)} - E_{ji}^{(k)} \right\| = O(\epsilon), \quad j = 1, 2, 3 \qquad (14.66)$$

or

$$\left\| E_{ji}^{(k+1)} - E_{ji} \right\| = O\left(\epsilon^k\right), \quad j = 1, 2, 3 \qquad (14.67)$$

The approximation

$$P_{ji}^{(k)}(\epsilon) = \mathbf{P}_{ji}(\epsilon) + \epsilon E_{ji}^{(k)}(\epsilon), \quad j = 1, 2, 3 \qquad (14.68)$$

will produce $O\left(\epsilon^{k+1}\right)$ approximation of the required solution P_i. Thus, having obtained P_i with the accuracy of $O\left(\epsilon^{k+1}\right)$, we get the same accuracy for the optimal trajectories and the approximate optimal control law. The price for this is that we have to solve k-times two reduced-order Lyapunov equations (14.62) at each discrete-time instant in the interval of interest.

14.6 Optimal Control of Weakly Coupled Bilinear Systems

The purpose of this section is to study the optimal control problem of weakly coupled bilinear systems with a quadratic performance criterion. We will study both the open-loop and closed-loop optimal control problems. The optimization of the time invariant bilinear weakly coupled system is considered in this section. The obtained results can be easily extended to the time varying case.

A sequence of linear state and costate equations is constructed such that the open-loop solution of the optimization problem is obtained in terms of the reduced-order subsystems. The obtained results utilize the recursive scheme for the optimal control of a general bilinear system with a quadratic performance criterion (Hofer and Tibken, 1988) and the time varying version of the reduced-order method for solving the linear-quadratic optimal open-loop weakly coupled control problem (Su and Gajic, 1991). This leads to the reduction in the size of the required computation and allows parallel processing of information.

The near-optimal closed-loop control is obtained in the form of a linear feedback law, with the feedback gains calculated from two reduced-order independent time varying linear-quadratic optimal control problems. The obtained results are based on the idea of the recursive reduced-order scheme for solving the algebraic Riccati equation for weakly coupled systems (Gajic et al., 1990) and the recursive scheme for the optimal control of a general bilinear system with a quadratic performance criterion (Cebuhar and Constanza, 1984). An algorithm which produces an arbitrary degree of accuracy for the closed-loop feedback control is derived. The results are demonstrated on the bilinear model of a paper making machine.

14.6.1 Open-Loop Control of Weakly Coupled Bilinear Systems

In this section, we exploit the iterative scheme (14.8)-(14.11), comprising a sequence of linear two-point boundary value problems, in order to derive the solution for the optimal open-loop control of weakly coupled bilinear systems. The solution is obtained in the spirit of the general theory of small parameter control problems, namely, the problem is decomposed into two reduced-order subproblems. The open-loop optimal control of weakly coupled linear systems has been studied in (Su and Gajic, 1991). The study of (Su and Gajic, 1991) is limited to the time invariant systems. In this section, we show that following the ideas of (Su and Gajic, 1991) we are able to handle in the same manner the time varying weakly coupled two-point boundary value problem.

The weakly coupled bilinear control system under consideration is represented by

$$
\begin{bmatrix} \dot{y}_1 \\ \dot{y}_2 \end{bmatrix} = \begin{bmatrix} A_1 & \epsilon A_2 \\ \epsilon A_3 & A_4 \end{bmatrix} \begin{bmatrix} y_1 \\ y_2 \end{bmatrix} + \begin{bmatrix} B_1 & \epsilon B_2 \\ \epsilon B_3 & B_4 \end{bmatrix} \begin{bmatrix} u_1 \\ u_2 \end{bmatrix} +
$$

$$+\left\{\begin{bmatrix} y_1 \\ y_2 \end{bmatrix}\begin{bmatrix} N_a & \epsilon N_b \\ \epsilon N_c & N_d \end{bmatrix}\right\}\begin{bmatrix} u_1 \\ u_2 \end{bmatrix}, \qquad \begin{bmatrix} y_1(t_0) \\ y_2(t_0) \end{bmatrix} = \begin{bmatrix} y_1^0 \\ y_2^0 \end{bmatrix} \qquad (14.69)$$

where $y_1 \in \Re^{n_1}$, $y_2 \in \Re^{n_2}$, $u_i \in \Re^{m_i}$, $i = 1, 2$, and ϵ is a small coupling parameter, with

$$\left\{\begin{bmatrix} y_1 \\ y_2 \end{bmatrix}\begin{bmatrix} N_1 & \epsilon N_2 \\ \epsilon N_3 & N_4 \end{bmatrix}\right\} = \sum_{i=1}^{n_1} y_{1i} \begin{bmatrix} N_{ai} & N_{bi} \\ N_{ci} & N_{di} \end{bmatrix}$$

$$+ \sum_{j=n_1+1}^{n_1+n_2} y_{2(j-n_1)} \begin{bmatrix} N_{aj} & N_{bj} \\ N_{cj} & N_{dj} \end{bmatrix} \qquad (14.70)$$

where $N_{ai} \in \Re^{n_1 \times m_1}$, $N_{bi} \in \Re^{n_1 \times m_2}$, $N_{ci} \in \Re^{n_2 \times m_1}$, $N_{di} \in \Re^{n_2 \times m_2}$.
A quadratic cost functional associated with (14.69) has the form

$$J = \frac{1}{2} \int_{t_0}^{t_f} \left(\begin{bmatrix} y_1 \\ y_2 \end{bmatrix}^T Q \begin{bmatrix} y_1 \\ y_2 \end{bmatrix} + \begin{bmatrix} u_1 \\ u_2 \end{bmatrix}^T R \begin{bmatrix} u_1 \\ u_2 \end{bmatrix} \right) dt$$

$$+ \frac{1}{2} \begin{bmatrix} y_1(t_f) \\ y_2(t_f) \end{bmatrix}^T F \begin{bmatrix} y_1(t_f) \\ y_2(t_f) \end{bmatrix} \qquad (14.71)$$

with Q, R, F having the weak coupling structures, that is

$$Q = \begin{bmatrix} Q_1 & \epsilon Q_2 \\ \epsilon Q_2^T & Q_3 \end{bmatrix}, \quad R = \begin{bmatrix} R_1 & 0 \\ 0 & R_2 \end{bmatrix}, \quad F = \begin{bmatrix} F_1 & \epsilon F_2 \\ \epsilon F_2^T & F_3 \end{bmatrix} \qquad (14.72)$$

In the following, we will utilize the recursive scheme (14.8)-(14.11) in order to find the optimal open-loop control law for the weakly coupled bilinear-quadratic optimal control problem represented by (14.69)-(14.72) in terms of the reduced-order subsystems.

It can be shown that the system of equations (14.8) preserves the weak coupling structure. Namely, the use of (14.69)-(14.72) in (14.4)-(14.5) and (14.8)-(14.11) produces

$$\begin{bmatrix} \dot{y}_1 \\ \dot{y}_2 \end{bmatrix}^{(k+1)} = \tilde{A}^{(k)} \begin{bmatrix} y_1 \\ y_2 \end{bmatrix}^{(k+1)} - \tilde{B}^{(k)} R^{-1} \tilde{B}^{(k)T} \begin{bmatrix} q_1 \\ q_2 \end{bmatrix}^{(k+1)}$$

$$\begin{bmatrix} \dot{q}_1 \\ \dot{q}_2 \end{bmatrix}^{(k+1)} = -\tilde{Q}^{(k)} \begin{bmatrix} y_1 \\ y_2 \end{bmatrix}^{(k+1)} - \tilde{A}^{(k)^T} \begin{bmatrix} q_1 \\ q_2 \end{bmatrix}^{(k+1)} \qquad (14.73)$$

where $q_1 \in \Re^{n_1}$ and $q_2 \in \Re^{n_2}$ are costate vectors corresponding, respectively, to the state variables y_1 and y_2. The time varying matrices in (14.73) are given by

$$\tilde{A}^{(k)} = \begin{bmatrix} \tilde{A}_1^{(k)} & \epsilon\tilde{A}_2^{(k)} \\ \epsilon\tilde{A}_3^{(k)} & \tilde{A}_4^{(k)} \end{bmatrix}, \qquad \tilde{Q}^{(k)} = \begin{bmatrix} \tilde{Q}_1^{(k)} & \epsilon\tilde{Q}_2^{(k)} \\ \epsilon\tilde{Q}_2^{(k)^T} & \tilde{Q}_3^{(k)} \end{bmatrix}$$

$$(14.74)$$

$$\tilde{S}^{(k)} = \tilde{B}^{(k)}R^{-1}\tilde{B}^{(k)^T} = \begin{bmatrix} \tilde{S}_1^{(k)} & \epsilon\tilde{S}_2^{(k)} \\ \epsilon\tilde{S}_2^{(k)^T} & \tilde{S}_3^{(k)} \end{bmatrix}$$

Note that partitions defined in (14.74) have to be performed by a computer only, in the process of computations, and there is no need for the corresponding analytical expressions. After some algebra the state-costate equations (14.73) can be written in the form

$$\begin{bmatrix} \dot{y}_1^{(k+1)} \\ \dot{q}_1^{(k+1)} \\ \dot{y}_2^{(k+1)} \\ \dot{q}_2^{(k+1)} \end{bmatrix} = \begin{bmatrix} \tilde{T}_1^{(k)} & \epsilon\tilde{T}_2^{(k)} \\ \epsilon\tilde{T}_3^{(k)} & \tilde{T}_4^{(k)} \end{bmatrix} \begin{bmatrix} y_1^{(k+1)} \\ q_1^{(k+1)} \\ y_2^{(k+1)} \\ q_2^{(k+1)} \end{bmatrix} \qquad (14.75)$$

The time varying matrices $\tilde{T}_i^{(k)}$ introduced in (14.75) are given by

$$\tilde{T}_1^{(k)} = \begin{bmatrix} \tilde{A}_1^{(k)} & -\tilde{S}_1^{(k)} \\ -\tilde{Q}_1^{(k)} & -\tilde{A}_1^{(k)^T} \end{bmatrix}, \qquad \tilde{T}_2^{(k)} = \begin{bmatrix} \tilde{A}_2^{(k)} & -\tilde{S}_2^{(k)} \\ -\tilde{Q}_2^{(k)} & -\tilde{A}_3^{(k)^T} \end{bmatrix}$$

$$\tilde{T}_3^{(k)} = \begin{bmatrix} \tilde{A}_3^{(k)} & -\tilde{S}_2^{(k)^T} \\ -\tilde{Q}_2^{(k)^T} & -\tilde{A}_2^{(k)^T} \end{bmatrix}, \qquad \tilde{T}_4^{(k)} = \begin{bmatrix} \tilde{A}_4^{(k)} & -\tilde{S}_2^{(k)} \\ -\tilde{Q}_3^{(k)} & -\tilde{A}_4^{(k)^T} \end{bmatrix}$$

$$(14.76)$$

The expression for the boundary conditions is changed due to an interchange of rows corresponding to $q_1^{(k+1)}$ and $y_2^{(k+1)}$, which modifies matrices defined in (14.9) as follows

$$U_1 \begin{bmatrix} w^{(k+1)}(t_0) \\ \lambda^{(k+1)}(t_0) \end{bmatrix} + V_1 \begin{bmatrix} w^{(k+1)}(t_f) \\ \lambda^{(k+1)}(t_f) \end{bmatrix} = c_1 \qquad (14.77)$$

with a new notation introduced as

$$\begin{bmatrix} y_1^{(k+1)} \\ q_1^{(k+1)} \end{bmatrix} = w^{(k+1)}, \quad \begin{bmatrix} y_2^{(k+1)} \\ q_2^{(k+1)} \end{bmatrix} = \lambda^{(k+1)} \qquad (14.78)$$

and

$$U_1 = \begin{bmatrix} I_{n_1} & 0 & 0 & 0 \\ 0 & 0 & 0 & 0 \\ 0 & 0 & I_{n_2} & 0 \\ 0 & 0 & 0 & 0 \end{bmatrix}; \quad c_1 = \begin{bmatrix} y_1^0 \\ 0 \\ y_2^0 \\ 0 \end{bmatrix}$$

$$V_1 = \begin{bmatrix} 0 & 0 & 0 & 0 \\ -F_1 & I_{n_1} & -\epsilon F_2 & 0 \\ 0 & 0 & 0 & 0 \\ -F_2^T & 0 & -F_3 & I_{n_2} \end{bmatrix}$$

$$(14.79)$$

In order to obtain the decoupled subsystems from (14.75), we apply the transformation of (Gajic and Shen, 1989) given by

$$\begin{bmatrix} \eta^{(k+1)} \\ \xi^{(k+1)} \end{bmatrix} = \mathbf{T}_2^{(k)}(t,\epsilon) \begin{bmatrix} w^{(k+1)} \\ \lambda^{(k+1)} \end{bmatrix} \qquad (14.80)$$

with

$$\mathbf{T}_2^{(k)}(t,\epsilon) = \begin{bmatrix} I_1 & -\epsilon L^{(k)} \\ \epsilon H^{(k)} & I_2 - \epsilon^2 H^{(k)} L^{(k)} \end{bmatrix}$$

$$(14.81)$$

$$\mathbf{T}_2^{(k)-1}(t,\epsilon) = \begin{bmatrix} I_1 - \epsilon^2 H^{(k)} L^{(k)} & \epsilon L^{(k)} \\ -\epsilon H^{(k)} & I_2 \end{bmatrix}$$

where I_1 and I_2 are identity matrices of order $2n_1$ and $2n_2$, respectively. The matrices $L^{(k)}$ and $H^{(k)}$ are the solutions of the following nonlinear differential equations

$$\dot{L}^{(k)} = \widetilde{T}_1^{(k)} L^{(k)} - L^{(k)} \widetilde{T}_4^{(k)} + \widetilde{T}_2^{(k)} - \epsilon^2 L^{(k)} \widetilde{T}_3^{(k)} L^{(k)}$$
$$\dot{H}^{(k)} = H^{(k)} \left(\widetilde{T}_1^{(k)} - \epsilon^2 L^{(k)} \widetilde{T}_3^{(k)} \right) - \left(\widetilde{T}_4^{(k)} + \epsilon^2 \widetilde{T}_3^{(k)} L^{(k)} \right) H^{(k)} + \widetilde{T}_3^{(k)}$$

$$(14.82)$$

The initial conditions for differential equations (14.82) are arbitrary (Qureshi and Gajic, 1991). The existence of the bounded solutions of (14.82) for sufficiently small values ϵ is established in (Qureshi and Gajic, 1991; Qureshi, 1992).

The transformation (14.80) applied to the system (14.75) produces two completely decoupled subsystems

$$\dot{\eta}^{(k+1)} = \left(\widetilde{T}_1^{(k)} - \epsilon^2 L^{(k)} \widetilde{T}_3^{(k)} \right) \eta^{(k+1)} \qquad (14.83)$$

422

$$\dot{\xi}^{(k+1)} = \left(\tilde{T}_4^{(k)} - \epsilon^2 \tilde{T}_3^{(k)} L^{(k)} \right) \xi^{(k+1)} \tag{14.84}$$

Consequently, the change of variables transforms the boundary conditions

$$U_2 \begin{bmatrix} \eta^{(k+1)}(t_0) \\ \xi^{(k+1)}(t_0) \end{bmatrix} + V_2 \begin{bmatrix} \eta^{(k+1)}(t_f) \\ \xi^{(k+1)}(t_f) \end{bmatrix} = c_1 \tag{14.85}$$

where

$$U_2 = U_1 T_2^{(k)^{-1}}(t_0, \epsilon), \qquad V_2 = V_1 T_2^{(k)^{-1}}(t_f, \epsilon) \tag{14.86}$$

The solutions of differential equations (14.83) and (14.84) are

$$\begin{aligned} \eta^{(k+1)}(t) &= \Phi^{(k)}(t, t_0, \epsilon)\, \eta^{(k+1)}(t_0) \\ \xi^{(k+1)}(t) &= \Psi^{(k)}(t, t_0, \epsilon)\, \xi^{(k+1)}(t_0) \end{aligned} \tag{14.87}$$

where $\Phi^{(k)}(t, t_0, \epsilon)$ and $\Psi^{(k)}(t, t_0, \epsilon)$ are the transition matrices of (14.83) and (14.84), respectively. The initial conditions $\eta^{(k+1)}(t_0)$ and $\xi^{(k+1)}(t_0)$ have to be determined. Substitution of (14.87) into (14.85) yields

$$\Delta^{(k+1)}(\epsilon) \begin{bmatrix} \eta^{(k+1)}(t_0) \\ \xi^{(k+1)}(t_0) \end{bmatrix} = c_1 \tag{14.88}$$

where

$$\Delta^{(k)}(\epsilon) = U_2(\epsilon) + V_2(\epsilon) \begin{bmatrix} \Phi^{(k)}(t_f, t_0, \epsilon) & 0 \\ 0 & \Psi^{(k)}(t_f, t_0, \epsilon) \end{bmatrix} \tag{14.89}$$

If $\Delta^{(k+1)^{-1}}(\epsilon)$ exists then the solution of (14.88) will be

$$\begin{bmatrix} \eta^{(k+1)}(t_0) \\ \xi^{(k+1)}(t_0) \end{bmatrix} = \Delta^{(k+1)^{-1}}(\epsilon)\, c_1 \tag{14.90}$$

Note that as $\epsilon \to 0$

$$\left\{ T_2^{(k)}(t, 0) \right\}^{-1} = \begin{bmatrix} I_1 & 0 \\ 0 & I_2 \end{bmatrix} = I \tag{14.91}$$

After partitioning the transition matrices $\Phi^{(k)}(t, t_0, 0)$ and $\Psi^{(k)}(t, t_0, 0)$ as

$$\Phi^{(k)}(t, t_0, 0) = \begin{bmatrix} \Phi_{11}^{(k)}(t, t_0, 0) & \Phi_{12}^{(k)}(t, t_0, 0) \\ \Phi_{21}^{(k)}(t, t_0, 0) & \Phi_{22}^{(k)}(t, t_0, 0) \end{bmatrix}$$

$$\Psi^{(k)}(t, t_0, 0) = \begin{bmatrix} \Psi_{11}^{(k)}(t, t_0, 0) & \Psi_{12}^{(k)}(t, t_0, 0) \\ \Psi_{21}^{(k)}(t, t_0, 0) & \Psi_{22}^{(k)}(t, t_0, 0) \end{bmatrix}$$

and after some algebra, the matrix $\Delta^{(k+1)}(\epsilon)$ is obtained in the form

$$\Delta^{(k+1)}(\epsilon) = \begin{bmatrix} I_{n_1} & 0 & 0 & 0 \\ 0 & \Delta_{22}^{(k+1)}(0) & 0 & 0 \\ 0 & 0 & I_{n_2} & 0 \\ 0 & 0 & 0 & \Delta_{44}^{(k+1)}(0) \end{bmatrix} + O(\epsilon) \quad (14.92)$$

where

$$\begin{aligned} \Delta_{22}^{(k+1)}(0) &= \Phi_{22}^{(k)}(t_f, t_0, 0) - F_1 \Phi_{12}^{(k)}(t_f, t_0, 0) \\ \Delta_{44}^{(k+1)}(0) &= \Psi_{22}^{(k)}(t_f, t_0, 0) - F_3 \Psi_{12}^{(k)}(t_f, t_0, 0) \end{aligned} \quad (14.93)$$

Since the matrices $\Phi_{22}^{(k)}(t_f, t_0, 0) - F_1 \Phi_{12}^{(k)}(t_f, t_0, 0)$ and $\Psi_{22}^{(k)}(t_f, t_0, 0) - F_3 \Psi_{12}^{(k)}(t_f, t_0, 0)$ are nonsingular (Kirk, 1970; page 211), so does $\Delta^{(k+1)}(\epsilon)$ for sufficiently small values of ϵ, with $0 < \epsilon \le \epsilon_1$ and ϵ sufficiently small. Thus, in summary, we have established the following theorem.

Theorem 14.2 *Let the problem matrices be continuous functions of t on the time interval $t_0 \le t \le t_f$, then for all sufficiently small ϵ the boundary value problem (14.83)–(14.86) has the solution given by*

$$\begin{bmatrix} \eta^{(k+1)}(t, \epsilon) \\ \xi^{(k+1)}(t, \epsilon) \end{bmatrix} = \begin{bmatrix} \Phi^{(k+1)}(t, t_0; \epsilon) & 0 \\ 0 & \Psi^{(k+1)}(t, t_0, \epsilon) \end{bmatrix} \Delta^{(k+1)-1}(\epsilon) c_1$$

Consequently, the solution of the original boundary problem (14.75)-(14.79) is obtained from (14.80)

$$\begin{bmatrix} w^{(k+1)}(t, \epsilon) \\ \lambda^{(k+1)}(t, \epsilon) \end{bmatrix} = \{ T_2(t, \epsilon) \}^{-1} \begin{bmatrix} \eta^{(k+1)}(t, \epsilon) \\ \xi^{(k+1)}(t, \epsilon) \end{bmatrix} \quad (14.94)$$

so that the required variables $y_1^{(k+1)}$ and $y_2^{(k+1)}$ are obtained by partitioning the vectors $w^{(k+1)}$ and $\lambda^{(k+1)}$ according to (14.78). The same hold for the costate variables, $q_1^{(k+1)}$ and $q_2^{(k+1)}$, that is, they are obtained form (14.80) and (14.78). ◇

Having obtained the approximate state trajectories $y_1^{(k+1)}$ and $y_2^{(k+1)}$ and the approximate costate trajectories $q_1^{(k+1)}$ and $q_2^{(k+1)}$, the approximate optimal open-loop control can be expressed as

$$u^{(k+1)}(t) = -R^{-1}\left(B + \left\{\begin{bmatrix} y_1^{(k+1)}(t) \\ y_2^{(k+1)}(t) \end{bmatrix} N\right\}\right)^T \begin{bmatrix} q_1^{(k+1)}(t) \\ q_2^{(k+1)}(t) \end{bmatrix} \quad (14.95)$$

The main problem that we are faced with in the presented method is the problem of finding the transition matrices $\Phi^{(k)}(t, t_0, \epsilon)$ and $\Psi^{(k)}(t, t_0, \epsilon)$ of the corresponding time varying systems. One way to overcome this problem is to study the optimal open-loop control of weakly coupled system in the discrete time domain. Research in that direction is underway.

14.6.2 Closed-Loop Control of Weakly Coupled Bilinear Systems

For the weakly coupled sequence of the linear systems (14.41), which approximate the solution of the bilinear-quadratic optimal control problem (14.69)-(14.72), the matrices $B_i(t), S_i(t), P_i(t)$ in equations (14.41)-(14.42) can be partitioned as

$$B_i(t) = \begin{bmatrix} B_{1i}(t) & \epsilon B_{2i}(t) \\ \epsilon B_{3i}(t) & B_{4i}(t) \end{bmatrix}, \quad S_i(t) = \begin{bmatrix} S_{1i}(t) & \epsilon S_{2i}(t) \\ \epsilon S_{2i}^T(t) & S_{3i} \end{bmatrix}$$

$$P_i(t) = \begin{bmatrix} P_{1i}(t) & \epsilon P_{2i}(t) \\ \epsilon P_{2i}^T(t) & P_{3i}(t) \end{bmatrix}$$

$$(14.96)$$

Partitioning (14.42) according to (14.96) and setting $\epsilon = 0$, we get an $O(\epsilon^2)$ approximation of (14.42) in terms of the reduced-order, decoupled algebraic Riccati equations

$$\begin{aligned}
&\mathbf{P}_{1i}(t) A_1 + A_1^T \mathbf{P}_{1i}(t) + Q_1 - \mathbf{P}_{1i}(t) S_{1i}(t) \mathbf{P}_{1i}(t) = 0 \\
&\mathbf{P}_{3i}(t) A_4 + A_4^T \mathbf{P}_{3i}(t) + Q_3 - \mathbf{P}_{3i}(t) S_{3i}(t) \mathbf{P}_{3i}(t) = 0 \\
&\mathbf{P}_{2i}(t)(A_4 - S_{3i}(t) \mathbf{P}_{3i}(t)) + (A_1 - S_{1i}(t) \mathbf{P}_{1i}(t))^T \mathbf{P}_{2i}(t) \\
&+ \mathbf{P}_{1i}(t) A_2 + A_2^T \mathbf{P}_{3i}(t) + Q_2 - \mathbf{P}_{1i}(t) S_{si}(t) \mathbf{P}_{3i}(t) = 0
\end{aligned} \quad (14.97)$$

The unique positive semidefinite solution of (14.97) exists under the following assumption.
Assumption 14.4 The triples $\left(A_1, B_{1i}(t), \sqrt{Q_1}\right)$ and $\left(A_4, B_{4i}(t), \sqrt{Q_3}\right)$ are stabilizable-detectable for every t.

$\triangle$

Corresponding solution, $\mathbf{P}_i(t)$, defined as

$$\mathbf{P}_i(t) = \begin{bmatrix} \mathbf{P}_{1i}(t) & \epsilon\mathbf{P}_{2i}(t) \\ \epsilon\mathbf{P}_{2i}^T(t) & \mathbf{P}_{3i}(t) \end{bmatrix} \qquad (14.98)$$

is $O\left(\epsilon^2\right)$ close to the optimal one, $P_i(t)$. An $O\left(\epsilon^2\right)$ perturbation made in the iterative scheme (14.41)-(14.43) propagates into next iteration, but due to the continuous dependence of the solution of the sequence of linear differential equations with respect to a perturbation in the system coefficients, the presented method produces

$$x_i(t) = \mathbf{x}_i(t) + O\left(\epsilon^2\right), \quad i = 1, 2, \dots \quad \forall t \geq 0 \qquad (14.99)$$

and

$$u_i^{app}(t) = \mathbf{u}_i^{app}(t) + O\left(\epsilon^2\right), \quad i = 1, 2, \dots \quad \forall t \geq 0 \qquad (14.100)$$

where

$$\dot{\mathbf{x}}_i(t) = A\mathbf{x}_i(t) + \mathbf{B}_i(t)\, \mathbf{u}_i^{app}(t) \qquad (14.101)$$

$$\mathbf{u}_i^{app}(t) = -R^{-1}\mathbf{B}_i(t)\,\mathbf{P}_i(t)\,\mathbf{x}_i(t) \qquad (14.102)$$

with

$$\mathbf{B}_i(t) = B + \{\mathbf{x}_{i-1}(t)\,N\} \qquad (14.103)$$

If one intends to improve the accuracy of the solution of the Riccati equation (14.42), one can use (in the last iteration with respect to i only) an iterative refinement of (Gajic et al., 1990; Shen and Gajic, 1990a). Define the approximations of the required solution of (14.42) as

$$P_{ji}^{(k)}(t, \epsilon) = \mathbf{P}_{ji}(t, \epsilon) + \epsilon^2 E_{ji}^{(k)}(t, \epsilon) \qquad j = 1, 2, 3 \qquad (14.104)$$

Then, the recursive reduced-order scheme for the error equations are obtained as (Gajic et al., 1990)

$$\begin{aligned}
E_{1i}^{(k+1)}\Delta_{1i} + \Delta_{1i}^T E_{1i}^{(k+1)} &= M_{1i}^{(k)} \\
E_{3i}^{(k+1)}\Delta_{4i} + \Delta_{4i}^T E_{3i}^{(k+1)} &= M_{3i}^{(k)} \\
E_{2i}^{(k+1)}\Delta_{4i} + \Delta_{1i}^T E_{2i}^{(k+1)} + E_{1i}^{(k+1)}\Delta_{2i} + \Delta_{3i}^T E_{3i}^{(k+1)} &= M_{2i}^{(k,k+1)}
\end{aligned}$$
$$(14.105)$$

for $k = 0, 1, 2, \dots$, and $E_{1i}^{(0)} = 0$, $E_{2i}^{(0)} = 0$, $E_{3i}^{(0)} = 0$, where the matrices Δ_{ij}, $j = 1, \dots, 4$, and $M_{1i}^{(k)}$, $M_{2i}^{(k,k+1)}$, $M_{3i}^{(k)}$ are given by

$$\Delta_{1i} = A_1 - S_{1i}\mathbf{P}_{1i}, \quad \Delta_{2i} = A_2 - S_{1i}\mathbf{P}_{2i} - S_2\mathbf{P}_{3i}$$
$$\Delta_{4i} = A_4 - S_{3i}\mathbf{P}_{3i}, \quad \Delta_{3i} = A_3 - S_{3i}\mathbf{P}_{2i}^{\mathbf{T}} - S_2^{\mathbf{T}}\mathbf{P}_{1i} \tag{14.106}$$

and

$$M_{1i}^{(k)} = P_{2i}^{(k)} S_{2i}^T P_{1i}^{(k)} + P_{1i}^{(k)} S_{2i} P_{2i}^{(k)^T} + P_{2i}^{(k)} S_{3i} P_{2i}^{(k)^T}$$
$$\quad - P_{2i}^{(k)} A_3 - A_3^T P_{2i}^{(k)^T} - \epsilon^2 E_{1i}^{(k)} S_{1i} E_{1i}^{(k)}$$

$$M_{3i}^{(k)} = P_{3i}^{(k)} S_{2i}^T P_{2i}^{(k)} + P_{2i}^{(k)^T} S_{2i} P_{3i}^{(k)} + P_{2i}^{(k)^T} S_{1i} P_{2i}^{(k)}$$
$$\quad + P_{3i}^{(k)} S_{2i}^T P_{3i}^{(k)} - P_{2i}^{(k)^T} A_2 - A_2^T P_{2i}^{(k)} + \epsilon^2 E_{3i}^{(k)} S_{3i} E_{3i}^{(k)}$$

$$M_{2i}^{(k,k+1)} = P_{2i}^{(k)} S_{2i}^T P_{2i}^{(k)} + \epsilon^2 E_{1i}^{(k+1)} S_{1i} E_{2i}^{(k)} + \epsilon^2 E_{2i}^{(k)} S_{3i} E_{3i}^{(k+1)}$$
$$\quad + \epsilon^2 E_{1i}^{(k+1)} S_{2i} E_{3i}^{(k+1)} \tag{14.107}$$

Note that under Assumption 14.4 both Δ_{1i} and Δ_{4i} are stable matrices.

It can be shown that the rate of convergence of (14.104)-(14.107) is $O\left(\epsilon^2\right)$ (Gajic et al., 1990), that is

$$\left\| E_{ji}^{(k+1)} - E_{ji}^{(k)} \right\| = O\left(\epsilon^2\right), \quad i = 0, 1, 2, ...; \ j = 1, 2, 3; \ k = 0, 1, ... \tag{14.108}$$

which implies

$$\left\| P_{ji} - P_{ji}^{(k)} \right\| = O\left(\epsilon^{2(k+1)}\right), \quad i = 0, 1, 2, ...; \ j = 1, 2, 3; \ k = 0, 1, ... \tag{14.109}$$

Having obtained $P_i(t)$ with the accuracy of $O\left(\epsilon^{2(k+1)}\right)$ produces the same accuracy for the approximations of the optimal state trajectories and optimal control laws.

14.7 Case Study: A Paper Making Machine

In order to demonstrate the efficiency of the proposed method for the near-optimal closed-loop control of singularly perturbed bilinear systems we have run a fourth-order real world example, a paper making machine control problem (Ying et al., 1992). The bilinear mathematical model of this system is formulated according to (14.1) and (14.3) as

$$A = \begin{bmatrix} -1.93 & 0 & 0 & 0 \\ .394 & -0.426 & 0 & 0 \\ 0 & 0 & -0.63 & 0 \\ 0.095 & -0.103 & 0.413 & -0.426 \end{bmatrix}, \quad B = \begin{bmatrix} 1.274 & 1.274 \\ 0 & 0 \\ 1.34 & -0.65 \\ 0 & 0 \end{bmatrix}$$

$$N_1 = \begin{bmatrix} 0 & 0 \\ 0 & 0 \\ .755 & .366 \\ 0 & 0 \end{bmatrix}, N_2 = N_4 = \begin{bmatrix} 0 & 0 \\ 0 & 0 \\ 0 & 0 \\ 0 & 0 \end{bmatrix}, N_3 = \begin{bmatrix} 0 & 0 \\ 0 & 0 \\ -.718 & -.718 \\ 0 & 0 \end{bmatrix}$$

Weighting matrices Q and R chosen as

$$Q = \begin{bmatrix} 1 & 0 & 0.13 & 0 \\ 0 & 1 & 0 & 0.09 \\ 0.13 & 0 & 0.1 & 0 \\ 0 & 0.09 & 0 & 0.2 \end{bmatrix}, \quad R = \begin{bmatrix} 1 & 0 \\ 0 & 1 \end{bmatrix}$$

Note that the matrices B, N_1, and N_2 have no weakly coupled forms. However, it has been shown in (Skataric et al., 1991) that the classes of linear-quadratic optimal control problems having weakly coupled system matrix and strongly coupled input matrix can be studied as the weakly coupled linear-quadratic optimal control problems by assuming the special form for the state penalty matrix. Small perturbation parameter is $\epsilon = 0.1$. Simulation results, obtained by using the MATLAB package, are presented in Figures 14.5-14.10. Figures 14.5-14.10 represent the approximate and optimal trajectories and the approximate and the optimal controls. The optimal ones are represented by the solid lines. It can be seen from the these plots that the approximate trajectories and controls are very good approximations for the optimal ones. The number of iterations performed are $i = 3$ and $k = 1$, where i represents the number of linear time varying systems in the sequence defined by (14.41) and k represents the number of iterations performed to increase the accuracy as defined by (14.104).

14.8 Conclusion

The results of this chapter can be applied to the nonlinear singularly perturbed and weakly coupled systems after they have been bilinearized. On the contrary to the linearization procedure (where all nonlinear terms are neglected), the system bilinearization preserves the nonlinear terms representing the product of the state and control variables. Through

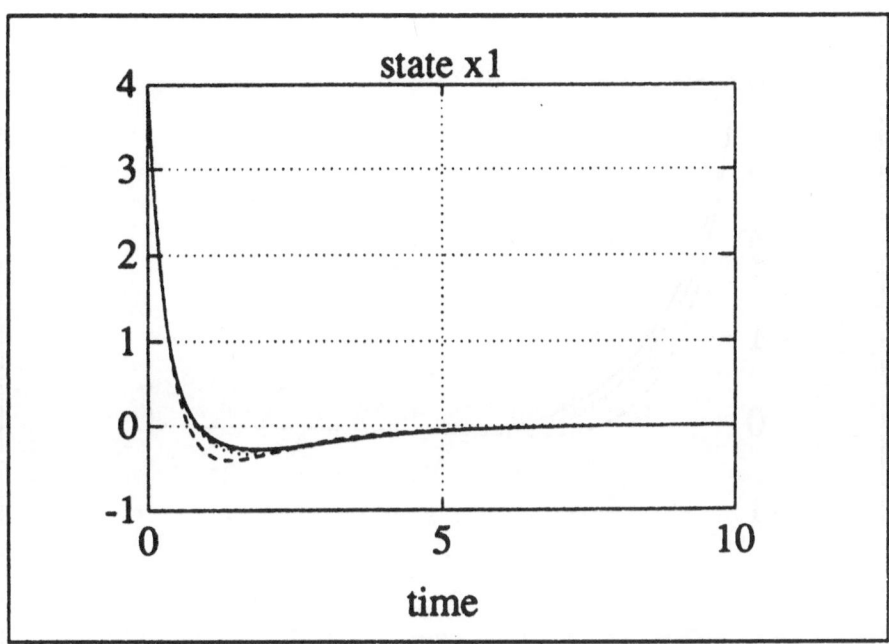

Figure 14.5: Optimal and approximate trajectories for x_1

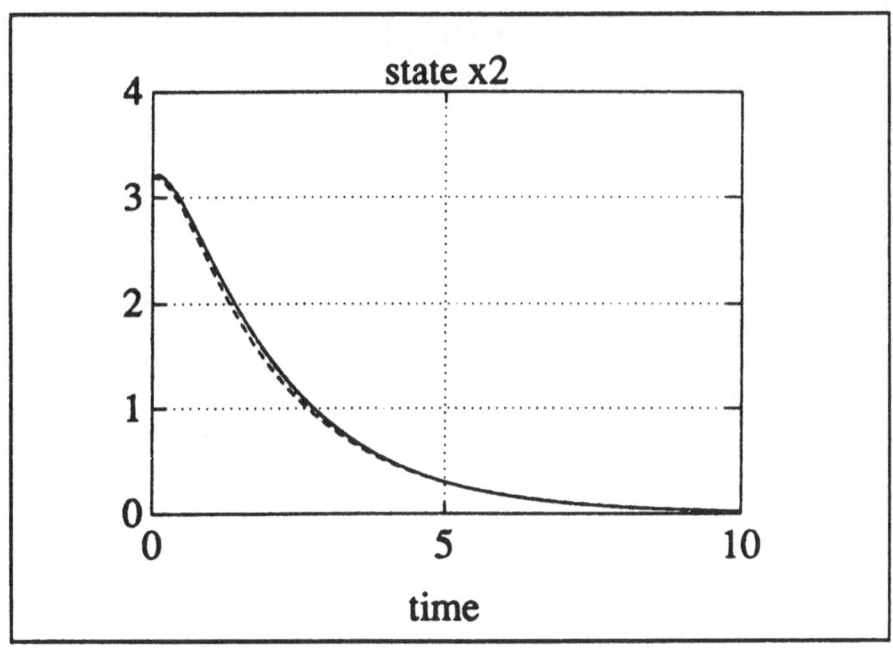

Figure 14.6: Optimal and approximate trajectories for x_2

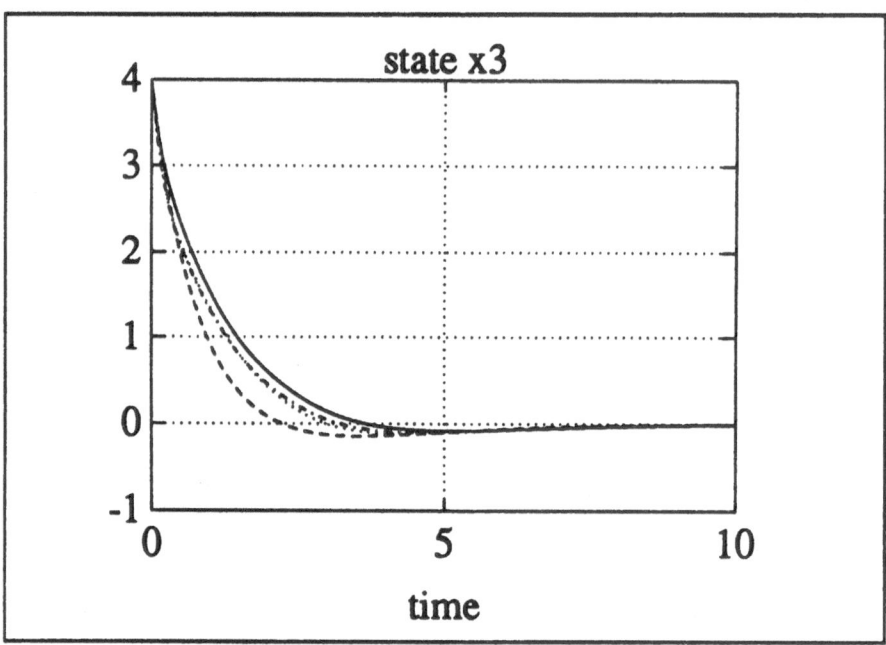

Figure 14.7: Optimal and approximate trajectories x_3

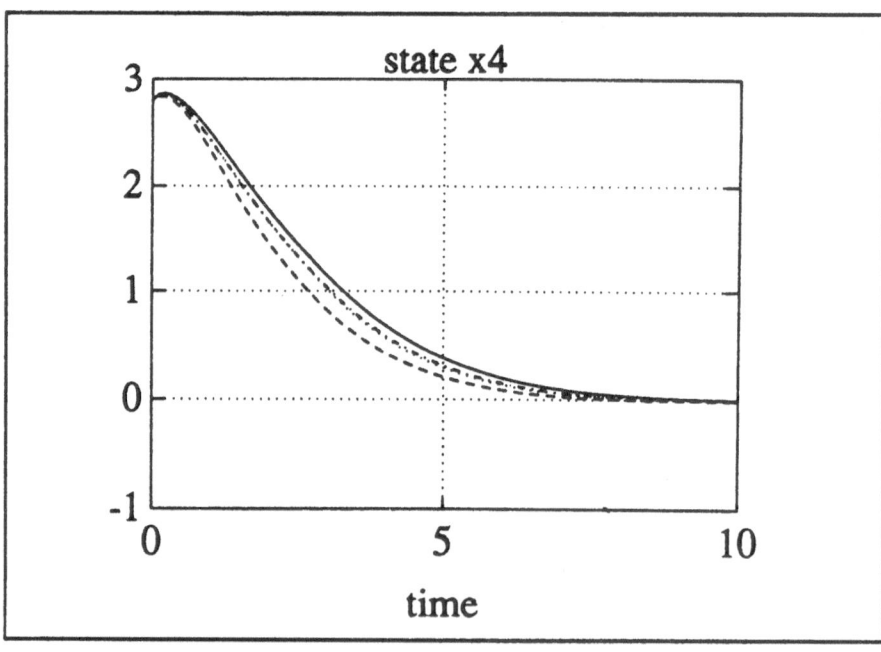

Figure 14.8: Optimal and approximate trajectories for x_4

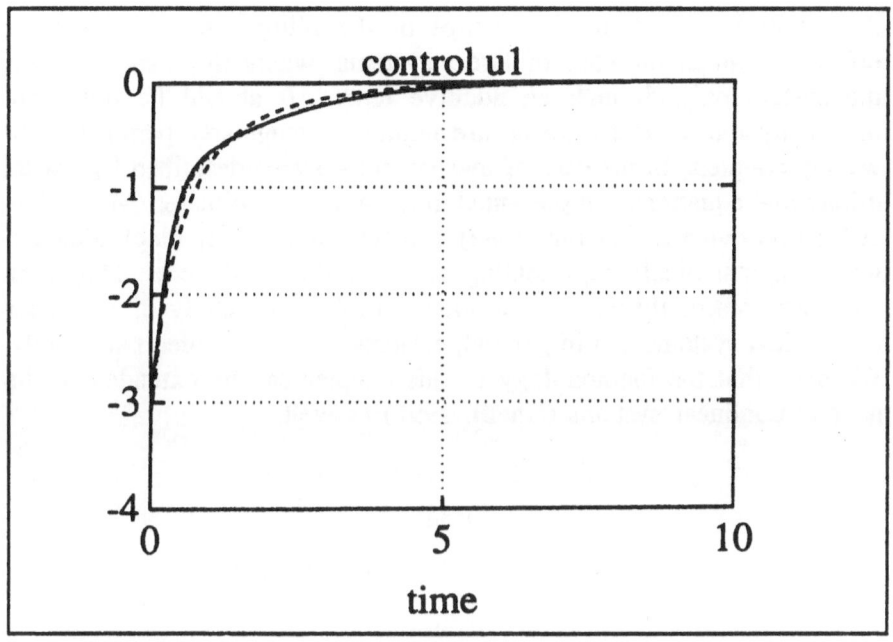

Figure 14.9: Optimal and approximate trajectories for control u_1

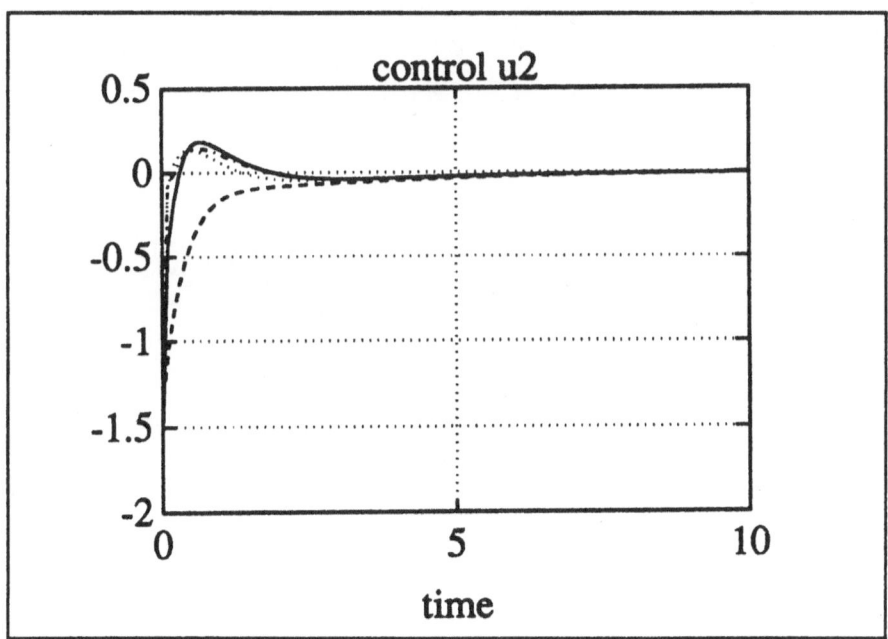

Figure 14.10: Optimal and approximate trajectories for control u_2

this multiplicative term the control of the bilinear systems is more effective than in the case of linear systems, where the control effects the system only through an additive term. It should be point out that many mechanical systems are nonlinear, singularly perturbed and weakly coupled. In the case of mechanical systems described by partial differential equations and presented in the modal coordinates (Meirovich, 1967; Meirovich and Baruh, 1983), the system matrix is block diagonal with diagonal blocks representing second order oscillators. The weak coupling control theory is a promising tool in the study of nonlinear mechanical systems and in general, systems with distributed parameters. We hope that the methodology of this chapter can be extended to the general nonlinear systems (Khalil, 1992) as well.

List of References

Ackerson, G. and K. Fu, "On the state estimation in switching environments," *IEEE Trans. Automatic Control*, AC-15(1970): 10-17.

Aganovic, Z., *Optimal Control of Singularly Perturbed Bilinear Systems*, Ph.D. Dissertation, Rutgers University, 1992.

Aganovic, Z. and Z. Gajic, "Composite near-optimal control of singularly perturbed bilinear systems," *Proc. Conf. Information Sciences and Systems*, Baltimore, USA, (1991a): 553-557.

Aganovic, Z. and Z. Gajic, "Optimal open-loop control of singularly perturbed bilinear systems," *Proc. Conf. Advances in Communication and Control*, Victoria, Canada, (1991b): 285-294.

Aldhaheri, R. and H. Khalil, "Aggregation of the policy iteration method for nearly completely decomposable Markov chains," *IEEE Trans. Automatic Control*, AC-36(1991): 178-187.

Anderson, B. and J. Moore, *Optimal Control — Linear Quadratic Methods*, Prentice-Hall, Englewood Cliffs, 1990.

Anderson, P. and A. Fouad, *Power System Control and Stability*, Iowa State University Press, Ames, 1984.

Arabacioglu, M., M. Sezer, and O. Oral, "Overlapping decomposition of large scale systems into weakly coupled subsystems," in *Computational and Combinatorial Methods in System Theory*, C. Byrnes and A. Lindquist, eds. North-Holland, Amsterdam, (1986): 135-147.

Arkun, Y. and S. Ramakrishnan, "Bounds of the optimum quadratic cost of structure constrained regulators," *IEEE Trans. Automatic Control*, AC-28(1983): 924-927.

Arnautovic, D., *Multivariable Voltage Regulator Synthesis in Multimachine Power Systems by Projective Control Method*, Ph.D. Dissertation, University of Belgrade, 1988.

Arnautovic, D. and J. Medanic, "Design of decentralized multivariable excitation controllers in multimachine power systems by projective controls," *IEEE Trans. Energy Conversion*, 2(1987): 598-604.

Arnautovic, D. and J. Medanic, "The sequential design of different multivariable excitation controllers in multimachine power systems," *Electric Power Systems Research*, 18(1990): 37-46.

Arnautovic, D. and D. Skataric, "Suboptimal design of hydroturbine governors," *IEEE Trans. Energy Conversion*, 6(1991): 438-444.

433

BIBLIOGRAPHY

Avramovic, B., "Subspace iteration approach to the time scale separation," *Proc. Conf. on Decision and Control*, Fort Lauderdale, U.S.A., (1979): 684-687.

Avramovic, B., P. Kokotovic, J. Winkelman, and J. Chow, "Area decomposition for electromechanical models of power systems," *Automatica*, 16(1980): 637-648.

Asamoah, F. and M. Jamshidi, "Stabilization of a class of singularly perturbed bilinear systems," *Int. J. Control*, 46(1983): 1589-1594.

Bar-Ness, Y. and A. Halbersberg, "Solution of the singular discrete regulator problem using eigenvector methods," *Int. J. Control*, 31 (1980): 615-625.

Baruh, H. and K. Choe, "Sensor placement in structural control," *AIAA J. Guidance, Dynamics and Control*, 13(1990): 524-533.

Basar, T., "A counter example in linear-quadratic games: existence of non-linear Nash strategies," *J. Optimization Theory and Applications*, 14(1974): 425-430.

Beale, S. and B. Shafai, "Robust control system design with a proportional-integral observer," *Int. J. Control*, 50(1989): 97-111.

Belanger, P. and T. McGillivray, "Computational experience with the solution of the matrix Lyapunov equation," *IEEE Trans. Automatic Control*, AC-21(1976): 799-800.

Bertrand, P., "A homotopy algorithm for solving coupled Riccati equations," *Optimal Control Appl. & Methods*, 6(1985): 351-357.

Bertsekas, D. and J. Tsitsiklis, *Parallel and Distributed Computation: Numerical Methods*, Prentice-Hall, Englewood Cliffs, 1989.

Bertsekas, D. and J. Tsitsiklis, "Some aspects of parallel and distributed iterative algorithms — A survey," *Automatica*, 27(1991): 3-21.

Blankenship, G., "Singularly perturbed difference equations in optimal control problems," *IEEE Trans. Automatic Control*, AC-26(1981): 911-917.

Bhattacharyya, S., A. Del Nero Gomes, and J. Howze, "The structure of robust disturbance rejection control," *IEEE Trans. Automatic Control*, AC-28(1983): 874-881.

Bittanti, S., A. Laub, and J. C. Willems, *The Riccati Equation*, Springer-Verlag, Berlin — New York, 1991.

Bruni, C., G. DiPillo, and G. Koch, "Bilinear systems: An appealing class of "nearly linear" systems in theory and applications," *IEEE Trans. Aut. Control*, AC-19(1974): 334-348.

Butuzov, V. and A. Vasileva, "Differential and difference equation systems with a small parameter for the case in which the unperturbed (singular) system is in the spectrum," *J. Differential Equations*, 6(1971): 499-510.

Calise, A. and D. Moerder, "Optimal output feedback design of systems with ill-conditioned dynamics," *Automatica*, 21(1985): 271-276.

Calvet, J. and A. Titli, "Overlapping vs partitioning in block-iteration methods: Application in large scale system theory," *Automatica*, 25(1989): 137-145.

Cebuhar, W. and V. Constanza, "Approximation procedures for the optimal control of bilinear and nonlinear systems," *J. Optimization Theory and Applications*, 43(1984): 615-627.

Chang, K., "Singular perturbations of a general boundary value problem," *SIAM J. Math. Anal.*, 3(1972): 520-526.

Chemouil, P. and A. Wahdam, "Output feedback control of systems with slow and fast modes," *Large Scale Systems*, 1(1980): 257-264.

Chen, C., *Linear System Theory and Design*. Holt, Rinehart and Winston, New York, 1984.

Chow, J., *Time Scale Modeling of Dynamic Networks*. Springer-Verlag, Lecture Notes in Control and Information Sciences, 47, New York, 1982.

Chow, J. and P. Kokotovic, "A decomposition of near-optimum regulators for systems with slow and fast modes," *IEEE Trans. Automatic Control*, AC-21(1976): 701-705.

Chow, J. and P. Kokotovic, "Near-optimal feedback stabilization of a class of nonlinear singularly perturbed systems," *SIAM J. Control and Optimization*, 16(1978a): 756-770.

Chow, J. and P. Kokotovic, "Two-time scale feedback design of a class of nonlinear systems," *IEEE Trans. Automatic Control*, AC-23(1978b): 438-443.

Chow, J. and P. Kokotovic, "A two-stage Lyapunov-Bellman feedback design of a class of nonlinear systems," *IEEE Trans. Automatic Control*, AC-26(1981): 656-663.

Chow, J. and P. Kokotovic, "Sparsity and time scales," *Proc. American Control Conference*, San Francisco, U.S.A., (1983): 656-661.

Cruz, J. and C. Chen, "Series solution of two-person, nonzero-sum, linear quadratic differential games," *J. Optimization Theory and Applications*, 7(1971): 240-257.

Delacour, J., M. Darwish, and J. Fantin, "Control strategies of large scale power systems," *Int. J. Control*, 27(1978): 753-767.

Delebecque, F. and J. Quadrat, "Optimal control of Markov chains admitting strong and weak interconnections," *Automatica*, 17(1981): 281-296.

Delebecque, F., J. Quadrat, and P. Kokotovic, "A unified view of aggregation and coherency in networks and Markov chains," *Int. J. Control*, 40(1984): 939-952.

De Vlieger, J., H. Verbruggen, and P. Bruijn, "A time-optimal control algorithm for digital computer control," *Automatica*, 18(1982): 239-244.

Dorato, P. and A. Levis, "Optimal linear regulators: the discrete time case," *IEEE Trans. Automatic Control*, AC-16(1970): 613-620.

Elgard, I. and E. Fosha, "Optimum megawatt-frequency control of multiarea electric energy systems," *IEEE Trans. Power Apparatus and Systems*, PAS-89(1970): 556-563.

Elliot, J., "NASA's advanced control law program for the F-8 digital fly-by-wire aircraft," *IEEE Trans. Automatic Control*, 22(1977): 753-757.

Ermer, C. and V. Vandelinde, "Output feedback gains for a linear discrete stochastic control problem," *IEEE Trans. Automatic Control*, 18(1973): 154-157.

Espana, M. and I. Landau, "Reduced order bilinear models for distillation columns," *Automatica*, 14(1978): 345-355.

Figalli, G., M. Cava, and L. Tomasi, "An optimal feedback control for a bilinear model of induction motor drives," *Int. J. of Control*, 39(1984): 1007-1016.

Fosha, E. and I. Elgard, "The megawatt-frequency control problem: A new approach via optimal control theory," *IEEE Trans. Power Apparatus and Systems*, PAS-89(1970): 563-578.

Fossard, A. and J. Magni, "Frequential analysis of singularly perturbed systems with state or output control," *Large Scale Systems*, 1(1980): 223-228.

Francis, B. and A. Glover, "Bounded peaking in the optimal linear regulator with cheap control," *IEEE Trans. Automatic Control*, AC-23(1978): 608-617.

Francis, B., "The optimal linear quadratic time invariant regulator with cheap control," *IEEE Trans. Automatic Control*, AC-24(1979): 616-621.

Gajic, Z., "Numerical fixed point solution of linear quadratic gaussian control problem for singularly perturbed systems," *Int. J. Control*, 43(1986): 373-387.

Gajic, Z., "The existence of a unique and bounded solution of the algebraic Riccati equation of the multimodel estimation and control problems," *Systems & Control Letters*, 10(1988): 185-190.

Gajic, Z. and H. Khalil, "Multimodel strategies under random disturbances and imperfect partial observations," *Automatica*, 22(1986): 121-125.

Gajic, Z. and T-Y. Li, "Simulation results for two new algorithms for solving coupled algebraic Riccati equations," *Third Int. Symposium on Differential Games*, Sophia Antipolis, France, (1988).

Gajic, Z., D. Petkovski, and N. Harkara, "The recursive algorithm for the optimal static output feedback control problem of linear singularly perturbed systems," *IEEE Trans. Automatic Control*, AC-34(1989): 465-468.

Gajic, Z., D. Petkovski, and X. Shen, *Singularly Perturbed and Weakly Coupled Linear Control Systems — A Recursive Approach*, Springer-Verlag, Lecture Notes in Control and Information Sciences, 140, New York, 1990.

Gajic, Z., B. Petrovic, and N. Rayavarupu, *Recursive Methods for Weakly Coupled and Singularly Perturbed Linear Estimation and Control Problems*, Technical Report, Rutgers University, 1987.

Gajic, Z. and X. Shen, "Decoupling transformation for weakly coupled linear systems," *Int. J. Control*, 50(1989): 1517-1523.

Gajic, Z. and X. Shen, "Parallel reduced-order controllers for stochastic linear singularly perturbed discrete systems," *IEEE Trans. Automatic Control*, AC-35(1991a): 87-90.

Gajic, Z. and X. Shen, "Study of the discrete singularly perturbed linear-quadratic control problem by a bilinear transformation," *Automatica*, 27(1991b): 1025-1028.

Gajic, Z. and D. Skataric, "Singularly perturbed weakly coupled linear control systems," *Proc. European Control Conf.*, Grenoble, France, (1991): 1607-1612.

Gardner, B. and J. Cruz, "Well-posedness of singularly perturbed Nash games," *J. Franklin Institute*, 306 (1978): 355-374.

Geromel, J. and P. Peres, "Decentralized load-frequency control," *Proc. IEE*, Part D., 132(1985): 225-230.

Gomathi, K., S. Prabhu, and M. Pai, "A suboptimal controller for minimum sensitivity of closed-loop eigenvalues to parameter variations," *IEEE Trans. Automatic Control*, AC-25(1980): 587-588.

Grodt, T. and Z. Gajic, "The recursive reduced order numerical solution of the singularly perturbed matrix differential Riccati equation," *IEEE Trans. Automatic Control*, AC-33 (1988): 751-754.

Guillen, J. and M. Armada, "Singular perturbation method for order reduction of large scale bilinear dynamical systems," *Proc. IFAC Symp. Large-Scale Systems Theory and Appl.*, Toulouse, (1980): 229-236.

Haddad, A., "Linear filtering of singularly perturbed systems," *IEEE Trans. Automatic Control*, AC-21(1976): 515-519.

Haddad, A. and P. Kokotovic, "Stochastic control of linear singularly perturbed systems," *IEEE Trans. Automatic Control*, AC-22(1977): 815-821.

Halyo, N. and J. Broussard, "A convergent algorithm for the stochastic infinite-time discrete optimal output feedback problem," *Proc. American Control Conf.*, Charlottesville, U.S.A., (1981): WA-1E.

Harkara, N., D. Petkovski, and Z. Gajic, "The recursive algorithm for the optimal static output feedback control problem of linear weakly coupled systems," *Int. J. Control*, 50(1989): 1-11.

Harvey, H. and G. Stein, "Quadratic weights for asymptotic regulator properties," *IEEE Trans. Automatic Control*, AC-23(1978): 378-387.

Hemker, P., "Numerical aspects of singular perturbation problems", in *Asymptotic Analysis II — Surveys and New Trends*, F. Verhulst, ed., Springer-Verlag, Berlin, Lecture Notes in Mathematics 985(1983): 267-287.

Hill, D., *Experiments in Computational Matrix Algebra*, Random House, New York, 1988.

Hopkins, W., J. Medanic, and W. Perkins, "Output feedback pole placement in the design of suboptimal linear quadratic regulators," *Int. J. Control*, 34(1981): 593-612.

Hoppensteadt, F. and W. Miranker, "Multitime methods for systems of difference equations," *Studies Appl. Math.*, 56(1977): 273-289.

Hofer, E. and B. Tibken, "An iterative method for the finite-time bilinear quadratic control problem," *J. Optimization Theory and Applications*, 57(1988): 411-427.

Hogan, S. and Z. Gajic, "Stochastic output feedback control of quasi-weakly coupled linear discrete systems," *Proc. American Control Conf.*, Chicago, USA, (1992): 1107-1111.

Hsieh, C., R. Skelton, and F. Dampa, "Minimum energy controllers with inequality constraints on output variances," *Optimal Control Appl. & Methods*, 10(1989): 347-366.

Huey, M., *A Reduced-Order Parallel Algorithm for Solving High-Gain and Cheap Control Problems*, M. S. Thesis, Rutgers University, 1992.

Ikeda, M. and D. Siljak, "Overlapping decompositions, expansions, and contractions of dynamical systems," *Large Scale Systems*, 1(1980): 29–38.

Ishimatsu, T., A. Mohri, and M. Takata, "Optimization of weakly coupled systems by a two-level method," *Int. J. Control*, 22(1975): 877–882.

Jamshidi, M., "An overview on the solution of the algebraic matrix Riccati equation and related problems," *Large Scale Systems*, 1(1980): 167–192.

Jameson, A. and R. O'Malley, "Cheap control of the time-invariant regulator", *Applied Mathematics and Optimization*, 1(1975): 337–354.

Kailath, T., *Linear Systems*, Prentice Hall, Englewood Cliffs, 1980.

Kalman, R., "Contributions to the theory of optimal control," *Bol. Soc. Mat. Mex.*, (1990): 102–119.

Kando, H., T. Iwazumi, and H. Ukai, "Singular perturbation modeling of large-scale systems with multi-time scale property," *Int. J. Control*, 48(1988): 2361–2387.

Kato, T., *Perturbation Theory of Linear Operators*, Springer-Verlag, New York, 1980.

Kaszkurewicz, E., A. Bhaya, and D. Siljak, "On the convergence of parallel asynchronous block-iterative computations," *Linear Algebra and Its Applications*, 131(1990): 139–160.

Katzberg, J., "Structured feedback control of discrete linear stochastic systems with quadratic cost," *IEEE Trans. Automatic Control*, AC–22(1977): 232–236.

Kautsky, J., N. Nichols, and P. Van Douren, "Robust pole assignment in linear state feedback," *Int. J. Control*, 41(1985): 1129–1155.

Keel, L. and S. Bhattacharyya, "A matrix equation approach to the design of low-order regulators," *SIAM J. Matrix Anal. Appl.*, 11(1990): 180–199.

Kenney, C. and R. Leipnik, "Numerical integration of the differential matrix Riccati equation," *IEEE Trans. Automatic Control*, AC–30(1985): 962–970.

Khalil, H., "Approximation of Nash strategies," *IEEE Trans. Automatic Control*, AC–25(1980a): 247–250.

Khalil, H., "Multi-model design of a Nash strategy," *J. Optimization Theory and Applications*, 31(1980b): 553–564.

Khalil, H., "On the robustness of output feedback control methods to modeling errors," *IEEE Trans. Automatic Control*, AC-26(1981): 524–526.

Khalil, H., "Output feedback control of linear two-time scale systems," *IEEE Trans. Automatic Control*, **AC-32**(1987): 784–792.

Khalil, H., "Feedback control of nonstandard singularly perturbed systems," *IEEE Trans. Automatic Control*, **AC-34**(1989): 1052–1060.

Khalil, H., *Nonlinear Systems*, Macmillan, New York, 1992.

Khalil, H. and Z. Gajic, "Near-optimum regulators for stochastic linear singularly perturbed systems," *IEEE Trans. Automatic Control*, **AC-29**(1984): 531–541.

Khalil, H. and P. Kokotovic, "Control strategies for decision makers using different models of the same system," *IEEE Trans. Automatic Control*, **AC–23**(1978): 289–298.

Khalil, H. and P. Kokotovic, "Control of linear systems with multiparameter singular perturbations," *Automatica*, **15**(1979a): 197–207.

Khalil, H. and P. Kokotovic, "D-stability and multiparameter singular perturbations," *SIAM J. Control and Optimization*, **17**(1979b): 56–65.

Khalil, H. and P. Kokotovic, "Feedback and well-posedness of singularly perturbed Nash games," *IEEE Trans. Automatic Control*, **AC-24**(1979c): 699–708.

Khorasani, K. and M. Azimi-Sadjadi, "Feedback control of two-time scale block implemented discrete-time systems," *IEEE Trans. Automatic Control*, **AC-32**(1987): 69–73.

Kirk, K., *Optimal Control Theory*, Prentice-Hall, Englewood Cliffs, 1970.

Kleinman, D., "On an iterative technique for Riccati equation computations," *IEEE Trans. Automatic Control*, **AC-13**(1968): 114–115.

Kokotovic, P., J. Allemong, J. Winkelman, and J. Chow, "Singular perturbations and iterative separation of the time scales," *Automatica*, **16**(1980): 23–33.

Kokotovic, P. and J. Cruz, "An approximation theorem for linear optimal regulators," *J. Math. Anal. Appl.*, **27**(1969): 249–252.

Kokotovic, P. and H. Khalil, *Singular Perturbations in Systems and Control*, IEEE Press, New York, 1986.

Kokotovic, P., H. Khalil, and J. O'Reilly, *Singular Perturbation Methods in Control: Analysis and Design*, Academic Press, Orlando, 1986.

Kokotovic, P., W. Perkins, J. Cruz, and G. D'Ans, " ϵ—coupling approach for near-optimum design of large scale linear systems," *Proc. IEE*, Part D., **116**(1969): 889–892.

Kokotovic, P. and G. Singh, "Optimization of coupled nonlinear systems," *Int. J. Control*, **14**(1971): 51–64.

Kokotovic, P. and R. Yackel, "Singular perturbation of linear regulators: basic theorems," *IEEE Trans. Automatic Control,* **AC-17**(1972): 29-37.

Kondo, R. and K. Furuta, "On the bilinear transformation of Riccati equations," *IEEE Trans. Automatic Control,* **AC-31**(1986): 50-54.

Kucera, V., "A contribution to matrix quadratic equations," *IEEE Trans. Automatic Control,* **AC-17**(1972), 344-347.

Kurtaran, B., "Suboptimal control for discrete linear constant stochastic systems," *IEEE Trans. Automatic Control,* **AC-20**(1975): 423-425.

Kurtaran, B. and M. Sidar, "Optimal instantaneous output feedback controllers for linear stochastic systems," *Int. J. Control,* **19**(1974): 797-816.

Kwakernaak, H. and R. Sivan, *Linear Optimal Control Systems,* Wiley-Interscience, New York, 1972.

Lancaster, P. and M. Tismenetsky, *The Theory of Matrices,* Academic Press, Orlando, 1985.

Lapidus, L. and N. Amundson, "Stagewise absorption and extraction equilibrium," *Ind. Engng. Chem.,* **42**(1950): 1071-1076.

Lapidus, L., E. Shapiro, S. Shapiro, and R. Stillman, "Optimization of process performance," *AIChEJ,* **7**(1961): 288-294.

Lewis, F., *Optimal Control,* Wiley, New York, 1986.

Levine, W. and M. Athans, "On the determination of the optimal constant output feedback gains for linear multivariable systems," *IEEE Trans. Automatic Control,* **AC-15**(1970): 44-48.

Levine, W., T. Johnson, and M. Athans, "Optimal limited state variable feedback controllers for linear systems," *IEEE Trans. Automatic Control,* **AC-16**(1971): 785-793.

Litkouhi, B., *Sampled-Data Control of Systems with Slow and Fast Modes,* Ph.D. Dissertation, Michigan State University, 1983.

Litkouhi, B. and H. Khalil, "Infinite-time regulators for singularly perturbed difference equations," *Int. J. Control,* **39**(1984): 587-598.

Litkouhi, B. and H. Khalil, "Multirate and composite control of two-time-scale discrete systems," *IEEE Trans. Automatic Control,* **AC-30**(1985): 587-598.

Luenberger, *Optimization by Vector Space Methods,* Wiley, New York, 1968.

Mahmoud, M., "A quantitative comparison between two decentralized control approaches," *Int. J. Control,* **28**(1978): 261-275.

Mahmoud, M., "Order reduction and control of discrete systems," *Proc. IEE,* Part D., **129**(1982): 129-135.

Mahmoud, M., "Stabilization of discrete systems with multiple-time scales," *IEEE Trans. Automatic Control*, **AC-31**(1986): 159–162.

Mahmoud, M., Y. Chen, and M. Singh, "Discrete two-time-scale systems," *Int. J. Systems Science*, **17**(1986): 1187–1207.

Makila, P. and H. Toivonen, "Computational methods for parametric LQ problems — A survey," *IEEE Trans. Automatic Control*, **AC-32**(1987): 658–671.

Medanic, J., "Design of low order optimal dynamic regulators for linear time invariant systems," *Proc. Conf. Information Sciences and Systems*, Baltimore, USA, (1979): 97–102.

Medanic, J., "Geometric properties and invariant manifolds of the Riccati equation," *IEEE Trans. Automatic Control*, **AC-27**(1982): 670–677.

Medanic, J. and B. Avramovic, "Solution of load-flow problems in power systems by ϵ—coupling method," *Proc. IEE*, Part D., **122**(1975): 801–805.

Medanic, J. and Z. Uskokovic, "The design of optimal output regulators for linear multivariable systems with constant disturbances," *Int. J. Control*, **37**(1983): 809–830.

Medanic, J., D. Petranovic, and N. Gluhajic, "The design of output regulators for discrete-time linear systems by projective controls," *Int. J. Control*, **41**(1985): 615–639.

Medanic, J. and Z. Uskokovic, "Design of coupled decentralized output regulators," *Int. J. Control*, **47**(1988): 1771–1794.

Medanic, J., H. Tharp, and W. Perkins, "Pole placement by performance criterion modification," *IEEE Trans. Automatic Control*, **AC-33**(1988): 469–472.

Meirovich, L., *Analytical Methods in Vibrations*, MacMillan, New York, 1967.

Meirovich, L. and H. Baruh, "On the problem of observation spillover in self-adjoint distributed parameter system," *J. Optimization Theory and Applications*, **39**(1983): 269–291.

Mendel, J., "A concise derivation of optimal constant limited feedback gains," *IEEE Trans. Automatic Control*, **AC-19**(1974): 447–448.

Meo, J., J. Medanic, and W. Perkins, "Design of digital PI+dynamic controllers using projective controls," *Int. J. Control*, **43**(1986): 539–559.

Miranker, W., *Numerical Methods for Stiff Equations*, D. Reidel Publishing Company, Dordrecht, Holland, 1981.

Moerder, D. and A. Calise, "Convergence of a numerical algorithm for calculating optimal output feedback gains," *IEEE Trans. Automatic Control*, AC-30(1985a): 900–903.

Moerder, D. and A. Calise, "Two-time scale stabilization of systems with output feedback," *AIAA J. Guidance, Dynamics and Control*, 8(1985b): 731–736.

Moerder, D. and A. Calise, "Near-optimal output feedback regulation of ill conditioned linear systems," *IEEE Trans. Automatic Control*, AC-33(1988): 463–466.

Mohler, R., "Natural bilinear control processes," *IEEE Trans. Syst. Sci. Cyber.*, SSC-6(1970): 192–197.

Mohler, R., *Bilinear Control Processes*, Academic Press, New York, 1973.

Mohler, R., *Nonlinear Systems — Applications to Bilinear Control*, Prentice-Hall, Englewood Cliffs, 1991.

Mohler, R. and W. Kolodziej, "An overview of bilinear systems theory and applications," *IEEE Trans. Syst. Sci. Cyber.*, SSC-10(1980): 683–688.

Molen, C. and C. Van Loan, "Nineteen dubious ways to compute the exponential of a matrix," *SIAM Review*, 20(1978): 801–836.

Murata, S., Y. Ando, and M. Suzuki, "Design of high gain regulator by the multiple time scale approach," *Automatica*, 26(1990): 585–591.

Naidu, D., *Singular Perturbation Methodology in Control Systems*, IEE Press, London, 1988.

Naidu, D. and A. Rao, *Singular Perturbation Analysis of Discrete Control Systems*, Lecture Notes in Mathematics, 1154, Springer-Verlag, Berlin, 1985.

Ohta, Y. and D. Siljak, "Overlapping block diagonal dominance and existence of Lyapunov functions," *J. Math. Anal. Appl.*, 112(1985): 396–410.

Oloomi, H. and M. Sawan, "The observer-based controller design of discrete-time singularly perturbed system," *IEEE Trans. Automatic Control*, AC-32(1987): 246–248.

O'Malley, R., *Introduction to Singular Perturbations*, Academic Press, New York, 1974a.

O'Malley, R., "Boundary layer methods for certain nonlinear singularly perturbed optimal control problems," *J. Math. Anal. Appl.*, 45(1974b): 468–484.

O'Malley, R., "A more direct solution of the nearly singular linear regulator problem," *SIAM Journal on Control and Optimization*, **14**(1976): 1063–1077.

O'Malley, R., *Singular Perturbation Methods for Ordinary Differential Equations*, Springer-Verlag, New York, 1991.

O'Malley, R. and A. Jameson, "Singular perturbations and singular arcs — part I," *IEEE Trans. Automatic Control*, **20**(1975): 218–226.

O'Malley, R. and A. Jameson, "Singular perturbations and singular arcs — part II," *IEEE Trans. Automatic Control*, **22**(1977): 328–337.

O'Reilly, J., "Partial cheap control of the time-invariant regulator," *Int. J. Control*, **37**(1983): 909–927.

Ortega, J. and W. Rheinboldt, *Iterative Solution of Nonlinear Equations In Several Variables*, Academic Press, New York, 1970.

Othman, H., N. Khraishi, and M. Mahmoud, "Discrete regulators with time scale separation," *IEEE Trans. Automatic Control*, **AC-30**(1985): 293–297.

Ozguner, U., "Near-optimal control of composite systems: the multi time-scale approach," *IEEE Trans. Automatic Control*, **AC-24**(1979): 652–655.

Ozguner, U. and W. Perkins, "A series solution to the Nash strategies for large scale interconnected systems," *Automatica*, **13**(1979): 313–315.

Pappas, T., A. Laub, and N. Sandell, "On the numerical solution of the discrete-time algebraic Riccati equation," *IEEE Trans. Automatic Control*, **AC-25**(1980): 631–641.

Papavassilopoulos, G., J. Medanic, and J. Cruz, "On the existence of Nash strategies and solutions to coupled Riccati equations in linear-quadratic games," *J. Optimization Theory and Applications*, **28**(1979): 49–75.

Papavassilopoulos, G. and P. Olsder, "On the linear-quadratic closed-loop, no memory Nash games," *J. Optimization Theory and Applications*, **42**(1984): 551–560.

Patniak, P., N. Viswanadham, and I. Sarma, "Computer control algorithms for a tubular ammonia reactor," *IEEE Trans. Automatic Control*, **AC-25**(1980): 642–651.

Petersen, I., "Linear quadratic differential games with cheap control," *Systems & Control Letters*, **8**(1986): 181–188.

Petkov, P., N. Christov, and M. Konstantinov, "A computational algorithm for pole placement assignment of linear multi-input systems," *IEEE Trans. Automatic Control*, **AC-31**(1986): 1044–1047.

Petkovski, D., "Design of decentralized proportional-plus-integral controllers for multivariable systems," *Computer and Chemical Engineering*, 5(1981): 51-56.

Petkovski, D. and M. Rakic, "On the calculation of optimum feedback gains for output constrained regulators," *IEEE Trans. Automatic Control*, 23(1978): 760.

Petkovski, D. and M. Rakic, "A series solution of feedback gains for output constrained regulators," *Int. J. Control*, 30(1979): 661-669.

Petrovic, B. and Z. Gajic, "The recursive solution of linear quadratic Nash games for weakly interconnected systems," *J. Optimization Theory and Applications*, 56(1988): 463-477.

Phillips, R., "Reduced order modeling and control of two-time scale discrete control systems," *Int. J. Control*, 31(1980): 761-780.

Phillips, R. and P. Kokotovic, "A singular perturbation approach to modeling and control of Markov chains," *IEEE Trans. Automatic Control*, AC-26(1981): 1087-1094.

Power, H., "Equivalence of Lyapunov matrix equations for continuous and discrete systems," *Electronic Letters*, 3(1967): 83.

Priel, B. and U. Shaked, " "Cheap" optimal control of discrete single input single output systems," *Int. J. Control*, 38(1983): 1087-1113.

Qureshi, M., *Parallel Algorithms for Discrete Singularly Perturbed and Weakly Coupled Filtering and Control Problems*, Ph.D. Dissertation, Rutgers University, 1992.

Qureshi, M. and Z. Gajic, "Boundary value problem of linear weakly coupled systems," *Proc. Allerton Conf. on Communication, Control and Computing*, Urbana, U.S.A., (1991): 455-462.

Qureshi, M., X. Shen, and Z. Gajic, "Open-loop control of linear singularly perturbed discrete systems," *Proc. Conf. Information Sciences and Systems*, Baltimore, USA, (1991): 151-155.

Qureshi, M. and Z. Gajic, "A new version of the Chang transformation," *IEEE Trans. Automatic Control*, AC-37(1992): 800-801.

Qureshi, M., X. Shen, and Z. Gajic, "Output feedback control of discrete linear singularly perturbed stochastic systems," *Int. J. Control*, 55(1992): 361-371.

Ramaker, R., J. Medanic, and W. Perkins, "Strictly proper projective controllers for disturbance attenuation," *Int. J. Control*, 52(1990): 963-982.

Saberi, A. and H. Khalil, "Quadratic-type Lyapunov functions for singularly perturbed systems," *IEEE Trans. Automatic Control*, **AC-29**(1984): 542-550.

Saberi, A. and H. Khalil, "Stabilization and regulation of nonlinear singularly perturbed systems — composite control," *IEEE Trans. Aut. Control*, **AC-30**(1985): 739-747.

Saberi, A. and P. Sannuti, "Cheap control problem of a linear uniform rank system: design by composite control," *Automatica*, **22**(1986): 757-759.

Saberi, A. and P. Sannuti, "Cheap and singular controls for linear quadratic regulators," *IEEE Trans. Automatic Control*, **AC-32**(1987): 208-219.

Sage, A. and C. White, *Optimum Systems Control*, Prentice-Hall, Englewood Cliffs, 1977.

Salgado, M., R. Middleton, and G. C. Goodwin, "Connection between continuous and discrete Riccati equation with applications to Kalman filtering," *Proc. IEE*, Part D., **135**(1988): 28-34.

Saksena, V., and J. Cruz, "A multimodel approach to stochastic Nash games," *Automatica*, **17**(1981a): 295-305.

Saksena, V., and J. Cruz, "Nash strategies in decentralized control of multiparameter singularly perturbed large scale systems," *Large Scale Systems*, **2**(1981b): 219-234.

Saksena, V., and T. Basar, "A multimodel approach to stochastic team problems," *Automatica*, **18**(1982): 713-720.

Saksena, V., J. Cruz, W. Perkins, and T. Basar, "Information induced multimodel solution in multiple decision maker problems," *IEEE Trans. Automatic Control*, **AC-28**(1983): 716-728.

Sannuti, P., "Direct singular perturbation analysis of high-gain and cheap control problems," *Automatica*, **19**(1983): 41-51.

Sannuti, P. and H. Wason, "Multiple time-scale decomposition of cheap control problems," *IEEE Trans. Automatic Control*, **30**(1985): 633-644.

Sasagawa, T., "On the finite escape phenomena for matrix Riccati equation," *IEEE Trans. Automatic Control*, **AC-27**(1982): 977-979.

Sesak, J., P. Likins, and T. Coradetti, "Flexible spacecraft control by model error sensitivity suppression," *J. of the Astronautical Sciences*, **27**(1979): 131-156.

Sen, S. and K. Datta, "Singular perturbation analysis of discrete cheap control problem," *Int. J. Systems Science*, **23**(1992): 57-70.

Sezer, M. and D. Siljak, "Nested ε—decomposition and clustering of complex systems," *Automatica*, **22**(1986): 321-331.

Sezer, M. and D. Siljak, "Nested epsilon decomposition of linear systems: Weakly coupled and overlapping blocks," *SIAM J. Matrix Anal. Appl.*, **12**(1991): 521-533.

Shapiro, E., D. Fredericks, and R. Roony, "Suboptimal constant output feedback and its application to modern flight control system design," *Int. J. Control*, **33**(1981): 505-517.

Shen, X., *Near-Optimum Reduced-Order Stochastic Control of Linear Discrete and Continuous Systems with Small Parameters*, Ph. D. Dissertation, Rutgers University, 1990.

Shen, X., "Solution of the singularly perturbed matrix difference Riccati equation," *Int. J. Systems Science*, **23**(1992): 403-410.

Shen, X. and Z. Gajic, "Near-optimum steady state regulators for stochastic linear weakly coupled systems," *Automatica*, **26**(1990a): 919-923.

Shen, X. and Z. Gajic, "Optimal reduced-order solution of the weakly coupled discrete Riccati equation," *IEEE Trans. Automatic Control*, **AC-35**(1990b): 600-602.

Shen, X. and Z. Gajic, "Approximate parallel controllers for discrete weakly coupled linear stochastic systems," *Optimal Control Appl. & Methods*, **11**(1990c): 345-354.

Shen, X., Z. Gajic, and D. Petkovski, "Parallel reduced-order algorithms for Lyapunov equations of large scale linear systems," *Proc. IMACS Symp. MCTS*, Lille, France, (1991a): 697-702.

Shen, X., Z. Gajic, and M. Qureshi, "Reduced-order solution of the weakly coupled matrix difference Riccati equation," *Proc. American Control Conf.*, Boston, U.S.A., (1991b): 2437-2438.

Shien, L. and Y. Tsay, "Transformations of a class of multivariable control systems to block companion forms," *IEEE Trans. Automatic Control*, **AC-27**(1982): 199-202.

Siljak, D., *Large-Scale Dynamic Systems: Stability and Structure*, North Holland, New York, 1978.

Siljak, D., *Decentralized Control of Complex Systems*, Academic Press, Cambridge, 1991.

Singh, N., Y. Singh, and S. Ahson, "Singular perturbation approach to near-optimum control with a prescribed degree of stability," *Int. J. Systems Science*, **18**(1987): 1703-1709.

Skataric, D., *Optimal Hydropower Plant Control in Isolated and Parallel Operation*, M.S. Thesis, University of Belgrade, 1989.

Skataric, D., *Quasi Singularly Perturbed and Weakly Coupled Linear Control Systems*, Ph. D. Dissertation, University of Novi Sad, 1992.

Skataric, D. and Z. Gajic, "Linear control of nearly singularly perturbed hydro power plants," *Automatica*, **28**(1992): 159-163.

Skataric, D., Z. Gajic, and D. Arnautovic, "Optimal reduced-order controllers for nearly weakly coupled linear systems," *Proc. Yugoslav Conf. ETAN*, Zagreb, Croatia, **VII**(1990): 27-33.

Skataric, D., Z. Gajic, and D. Petkovski, "Reduced-order solution for a class of linear quadratic optimal control problems," *Proc. Allerton Conf. on Communication, Control and Computing*, Urbana, U.S.A., (1991): 440-447.

Smith, D., "Decoupling and order reduction via the Riccati transformation," *SIAM Review*, **29**(1987): 91-113.

Sontag, E., *Mathematical Control Theory — Deterministic Finite Dimensional Systems*, Springer-Verlag, New York, 1990.

Srikant, R. and T. Basar, "Optimal solutions in weakly coupled multiple decision maker Markov chains with nonclassical information," *Proc. Conf. on Decision and Control*, Tampa, U.S.A., (1989): 168-173.

Srikant, R. and T. Basar, "Iterative computation of noncooperative equilibria in nonzero-sum differential games with weakly coupled players," *J. Optimization Theory and Applications*, **71**(1991): 137-168.

Srikant, R. and T. Basar, "Asymptotic solutions of weakly coupled stochastic teams with nonclassical information," *IEEE Trans. Automatic Control*, **AC-37**(1992a): 163-173.

Srikant, R. and T. Basar, "Sequential decomposition and policy iteration schemes for M-player games with partial weak coupling," *Automatica*, **28**(1992b): 95-105.

Starr, A. and Y. Ho, "Nonzero-sum differential games", *J. Optimization Theory and Applications*, **3**(1969): 49-79.

Stewart, G., *Introduction to Matrix Computation*, Academic Press, New York, 1973.

Su, W., *Contributions to the Open and Closed Loop Control Problems of Linear Weakly Coupled and Singularly Perturbed Systems*, M.S. Thesis, Rutgers University, 1990.

Su, W. and Z. Gajic, "Reduced-order solution to the finite time optimal control problems of linear weakly coupled systems," *IEEE Trans. Automatic Control*, **AC-36**(1991): 498-501.

Su, W. and Z. Gajic, "Decomposition method for solving weakly coupled algebraic Riccati equation," *AIAA J. Guidance, Dynamics and Control*, **15**(1992): 536–538.

Su, W., Z. Gajic, and X. Shen, "The recursive reduced-order solution of an open-loop control problem of linear singularly perturbed systems," *IEEE Trans. Automatic Control*, **AC-37**(1992a): 279–281.

Su, W., Z. Gajic, and X. Shen, "The exact slow-fast decomposition of the algebraic Riccati equation of singularly perturbed systems," *IEEE Trans. Automatic Control*, **AC-37**(1992b): 1456–1459.

Suzuki, M., "Composite control of singularly perturbed systems," *IEEE Trans. Automatic Control*, **AC-26**(1981): 505–507.

Teneketzis, D. and N. Sandell, "Linear regulator design for stochastic systems by multiple time-scale method," *IEEE Trans. Automatic Control*, **AC-22**(1977): 615–621.

Toivonen, H., "A globally convergent algorithm for the optimal constant output feedback problem," *Int. J. Control*, **41**(1985): 1589–1599.

Toivonen, H. and P. Makila, "Newton's method for solving parametric linear quadratic control problems," *Int. J. Control*, **46**(1987): 897–911.

Tharp, H., "Optimal pole-placement in discrete systems," *IEEE Trans. Automatic Control*, **AC–37**(1992): 645–648.

Tzafestas, S. and K. Anagnostou, "Stabilization of singularly perturbed strictly bilinear systems," *IEEE Trans. Automatic Control*, **AC–29**(1984a): 943–946.

Tzafestas, S. and K. Anagnostou, "Stabilization of ϵ-coupled bilinear systems using state feedback control," *Int. J. System Science*, **15**(1984b): 639–646.

Washburn, H. and J. Mendel, "Multistage estimation of dynamical and weakly coupled systems in continuous-time linear systems," *IEEE Trans. Automatic Control*, **AC-25**(1980): 71–76.

West, P., S. Bingulac, and W. Perkins, "L-A-S: A computer aided control system design language," in *Computer-Aided Systems Engineering*, M. Jamshidi and C. Herget, eds., North Holland, Amsterdam, (1985): 243–261.

Wilde, R. and P. Kokotovic, "A dichotomy in linear control theory," *IEEE Trans. Automatic Control*, **AC-17**(1972): 382–383.

Wilde, R. and P. Kokotovic, "Optimal open- and closed-loop control of singularly perturbed linear systems," *IEEE Trans. Automatic Control*, **AC-18**(1973): 616–625.

Wiener, N., *Cybernetics*, MIT Press, Cambridge, 1948.

Wonham, W., "On a matrix Riccati equation of stochastic control," *SIAM J. Control*, **6**(1968): 681–697.

Yackel, R. and P. Kokotovic, "A boundary layer method for the matrix Riccati equation," *IEEE Trans. Automatic Control*, **AC-17**(1973): 17–24.

Ying, Y., M. Rao, and X. Shen, "Bilinear decoupling control and its industrial application," *Proc. American Control Conf.*, Chicago, USA, (1992): 1163–1167.

Young, K., P. Kokotovic, and V. Utkin, "A singular perturbation analysis of high-gain feedback systems," *IEEE Trans. Automatic Control*, **22**(1977): 931–937.

Zangwill, W. and C. Garcia, *Pathways to Solutions, Fixed Points and Equilibria*, Prentice-Hall, Englewood Cliffs, 1981.

Zecevic, A. and D. Siljak, "A block-parallel Newton method via overlapping decompositions," *Proc. American Control Conf.*, Chicago, USA, (1992): 1653–1659.

Zhuang, J. and Z. Gajic, "Stochastic multimodel strategy with perfect measurements," *Control — Theory and Advanced Technology*, **7**(1991): 173–182.

Index